PESTICIDES IN THE ENVIRONMENT

PESTICIDES IN THE ENVIRONMENT

Impact, Assessment, and Remediation

Edited by

ANKET SHARMA

Texas Tech University, Lubbock, TX, United States; Zhejiang A&F University, Hangzhou, Zhejiang, China

VINOD KUMAR

Department of Botany, Government Degree College, Ramban, Jammu & Kashmir, India

BINGSONG ZHENG

Zhejiang A&F University, Hangzhou, Zhejiang, China

ELSEVIER

Elsevier
Radarweg 29, PO Box 211, 1000 AE Amsterdam, Netherlands
The Boulevard, Langford Lane, Kidlington, Oxford OX5 1GB, United Kingdom
50 Hampshire Street, 5th Floor, Cambridge, MA 02139, United States

ISBN: 978-0-323-99427-9

For Information on all Elsevier publications visit our website at https://www.elsevier.com/books-and-journals

Publisher: Candice Janco
Editorial Project Manager: Aera Gariguez/Namrata Lama
Production Project Manager: Sruthi Satheesh
Cover Designer: Christian Bilbow

Typeset by Aptara, New Delhi, India

Contents

Contributors ix

1. Impact of pesticides on aquatic invertebrates and vertebrates: Recent updates and future perspectives **1**

Megha Andotra, Simran Sharma, Ankeet Bhagat, Sunil Sharma and Owias Iqbal Dar

1.1 Introduction 1
1.2 Pesticide toxicity to aquatic invertebrates 2
1.3 Pesticide toxicity to aquatic vertebrates 3
1.4 Conclusion 12
References 12

2. Pesticide types and mechanism of their action in nontarget organisms **15**

Arun Dev Singh, Jaspreet Kour, Shalini Dhiman, Kamini Devi, Neerja Sharma, Pardeep Kumar, Deepak Kumar, Mohd. Ali, Parkirti, Puja Ohri, Amrit Pal Singh and Renu Bhardwaj

2.1 Introduction 15
2.2 Pesticides: types and classification 16
2.3 Pesticide toxicity and their mechanism of action in plants 17
2.4 Pesticide toxicity and their mechanism of action in humans 19
2.5 Pesticide toxicity and their mechanism of action in honeybees 21
2.6 Pesticide toxicity and their mechanism of action in soil microbes 29
2.7 Pesticide toxicity and their mechanism of action in the aquatic ecosystem (fish) 31
2.8 Pesticide detoxification mechanism in plants 32
2.9 Conclusion and future prospects 37
References 37

3. Pesticide biology in soil: Sorption, leaching, and accumulation **49**

Sandeep, Sonika Sharma, Ashish Sharma, Raj Bala, Indu Sharma and Ashutosh Sharma

3.1 Introduction 49
3.2 Release of pesticides in soil during agricultural activities 52
3.3 Pesticide sorption by soil organic matter 53
3.4 Effect of pesticide on soil process 53
3.5 Leaching of pesticides in soil 55
3.6 Impact of pesticides on the soil biology 56
3.7 Accumulation of pesticides in soil and its remediation 56

3.8 Conclusion and future prospects 61
Acknowledgment 62
Conflicts of interest 62
References 62

4. Pesticide biology in plants: Plant uptake, translocation, and accumulation 67
Harleen Kaur, Ravneet Kaur, Swapnil Singh, Neetu Jagota and Ashish Sharma

4.1 Introduction 67
4.2 Effect of pesticides on plants 68
4.3 Plant uptake 68
4.4 Translocation of pesticides 76
4.5 Storage and metabolism of pesticides 78
4.6 Conclusion 80
References 81

5. Impact of pesticide application on the food chain and food web 87
Shalini Dhiman, Jaspreet Kour, Arun Dev Singh, Kamini Devi, Raman Tikoria, Mohd. Ali, Deepak Kumar, Puja Ohri and Renu Bhardwaj

5.1 Introduction 87
5.2 Pesticide types and occurrences in the environment 88
5.3 Pesticide transfer: from environment to food (food chain/web) 91
5.4 Pesticide residues impact 96
5.5 Conclusion 108
References 109

6. Methods for pesticide residue analysis: Recent advancements 119
Antul Kumar, Himani Gautam, Anuj Choudhary, Harmanjot Kaur, Sahil Mehta, V. Mohan Murali Achary, Bindu Yadav and Om Prakash Narayan

6.1 Introduction 120
6.2 Method validation 121
6.3 Different parameters of method validation 121
6.4 Detection techniques for pesticide residual analysis 125
6.5 Instrumental detection methods 137
6.6 Advanced extraction and detection methods 139
6.7 Residue analysis in various systems 140
6.8 Conclusion and future prospective 143
References 145

7. Impact of pesticide application: Positive and negative side 155
Shveta Saroop and Sonam Tamchos

7.1 Introduction 155

7.2 Positive side of pesticide application 157
7.3 Negative side of pesticide application 162
7.4 Conclusion and future prospects 172
References 173

8. Pesticides-mediated ROS generation in plants **179**

Ashutosh Sharma, Sanjay Kumar Sharma, Narayan Singh, Vasudha Maurya, Sukhwinder Kaur, Rahul Kumar and Indu Sharma

8.1 Introduction 179
8.2 Various classes of pesticides and their comparison in terms of ROS generation 181
8.3 The pesticide-induced oxidative burst 183
8.4 Effect of pesticides on ROS-generating systems 186
8.5 Effect of pesticides on ROS scavenging systems 193
8.6 Various methods to reduce pesticide-induced oxidative stress in plants 194
8.7 Conclusion and future prospects 197
Conflicts of interest 198
Acknowledgment 198
References 198

9. Pesticide impacts on humans **203**

Raman Jasrotia, Seema Langer and Menakshi Dhar

9.1 Introduction 203
9.2 History of pesticides 204
9.3 Classification of pesticides 205
9.4 Pyrethroids 209
9.5 Importance of pesticides 211
9.6 Pesticides and humans 212
9.7 Conclusion 217
Acknowledgment 217
References 217

10. Role of microbes in pesticide bioremediation: Recent advances and biotechnological implications **223**

Roohi Sharma and Shveta Saroop

10.1 Pesticides as an environmental toxic and bio-remediation strategy thereof 223
10.2 Microbial-assisted bioremediation 225
10.3 Enzymatic degradation pathway of pesticides by microbes 229
10.4 Bioremediation processes 235
10.5 Recent advances and biotechnological implications 238
10.6 Conclusion 240
References 240

11. Impact of nanopesticides in the environment: Solutions, threats, and opportunities **251**

Romica Verma and Shveta Saroop

11.1 Introduction 251
11.2 Health impacts 266
11.3 Emerging trends and future perspectives 266
11.4 Agricultural applications of stimuli responsive systems 274
11.5 Temperature-sensitive systems 275
11.6 Conclusion and future perspectives 277
References 277

12. Plant hormones (brassinosteroids, salicylates, and jasmonates) as potential mitigators of pesticide and herbicide stress in plants **293**

Barket Ali

12.1 Introduction 293
12.2 Effect of brassinosteroids on the herbicide and pesticide stress 295
12.3 Effect of salicylates on the herbicide and pesticide stress 296
12.4 Effect of jasmonates on herbicide and pesticide stress 299
12.5 Conclusion 300
References 301

13. Alternatives to chemical pesticides: Current trends and future implications **307**

Sumit Singh, Shivali Puri and Satwinder Kaur Sohal

13.1 Introduction 307
13.2 Advantages of using biopesticides over conventional chemical pesticides 308
13.3 Adverse effects of chemical pesticides on human health 309
13.4 Types of biopesticides and their modes of action 310
13.5 Current scenario of biopesticides 321
13.6 Limitations and challenges in the promotion, commercialization, and consumption of biopesticides 322
13.7 Way forward: Future prospective to be considered 324
13.8 Policy recommendations 325
13.9 Conclusion 326
References 326

Index *335*

Contributors

V. Mohan Murali Achary
International Centre for Genetic Engineering and Biotechnology, New Delhi, India

Mohd. Ali
Department of Zoology, Guru Nanak Dev University, Amritsar, Punjab, India

Barket Ali
Department of Botany, Government Degree College, Kilhotran, Doda, Jammu and Kashmir, India

Megha Andotra
Aquatic Toxicology Lab, Department of Zoology, Guru Nanak Dev University, Amritsar, Punjab, India

Raj Bala
Department of Botany & Environment Studies, DAV University, Jalandhar, Punjab, India

Ankeet Bhagat
Aquatic Toxicology Lab, Department of Zoology, Guru Nanak Dev University, Amritsar, Punjab, India

Renu Bhardwaj
Department of Botanical and Environmental Sciences, Guru Nanak Dev University, Amritsar, Punjab, India

Anuj Choudhary
Department of Botany, Punjab Agricultural University, Ludhiana, Punjab, India

Owias Iqbal Dar
Department of Biosciences, University Institute of Biotechnology (UBIT), Chandigarh University, Mohali, Punjab, India

Kamini Devi
Department of Botanical and Environmental Sciences, Guru Nanak Dev University, Amritsar, Punjab, India

Menakshi Dhar
Department of Zoology, University of Jammu, Jammu, Jammu and Kashmir, India

Shalini Dhiman
Department of Botanical and Environmental Sciences, Guru Nanak Dev University, Amritsar, Punjab, India

Himani Gautam
Department of Entomology, Dr. YS Parmar University of Horticulture and Forestry, Nauni Solan, Himachal Pradesh, India

Neetu Jagota
Department of Botany & Environment Studies, DAV University, Jalandhar, Punjab, India

Raman Jasrotia
Department of Zoology, University of Jammu, Jammu, Jammu and Kashmir, India

Harleen Kaur
Department of Botany & Environment Studies, DAV University, Jalandhar, Punjab, India

Ravneet Kaur
Department of Botany & Environment Studies, DAV University, Jalandhar, Punjab, India

Harmanjot Kaur
Department of Botany, Punjab Agricultural University, Ludhiana, Punjab, India

Sukhwinder Kaur
Department of Life Sciences, University Institute of Sciences, Sant Baba Bhag Singh University, Jalandhar, Punjab, India

Jaspreet Kour
Department of Botanical and Environmental Sciences, Guru Nanak Dev University, Amritsar, Punjab, India

Pardeep Kumar
Department of Botanical and Environmental Sciences, Guru Nanak Dev University, Amritsar, Punjab, India

Deepak Kumar
Department of Zoology, Guru Nanak Dev University, Amritsar, Punjab, India

Antul Kumar
Department of Botany, Punjab Agricultural University, Ludhiana, Punjab, India

Rahul Kumar
Faculty of Agricultural Sciences, DAV University, Jalandhar, Punjab, India

Seema Langer
Department of Zoology, University of Jammu, Jammu, Jammu and Kashmir, India; Institute of Human Genetics, University of Jammu, Jammu, Jammu and Kashmir, India

Vasudha Maurya
Faculty of Agricultural Sciences, DAV University, Jalandhar, Punjab, India; Department of Biotechnology, DAV University, Jalandhar, Punjab, India

Sahil Mehta
Department of Botany, Hansraj College, University of Delhi, Delhi, India

Om Prakash Narayan
College of Liberal Arts and Sciences, University of Florida, Gainesville, FL, United States

Puja Ohri
Department of Zoology, Guru Nanak Dev University, Amritsar, Punjab, India

Parkirti
Department of Zoology, Guru Nanak Dev University, Amritsar, Punjab, India

Shivali Puri
Department of Zoology, DAV University, Jalandhar, Punjab, India

Sandeep
Department of Botany & Environment Studies, DAV University, Jalandhar, Punjab, India; Faculty of Agricultural Sciences, DAV University, Jalandhar, Punjab, India

Shveta Saroop
Department of Botany, Government Degree College, Kathua, Jammu and Kashmir, India

Simran Sharma
Aquatic Toxicology Lab, Department of Zoology, Guru Nanak Dev University, Amritsar, Punjab, India

Sunil Sharma
Aquatic Toxicology Lab, Department of Zoology, Guru Nanak Dev University, Amritsar, Punjab, India

Neerja Sharma
Department of Botanical and Environmental Sciences, Guru Nanak Dev University, Amritsar, Punjab, India

Sonika Sharma
Faculty of Agricultural Sciences, DAV University, Jalandhar, Punjab, India

Indu Sharma
Department of Life Sciences, University Institute of Sciences, Sant Baba Bhag Singh University, Jalandhar, Punjab, India

Ashish Sharma
Department of Botany & Environment Studies, DAV University, Jalandhar, Punjab, India

Ashutosh Sharma
Faculty of Agricultural Sciences, DAV University, Jalandhar, Punjab, India

Sanjay Kumar Sharma
Department of Life Sciences, University Institute of Sciences, Sant Baba Bhag Singh University, Jalandhar, Punjab, India

Roohi Sharma
Department of Botany, University of Jammu, Jammu, Jammu and Kashmir, India

Amrit Pal Singh
Department of Pharmaceutical Sciences, Guru Nanak Dev University, Amritsar, Punjab, India

Arun Dev Singh
Department of Botanical and Environmental Sciences, Guru Nanak Dev University, Amritsar, Punjab, India

Swapnil Singh
Department of Botany & Environment Studies, DAV University, Jalandhar, Punjab, India

Narayan Singh
Faculty of Agricultural Sciences, DAV University, Jalandhar, Punjab, India; Department of Biotechnology, DAV University, Jalandhar, Punjab, India

Sumit Singh
Department of Zoology, Guru Nanak Dev University, Amritsar, Punjab, India

Satwinder Kaur Sohal
Department of Zoology, Guru Nanak Dev University, Amritsar, Punjab, India

Sonam Tamchos
Regional Ayurveda Research Institute, Jammu (CCRAS, Ministry of AYUSH, Government of India), Jammu and Kashmir, India

Raman Tikoria
Department of Zoology, School of Bioengineering and Biosciences, Lovely Professional University, Phagwara, Punjab, India

Romica Verma
Department of Botany, University of Jammu, Jammu, Jammu and Kashmir, India

Bindu Yadav
School of Environmental Sciences, Jawaharlal Nehru University, New Delhi, India; Department of Agricultural & Biological Engineering, University of Florida, Gainesville, FL, United States

CHAPTER 1

Impact of pesticides on aquatic invertebrates and vertebrates: Recent updates and future perspectives

Megha Andotra[a,*], Simran Sharma[a,*], Ankeet Bhagat[a], Sunil Sharma[a] and Owias Iqbal Dar[b,*]

[a]Aquatic Toxicology Lab, Department of Zoology, Guru Nanak Dev University, Amritsar, Punjab, India
[b]Department of Biosciences, University Institute of Biotechnology (UBIT), Chandigarh University, Mohali, Punjab, India

1.1 Introduction

Environmental Protection Agency (EPA) considers pesticides to be a group of chemical substances which are extensively utilized to repel and control the population of pests. The use of pesticides has dramatically increased over the past years and has become an important part of agricultural practices. Classification of pesticides is predominantly based on the target organism, applications, and chemical nature. On the basis of the target organism pesticides can be cataloged as fungicides (that inhibit or kill fungus), insecticides (kill insects), rodenticides (chemicals that kill rodents), herbicides (pesticides that kill or control weeds), insect repellants (substances applied on cloth or skin to kill insects), and fumigants (gaseous substances used to control or kill pests like ticks, bedbugs). Based on applications pesticides can be categorized as domestic pesticides (used to kill insects like bacteria, cockroach, mice, protozoa, etc.), public health (kills vector that causes diseases) agriculture (used to protect weeds, insects, and crop pest) (Rani et al., 2021). Categorization on the basis of chemical nature includes pyrethroids, organochlorines, carbamates, organophosphates, phenyl amides (acetamides, acylanalides, carbanilates, and toluidines) trazines, phenoxyalkonates, benzoic acid derivatives, phthalimide derivatives, benzonitriles, and dipyrids (Jayaraj et al., 2016). Continuous and repetitive use of pesticides leads to the deprivation of biodiversity. In 2008 a report by Green Peace showed that India becomes one of the major industries in the production of 90,000 metric tons of pesticides in Asia (Kumarasamy et al., 2012). Plants uptake only a small portion of the pesticide and the majority portion enters into the air, water, or seeps deep into the soil (Aktar et al., 2009). According to Eleftherohorinos (2008), the contamination of

* Authors contributed equally.

Pesticides in the Environment: Impact, Assessment, and Remediation
DOI: https://doi.org/10.1016/B978-0-323-99427-9.00002-1

pesticides relies mainly on interconnection between soil persistence and soil adsorption, soil factors (soil fauna, soil moisture, soil microflora, inorganic surfaces, pH, organic components) the climatic variation, the plant species, the physicochemical properties (solubility, vapor pressure, stability, pKa). Overutilization of pesticides results in extreme outcomes for soil fauna which in turn can lead to depletion in the number of beneficial microbes that can hold onto different nutrients (Savonen, 1997). As stated by U.S. Environmental Protection Agency, up to 46 pesticides were reported in the groundwater of 26 states (Trautmann et al., 1989). They are deliberately designed to cause detrimental effects to specific species or groups of organisms making them apt for their use as pest control agents but sometimes they can also cause harmful effects in nontarget organisms. In a report by the World Health Organization approximately 3 million instances of poisoning by pesticide and 220,000 deaths were tolled each year majorly in developing countries. Invertebrates are devasted by pesticide exposure. The number of taxa and total invertebrate abundance were greatly affected (Ward et al., 1995). Growth rate, survival, and reproduction were also affected by pesticides (Duch et al., 2021; Theys et al., 2021). Pesticides affect vertebrates by increasing mortality, decreasing sperm count, increasing apoptosis, inhibition of antioxidant enzyme activity, diminishing swimming speed, reactive oxygen species (ROS) induced damage, decreasing assay of acetylcholinesterase (AChE) activity, alterations in the histopathology of the liver, kidney, testis, and ovary, and changes in serum glucose level (Bharti & Rasool, 2021; Deka & Mahanata, 2012; Stara et al., 2018).

1.2 Pesticide toxicity to aquatic invertebrates

The route of pesticide subjection to aquatic invertebrates varies from passing across the surface of body walls to intake from pore water, across respiratory surfaces (Chapman & Mason, 1993). The toxicity as well as bioavailability of pesticides depend mostly on whether the pesticide is dissolved in water or adsorbed on sediment. The quantity and quality of organic matter decide the sorption capacity of sediment (Davies & Caseley, 1999; Diepens et al., 2014; Landrum et al., 1996). For example, the interaction of contaminants with food sources makes them available to organisms resulting in their uptake and eliciting toxic effects in them (Lotufo et al., 2001; Schlekat et al., 2000) whereas pesticides' interaction with humic acid layering on sediments make them biologically unavailable to organisms (Decho & Luoma, 1994). According to DeLorenzo et al. (2001), pesticides affect photosynthesis, respiration, and biosynthetic reactions as well as cell division, growth, and molecular composition in microorganisms. Pesticides also affected growth rate, feeding rate, and survival in midges and crustaceans (Stoughton et al., 2008; Theys et al., 2021). Reproductive problems were also reported in water fleas (Duchet et al., 2021). Also, in crustaceans reduced heartbeat, locomotion activity, and gill ventilation were observed (Siregar et al., 2021). A summary of toxicity of various pesticides on different Invertebrates is given in Table 1.1.

1.3 Pesticide toxicity to aquatic vertebrates

The deterioration of the aquatic environment occurs as a result of the indiscriminate use of pesticides, weedicides, herbicides and other related chemicals. The introduction of these chemicals leads to the mortality of thousands of aquatic animals all over the United States (Helfrich et al., 2009). These aquatic animals include fish, frogs, water birds, mussels, turtles, and other aquatic living organism (Helfrich et al., 2009; Lakhani, 2015). A number of effects are seen in fish on coming in contact with pesticides and some of them are restlessness, loss of balance, rapid body movement, uncontrollable muscle contractions, problem in respiration, excess mucous secretion, and change in coloration (Khan & Law, 2005). It is assumed that these chemicals are the main culprit behind the reduced individuals of amphibians, as they are very sensitive organisms (Muths et al., 2006). Atrazine an herbicide is highly poisonous or toxic to many species of amphibians and fishes as it negatively affects the immune system of these organisms (Forson & Storfer, 2006; Rohr et al., 2003). A similar effect has been shown by carbaryl which is also highly toxic to many amphibian species, and there is a high mortality rate of tadpoles and juvenile frogs because of the excessive use of herbicide known as glyphosate (Relyea, 2005). These toxicants have the potential of blocking the action of hormones in reptiles, amphibians, and fishes and cause reproductive malfunctioning and abnormal development. They also impact the secretion rate of hormones or their clearance from the body by targeting the tissues through hormone receptors or nonreceptors (Khan & Law, 2005).

On treatment of mosquito fish (*Gambusia affinis*) with LC_{50} concentration 297 μg/L of chlorpyrifos for 96 hours, diffidence in antioxidant enzyme activities like catalase-CAT (77%), superoxide dismutase-SOD (71%), and glutathione reductase-GR (65%), maximal reduction in swimming speed (84% ± 5.12%) and amplification of lipid peroxidation was observed that indicates that ROS-induced damage may be a major toxic outcome of chlorpyrifos (Kavitha & Rao, 2008).For the evaluation of the toxic effects of the vinclozolin (1.0, 10.0, and 100.0 μg per milligram fodder) and p,p′-DDE (0.1, 1.0, or 10.0 μg per milligram fodder), an experiment was done on Guppy fish wild-type (*Poecilia reticulate)* in Denmark. A 15% and 70% mortality of fish was observed on exposure to 100 μg/mg of vinclozolin and 10 μg/mg of p,p′-DDE respectively. The most conspicuous effect was decrease in sperm count when treated with largest concentration of vinclozolin but increase in number of sperm cells in comparison to control on exposure of 0.1 μg/mg of p,p′-DDE. Furthermore, the weight of testes is about 2.8% of the body weight in the control group and Gonadosomatic index (GSI) values between 1.7% and 2.0% of specified fishes exposed to the two chemicals (Baatrup & Junge, 2001).

Table 1.1 Toxicity of various pesticides to different aquatic invertebrates.

Test organisms	Pesticide	Concentration	Response	References
Chironomus riparius	^{14}C-lindane fate, distribution and toxicity with respect to varying sediment organic content (10, 20, and 40% OM of dry wt.), a density gradient of C. riparius (0, 2000, 6000, and 18,000 ind/m^2)	0 (controls), 7.5, 22.5, 68, and 102 μg/L.	*Chironomus* densities affected both larval mortality, development and partitioning of lindane between sediment, interstitial water and overlying water whereas sediment organic matter affected only larval development and sediment partitioning, larval mortality and development were also affected by lindane.	Goedkoop & Peterson (2003)
Chironomus riparius	Lindane	0, 0.01, 0.1, 0.5, 0.75, 1 ppm for 48 hours.	Reduced fecundity, adult body size, effected larval behavior and delayed emergence time.	Hirthe et al. (2001)
Daphnia magna, Hyalella azteca, Chironomus tentans	Chlorpyrifos, aldicarb, and chlordane		Invertebrates tested were ≥200 times more sensitive compared to vertebrate fish *P. promelas* to chlorpyrifos, *H. azteca* and *D. magna* was more sensitive to chlorpyrifos, *C. tentans* was most sensitive for both aldicarb and chlordane respectively.	Moore et al. (1998)
74 nonchironomid and 24 chironomid taxa	Chlorpyrifos in an outdoor artificial stream system	0.1 μg/L and 5 μg/L for 21 days.	Significant reduction in total invertebrate abundance and number of taxa at both doses. Shannon-Weaver diversity reduced at both doses whereas evenness increased at high doses.	Ward et al. (1995)

(continued on next page)

Table 1.1 Toxicity of various pesticides to different aquatic invertebrates—cont'd

Test organisms	Pesticide	Concentration	Response	References
Chironomus tentans and *Hyalella azteca*	Imidacloprid, Admire LC_{50} 65.43 and 17.44 μg/L for *H. azteca* respectively and 5.75 and 5.40 μg/L for for *C. tentans* respectively	28-day chronic tests.	Reduction in invertebrate growth and survival was seen but organisms are able to recover from short-term pulse exposure to similar imidacloprid concentrations if the stressor is removed after four days.	Stoughton et al. (2008)
Neocaridina denticulata	Imidacloprid 0.03125, 0.0625, 0.125, 0.25, 0.5, and 1 ppm	24, 48, and 72 hours.	reduced heartbeat, gill ventilation, reduction of locomotor activity.	Siregar et al. (2021)
Ceriodaphnia dubia and *Daphnia magna*	Imidacloprid (0.256 μg/L), clothianidin (3.11 μg/L), and thiamethoxam (1.49 μg/L) and their mixture	Chronic tests were performed.	Survival of both species not affected, Reproduction affected in *C. dubia* in mixture only but in *D. magna* significant reduction in reproduction at all concentrations.	Duchet et al. (2021)
Asellus aquaticus	Chlorpyrifos 0.2 μg/L at 22 °C and in presence of DTF	10 days	Growth rate significantly reduced at both conditions and feeding rate was affected at high temperature.	Theys et al. (2021)

A study on the exposure of crayfish (*Cherax destructor*) to environmental atrazine concentration (ATRenv) and 10% of the 96-hours LC_{50} (ATR10%) reported that the exposed fishes were less agile than control fishes although food consumption was comparable. Twelve percent of fishes died during the exposure period in the ATR10% group. A significant increase in glucose (74%), alkaline phosphatase (278%), lactate (96%), and ammonia levels (43%) was seen in the hemolymph of the ATR10% exposure group. In the ATR10% group after 7 days, activity of catalase was 66% lower in muscle, and about 88% elevated in the same group for hepatopancreas after 14 days, in comparison to control. The superoxide dismutase activity was decreased by 47% and 50% in hepatopancreas at ATRenv and ATR10% after an exposure duration of 14 days respectively. Gills of crayfish revealed hypertrophy of cells present on the margin of gill fragments, detachment of the membrane surface, disintegration with pseudocyst formation, breakdown of the epithelial layer, and eosinophil granulated plasma in a larger number of elements in both ATR exposure groups (ATRenv and ATR10%) (Stara et al., 2018).

Wang et al. (2018) evaluated the toxicity of cyprodinil (CYP) and kresoxim-methyl (KRM) to various life stages of zebrafish (*Danio rerio*). After determination of the 96-hour LC_{50} values of CYP and KRM at various life stages, it was seen that these chemicals are fatal to fish and the juvenile fish was more sensitive to CYP and KRM, and the adult fish was more sensitive. Furthermore, KRM is more lethal than CYP for all the life stages of zebrafish. In zebrafish embryos, slight changes in the activity of caspase 3 in all the KRM, CYP, and MIX (KRM and CYP) exposures were observed, but caspase 3 activity was significantly elevated for the mean and high concentrations of CYP in comparison to control. In addition, a significant reduction in caspase 3 activity was noticed at the mean concentration of MIX exposure when compared with the parallel CYP exposure. The activity of Cu/Zn-superoxide dismutase was significantly diminished in low concentrations of MIX exposure and all the CYP and KRM exposure groups excluding low concentrations of KRM exposure in comparison to control whereas significant elevation was detected in the mean and high concentrations of MIX exposure when compared with the parallel CYP or KRM exposure. A significant elevation in glutathione-s-transferase level in zebrafish embryos on exposure to a high concentration of CYP and KRM was reported and slight changes were seen in almost all exposures after 96 hours compared with the control. The activity of catalase was not significantly different for every exposure in comparison to the control but significant elevation was seen in the mean and high concentrations of MIX exposure when compared with the parallel CYP and KRM exposures or the control. However, carboxyl esterase activity in the embryos of zebrafish was significantly reduced in all the exposures in comparison to the control.

Exposure to carbamate and organophosphorus insecticides to Rusty blenny (*Parablennius sanguinolentus)* along the Western Calabrian coast results in the reduction of AChE and BChE activity in the brain (Parrino et al., 2020) and a similar result was

Table 1.2 Toxicity of various pesticides on different aquatic vertebrates.

Chemicals	Test organisms (common name and scientific name)	Concentration and duration of exposure	Significant findings	References
Vinclozolin	Guppy fish wild-type (*Poecilia reticulate*)	1.0, 10.0, and 100.0 μg per mg fodder for 30 days	15% of the fish died during the exposure; reduced the number of sperm cells, Vinclozolin treated groups at 100 μg/mg contained 3–4 million sperm cells; whereas gonadosomatic index (GSI) value lies between 1.7% and 2.0%.	Baatrup & Junge (2001)
p,p′-DDE	Guppy fish wild-type (*Poecilia reticulate*)	0.1, 1.0, or 10.0 μg per mg fodder for 30 days	70% of the fish dies during the treatment; p,p′-DDE at 0.1 μg/mg, sperm cell count exceeded that of control fish; whereas GSI value lies between 1.7% and 2.0%.	Baatrup & Junge (2001)
Atrazine	Cray fish (*Cherax destructo*)	Within ± 5% of the intended concentration for 14 days	In hepatopancreas, SOD activity by 25% at ATRenv and 35% at ATR10% during 7-day exposure and after 14 days exposure period, there was up to 47% at ATRenv and 50% at ATR 10%; Hypertrophy of cells on the periphery of fragments, separation of the membrane surface, fragmentation with pseudocyst formation and disintegration of epithelium.	Stara et al. (2018)
Cyprodinil and kresoxim-methyl	Zebrafish (*Danio rerio*)	96 hours	According to 96-hour LC_{50} values of cyprodinil and kresoxim-methyl, the sensitivity of *D. rerio* to CYP is maximum for juvenile fish followed by embryo, larvae and adult fish while sensitivity to KRM was highest for adult fish and then juvenile fish, larvae and embryo; elevation of caspase 3 in response to mean and high concentrations of CYP exposure; significant decline of caspase 3 activity on exposure of mean concentration of MIX exposure compared with the parallel CYP exposure.	Wang et al. (2018)

(*continued on next page*)

Table 1.2 Toxicity of various pesticides on different aquatic vertebrates—cont'd

Chemicals	Test organisms (common name and scientific name)	Concentration and duration of exposure	Significant findings	References
Chlorpyrifos	Mosquito fish (*Gambusia affinis*)	297 μg/L for 96 hours	Reduction in the swimming speed by 84% ± 5.12%; the exposed fish exhibited significant inhibition in antioxidant enzyme activities like CAT (77%), SOD (71%), and GR (65%); the increased lipid peroxidation in the present study suggests that ROS-induced damage may be one of the main toxic effect of chlorpyrifos.	Kavitha & Rao (2008)
Carbamate and organophosphorus insecticides	Rusty blenny *(Parablenniussan-guinolentus)*	1.0 μg/L; 0.14 μg/L	Decrease in AChE and BChE activity in the brain.	Parrino et al. (2020)
Atrazine and chlorpyrifos	Fathead minnow (*Pimephales promelas*), Bluegill (*Lepomis-macrochirus*), Green frog (*Rana clamitans*), African clawed frog (*Xenopus laevis*)	0, 100, and 1000 μg/L; EC10, EC25, EC50, EC75, and EC95 for each species with atrazine for 48 hours and for next 48 hours with atrazine alone or mixture of both.	AChE activity lowered in *P. promelas* and *X. laevis.*	Wacksman et al. (2006)
Atrazine	South American toad (*Rhinellaarenarum*)	1, 2.5, 5, 7.5, 10, 15, 20, 25 and 30mg/L (Stage 4 and Stage 25 larvae); 0.1, 1, 5, 10, 15 and 20mg/L (Stage 38-39 larvae) for 21 days.	Stage 25 larvae mortality was 78.3% and 31.8% with 20 mg/L and 30mg/L of atrazine, respectively. 100% mortality with 15 mg/L and 20 mg/L of atrazine; 50% mortality with 10 mg/L.	Brodeur et al. (2009)

(continued on next page)

Endosulfan	Male Cichlid (*Cichlasomadimerus)*	0.7 µg/L (as active ingredient and commercial formulatio) for 14 days.	Low HSI, Bioaccumulation of α-endosulfan in gills > brain > muscle with active ingredient (A.I.) and no detectable amount with commercial formulation (C.F.). For β-endosulfan, bioaccumulation due to A.I. was gills > brain > muscle > liver while due to C.F., it was liver > gills > brain > muscle. High level of endosulfan-sulfate was observed with both the exposures. Amount of lipid percentage and endosulfan-sulfate had a positive correlation when exposed to A.I. but not with C.F. in terms of bioaccumulation factors, muscle had the lowest while liver had the highest factors in both the exposures.	Cuña et al. (2020)
Malathion	Asian stinging catfish *(Heteropneustesfossilis)*	0.2ppb (LC_{50}) for 10, 20, and 30 days.	In liver, swollen and dissociated hepatocytes, granular cytoplasm, pycnotic nuclei, increase in fibro connective tissue, rupture of capillary endothelium, acute and extensive necrosis of liver cells was observed. In kidney, disintegrated renal tubules, ruptured blood vessels, pyknosis, apoptotic and necrotic cells, degeneration of hematopoietic cells and haemmorhage was recorded. In ovary, adhesion of primary follicles, cytoplasmic retraction, increased atretic oocytes, destruction of the ovigerous lamellae and vitellogenic membrane, increased intrafollicular spaces, vacuolated and necrotic cytoplasm, extrusion of karyoplasm and embedded nucleoli into the surrounding cytoplasm in oocyte were observed.	Deka & Mahanata (2012)

(continued on next page)

Table 1.2 Toxicity of various pesticides on different aquatic vertebrates—cont'd

Chemicals	Test organisms (common name and scientific name)	Concentration and duration of exposure	Significant findings	References
Atrazine	Broad-banded grassland frog (*Ptychaden-abibroni)* at different developmental stages (1, 2, 3, and 4 weeks)	0, 200, 400, 600 μg/L for 96 hours and 0, 3, 30, 100 μg/L for 28 days.	Mortality %: 600 μg/L > 400 μg/L > 200 μg/L> 0 μg/L. Mortality %: 1 week > 2week > 3 week > 4 week. Glycogen concentration: 600 μg/L > 400 μg/L > 200 μg/L > 0 μg/L. Glycogen concentration: 1 week> 2week > 3 week > 4 week.	Ezemonye & Tongo (2009)
Endosulfan	Male Indus lotak (*Cyprinion watsoni)*	0.5 ppb and 1 ppb for 30 days.	During spawning and quiescent periods, mortality increases with dose. A highly significant increase in Fulton's condition factor was observed in 1 ppb exposure group during both periods but only during quiescent period during 0.5 ppb exposure. A highly significant decrease in testicular testosterone in 0.5 ppb and 1 ppb exposure during both periods. In 0.5 ppb testes of both periods, increase in interlobular areas, fewer, smaller, and loosely arranged spermatogonia, clumped spermatocytes and increase in spermatid number was observed. During both periods, in 1 ppb exposed testes, more interlobular areas, absence of spermatogonia, clumped spermatocytes, thinning or rupture of interlobular wall, disintegrated spermatogonia and absence of primary and secondary spermatocytes was observed.	Ul Islam et al. (2017)

(continued on next page)

Malathion	Snake-headed fish (*Channa punctatus)*	1/20th of 96-hour LC_{50} (0.4 mg/L) for 1, 4, 8, and 12 days.	K and organosomatic indices (HSI & KSI) reduced; weight of liver and renal tissue declined significantly with increase in exposure period. Serum glucose level increased while total protein content decreased with exposure duration. Transaminases and ALP enzyme activity in fish serum decreased in time dependent manner. SOD activity in fish liver and kidney; CAT activity in liver and kidney of fish; LPO in kidney were increased steadily during exposure while LPO in liver first increased significantly after 8 days but declined marginally after 12 days. Calcium concentration remains increased throughout the exposure duration. Urea, urea nitrogen, and creatinine decreased after 4 days but increased during the remaining exposure duration. In liver, after 4 days exposure, dilated sinusoids, hypertrophied sinusoids, congested RBCs along the central vein, pyknotic nuclei, vacuolization in hepatocytes, rupture of hepatocyte membrane and release of cytoplasmic content was observed but after 8 and 12 days, congestion of erythrocytes along hepatic portal vein, liver cell necrosis and fibrosis of sinusoids was noticed. After 4 days of exposure in kidney, glomeruli and renal tubules shrunk in size. Detachment of epithelium layer from basement membrane, degeneration and hypertrophy of renal tubules and vacuolization in renal cells was recorded following 8 days of exposure. After 12 days, renal tubules' lumen shrunk, dilation of spaces in Bowman's capsule, atrophy of renal tubules and its epithelial lining was found.	Bharti & Rasool (2021)

observed when Fathead minnow (*Pimephales promelas)*was exposed to atrazine (0, 100, and 1000 μg/L) for first 48 hours and a mixture of atrazine and chlorpyrifos (EC10, EC25, EC50, EC75, and EC95) or atrazine alone for the next 48 hours (Wacksman et al., 2006). In male chlid, low HSI, high endosulfan-sulfate, and bioaccumulation was observed for α- and β-endosulfan with active ingredient but with commercial formulation, bioaccumulation of only β-endosulfan was seen on exposure to 0.7 μg/L of endosulfan as an active ingredient and commercial formulation for 14 days. Endosulfan as an active ingredient has a positive correlation with the amount of lipids. The muscle had the lowest while the liver had the highest bioaccumulation factors with both exposures (Cuña et al., 2020). An exposure of 0.2 ppb (LC_{50}) malathion for 10, 20, and 30 days to *Heteropneustes fossilis* results in significant histopathological alterations in the liver, kidney, and ovary (Deka & Mahanata, 2012).

Exposure of 0.5 ppb and 1 ppb of endosulfan to *Cyprinion watsoni* during their spawning and quiescent periods show a dose-dependent increase in mortality, the decline in testicular testosterone, and a significant elevation in Fulton's condition factor in 1 ppb exposure during both periods and 0.5 ppb during the quiescent period and extensive histological changes in testes (Ul Islam et al., 2017. Malathion exposure of 1/20th of 96-hour LC_{50} (0.4 mg/L) for 1, 4, 8, and 12 days to *Channa punctatus* results in a reduction in hepatosomatic indices (HSI) and kidney somatic indices (KSI), reduction in liver and renal tissue weight, increase in serum glucose and decline in total protein with increasing exposure, decrease in activity of transaminases and alkaline phosphatase-ALP enzyme in serum, increase in activity of SOD and CAT in liver and kidney, increase in lactoperoxidase-LPO activity in kidney, high calcium, alterations in urea, urea nitrogen and creatinine, and significant differences in histology of liver and kidney (Bharti & Rasool, 2021). A Summary of experimental studies based of effects of pesticide exposure to different aquatic vertebrates is provided in Table 1.2.

1.4 Conclusion

In conclusion, it is evident that pesticides have a detrimental impact on aquatic vertebrates as well as invertebrates. The negative impact of pesticides on aquatic life extends beyond individual organisims, posing a threat to the overall health and sustainability of aquatic ecosystem. Therefore, there is a need to standardize the ways for properly disposing off these hazardous contaminants to make environment eco-friendly for all aquatic organisms.

References

Aktar, M. W., Paramasivam, M., Sengupta, D., Purkait, S., Ganguly, M., & Banerjee, S. (2009). Impact assessment of pesticide residues in fish of Ganga river around Kolkata in West Bengal. *Environmental Monitoring and Assessment, 157*(1), 97–104.

Baatrup, E., & Junge, M. (2001). Antiandrogenic pesticides disrupt sexual characteristics in the adult male guppy *Poecilia reticulata*. *Environmental Health Perspectives, 109*(10), 1063–1070.

Bharti, S., & Rasool, F. (2021). Analysis of the biochemical and histopathological impact of a mild dose of commercial malathion on *Channa punctatus* (Bloch) fish. *Toxicology Reports, 8*, 443–455.

Brodeur, J. C., Svartz, G., Perez-Coll, C. S., Marino, D. J., & Herkovits, J. (2009). Comparative susceptibility to atrazine of three developmental stages of *Rhinella arenarum* and influence on metamorphosis: Non-monotonous acceleration of the time to climax and delayed tail resorption. *Aquatic Toxicology, 91*(2), 161–170.

Chapman, P. J., & Mason, R. D. (1993). British and European community regulations and registration requirements for non-pesticidal co-formulants in pesticides and for adjuvants. *Pesticide Science, 37*(2), 167–171.

Cuña Da, R. H., Nostro Lo, F. L., Shimabukuro, V., Ondarza, P. M., & Miglioranza, K. S. B. (2020). Bioaccumulation and distribution behavior of endosulfan on a cichlid fish: Differences between exposure to the active ingredient and a commercial formulation. *Environmental Toxicology and Chemistry, 39*(3), 604–611.

Davies, J., & Caseley, J. C. (1999). Herbicide safeners: A review. *Pesticide Science, 55*(11), 1043–1058.

Decho, A. W., & Luoma, S. N. (1994). Humic and fulvic acids: Sink or source in the availability of metals to the marine bivalves *Macoma balthica* and *Potamocorbula amurensis*? *Marine Ecology Progress Series, 108*, 133.

DeLorenzo, M. E., Scott, G. I., & Ross, P. E. (2001). Toxicity of pesticides to aquatic microorganisms: A review. *Environmental Toxicology and Chemistry: An International Journal, 20*(1), 84–98.

Deka, S., & Mahanta, R. (2012). A study on the effect of organophosphorus pesticide malathion on hepato-renal and reproductive organs of *Heteropneustes fossilis* (Bloch). *Science, 1*(1), 1–13.

Diepens, N. J., Pfennig, S., Van den Brink, P. J., Gunnarsson, J. S., Ruepert, C., & Castillo, L. (2014). Effect of pesticides used in banana and pineapple plantations on aquatic ecosystems in Costa Rica. *Journal of Environmental Biology, 35*(sp. issue), 73–84.

Duchet, C., Mitchell, C. J., McIntyre, J. K., & Stark, J. D. (2021). Chronic toxicity of three neonicotinoid insecticides and their mixture on two daphniid species: *Daphnia magna* and *Ceriodaphnia dubia*. *BioRxiv, 2021*, 2021-04.

Eleftherohorinos, I. G. (2008). Weed science: weeds, herbicides, environment, and methods for weed management. *AgroTypos, Athens, Greece*, 408.

Ezemonye, L. I. N., & Tongo, I. (2009). Lethal and sublethal effects of atrazine to amphibian larvae. *Jordan Journal of Biological Sciences, 2*(1), 29–36.

Forson, D. D., & Storfer, A. (2006). Atrazine increases rana virus susceptibility in the tiger salamander, *Ambystoma tigrinum*. *Ecological Applications, 16*(6), 2325–2332.

Goedkoop, W., & Peterson, M. (2003). The fate, distribution, and toxicity of lindane in tests with *Chironomus riparius*: Effects of bioturbation and sediment organic matter content. *Environmental Toxicology and Chemistry: An International Journal, 22*(1), 67–76.

Helfrich, L. A., Weigmann, D. L., Hipkins, P. A., & Stinson, E. R. (2009). *Pesticides and aquatic animals: A guide to reducing impacts on aquatic systems. Virginia Polytechnic Institute and State University, 24*.

Hirthe, G., Fisher, T. C., Crane, M., & Callaghan, A. (2001). Short-term exposure to sub-lethal doses of lindane affects developmental parameters in *Chironomus riparius* Meigen, but has no effect on larval glutathione-S-transferase activity. *Chemosphere, 44*(4), 583–589.

Jayaraj, R., Megha, P., & Sreedev, P. (2016). Organochlorine pesticides, their toxic effects on living organisms and their fate in the environment. *Interdisciplinary Toxicology, 9*(3–4), 90.

Kavitha, P., & Rao, J. V. (2008). Toxic effects of chlorpyrifos on antioxidant enzymes and target enzyme acetylcholinesterase interaction in mosquito fish, *Gambusia affinis*. *Environmental Toxicology and Pharmacology, 26*(2), 192–198.

Khan, M. Z., & Law, F. C. (2005). Adverse effects of pesticides and related chemicals on enzyme and hormone systems of fish, amphibians and reptiles: A review. *Proceedings of the Pakistan Academy of Sciences, 42*(4), 315–323.

Kumarasamy, P., Govindaraj, S., Vignesh, S., Rajendran, R. B., & James, R. A. (2012). Anthropogenic nexus on organochlorine pesticide pollution: A case study with Tamiraparani river basin, South India. *Environmental Monitoring and Assessment, 184*(6), 3861–3873.

Lakhani, L. (2015). How to reduce impact of pesticides in aquatic environment. *International Journal of Research-GRANTHAALAYAH, 3*(9SE), 1–5.
Landrum, P. F., Harkey, G. A., & Kukkonen, J. (1996). Evaluation of organic contaminant exposure in aquatic organisms: The significance of bioconcentration. *Ecotoxicology: A Hierarchical Treatment, 2*, 85.
Lotufo, G. R., Landrum, P. F., & Gedeon, M. L. (2001). Toxicity and bioaccumulation of DDT in freshwater amphipods in exposures to spiked sediments. *Environmental Toxicology and Chemistry: An International Journal, 20*(4), 810–825.
Moore, M. T., Huggett, D. B., Gillespie Jr, W. B., Rodgers Jr, J. H., & Cooper, C. M (1998). Comparative toxicity of chlordane, chlorpyrifos, and aldicarb to four aquatic testing organisms. *Archives of Environmental Contamination and Toxicology, 34*(2), 152–157.
Muths, E., Gallant, A. L., Grant, E. H. C., Battaglin, W. A., Green, D. E., Staiger, J. S., Walls, S. C., Gunzburger, M,S., & Kearney, R. F. (2006). *The amphibian research and monitoring initiative (ARMI): 5-year report [No. 2006-5224]*. United States Geological Survey.
Parrino, V., Minutoli, R., Paro, G. L., Surfaro, D., & Fazio, F. (2020). Environmental assessment of the pesticides in *Parablennius sanguinolentus* along the Western Calabrian coast (Italy). *Regional Studies in Marine Science, 36*, 101297.
Rani, L., Thapa, K., Kanojia, N., Sharma, N., Singh, S., Grewal, A. S., Srivastav, A. L., & Kaushal, J. (2021). An extensive review on the consequences of chemical pesticides on human health and environment. *Journal of Cleaner Production, 283*, 124657.
Relyea, R. A. (2005). The lethal impact of Roundup on aquatic and terrestrial amphibians. *Ecological Applications, 15*(4), 1118–1124.
Rohr, J. R., Elskus, A. A., Shepherd, B. S., Crowley, P. H., McCarthy, T. M., Niedzwiecki, J. H., Sager, T., Sih, A., & Palmer, B. D. (2003). Lethal and sublethal effects of atrazine, carbaryl, endosulfan, and octylphenol on the streamside salamander (*Ambystoma barbouri*). *Environmental Toxicology and Chemistry: An International Journal, 22*(10), 2385–2392.
Savonen, C. (1997). *Soil microorganisms object of new OSU service*. Good Fruit Grower. http://www.goodfruit.com/archive/1995/6other.html.
Schlekat, C. E., Dowdle, P. R., Lee, B. G., Luoma, S. N., & Oremland, R. S. (2000). Bioavailability of particle-associated Se to the bivalve *Potamocorbula amurensis*. *Environmental Science & Technology, 34*(21), 4504–4510.
Siregar, P., Suryanto, M. E., Chen, K. H. C., Huang, J. C., Chen, H. M., Kurnia, K. A., Santoso, F., Hussain, A., Hieu, B. T. N., Saputra, F., Audira, G., Roldan, M. J. M., Fernandez, R. A., Macabeo, A. P. G., Lai, H. T., & Hsiao, C. D. (2021). Exploiting the freshwater shrimp *Neocaridina denticulata* as aquatic invertebrate model to evaluate nontargeted pesticide induced toxicity by investigating physiologic and biochemical parameters. *Antioxidants, 10*(3), 391.
Stara, A., Kouba, A., & Velisek, J. (2018). Biochemical and histological effects of sub-chronic exposure to atrazine in crayfish *Cherax destructor*. *Chemico-Biological Interactions, 291*, 95–102.
Stoughton, S. J., Liber, K., Culp, J., & Cessna, A. (2008). Acute and chronic toxicity of imidacloprid to the aquatic invertebrates Chironomus tentans and Hyalella azteca under constant-and pulse-exposure conditions. *Archives of Environmental Contamination and Toxicology, 54*(4), 662–673.
Theys, C., Verheyen, J., Tüzün, N., & Stoks, R. (2021). Higher mean and fluctuating temperatures jointly determine the impact of the pesticide chlorpyrifos on the growth rate and leaf consumption of a freshwater isopod. *Chemosphere, 273*, 128528.
Trautmann, N. M., Porter, K. S., & Wagenet, R. J. (1989). Pesticides and groundwater: A guide for the pesticide user. In: *Natural resources cornell cooperative extension*. Cornell University. http://psep.cce.cornell.edu/facts-slides-self/facts/pest-gr-gud-grw89.aspx.
Ul Islam, F., Jalali, S., Shafqat, M. N., & Shah, S. T. A. (2017). Endosulfan is toxic to the reproductive health of male freshwater fish, *Cyprinion watsoni*. *The Science of Nature, 104*(11–12), 104.
Wacksman, M. N., Maul, J. D., & Lydy, M. J. (2006). Impact of atrazine on chlorpyrifos toxicity in four aquatic vertebrates. *Archives of Environmental Contamination and Toxicology, 51*(4), 681–689.
Wang, Y., Dai, D., Yu, Y., Yang, G., Shen, W., Wang, Q., Weng, H., & Zhao, X. (2018). Evaluation of joint effects of cyprodinil and kresoxim-methyl on zebrafish, *Danio rerio*. *Journal of Hazardous Materials, 352*, 80–91.
Ward, S., Arthington, A. H., & Pusey, B. J. (1995). The effects of a chronic application of chlorpyrifos on the macroinvertebrate fauna in an outdoor artificial stream system: species responses. *Ecotoxicology and Environmental Safety, 30*(1), 2–23.

$-160-

CHAPTER 2

Pesticide types and mechanism of their action in nontarget organisms

Arun Dev Singh[a], Jaspreet Kour[a], Shalini Dhiman[a], Kamini Devi[a], Neerja Sharma[a], Pardeep Kumar[a], Deepak Kumar[b], Mohd. Ali[b], Parkirti[b], Puja Ohri[b], Amrit Pal Singh[c] and Renu Bhardwaj[a]
[a]Department of Botanical and Environmental Sciences, Guru Nanak Dev University, Amritsar, Punjab, India
[b]Department of Zoology, Guru Nanak Dev University, Amritsar, Punjab, India
[c]Department of Pharmaceutical Sciences, Guru Nanak Dev University, Amritsar, Punjab, India

2.1 Introduction

Global population explosion has emerged as a challenge for the climate and thus put pressure on the coupled abiotic and biotic factors. However, to maintain such populations higher yields are supposed to be achieved. Thus, modern agricultural practices have been adopted which require the introduction of chemical pesticides such as organochlorine, pyrethroids, organophosphate, carbamates, benzoic acid herbicides, phenoxy herbicides, triazines, and urea to eliminate all kinds of threats (Fatma et al., 2018; Mieldažys et al., 2016). Pesticides are specific compositions of certain chemicals that are specifically made to eliminate, or lowering insect-pests involved in the destruction of field crops and low yields. These chemicals act against a variety of insect pests, and other pathogens like bacteria, fungi, nematodes, mollusks, and multiple weeds that exert pressure and act as competitors with humans for food resources and spread diseases. These chemicals are also used to eliminate mosquitoes, induce defoliants, as surface disinfectants, etc. On the contrary, these toxic chemicals show harmful impacts on nontarget organisms such as humans, fishes, beneficial microbes, and plants, thus required to be used with great care and disposed of properly. However, these chemical constituents are generally classified as insecticides, fungicides, herbicides, rodenticides, etc. Different pesticides show variability in their physical and chemical compositions and thus can further be classified accordingly (Kaur et al., 2019; Yadav & Devi, 2017). However, Drum (1980) suggested three bases for the classification of different pesticides either on the bases of their chemical composition, their mode of entry, and the target organisms. High loads of pesticide applications not only kill the target organisms but lead to detrimental impacts on the nontarget organisms as well as the host plant itself. In plants, the pesticidal residue interferes with the normal metabolic pathways of the host plants and other biomolecule activity which however leads to morphotoxicity, cytotoxicity, phototoxicity, and genotoxicity in plants. These toxic impacts lead to retarded growth, abnormal cell cycle, curling of leaves, damage in

Pesticides in the Environment: Impact, Assessment, and Remediation
DOI: https://doi.org/10.1016/B978-0-323-99427-9.00012-4

photosystems, chlorosis, necrosis, lowering of germination potential and reactive oxygen species (ROS) generation in the cellular environment, etc. (Kilic et al., 2015; Shakir et al., 2016). In the case of other nontarget organisms, humans get exposed to these toxic chemicals through dust, polluted air, dermal exposure, and mainly through the consumption of infected water and food crops. On the basis of its doses and route of exposure, pesticide impacts can be grouped into chronic and acute toxicity in humans. Chronic toxicity includes nerve disorders, birth disorders, genetic changes, tumors, reproductive effects, endocrine disruption, etc. (Moreira et al., 2021; Sabarwal et al., 2018) whereas the acute effects include dermal infections, eye and skin irritation, headache, dizziness, nausea, cramps, panic attacks, and even the death of the patient could occur (Kalyabina et al., 2021; Yadav & Devi, 2017). Also, some other nontarget insect pests such as pollinators like honeybees, are found to have declined in their populations due to inadequate applications of these pesticides and thus can hamper the normal ecosystem process (Ward et al., 2022). Application of these pesticides in field crops will lead to multifold accumulation of these toxic chemicals into the soils which furthermore leads to the death and destruction of soil microflora and however compromises the soil fertility and other biogeochemical cycles such as nitrogen fixation and mineralization processes of the soil ecosystem (Duke, 2021; Yadav & Devi, 2017). On the other hand, pesticide runoff and washed-away operations led to their entry into the aquatic environment and however assist the entry of toxic residues into the aquatic animals through the food chains (Yadav & Devi, 2017). However, plants' defense machinery gets activated under these stressful conditions and mediates the degradation or detoxification of these pesticidal residues through different enzymatic actions. Also, the detoxification mechanism involves the signaling transduction mechanism of multiple phytohormones such as brassinostroids, salicylic acid (SA), jasmonic acid (JA), etc. (Lu et al., 2018; Ma et al., 2018; Yu et al., 2022).

2.2 Pesticides: types and classification

Pesticides are chemicals that are used to protect crops, maintain yields, and control pests (Mieldažys et al., 2016). Pesticides are classified on the bases of various criteria. For classifying pesticides, the most commonly used criteria include (Botitsi et al., 2017).

- Classification of pesticides based on the target organism (Yadav et al., 2017). Pesticides are categorized based on their target pest and are named with the suffix "-cide" which means "killer" in Latin. This naming convention reflects their intended activity of killing the targeted pests (Fishel & Ferrell, 2019).
- Classification of pesticides based on origin (Biondi et al., 2012). They are divided into two categories:
    1. Biopesticides are derived from biological sources (Amin et al., 2021). These are microbial pesticides, biochemical pesticides, and plant-derived pesticides (Souto et al., 2021).

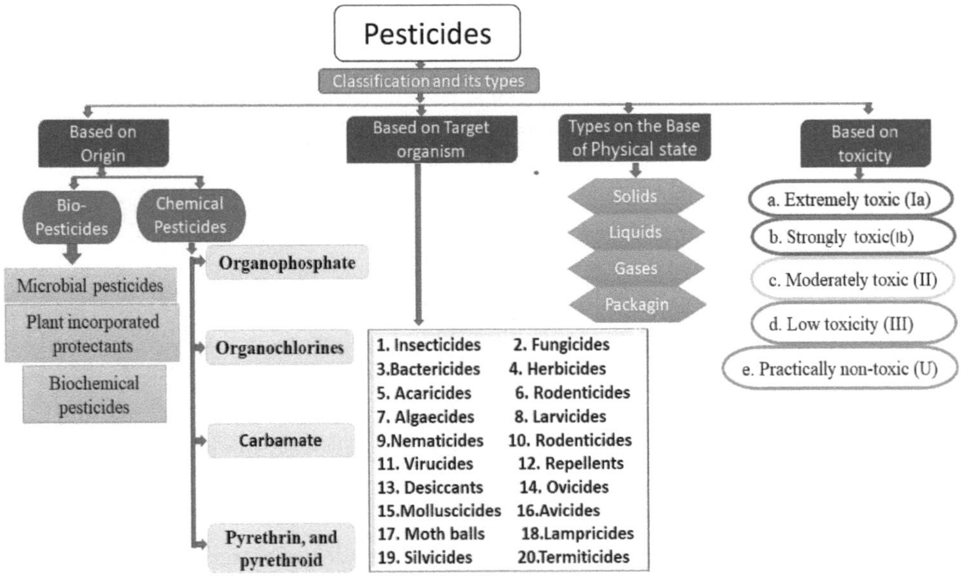

Figure 2.1 Flowchart represents the classification and types of pesticides.

2. Chemical pesticides are determined by their chemical composition (Hoffman et al., 2002). This classification provides information on the efficacy and chemical properties of pesticides. Depending on their chemical makeup, pesticides may be divided into four major categories. They include carbamates, pyrethrin, and pyrethroids, as well as organochlorines, organophosphorus, and carbamates (Zacharia, 2011).

- Classification based on the toxicity of pesticides (Eddleston et al., 2002). The pesticides are classified on the bases of toxicity from extremely hazardous pesticides to partially hazardous (WHO, 2009).
- Type on the bases of physical state (Radosevich et al., 2007). Pesticides can be classified based on their physical state, including whether they are solid, liquid, or gas. These kinds of classification of pesticides are shown in Fig. 2.1.

2.3 Pesticide toxicity and their mechanism of action in plants

Various reports revealed the adverse effects at the cellular or genomic level, genotoxicity, cytotoxic, aneuploidy, DNA damage, high frequencies of mutation, and alteration in the photosynthetic and antioxidant defense systems with increased rates of pesticide usage (Anitha & Savitha, 2013; Fatma et al., 2018; Horbowicz et al., 2013). Although the mechanism of action of pesticides in plants is still unclear. Beyond the allowed limit of pesticide dose, it severely affects the physiological, morphological, and molecular processes at various phases of plant life (Shakir et al., 2016). Extravagantly recurrent application of these pesticides reduced the germination and also retarded the growth of reproductive and vegetative organs of several crops (Kilic et al., 2015). The morphological signs that are used to assess pesticide toxicity include retard growth, poor germination and yield, yellowing, and curling of leaves, etc. the exogenous supplementation of insecticides like dimethon-S-methyl, pirimicarb, dicofol, and dimethoate affects the leaf area/number, photosynthetic machinery, and other molecular changes in tamarind plants (wild) (Saladin & Clément, 2005). Another study revealed the negative effect of fungicides (pyrimethanil and fludioxonil) in grapevine plants showed lowered photosynthetic efficiency and also reduced photosynthetic content due to the overproduction of secondary metabolites (Saladin et al., 2003). The application of fungicides in vitro plants lowered the stomatal conductance, transpiration rate, and CO_2 absorption rate whereas, in the fruiting shoots, there is no alteration in the transpiration and absorption rates (Dias, 2012). The application of glufosinate (herbicide) to tomato plants initiated the accumulation of ammonium and the generation of ethylene (stress and senescence markers).

The pea plants were treated with imazethapyr which increased the content of free amino acid but reduced the content of protein (hydrolysis of protein) (Royuela et al., 2003). It was studied that the nitrogen and carbon metabolism exhibited irregular pathways but the deviations in the metabolic rate depended on the extent of pesticide contact and crop type. It was reported that the application of sulfentrazone, mancozeb, and imidacloprid on *Allium cepa* had toxic effects both at the cellular and genomic level by triggering chromosomal aberrations, chromosome looping by spindle deactivation, and anomalous DNA compression resulting in reduced mitotic index (Bianchi et al., 2016; Fatma et al., 2018).

Pesticides induced the production of ROS which was regulated by a highly controlled mechanism that maintained the balance between the formation of ROS and their reclamation during growth. The overdose of pesticides leads to induced stress in targeted and nontargeted plants by varying the physicochemical and metabolic responses. It was recently reported that the application of pesticides including imidacloprid, alphacypermethrin, and emamectin benzoate to tomato saplings caused cell injury, alteration in the antioxidants like **s**uperoxide dismutase (SOD), peroxidase (POD), and catalase (CAT) activity, disruption in cell viability, and also increased the glutathione reductase activity that led to destroying the biomolecules, cell membrane, nucleic acids, etc. (Ahmad et al., 2018). Fatma et al. (2018) studied the dose-dependent approach of pesticides (chlorpyrifosin and mancozeb) which enhanced the activities of defensive enzymes in *Allium sp.* but beyond the particular limit this led to morpho-toxicity of pesticides.

The pesticides also intervened with the photosystem by binding with plastoquinone (D1 protein) resulting in hampering the electron transport and blocking the PQH_2. Paraquat (herbicide) triggers the generation of free radical species ($O_2\cdot^-$) by interfering in the photosystem. These reactive molecules disrupted the electron gradient structure,

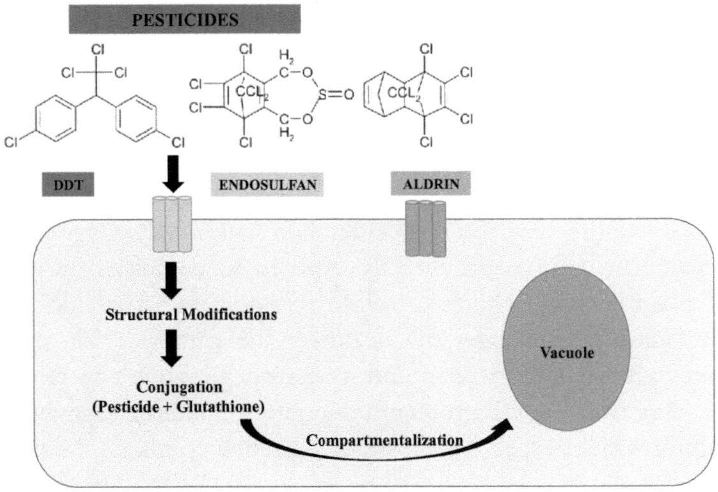

Figure 2.2 Mechanism of pesticides phytoremediation inside plant cells. Reproduced from Kamal et al. (2020).

hence, disturbing the transportation of electrons to the end molecule and impeding the route of energy production (Kamal et al., 2020; Laskay & Lokas, 2011). Among defensive enzymes, glutathione-S-transferase (GST), a multifunctional enzyme that regulates the ascorbate–glutathione cycle further controls the herbicide's conversions to glutathione (GSH) resulting in increasing pesticide tolerance. Singh and Prasad (2018) reported the application of prometryn and glyphosate induced the activity of GST in *Triticum aestivum* and *Arachis hypogaea* plants respectively. Furthermore, proline, an important osmolyte, is considered a stress marker as its level is high during a stressed environment (Ahmad et al., 2017).

Phytoremediation had been studied as the most effective, advanced, and cost-effective method used to reduce pesticide residue via plant rhizosphere system. Though the entire procedure is intricate and multistep and covers several biophysiological and chemical responses in an organized manner such as hydrolysis, reduction, and chemical conjugation; the physical reorganization takes place equally on both sides of the plant (Hoagland et al., 2001). The reaction process catalyzed in the cell during phytoremediation is exemplified in Fig. 2.2. Detailed mode of action is yet to be explored.

2.4 Pesticide toxicity and their mechanism of action in humans

Pesticides are the compounds that are most strictly regulated because of their intended harmful potential for living organisms. Pesticide residues can be transmitted across the supply chain for livestock products after being applied to croplands (Li et al., 2023). To avoid or minimize produce losses brought on by pests, pesticides are frequently employed in agricultural production. Moreover, they are employed to enhance the quality of produce, particularly its look, which is frequently important to consumers (Godage et al., 2023). Humans are usually exposed to pesticides from a variety of sources, such as nutrition, employment, or environmental exposure. Increasing research suggests that residents living away from agricultural areas are exposed to pesticides at levels that are higher than those of residents who live in agricultural areas (Ottenbros et al., 2023). Despite the fact that pesticides have many advantages for agriculture, working on treated fields or being directly exposed to pesticides during mixing and application offers substantial health risks to farmers. Moreover, pesticide residues in food and water may indirectly endanger the health of the general public (Philippe et al., 2021). Pesticides have both short-term and long-term negative effects on a variety of organisms, including neurological problems, respiratory conditions, genetic damage, and reproductive abnormalities especially in underdeveloped countries, pesticide poisoning is a major cause of illness and fatality (Dar et al., 2023). Various studies have been conducted by worldwide scientists showing the adverse effects of pesticides. It has been demonstrated that pesticides that harm the nervous system, are genotoxic, and induce metabolic abnormalities (He & Liu, 2023; Macar et al., 2023; Zhao et al., 2023). Similarly, through interacting with plasma proteins, blood cells, and lipids in various organs and peripheral tissues, pesticides are typically transported throughout the organism (Bertol & Favretto, 2023; Kalyabina et al., 2021). The lipophilicity of substances affects how strongly molecules bind. Hence, information on the solubility of toxicants in water serves as the foundation for calculations of their bioaccumulation, biodegradation, hydrolysis, and adsorption rates (Kalyabina et al., 2021; Wang et al., 2023). Additionally, a balanced diet for humans must include plenty of vegetables to ensure that the necessary daily intake of important nutrients, vitamins, and minerals is met, although vegetables infected with pesticide residues could be dangerous to customers' acute and long-term health. Children and newborns are the most at risk and severely impacted target demographics (Meftaul et al., 2023).

2.4.1 Mechanism of action

Pesticides may have an impact on the metabolic processes and enzyme reactions involved in whole-cell activity (Bokade & Bajaj, 2023; Kalyabina et al., 2021). The primary component of the mitochondrial electron transport chain, nicotinamide adenine dinucleotide + hydrogen (NADH)-dehydrogenase, can practically be diminished in activity by a few pesticides. Oxidative stress and neurotoxicity may be mitigated by chlorpyrifos' poor NADH-dehydrogenase movement (Sule et al., 2022). Moreover, the production of ROS and responsive nitrogen species (RNS) by pesticides can lead to oxidative stress and damage to cell structures. Moreover, hepatotoxicity was caused by elevated ROS/RNS in vertebrates during the digestion and biotransformation of hazardous chemicals (Doroudian et al., 2022; Kalyabina et al., 2021). There are still few studies showing that chemical pesticides have negative effects on genetic and epigenetic modifications, specifically in people who are exposed to them by DNA methylation and miRNA expression, in addition to their effects on cells, metabolic systems, and other things (Teodoro et al., 2019; Wu et al., 2023). Overexpression of miRNAs and DNA methylation status is associated with important cell and molecular pathways that result in various clinical illnesses (Bhatti et al., 2023; Teodoro et al., 2019). Additionally, pesticides can behave as endocrine disrupting chemicals that can interfere with the production, secretion, transport, binding, action, or removal of endogenous hormones, which are necessary for the homeostasis of processes including development and reproduction. However, in spermatogenesis, the luteinizing hormone (LH) normally binds to the LH/choriogonadotropin receptor found on the membrane of Leydig cells. Protein kinase A (PKA) and cyclic AMP (cAMP) levels rise, which starts a cascade that results in testosterone production. Pesticide exposure affects the testosterone production cascade by lowering cAMP and PKA levels and decreasing the expression of genes that are steroidogenic. Furthermore, it results in increased ROS levels that aggravate mitochondrial damage is an acronym for luteinizing hormone (Moreira et al., 2021; Venkaiah et al., 2022). Similarly, due to their lipophilic nature, pesticides can reach the brain when ingested, absorbed through the skin, or inhaled (Khalil et al., 2022; Narasimhamurthy et al., 2022). When it enters the brain, it can target elements in a number of signaling pathways, including the cAMP response element binding protein (CREB) phosphorylation and gene expression pathways, the brain-derived neurotrophic factor (BDNF)/tropomyosin-related kinase B (TrkB) pathway, the rat sarcoma (RAS)/rapidly accelerated fibrosarcoma (RAF)/mitogen-activated protein kinase (MEK) pathway, the phosphoinositide 3-kinase (Pi3K)/protein kinase B (AKT) pathway, the phospholipase C (PLC)/protein kinase C (PKC) pathway, the calcium–calmodulin pathway, and the cAMP pathway. The numerous roles of CREB, including the control of neuronal plasticity and survival, can be impacted by changes in phosphorylation or gene expression (Narasimhamurthy et al., 2022) (Fig. 2.3).

2.5 Pesticide toxicity and their mechanism of action in honeybees

Crops are frequently plagued by a variety of pests and weeds, which lowers agricultural yields and quality (Kumar & Ohri, 2023; Sharma et al., 2023). To improve global crop production, modern agriculture increasingly relies on the use of agrochemicals including pesticides and fertilizers (Nguegang et al., 2022; Sharma et al., 2022). The usage of agrochemicals, however, is anticipated to treble by 2050, which would raise the dangers and negative impacts on nontarget organisms, including honeybees (Zhao et al., 2022).

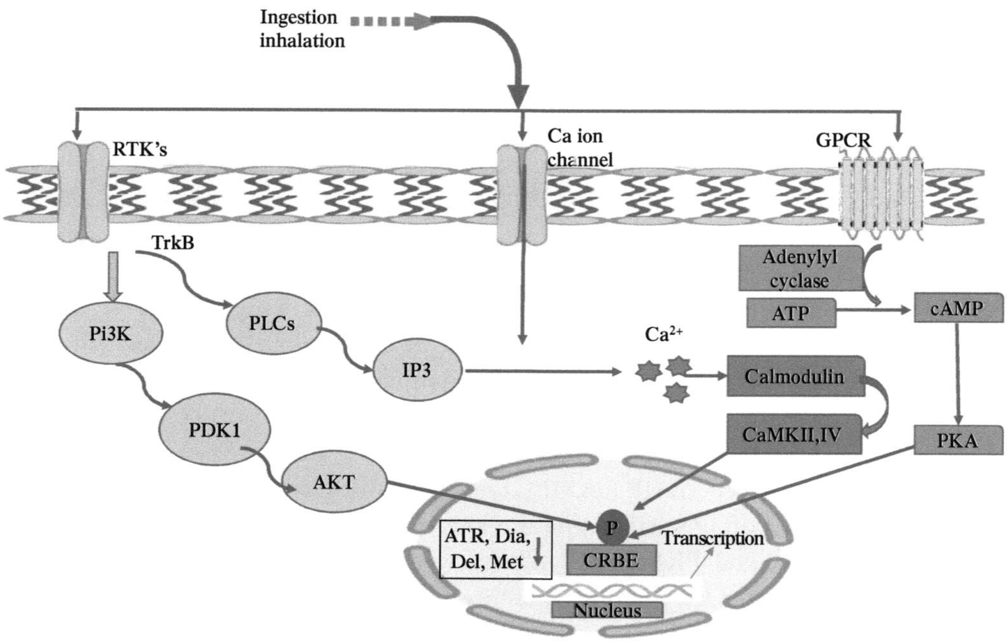

Figure 2.3 Mechanism of pesticides involved neurotoxicity.

Various research (Table 2.1) have demonstrated that pesticides have negative impacts on honeybee behavior and physiological health (Lv et al., 2023; Pervez & Manzoor, 2023; Yang et al., 2023). Honeybees that are out foraging can travel up to 15 km from the hive in search of nectar, pollen, and water, which are essential for sustaining a colony. Foragers are exposed to pesticides when they are foraging by a number of methods. Firstly, pesticides may be present in pollen and nectar from different crops, which could cause oral contact while honeybees are feeding (Hellström et al., 2023; Koech et al., 2023). Second, through spray drift or treatment seed, highly concentrated systemic pesticides may be released into surface water, guttation water, or air, further raising the danger of honeybee exposure to pesticides (Alsafran et al., 2022; Bonmatin et al., 2015; Ward et al., 2022). The situation is made worse by the fact that honeybees are exposed to pesticide contamination from the environment and bring pesticides in different forms back to the beehive, where the larvae, pupae, nurse bees, and queen are vulnerable (Van der Sluijs et al., 2013; Walsh et al., 2020). Interestingly, beekeeper-applied pesticides, such as acaricides, also result in detrimental effects on honeybees in addition to agricultural pesticides (Johnson et al., 2013; Tihelka, 2018). Both pesticides used by beekeepers and those used on farms can linger for several years in honeybees' wax combs, which may worsen their effects on honeybees (Lu et al., 2020). Fig. 2.4 illustrates various ways by which pesticides can be introduced to honeybees.

Table 2.1 Toxicity effect of different pesticides or their active ingredients in honeybees.

Pesticide/active ingredient	Honeybee species	Reported effect	References
Sulfoxaflor	*Apis mellifera*	Increased activity of alkaline phosphatase, acetylcholinesterase, and alkaline phosphatase.	Ahmed et al. (2023)
Glufosinate-ammonium and sulfoxaflor	*A. mellifera*	Affected immunocompetence, decreased life span, and altered gut bacterial composition of bees.	Castelli et al. (2023)
Sulfoxaflor	*A. mellifera*	Decreased food intake, body weight, and survival rate of bees.	Cheng et al. (2023)
Pyraclostrobin and chlorpyrifos	–	Downregulation of *ilp-1* that affects caste structure of colony.	Christen (2023)
λ-cyhalothrin, Acephate, Imidacloprid, and Carbaryl	*A. mellifera*	Reduced cell viability in a concentration-dependent manner and increased expression of stress response genes.	Goblirsch & Adamczyk (2023)
Azoxystrobin, boscalid, and pyraclostrobin	*A. cerana cerena*	Increased intercellular spaces and vacuolization in midgut epithelial cells and disrupted microbial balance in bee gut.	Huang et al. (2023)
Flupyradifurone	*A. mellifera*	Affected larval proteins expression.	Kablau et al. (2023)
Thiamethoxam and 7 other pesticides	*A. mellifera*	Exhibited feeding toxicity.	Li et al. (2023)
Twenty-one chemicals from different products	*A. mellifera*	Interfered with sense of smell.	Mayack (2023)
Clethodim and haloxyfop-p-methyl	*A. mellifera*	Reduced survival and body weight of bees, altered morphology and caused deleterious effects on hemocyte.	Mohamed et al. (2023)
Imidacloprid	*A. mellifera carnica*	Reduced levels of electrolytes and induced bioelement malnutrition.	Paleolog et al. (2023)
Boscalid	*A. mellifera*	Decreased number of spermatozoa stored in queen spermatheca and increased queen mortality indicating disrupted reproduction.	Pineaux et al. (2023)

(*continued on next page*)

Table 2.1 Toxicity effect of different pesticides or their active ingredients in honeybees—cont'd

Pesticide/active ingredient	Honeybee species	Reported effect	References
Azoxystrobin	*A. mellifera*	Induced morphological alterations such as intense cytoplasmic vacuolization and cell fragmentation.	Serra et al. (2023)
Difenoconazole, glyphosate and imidacloprid	–	Altered physiology of newly emerged bees.	Almasri et al. (2022)
Emamectin benzoate	*A. florea*	Induced 65%–100% mortality in a concentration dependent manner.	Anwar et al. (2022)
Sulfoxaflor	*A. mellifera*	Reduced daily flight activity and total number of flights.	Barascou et al. (2022)
Sulfoxaflor	*A. mellifera*	Affected homing ability of honeybees.	Capela et al. (2022)
Imidacloprid	*A. mellifera*	Induced autophagy and apoptosis in bee's midgut and altered ultrastructure including nuclear pyknosis, disorganization of the striated border, enlarged intercellular spaces and formation of cytoplasmic vacuoles.	Carneiro et al. (2022)
Sulfoxaflor	*A. mellifera*	Impaired learning and memory retrieval.	Cartereau et al. (2022)
Imidacloprid	*A. mellifera*	Reduced activity of glutathione S-transferase, acetylcholinesterase, and increased larval mortality.	Delkash-Roudsari et al. (2022)
Sulfoxaflor	*A. mellifera*	Reduced brood production, bee bread, and honey and impacted colony growth.	El-Din et al. (2022)
Cyantraniliprole and sulfoxaflor	*A. mellifera*	Lowered body weight of larvae and sulfoxaflor also induced deformation in wings of emerged bees.	Kim et al. (2022)
Imidacloprid	*A. mellifera*	Affected somatic muscle development, phototransduction, and behavior and altered gene expression in larvae, pupae and adults.	Chen et al. (2021)

(continued on next page)

Table 2.1 Toxicity effect of different pesticides or their active ingredients in honeybees—cont'd

Pesticide/active ingredient	Honeybee species	Reported effect	References
Thiamethoxam	*A. mellifera*	Increased juvenile hormone 3 in larvae, upregulated 140 differentially expressed genes and altered gustatory and odorant-related genes.	Li et al. (2021a)
Sulfoxaflor	*A. mellifera*	Disrupted larval development leading to eclosion failure, resulted death of adult, and induced oxidative stress in larvae as well as adults.	Li et al. (2021b)
Emamectin benzoate	*A. mellifera*	Caused more than 50% larval mortality in 6 days.	Ulziibayar et al. (2021)
S-dinotefuran	*A. cerana* and *A. mellifera*	Altered neuronal signaling.	Zhang et al. (2021)
Roundup®	*A. mellifera*	Increased food consumption and decreased survival rate.	Faita et al. (2020)
Imidacloprid	*A. cerana*	Upregulation of genes involved in mechanistic target of rapamycin (mTOR) signaling pathway and phenylalanine metabolism and decreased activity of detoxifying enzymes including carboxylesterase and cytochrome P450 monooxygenase.	Gao et al. (2020)
Flupyradifurone	*A. mellifera*	Induced cell swelling of brain cells, pycnotic nucleus, and premature onset of foraging.	Hesselbach et al. (2020)
Azadirachtin and chlorantraniliprole	*A. mellifera*	Both pesticides impaired ability to flight and azadirachtin also resulted in mortality of bees.	Naiara Gomes et al. (2020)
Acetamiprid	*A. mellifera*	Induced precocious foraging and shortened lifespan in worker bee.	Shi et al. (2020)
Chlorantraniliprole	*A. mellifera*	Significantly increased mortality, caused lethargic behavior, and reduced locomotion activity.	Williams et al. (2020)
Nitenpyram	*A. mellifera*	Reduced food consumption, survival, and altered gut microbiota.	Zhu et al. (2020)
Clothianidin	*A. mellifera*	Altered biosensory abilities.	Bartling et al. (2019)
Imidacloprid	*A. mellifera*	Altered synapse regulation, induced apoptosis, and oxidative stress.	Catae et al. (2019)

(continued on next page)

Table 2.1 Toxicity effect of different pesticides or their active ingredients in honeybees—cont'd

Pesticide/active ingredient	Honeybee species	Reported effect	References
Acetamiprid	*A. mellifera*	Increased mortality of honeybees, lowered flight intensity, colony intensity, and area covered by pupae.	Chen et al. (2019)
Clothianidin	*A. mellifera*	Decreased glycogen and lipid content in bees and reduced body weight.	Cook et al. (2019)
Boscalid	*A. mellifera*	Reduced wingbeat frequency and forager efficiency.	Liao et al. (2019)
Flupyradifurone	*A. mellifera*	Induced mortality and abnormal behavior.	Tosi & Nieh (2019)
Diflubenzuro	*A. mellifera*	Increased larval mortality.	Wade et al. (2019)
Coumaphos	*A. mellifera*	Increased mortality and downregulated catalase.	Gregorc et al. (2018)
Fupyradifurone	*A. mellifera*	Reduced taste and appetitive behavior.	Hesselbach et al. (2018)
Spinosad	*A. mellifera*	Altered behavior, increased resting time, and reduced walking velocity and distance.	Lopes et al. (2018)
Clothianidin	*A. mellifera*	Reduced body weight, increased mortality, and enhanced activity of cytochrome P450 monooxygenase.	Yao et al. (2018a)
Acephate	*A. mellifera*	Inhibited activity of acetylcholinesterase, glutathione-S-transferase and, esterase by 29.4, 20.4, and 79.9% respectively and reduced body weight.	Yao et al. (2018b)
Imidacloprid	*A. mellifera*	Impaired climbing ability and downregulated muscle related genes.	Wu et al. (2017)
Coumaphos and imidacloprid	*A. mellifera*	Decreased sperm viability in queen bees, decreased activity of development and immunity genes, and suppressed expression of antioxidative genes.	Chaimanee et al. (2016)
Coumaphos and prochloraz	*A. mellifera*	Downregulate immune-related genes in prepupae.	Cizelj et al. (2016)

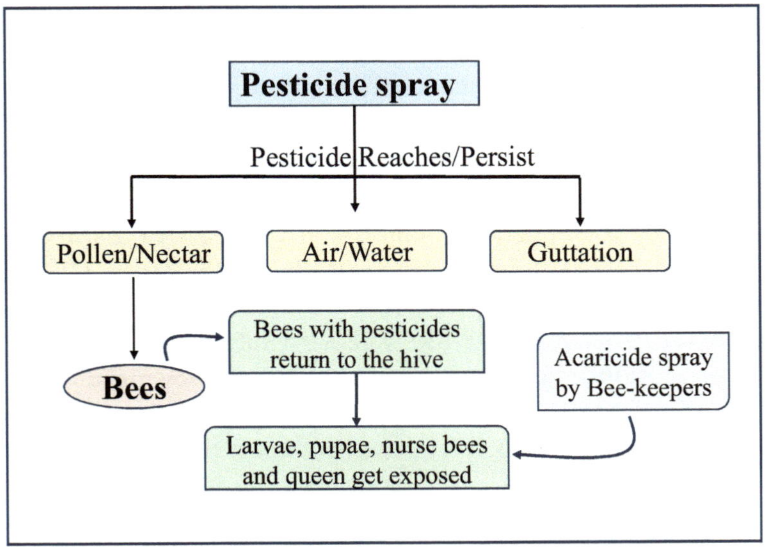

Figure 2.4 Illustration of various ways by which pesticides can be introduced to honeybees.

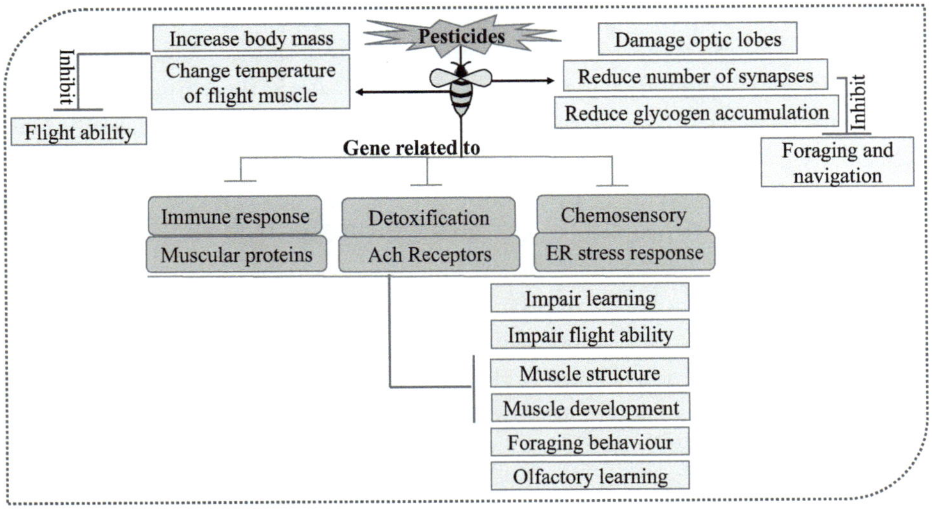

Figure 2.5 Different regulatory mechanisms by which pesticides induce toxicity to honeybees.

2.5.1 Mechanism of action

Different pesticides have different modes of action (Fig. 2.5) and their impact depends on the exposure time and the concentration. Some pesticides such as neonicotinoids and the byproducts of their breakdown operate as neuroactivators of nAChRs in insects (Stanley et al., 2016). Neonicotinoids deactivate neurons in the mushroom bodies of honeybees' brains and paralyze insects when they bind to the postsynaptic nicotinic receptors or nAChRs (Palmer et al., 2013). In addition, neonicotinoids' harmful impacts on honeybee health are evident in a variety of areas, including olfactory learning, navigation, flight prowess, and dancing circuits. In a study, a transcriptome analysis of honeybee brain samples was used to investigate the molecular mechanism by which imidacloprid, a neonicotinoid affects olfaction. The results showed that the expression levels of genes related to immune responses, (*ABAECIN, DEFENSIN1, APISIMIN,* and *PROPHENOLOXIDASE*) detoxification (*CYP6A2, CYP6A17, CYP9E2, and ES-TERASEA2*) and chemosensory responses (*DORANT-BINDING PROTEIN4 (OBP4), OBP16, OBP18, OBP21,* and *SLC18A2*), were all significantly downregulated by imidacloprid (Li et al., 2019). Further research revealed the downregulation of cAMP-dependent kinase and cAMP response element-binding protein honeybees treated with imidacloprid (Christen et al., 2016). It is interesting to note that a recent study found that honeybees treated with thiamethoxam were unable to recognize floral fragrances, while those exposed to clothianidin, imidacloprid, and dinotefuran took longer to recognize these differences (Mustard et al., 2020). In another study, it was found that *S*-dinotefuran exhibited 22.9-fold more olfactory toxicity in honeybees than *R*-dinotefuran at 48 hours however, *R*-dinotefuran altered gene regulation more than S-dinotefuran, according to high-throughput circular RNA sequencing of honeybee brain (Liu et al., 2019). Further, neonicotinoid buildup in the bee brain from eating over several days causes circadian rhythmicity in many individual bees to be disrupted, alters the timing of behavioral circadian rhythms in bees that stay rhythmic, and reduces sleep quality (Tackenberg et al., 2020). Similarly, in a study, bees that were exposed to imidacloprid residues began foraging earlier in adulthood, made fewer orientation flights, and had 28% less foraging flights overall (Colin et al., 2019). In another experiment, Jacob et al. (2019) tested the toxicity of different neonicotinoids including acetamiprid, thiacloprid, and imidacloprid, and found that all the tested pesticides had a negative impact on speed, continuous mobility, resting frequency, and length and distance traveled by a honeybee. In addition to these behaviors and physiological traits, neonicotinoids also affect honeybees' locomotion and sucrose responsiveness, which is a sign of foraging behavior, climbing prowess, and lipid and glycogen levels (Zhao et al., 2022).

Some pesticides such as chlorantraniliprole is reported to block the transcription of *VITELLOGENIN*, which has significant effects on honeybee foraging behaviors and lifespan. Moreover, the transcription of *RELISH* is upregulated while *DOMELESS* and *HOPSCOTCH* are downregulated by chlorantraniliprole, suggesting that the drug may have a negative impact on the immune system as well (Christen & Fent, 2017). However, a recent study indicated that honeybees exposed to sublethal amounts of spinosad had lower expression levels of acetylcholine receptors and genes involved in metabolism and oxidative phosphorylation. This suggests that spinosad has a negative impact on metabolism and the generation of energy (Christen et al., 2019). In addition, honeybees

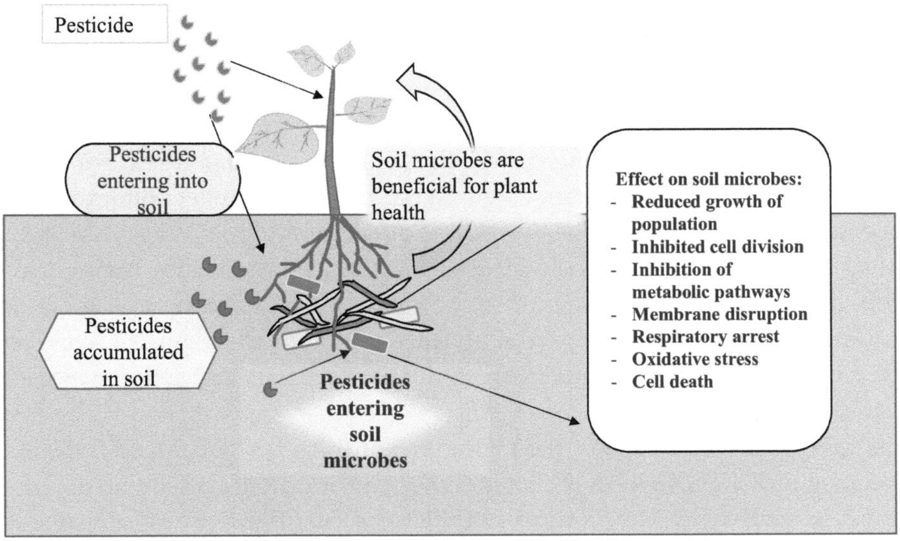

Figure 2.6 The figure represents the mechanism of action of pesticides on soil microorganisms.

die when exposed to spinosad doses that are field advised. In honeybees exposed to 50% and 5% lethal quantities of spinosad, behavioral activity is altered and movement velocity and distance are decreased (Lopes et al., 2018). Recently a study reported that *Apis mellifera CYP6AQ1, CYP9Q2,* and *CYP9Q3* can degrade flupyradifurone and transgenic flies having the expression of *A. mellifera* CYP9Q2 or CYP9Q3 exhibit much lower sensitivity to flupyradifurone, suggesting that these three *CYP* genes play crucial roles in pesticide tolerance (Haas et al., 2021). Therefore, different pesticides affect the survival of honeybees in various ways.

2.6 Pesticide toxicity and their mechanism of action in soil microbes

Pesticides are a group of chemicals that are used worldwide in the agricultural field. These are important constituents of modern agriculture, though they are designed to kill unwanted species their ill effects on various nontarget species cannot be neglected. These chemicals have a detrimental effect on the biodiversity of organisms (Kaur et al., 2019). Pesticidal application in agriculture increases crop productivity, and yield, protects crops from pests, and is more economical. The excess or repeated use of pesticides accumulates these chemicals in soil. This may affect soil microflora and physiochemical properties of soil including adsorption, degradation, and transportation (Fig. 2.6). It can alter soil fertility and disturb the soil ecosystem. It can also alter various biochemical reactions of soil like nitrification, ammonification, and nitrogen fixation (Yadav et al., 2017; Zikankuba et al., 2019). Pesticides along with affecting target organisms also interact with soil microorganisms. Soil microorganisms play an essential role in modulating soil fertility, nutrient uptake, maintaining soil structure, and maintaining nutrient cycle thus helping in the growth and development of plants. (Karpouzas et al., 2022). But these useful microorganisms of soil are at great risk due to excessive use of herbicides, insecticides, rodenticides, weedicides, and fungicides. These chemicals change the normal activities of enzymes, for example, buprofezin chemical cause an adverse effect on invertase in soil and chemicals like pentachlorophenol inhibits the growth of microorganism and affects microbial abundance in soil (Hashimi et al., 2020). The herbicides having a triazine group if applied for a longer duration can be hazardous. In the same way, continuous application of atrazine can reduce the abundance of soil microorganisms. While atrazine and metolachlor can reduce the richness of different bacteria and actinomycetes in soil (Meena et al., 2020).

The mechanism of action of pesticides on soil microbes like *Mesorhizobium* strain, *Azotobacter,* and *Bacillus* sp. mainly includes the entry of pesticidal chemicals inside the cell from the surrounding environment of the microbes. Transport of inorganic ions in excess amount across the cytoplasmic membrane of microorganisms results in higher toxicity and cause oxidative stress by obstructing electron flow through the electron transport chain, it also causes membrane damage, reduces the growth pattern of microbes, cause respiratory arrest, and disturb their internal metabolic activity by binding to sulfide and amino groups (Shahid et al., 2019). These chemicals in microbes also show adverse effects on cell metabolism, growth and cell division, light process, etc. The pesticides that inhibit cell metabolism are divided into seven groups including the acetyl-CoA inhibiting group, acetohydroxyacid synthase group, 5-enopyruvylshikimate-3-phosphate synthase inhibiting group, glutamine synthetase inhibiting group. While light process inhibiting groups involve both photosystem inhibiting pesticides, protoporphyrinogen oxidase inhibiting pesticides, phytoene dehydrogenase inhibiting pesticides, and deoxyxylulose 5-phosphate synthase (DXP) synthase inhibiting (Thiour-Mauprivez et al., 2019). Soil microbes have different tolerance capacities towards different pesticidal chemicals. For example, there was a reduction in microbial biomass carbon on excessive usage of organophosphate pesticide while actinomycete and fungal growth were found to be suppressed by biodegradation of isoproturon pesticide, while the fungicide Captan reduces total culturable fungi, nitrifying bacteria, and aerobic nitrogen-fixing bacteria and soil dominating bacteria and total culturable bacteria were also found to be affected by excessive use of insecticide imidacloprid. Other pesticidal chemicals like paraquat, atrazine, primeextra and glyphosate, ethion, carbofuran, and hexaconazole decline the population of soil bacteria (Tripathi et al., 2020). In a study, the effect of organochlorine pesticide on *E. cloacae* reveals that it can alter the surface morphological structures like the formation of disordered rod-shaped cells with disorganized cell structures (Shahid et al., 2021). A study on the effect of pyraclostrobin on a fungus found that this chemical can inhibit respiration in fungus by inhibiting mitochondrial cytochrome-b and c1.

It also inhibited dehydrogenase activity and glucosidase activity (Zhang et al., 2019). Another study reveals the effect of synthetic pesticides on *Mesorhizobium ciceri* affecting its permeability of the membrane, cell shape, growth, and antioxidant enzymes, thus reducing cell viability (Shahid et al., 2021). Studies show that in the soil bacterium Pseudomonas, glyphosate disturbs aromatic amino acid biosynthetic pathways and it also inhibits secondary metabolite precursor, tryptophan while the herbicide glyphosate is a common herbicide that has an adverse effect on soil microbes such as *Bradyrhizobium* sp. (Duke, 2021; Zaller et al., 2019). In one more study, the effect of pesticides was observed on surface morphology, cell viability, growth patterns, and growth regulators of *Azotobacter vinelandii*. A higher concentration of pesticide shows various ill effects including a decline in bacterial growth, damaged structure, broken cell membrane, respiratory arrest, disturbed metabolism, and impaired metabolic pathways (Shahid et al., 2019). Another study shows the effect of organophosphate pesticides on microorganisms producing siderophore, where siderophore production decreased when the dose of pesticides was increased in *Salmonella typhimurium, A. vinelandii, Rhizobium leguminosarum, Bacillus brevis, Pseudomonas fluorescens* (Kumar et al., 2019). Therefore, the main concern should be given to rational pesticide selection and evaluation of various strategies to remediate chemical toxicity.

2.7 Pesticide toxicity and their mechanism of action in the aquatic ecosystem (fish)

There are a number of negative consequences on aquatic creatures caused by the discharge of pesticidal chemicals from agro-industrial sources into the aquatic environment. Marine animals especially fish have been found to accumulate these toxic contaminants residues via two distinct pathways: direct ingestion from polluted water, and indirect ingestion by higher-level predators, i.e., by mainly entering the food chain (Sharma et al., 2020). According to Arias-Estévez et al. (2008), the residue left behind by pesticides can go from the fields into the water and the air, where it has a detrimental effect on the species that are present in those environments. According to M'Anampiu (2011), when aquatic organisms like fish absorb pesticides, it can have a negative impact not just on the fish's health but also on the quality of the flesh that humans consume when they eat the fish. Pesticide stress often inhibits aquatic animal enzymes, metabolic abnormalities, growth retardation, and fertility and longevity (Sharma et al., 2020). When the water is acidic, fish gills can be taken divalent ions of many toxic pollutants directly from the water (Haider & Inbaraj, 1986). Under laboratory conditions, pesticide exposure was reported to decrease protein levels in the fish brain, gill, muscle, and liver (Sharma et al., 2020; Tilak & Rao, 1991). Additionally, pesticides disrupt endocrine hormones, causing male fish to develop feminine traits (Sharma et al., 2020). These visible signs of developmental disturbance are linked to lower hatching rates, offspring viability, and adult fertility and sterility.

Chlorpyrifos, carbofuran (Carbofuran), pyrethroid, thiamethoxam, beta-cypermethrin, tetraconazole, azoxystrobin, bifenthrin, chlorfenapyr, acetamiprid, dimethoate, malathion, etc., are some of the most popular pesticides which were globally used. These pesticides are hazardous to the animal kingdom and cause various kinds of toxicity like neurotoxicity and genotoxicity (Table 2.2). Further after being distributed inside the organism tissue these pesticides get converted to more dangerous forms like sulfates and glucuronides, which have major deleterious effects that can further have various negative consequences for fish flesh quality and even threaten the survival of these aquatic species (Richterova & Svobodova, 2012). Additionally, these pesticide block axon ion channels. Therefore, aquatic species can experience osmotic stress from concentration gradient disruptions (Murthy et al., 2013). In addition to these aquatic creatures, amphibians are also susceptible to pesticide contamination (Mann et al., 2003). In Müller's clawed frog *Xenopus muelleri* from South Africa's lower Phongolo River floodplain, dichlorodiphenyltrichloroethane (DDT) bioaccumulation increased with time (Wolmarans et al., 2018). Pesticides also harm adult waterbird breeding and embryo development such as poor egg quality, mortality, incapacity of chicks to flourish, skeletal deformities, defective reproductive system and neurological system development, and hormonal mimicking of oestrogens affect embryos. These pesticidal chemicals can kill adult birds, diminish fertility, suppress egg development, thin eggshells, and disrupt incubation and chick rearing (Kadiru et al., 2022).

2.8 Pesticide detoxification mechanism in plants

Pesticide toxicity negatively affects plant growth and development due to oxidative damage of cells by the generated ROS. But under this stressful condition, the antioxidative defense system of plants gets activated which metabolizes the pesticide residues and improves the developmental and physiological functions of the plant's system. There are three phases in pesticide detoxification, i.e., phase I, phase II, and phase III.

Phase I detoxification of pesticides: In phase I, pesticides are enzymatically catalyzed to less hydrophobic metabolites by hydroxylation. Generally, cytochrome P450 monooxygenases (mixed-function oxidase system) and NADPH–cytochrome P450 reductase are involved in the transformation of pesticide compounds (Aigrain et al., 2010).

Carbon hydroxylation leads to the detoxification of pelargonic acid by CYP72A18 in *Oryza sativa,* diuron, fluometuron, linuron, and triasulfuron by CYP71A10 in *Glycine max* and triasulfuron and chlorsulfuron in *Triticum aestivum* (Imaishi & Matumoto, 2007; Ohkawa & Inui, 2014; Xiang et al., 2006). Some pesticides with chemical groups –NH_2/CO–NH. –RCOOR′ and P $(OR)_3$ are selectively modified by hydrolase enzyme (Cho & Kong, 2007; Wan et al., 2017; Zhou et al., 2015). It was shown that hydrolysates of pesticide amide bonds obtained from isoproturon have been recognized in rice and

Table 2.2 Toxicity effects of pesticides on aquatic creatures.

S. no.	Pesticide	Used globally as	Toxic effect on aquatic creatures	References
1.	Chlorpyrifos, thiamethoxam, beta-cypermethrin, tetraconazole, and azoxystrobin	Insecticide	The results showed that the most toxic pesticidal substances to *Gobiocypris rarus* were beta-cypermethrin, chlorpyrifos, and azoxystrobin. These insecticides were shown to be more toxic to larvae of a rare minnow than to embryos. Pesticide combinations harmed fish synergistically.	Yang et al. (2021)
2.	Cypermethrin (synthetic pyrethroid)	Insecticide	Chronic exposure of cypermethrin in *Catla catla* increased kidney stress enzyme activity after 15 days, then decreased up to 45 days. Histopathological examination reveals hematopoietic tissue proliferated and tubules and glomerulus degeneration. Cypermethrin could result in a variety of pathological abnormalities in the renal system and potentially interfering excretory system.	Sharma et al. (2021)
3.	Chlorpyrifos and bifenthrin	Insecticide	*Labeo rohita* DNA damage increased with time up to 56 days after being exposed to mixture of chlorpyrifos and bifenthrin doses, but then it went down slightly over the next 14 days.	Bano et al. (2021)
4.	Chlorfenapyr	Acaricide, insecticide	Comparative studies of the effects of chlorfenapyr, dimethoate, and acetamiprid under chronic exposure revealed notable differences in blood profile and disrupted thyroid profile *Cirrhinus mrigala*.	Ghayyur et al. (2021)
	Acetamiprid	Pesticide		
	Dimethoate	Broad range of insecticides, pesticides, and acaricides	Pesticide-treated fish had increased serum liver biomarker enzymes. All treated groups showed gill and liver histological changes.	

(*continued on next page*)

Table 2.2 Toxicity effects of pesticides on aquatic creatures—cont'd

S. no.	Pesticide	Used globally as	Toxic effect on aquatic creatures	References
5.	Malathion	Insecticide	Once inside a human body, it might stay there for two generations. After administering a low dose of malathion to *Channa punctatus* fish (Bloch), morphometric indices, biochemical parameters, and enzymatic levels, especially antioxidant processes, decreased with time.	Bharti & Rasool (2021); Kalantary et al. (2016)
6.	Carbofuran (carbofuran)	Nematicide, acaricide, and insecticide	Causes a reduction in the brain and muscle activity of the common craps.	Ensibi et al. (2012)
7.	Chlorpyrifos	Organophosphate insecticide	Causes neurotoxicity in zebrafishes.	Yen et al. (2011)
8.	Cypermethrin (pyrethroid)	Insecticide and pesticides	Reported to be extremely hazardous to marine invertebrates and fish.	Gautam & Gupta (2008)
9.	Cypermethrin	Insecticide and pesticides	Young brown trout (*Salmo trutta* L.) exposed to cypermethrin doses in a huge stream water aquarium showed disturbed reproductive capacity.	Jaensson et al. (2007)
10.	Pyrethroids	Insecticide	More hazardous to mammals and birds at lower temperatures whereas more than 100 times extra hazardous to fish due to insufficient hydrolytic enzymes.	Aydın et al. (2005)

wheat (Lu et al., 2015; Lu et al., 2020). Certain organophosphorus pesticides are detoxified by laccase enzymes that catalyze single electrons of metabolites/substrates by using O_2 (Majeau et al., 2010; Maruyama et al., 2006). It has been identified that herbicides atrazine and isoproturon are metabolized by two laccase–encoding genes (LOC_Os01g63180 and LOC_Os12g15680) in rice (Huang et al., 2016).

Phase II detoxification of pesticides: In Phase II detoxification, the enzymes catalyze the conjugation of pesticides with many endogenous compounds like reduced GSH, glycine sulfate, or glucuronic acid by adding polar groups in their structure. GSH is an important antioxidant which involved in enzymatic and nonenzymatic defense systems and protects the cell from oxidative damage generated by pesticides. In pesticide detoxification, GSH reacts with electrophilic molecules and replaces –H, –Cl, or –O–N=O groups (Lushchak, 2012).

Glutathione-S-transferases are involved in pesticide resistance by conjugation reaction with GSH to less hydrophobic compounds and generate water-soluble metabolites. Based on sequence similarity, plant-specific, and detoxification mechanism, GSTs can be divided into two classes, i.e., phi (F) and tau (U). In GST-catalyzed conjugation of pesticide reactions, GSH is added to an unsaturated bond/substitution reaction. In a substitution reaction, the halide is the leaving group which enhances the polarity and greatly reduces the toxicity effect of pesticide. It has been observed that some GSTUs and their clones (GmGSTU2–2, GmGSTU4–4, and GmGSTU10–10) detoxifying the chlorimuron ethyl herbicide and fluorodifen in *Glycine max* (Axarli et al., 2016; Cummins et al., 2011). The phi class GST (AtuGSTF2) was found to be effective against atrazine toxicity in *Amaranthus tuberculatus* (Evans et al., 2017).

Another phase II enzyme glycosyltransferases (GTs) catalyze the conjugation of pesticides with sugars and other chemicals. The most commonly observed GTs involved in pesticide conjugating mechanisms are planting UDP–glycosyltransferases (UGTs) which use UDP-glucose as a nucleotide sugar donor. Most of the GTs have been playing an important role against the toxic effects of insecticides, herbicides, and fungicides through O–glycosylation in plants (Tiwari et al., 2016). In grapes and strawberries, the less hydrophobic fungicide compounds which were hydroxylated in phase-I transformed into O–glycosylation products (Bauer et al., 2018). Detoxification of pesticides also occurs by the process of condensation in which amino/carboxylic groups are condensed with the metabolites of xenobiotics and form conjugates. Amino groups of pesticide isoproturon undergo condensation with amino groups like serine, aspartic acid, glutamate, alanine, and threonine and form 13 conjugates (Lu et al., 2020). Condensation depends on the enzymes and cofactors involved in pesticide metabolism. Pesticide metabolites containing the primary $-NH_2$ groups like sulfomides, hydrazines, or amino acids are also detoxified by the process of acetylation by enzymes acetyltransferases (ATs) (Levsen et al., 2005). Methylation of toxic metabolites by methyltransferases using S–adenosyl

L–methioninehas as a donor has been shown to be effective against atrazine toxicity in some plants (Zhang & Yang, 2021).

Phase III detoxification of pesticides: In phase III, ABC transporters (ATP–binding cassette type proteins) play a key role in detoxification mechanisms. ABC transporters are a group of integral membrane proteins forming channels for the transport of various molecules such as saccharides, amino acids, metals, xenobiotics, etc. (Zhang et al., 2018). Based on the protein size, alignment, sequence similarity, and existence of linker domains; ABC transporters are arranged in 13 subfamilies in plants. 6 ABC transporter genes were involved in the transportation of atrazine conjugated in *Medicago sativa* (Zhang et al., 2016). It has also been found that ABC transporter genes specifically for shoots, roots, and all tissues were induced by isoproturon in *Oryza sativa* (Lu et al., 2016).

2.8.1 Pesticide detoxification by some growth regulators

2.8.1.1 *Salicylic acid*

Plant hormone SA regulates various metabolic activities and stress conditions in plants. It was reported that SA help in the detoxification of napropamide and propazine residues in *Brassica napus* and *Zea mays* cells by enhancing their metabolism (Cui et al., 2010; Liang et al., 2012). The application of SA decreased oxygen accumulation and enhanced the antioxidant enzymatic activity in plants.

Two pathways are proposed for SA-induced pesticide detoxification mechanism in plants, i.e., PAL (phenylalanine ammonia lyase)—a responsive pathway and the ICS (isochorismate synthase)—IPL (isomerized pyruvate lyase) pathway. In this study, a set of mutants unable to synthesize SA is used to identify the role of SA in pesticide detoxification. Enhanced accretion of isoproturon residues was observed in plant cells of *Arabidopsis* with mutant lines pal1–1, pal1–2, and eps1–1, but the application of SA can restore the phenotype under pesticide stress (Lu et al., 2018).

2.8.1.2 *Jasmonic acid*

Growth regulator JA is a signaling molecule that regulates the secondary metabolism pathway under normal conditions and also modulates defense responses in plant cells. It was studied that JA improved the imazapic and isoproturon detoxification in tobacco and wheat plants (Kaya & Doganlar, 2016; Ma et al., 2018). Molecular studies have revealed that JA receptor CORONATINE INSENSITIVE 1 (COI1a) encoding an F–box protein can be induced by pesticide exposure. Overexpression of OsCOI1a showed resistance to atrazine toxicity in rice phenotypes which resulted in improved growth and reduced accumulation of atrazine residues in rice (Ma et al., 2021). This study confirmed that JA helps to remove the pesticide metabolites by dealkylation and glycosylation processes.

2.8.1.3 Brassinosteroids

This group of phytohormones regulates metabolic processes, and physiological, biochemical, and stress responses in plants. Reduction in insecticide imidacloprid and fungicide chlorothalonil residues has been observed by the application of brassinosteroids (BRs) in *Brassica juncea* and *Solanum lycopersicon* (Sharma et al., 2018; Ali et al., 2021). BRs mainly activate the plant antioxidant defense system and increase the activity of enzymes involved in phase I–III detoxifying processes. It has also been reported that nitric oxide and mitogen-activated protein kinase (MAPK) help in BR-induced pesticide detoxification (Yin et al., 2016).

2.9 Conclusion and future prospects

Insect pests are troubling the world ecosystem as well as are responsible for huge losses of yields in agricultural setups and stored crops. Thus, pesticides have been introduced to eliminate the losses and to fulfill the demands of the expanding populations. Whereas indiscriminate and uncontrolled applications of these chemicals are now considered to be a challenge not only for crop plants as well as it leads to toxic impacts on nontarget organisms such as humans, pollinators, aquatic fauna, and the important soil microflora. Thus, to maintain sustainable crop production and development new areas should be explored to get ecofriendly alternatives to these toxic synthetic pesticides. Natural plant products could be explored and however can become an alternative to synthetic toxicants. On the other hand, nanotechnology can be introduced to cope with these challenges, as it significantly lowers the amounts of agrochemicals required to eliminate insect pests.

References

Ahmad, F., Singh, A., & Kamal, A. (2017). Ameliorative effect of salicylic acid in salinity stressed *Pisum sativum* by improving growth parameters, activating photosynthesis and enhancing antioxidant defense system. *Bioscience Biotechnology Research Communications, 10*, 481–489.

Ahmad, F., Singh, A., & Kamal, A. (2018). Crosstalk of brassinosteroids with other phytohormones under various abiotic stresses. *Journal of Applied Biology and Biotechnology, 6*(1), 56–62.

Ahmed, F. S., Helmy, W. S., El-Din, H. S., & Al Naggar, Y. (2023). Chronic in-hive exposure to a field-relevant concentration of Closer^{TM} SC (24% sulfoxaflor) insecticide altered immunological and physiological markers of honey bee foragers (*Apis mellifera*). *Apidologie, 54*(1), 1–15. https://doi.org/10.1007/s13592-022-00987-6.

Aigrain, L., Pompon, D., Truan, G., & Moréra, S. (2010). Cloning, purification, crystallization and preliminary x–ray analysis of a chimeric NADPH–cytochrome P450 reductase. *Acta Crystallographica Section A: Foundations & Advances, 65*, 210–212.

Ali, M., Cheng, J. H., & Sun, D. W. (2021). Effect of plasma activated water and buffer solution on fungicide degradation from tomato (Solanum lycopersicum) fruit. *Food Chemistry, 350*, 129195.

Almasri, H., Liberti, J., Brunet, J. L., Engel, P., & Belzunces, L. P. (2022). Mild chronic exposure to pesticides alters physiological markers of honey bee health without perturbing the core gut microbiota. *Scientific Reports, 12*(1), 4281. https://doi.org/10.1038/s41598-022-08009-2.

Alsafran, M., Rizwan, M., Usman, K., Saleem, M. H., & Al Jabri, H. (2022). Neonicotinoid insecticides in the environment: A critical review of their distribution, transport, fate, and toxic effects. *Journal of Environmental Chemical Engineering, 10*, 108485. https://doi.org/10.1016/j.jece.2022.108485.

Amin, S., Solangi, A. R., Hassan, D., Hussain, N., Ahmed, J., & Baksh, H. (2021). Recent trends in development of nanomaterials based green analytical methods for environmental remediation. *Current Analytical Chemistry, 17*(4), 438–448.

Anitha, S. R., & Savitha, G. (2013). Impact of mancozeb stress on seedling growth, seed germination, chlorophyll and phenolic contents of rice cultivars. *International Journal of Science & Research, 4*(7), 292–296.

Anwar, M. I., Sadiq, N., Aljedani, D. M., Iqbal, N., Saeed, S., Khan, H. A. A., … Khan, K. A. (2022). Toxicity of different insecticides against the dwarf honey bee, *Apis florea* Fabricius (Hymenoptera: Apidae). *Journal of King Saud University-Science, 34*(1), 101712. https://doi.org/10.1016/j.jksus.2021.101712.

Arias-Estévez, M., López-Periago, E., Martínez-Carballo, E., Simal-Gándara, J., Mejuto, J. C., & García-Río, L. (2008). The mobility and degradation of pesticides in soils and the pollution of groundwater resources. *Agriculture, Ecosystems & Environment, 123*(4), 247–260.

Axarli, I., Muleta, A. W., Vlachakis, D., Kossida, S., Kotzia, G., & Maltezos, A. (2016). Directed evolution of tau class glutathione transferases reveals a site that regulates catalytic efficiency and masks co-operativity. *Biochemical Journal, 473*, 559–570.

Aydın, R., Köprücü, K., Dörücü, M., Köprücü, S. Ş., & Pala, M. (2005). Acute toxicity of synthetic pyrethroid cypermethrin on the common carp (*Cyprinus carpio L.*) embryos and larvae. *Aquaculture International, 13*, 451–458.

Bano, N., Nadeem, A., Maalik, S., Mushtaq, S., Iqbal, N., Khan, A. K., et al. (2021). Effect of pesticides on erythrocytes of indigenous fish *Labeo rohita*. *Journal of King Saud University-Science, 33*(7), 101586.

Barascou, L., Requier, F., Sené, D., Crauser, D., Le Conte, Y., & Alaux, C. (2022). Delayed effects of a single dose of a neurotoxic pesticide (sulfoxaflor) on honeybee foraging activity. *Science of the Total Environment, 805*, 150351. https://doi.org/10.1016/j.scitotenv.2021.150351.

Bartling, M. T., Vilcinskas, A., & Lee, K. Z. (2019). Sub-lethal doses of clothianidin inhibit the conditioning and biosensory abilities of the western honeybee *Apis mellifera*. *Insects, 10*(10), 340. https://doi.org/10.3390/insects10100340.

Bauer, A., Luetjohann, J., Rohn, S., Jantzen, E., & Kuballa, J. (2018). Development of a suspect screening strategy for pesticide metabolites in fruit and vegetables by UPLC–Q–TOF–MS. *Food Analytical Methods, 11*, 1–17.

Bertol, E., & Favretto, D. (2023). Pharmacokinetics: drug absorption, distribution, & elimination. In *Karch's drug abuse handbook* (pp. 133–178). CRC Press.

Bharti, S., & Rasool, F. (2021). Analysis of the biochemical and histopathological impact of a mild dose of commercial malathion on *Channa punctatus* (Bloch) fish. *Toxicology Reports, 8*, 443–455.

Bhatti, J. S., Khullar, N., Vijayvergiya, R., Navik, U., Bhatti, G. K., & Reddy, P. H. (2023). Mitochondrial miRNA as epigenomic signatures: Visualizing aging-associated heart diseases through a new lens. *Ageing Research Reviews, 86*, 101882.

Bianchi, J., Fernandes, T. C. C., & Marin-Morales, M. A. (2016). Induction of mitotic and chromosomal abnormalities on *Allium cepa* cells by pesticides imidacloprid and sulfentrazone and the mixture of them. *Chemosphere, 144*, 475–483.

Biondi, A., Desneux, N., Siscaro, G., & Zappalà, L. (2012). Using organic-certified rather than synthetic pesticides may not be safer for biological control agents: Selectivity and side effects of 14 pesticides on the predator *Orius laevigatus*. *Chemosphere, 87*(7), 803–812.

Bokade, P., & Bajaj, A. (2023). Molecular advances in mycoremediation of polycyclic aromatic hydrocarbons: Exploring fungal bacterial interactions. *Journal of Basic Microbiology, 63*, 239–256.

Bonmatin, J. M., Giorio, C., Girolami, V., Goulson, D., Kreutzweiser, D. P., Krupke, C., … Tapparo, A. (2015). Environmental fate and exposure; neonicotinoids and fipronil. *Environmental Science and Pollution Research, 22*, 35–67. https://doi.org/10.1007/s11356-014-3332-7.

Botitsi, H., Tsipi, D., & Economou, A. (2017). Current legislation on pesticides. In *Applications in high resolution mass spectrometry* (pp. 83–130). Elsevier.

Capela, N., Sarmento, A., Simões, S., Azevedo-Pereira, H. M., & Sousa, J. P. (2022). Sub-lethal doses of sulfoxaflor impair honey bee homing ability. *Science of the Total Environment, 837*, 155710. https://doi.org/10.1016/j.scitotenv.2022.155710.

Carneiro, L. S., Martinez, L. C., De Oliveira, A. H., Cossolin, J. F. S., De Resende, M. T. C. S., Gonçalves, W. G., Medeiros-Santana, L., & Serrão, J. E. (2022). Acute oral exposure to imidacloprid induces apoptosis and autophagy in the midgut of honey bee *Apis mellifera* workers. *Science of the Total Environment, 815*, 152847.

Cartereau, A., Pineau, X., Lebreton, J., Mathé-Allainmat, M., Taillebois, E., & Thany, S. H. (2022). Impairments in learning and memory performances associated with nicotinic receptor expression in the honeybee *Apis mellifera* after exposure to a sublethal dose of sulfoxaflor. *PLoS ONE, 17*(8), e0272514. https://doi.org/10.1371/journal.pone.0272514.

Castelli, L., Branchiccela, B., Zunino, P., & Antúnez, K. (2023). Insights into the effects of sublethal doses of pesticides glufosinate-ammonium and sulfoxaflor on honey bee health. *Science of the Total Environment*, 161331. https://doi.org/10.1016/j.scitotenv.2022.161331.

Catae, A. F., da Silva Menegasso, A. R., Pratavieira, M., Palma, M. S., Malaspina, O., & Roat, T. C. (2019). MALDI-imaging analyses of honeybee brains exposed to a neonicotinoid insecticide. *Pest Management Science, 75*(3), 607–615. https://doi.org/10.1002/ps.5226.

Chaimanee, V., Evans, J. D., Chen, Y., Jackson, C., & Pettis, J. S. (2016). Sperm viability and gene expression in honey bee queens (*Apis mellifera*) following exposure to the neonicotinoid insecticide imidacloprid and the organophosphate acaricide coumaphos. *Journal of Insect Physiology, 89*, 1–8. https://doi.org/10.1016/j.jinsphys.2016.03.004.

Chen, L., Yan, Q., Zhang, J., Yuan, S., & Liu, X. (2019). Joint toxicity of acetamiprid and co-applied pesticide adjuvants on honeybees under semifield and laboratory conditions. *Environmental Toxicology and Chemistry, 38*(9), 1940–1946. https://doi.org/10.1002/etc.4515.

Chen, Y. R., Tzeng, D. T., & Yang, E. C (2021). Chronic effects of imidacloprid on honey bee worker development: Molecular pathway perspectives. *International Journal of Molecular Sciences, 22*(21), 11835. https://doi.org/10.3390/ijms222111835.

Cheng, S., Dai, P., Li, R., Chen, Z., Liang, P., Xie, X., … Gao, X. (2023). The sulfoximine insecticide sulfoxaflor exposure reduces the survival status and disrupts the intestinal metabolism of the honeybee *Apis mellifera*. *Journal of Hazardous Materials, 442*, 130109. https://doi.org/10.1016/j.jhazmat.2022.130109.

Cho, H. Y., & Kong, K. H. (2007). Study on the biochemical characterization of herbicide detoxification enzyme, glutathione S–transferase. *Biofactors, 30*, 281–287.

Christen, V. (2023). Different effects of pesticides on transcripts of the endocrine regulation and energy metabolism in honeybee foragers from different colonies. *Scientific Reports, 13*(1), 1985. https://doi.org/10.1038/s41598-023-29257-w.

Christen, V., & Fent, K. (2017). Exposure of honey bees (*Apis mellifera*) to different classes of insecticides exhibit distinct molecular effect patterns at concentrations that mimic environmental contamination. *Environmental Pollution, 226*, 48–59. https://doi.org/10.1016/j.envpol.2017.04.003.

Christen, V., Krebs, J., Bünter, I., & Fent, K. (2019). Biopesticide spinosad induces transcriptional alterations in genes associated with energy production in honey bees (*Apis mellifera*) at sublethal concentrations. *Journal of Hazardous Materials, 378*, 120736. https://doi.org/10.1016/j.jhazmat.2019.06.013.

Christen, V., Mittner, F., & Fent, K. (2016). Molecular effects of neonicotinoids in honey bees (*Apis mellifera*). *Environmental Science and Technology, 50*(7), 4071–4081. https://doi.org/10.1021/acs.est.6b00678.

Cizelj, I., Glavan, G., Božič, J., Oven, I., Mrak, V., & Narat, M. (2016). Prochloraz and coumaphos induce different gene expression patterns in three developmental stages of the Carniolan honey bee (*Apis mellifera carnica* Pollmann). *Pesticide Biochemistry and Physiology, 128*, 68–75. https://doi.org/10.1016/j.pestbp.2015.09.015.

Colin, T., Meikle, W. G., Wu, X., & Barron, A. B. (2019). Traces of a neonicotinoid induce precocious foraging and reduce foraging performance in honey bees. *Environmental Science and Technology, 53*(14), 8252–8261. https://doi.org/10.1021/acs.est.9b02452.

Cook, S. C. (2019). Compound and dose-dependent effects of two neonicotinoid pesticides on honey bee (*Apis mellifera*) metabolic physiology. *Insects, 10*(1), 18. https://doi.org/10.3390/insects10010018.

Cui, J., Zhang, R., Wu, G. L., Zhu, H. M., & Yang, H. (2010). Salicylic acid reduces napropamide toxicity by preventing its accumulation in rapeseed (*Brassica napus* L.). *Archives of Environmental Contamination and Toxicology, 59*, 100–108.

Cummins, I., Dixon, D. P., Freitag, P. S., Skipsey, M., & Edwards, R. (2011). Multiple roles for plant glutathione transferases in xenobiotic detoxification. *Drug Metabolism Reviews, 43*, 266–280.

Dar, M. A., Hamid, B., & Kaushik, G. (2023). Temporal trends in the use and concentration of organophosphorus pesticides in Indian riverine water, toxicity, and their risk assessment. *Regional Studies in Marine Science, 59*, 102814.

Delkash-Roudsari, S., Goldansaz, S. H., Jahromi, K. T. Ashouri A., & Abramson, C. I (2022). Side effects of imidacloprid, ethion, and hexaflumuron on adult and larvae of honey bee *Apis mellifera* (Hymenoptera, Apidae). *Apidologie, 53*(1), 17. https://doi.org/10.1007/s13592-022-00910-z.

Dias, M. C. (2012). Phytotoxicity: An overview of the physiological responses of plants exposed to fungicides. *Journal of Botany, 2012*, 135479.

Domingues, C. E., Abdalla, F. C., Balsamo, P. J., Pereira, B. V., de Alencar Hausen, M., Costa, M. J., & Silva-Zacarin, E. C. (2017). Thiamethoxam and picoxystrobin reduce the survival and overload the hepato-nephrocitic system of the Africanized honeybee. *Chemosphere, 186*, 994–1005. https://doi.org/10.1016/j.chemosphere.2017.07.133.

Doroudian, A., Emadi, M., Hosseinzadeh, R., & Maghami, F. (2022). Biological and molecular effects of pesticides on human health. In *Pesticides-updates on toxicity, efficacy and risk assessment*. IntechOpen.

Drum, C. (1980). Soil chemistry of pesticides. PPG Industries Inc. USA.

Duke, S.O. (2021). Glyphosate: Uses other than in glyphosate-resistant crops, mode of action, degradation in plants, and effects on non-target plants and agricultural microbes. *Reviews of Environmental Contamination and Toxicology, 255*, 1–65, https://doi.org/10.1007/398_2020_53/COVER.

Eddleston, M., Karalliedde, L., Buckley, N., Fernando, R., Hutchinson, G., Isbister, G., ... Smit, L. (2002). Pesticide poisoning in the developing world: A minimum pesticides list. *The Lancet, 360*(9340), 1163–1167.

El-Din, H. S., Helmy, W. S., Al Naggar, Y., & Ahmed, F. S (2022). Chronic exposure to a field-realistic concentration of Closer® SC (24% sulfoxaflor) insecticide impacted the growth and foraging activity of honey bee colonies. *Apidologie, 53*(2), 22. https://doi.org/10.1007/s13592-022-00937-2.

Ensibi, C., Hernández-Moreno, D., Míguez Santiyán, M. P., et al. (2012). Effects of carbofuran and deltamethrin on acetylcholinesterase activity in brain and muscle of the common carp. *Environmental Toxicology, 29*, 386–393.

Evans, A. F., Jr, O'Brien, S. R., Ma, R., Hager, A. G., Riggins, C. W., Lambert, K. N., & Riechers, D. E. (2017). Biochemical characterization of metabolism–based atrazine resistance in Amaranthus tuberculatus and identification of an expressed GST associated with resistance. *Plant Biotechnology Journal, 15*, 1238–1249.

Faita, M. R., Cardozo, M. M., Amandio, D. T. T., Orth, A. I., & Nodari, R. O. (2020). Glyphosate-based herbicides and *Nosema sp.* microsporidia reduce honey bee (*Apis mellifera* L.) survivability under laboratory conditions. *Journal of Apicultural Research, 59*(4), 332–342. https://doi.org/10.1080/00218839.2020.1736782.

Fatma, F., Kamal, A., & Srivastava, A. (2018). Exogenous application of salicylic acid mitigates the toxic effect of pesticides in *Vigna radiata* (L.) Wilczek. *Journal of Plant Growth Regulation, 37*, 1185–1194.

Fatma, F., Verma, S., Kamal, A., & Srivastava, A. (2018). Phytotoxicity of pesticides mancozeb and chlorpyrifos: Correlation with the antioxidative defence system in *Allium cepa*. *Physiology and Molecular Biology of Plants, 24*, 115–123.

Fishel, F.M., & Ferrell, J.A. (2019). Managing pesticide drift. Agronomy Department, University of Florida. Publication #I232.

Gao, J., Jin, S. S., He, Y., Luo, J. H., Xu, C. Q., Wu, Y. Y., Hou, C.-S., Wang, Q., & Diao, Q. Y. (2020). Physiological analysis and transcriptome analysis of Asian honey bee (*Apis cerana cerana*) in response to sublethal neonicotinoid imidacloprid. *Insects, 11*(11), 753. https://doi.org/10.3390/insects11110753.

Gautam, P. P., & Gupta, A. K. (2008). Toxicity of cypermethrin to the juveniles of freshwater fish *Poecilia reticulata* (Peters) in relation to selected environmental variables. *Natural Product Radiance, 7*, 314–319.

Ghayyur, S., Khan, M. F., Tabassum, S., Ahmad, M. S., Sajid, M., Badshah, K., Khan, M. A., Saira, Ghayyur, S., Khan, N. A., Ahmad, B., & Qamer, S. (2021). A comparative study on the effects of selected pesticides on hemato-biochemistry and tissue histology of freshwater fish *Cirrhinus mrigala* (Hamilton, 1822). *Saudi Journal of Biological Sciences, 28*(1), 603–611.

Goblirsch, M., & Adamczyk, J. J. (2023). Using the honey bee (*Apis mellifera* L.) cell line AmE-711 to evaluate insecticide toxicity. *Environmental Toxicology and Chemistry, 42*(1), 88–99. https://doi.org/10.1002/etc.5500.

Godage, N. H., Cudjoe, E., Chau, T., & Gionfriddo, E. (2023). Quantitative determination of pesticides in human plasma using bio-SPME-LC–MS/MS: A robust tool to assess occupational exposure to pesticides. *Analytical and Bioanalytical Chemistry*, 1–12.

Gregorc, A., Alburaki, M., Rinderer, N., Sampson, B., Knight, P. R., Karim, S., & Adamczyk, J. (2018). Effects of coumaphos and imidacloprid on honey bee (Hymenoptera: Apidae) lifespan and antioxidant gene regulations in laboratory experiments. *Scientific Reports, 8*(1), 1–13. https://doi.org/10.1038/s41598-018-33348-4.

Haas, J., Zaworra, M., Glaubitz, J., Hertlein, G., Kohler, M., Lagojda, A., Lueke, B., Maus, C., Almanza, M.-T., Emyr Davies, T. G., Bass, C., & Nauen, R. (2021). A toxicogenomics approach reveals characteristics supporting the honey bee (*Apis mellifera* L.) safety profile of the butenolide insecticide flupyradifurone. *Ecotoxicology and Environmental Safety, 217*, 112247. https://doi.org/10.1016/j.ecoenv.2021.112247.

Haider, S., & Inbaraj, R. M. (1986). Relative toxicity of technical material and commercial formulation of malathion and endosulfan to a freshwater fish, *Channa punctatus* (Bloch). *Ecotoxicology and Environmental Safety, 11*, 347–351.

Hashimi, M. H., Hashimi, R., & Ryan, Q. (2020). Toxic effects of pesticides on humans, plants, animals, pollinators and beneficial organisms. *Asian Plant Research Journal, 5*(4), 37–47.

He, F., & Liu, R. (2023). Mechanistic insights into phenanthrene-triggered oxidative stress-associated neurotoxicity, genotoxicity, and behavioral disturbances toward the brandling worm (*Eisenia fetida*) brain: The need for an ecotoxicological evaluation. *Journal of Hazardous Materials, 450*, 131072.

Hellström, S., Strobl, V., Straub, L., Osterman, W. H., Paxton, R. J., & Osterman, J. (2023). Environmental and sustainability indicators. *Environmental and Sustainability Indicators, 18*, 100239. https://doi.org/10.1016/j.indic.2023.100239.

Hesselbach, H., & Scheiner, R. (2018). Effects of the novel pesticide flupyradifurone (Sivanto) on honeybee taste and cognition. *Scientific Reports, 8*(1), 1–8. https://doi.org/10.1038/s41598-018-23200-0.

Hesselbach, H., Seeger, J., Schilcher, F., Ankenbrand, M., & Scheiner, R. (2020). Chronic exposure to the pesticide flupyradifurone can lead to premature onset of foraging in honeybees *Apis mellifera*. *Journal of Applied Ecology, 57*(3), 609–618. https://doi.org/10.1111/1365-2664.13555.

Hoagland, R. E., Zablotowicz, Z. M., & Hall, J. C. (2001). Pesticide metabolism in plants and microorganisms: An overview. *Weed Science, 51*(4), 472–495.

Hoffman, D.J., Rattner, B.A., Burton Jr, G.A., & Cairns Jr, J. (Eds.). (2002). *Handbook of ecotoxicology*. CRC Press.

Horbowicz, M., Sempruch, C., Kosson, R., Koczkodaj, D., & Walas, D. (2013). Effect of fluazifop-p-butyl treatment on pigments and polyamines level within tissues of non-target maize plants. *Pesticide Biochemistry and Physiology, 107*(1), 78–85.

Huang, M., Dong, J., Yang, S., Xiao, M., Guo, H., Zhang, J., & Wang, D. (2023). Ecotoxicological effects of common fungicides on the eastern honeybee *Apis cerana cerana* (Hymenoptera). *Science of the Total Environment, 868*, 161637. https://doi.org/10.1016/j.scitotenv.2023.161637.

Huang, M. T., Lu, Y. C., Zhang, S., Luo, F., & Yang, H. (2016). Rice (*Oryza sativa*) laccases involved in modification and detoxification of herbicides atrazine and isoproturon residues in plants. *Journal of Agricultural and Food Chemistry, 64*, 6397–6406.

Imaishi, H., & Matumoto, S. (2007). Isolation and functional characterization in yeast of CYP72A18, a rice cytochrome P450 that catalyzes (ω–1)–hydroxylation of the herbicide pelargonic acid. *Pesticide Biochemistry and Physiology, 88*, 71–77.

Jacob, C. R., Malaquias, J. B., Zanardi, O. Z., Silva, C. A., Jacob, J. F., & Yamamoto, P. T. (2019). Oral acute toxicity and impact of neonicotinoids on *Apis mellifera* L. and *Scaptotrigona postica Latreille* (Hymenoptera: Apidae). *Ecotoxicology (London, England), 28*(7), 744–753. https://doi.org/10.1007/s10646-019-02070-w.

Jaensson, A., Scott, A. P., Moore, A., Kylin, H., & Olsén, K. H. (2007). Effects of a pyrethroid pesticide on endocrine responses to female odours and reproductive behaviour in male parr of brown trout (*Salmo trutta* L.). *Aquatic Toxicology, 81*(1), 1–9.

Johnson, R. M., Dahlgren, L., Siegfried, B. D., & Ellis, M. D. (2013). Acaricide, fungicide and drug interactions in honey bees (*Apis mellifera*). *PLoS ONE, 8*(1), e54092. https://doi.org/10.1371/journal.pone.0054092.

Kablau, A., Erler, S., Eckert, J. H., Pistorius, J., Sharbati, S., & Einspanier, R. (2023). Effects of flupyradifurone and two reference insecticides commonly used in toxicological studies on the larval proteome of the honey bee *Apis mellifera*. *Insects, 14*(1), 77. https://doi.org/10.3390/insects14010077.

Kadiru, S., Patil, S., & D'Souza, R. (2022). Effect of pesticide toxicity in aquatic environments: A recent review. *International Journal of Fisheries and Aquatic Studies, 10*(3), 113–118.

Kalantary, R. R., Azari, A., Esrafili, A., Yaghmaeian, K., Moradi, M., & Sharafi, K. (2016). The survey of Malathion removal using magnetic graphene oxide nanocomposite as a novel adsorbent: Thermodynamics, isotherms, and kinetic study. *Desalination and Water Treatment, 57*(58), 28460–28473.

Kalyabina, V. P., Esimbekova, E. N., Kopylova, K. V., & Kratasyuk, V. A. (2021). Pesticides: Formulants, distribution pathways and effects on human health–a review. *Toxicology Reports, 8*, 1179–1192.

Kamal, A., Ahmad, F., & Shafeeque, M. A. (2020). Toxicity of pesticides to plants and non-target organism: A comprehensive review. *Iranian Journal of Plant Physiology, 10*(4), 3299–3313.

Karpouzas, D. G., Vryzas, Z., & Martin-Laurent, F. (2022). Pesticide soil microbial toxicity: Setting the scene for a new pesticide risk assessment for soil microorganisms (IUPAC Technical Report). *Pure and Applied Chemistry, 94*, 1161–1194.

Kaur, R., Mavi, G. K., Raghav, S., & Khan, I. (2019). Pesticides classification and its impact on environment. *The International Journal of Current Microbiology and Applied Sciences, 8*(3), 1889–1897.

Kaya, A., & Doganlar, Z. B. (2016). Exogenous jasmonic acid induces stress tolerance in tobacco (*Nicotiana tabacum*) exposed to imazapic. *Ecotoxicology and Environmental Safety, 124*, 470–479.

Khalil, M., Iqbal, M., Turan, V., Tauqeer, H. M., Farhad, M., Ahmed, A., & Yasin, S. (2022). Household chemicals and their impact. In *Environmental micropollutants* (pp. 201–232). Elsevier.

Kilic, S., Duran, R. E., & Coskun, Y. (2015). Morphological and physiological responses of maize (*Zea mays* L.) seeds grown under increasing concentrations of chlorantraniliprole insecticide. *Polish Journal of Environmental Studies, 24*(3), 1069–1075.

Kim, J., Chon, K., Kim, B. S., Oh, J. A., Yoon, C. Y., & Park, H. H. (2022). Assessment of acute and chronic toxicity of cyantraniliprole and sulfoxaflor on honey bee (*Apis mellifera*) larvae. *Pest Management Science, 78*(12), 5402–5412. https://doi.org/10.1002/ps.7162.

Koech, S. J., Karanja, R. H., Kurgat, J. K., Mokaya, H. O., Dubois, T., & Lattorff, H. M. G. (2023). Pesticide contamination and their botanical sources in pollen loads collected by honeybees in Kenya: A spatio-temporal context. *Agriculture, Ecosystems and Environment, 343*, 108264. https://doi.org/10.1016/j.agee.2022.108264.

Kumar, D., & Ohri, P. (2023). Say "NO" to plant stresses: Unravelling the role of nitric oxide under abiotic and biotic stress. *Nitric Oxide, 130*(1), 36–57. https://doi.org/10.1016/j.niox.2022.11.004.

Kumar, V., Singh, S., & Upadhyay, N. (2019). Effects of organophosphate pesticides on siderophore producing soils microorganisms. *Biocatalysis and Agricultural Biotechnology, 21*, 101359.

Laskay, G., & Lakos, A. (2011). Counteracting the effects of paraquat on photosynthesis by chlorogenic acid. *Acta Biologica Szegediensis, 55*(1), 101–103.

Levsen, K., Schiebel, H. M., Behnke, B., Dötzer, R., Dreher, W., Elend, M., & Thiele, H. (2005). Structure elucidation of phase II metabolites by tandem mass spectrometry: An overview. *Journal of Chromatography A, 1067*, 55–72.

Li, H., Liu, S., Chen, L., Luo, J., Zeng, D., & Li, X. (2021a). Juvenile hormone and transcriptional changes in honey bee worker larvae when exposed to sublethal concentrations of thiamethoxam. *Ecotoxicology and Environmental Safety, 225*, 112744. https://doi.org/10.1016/j.ecoenv.2021.112744.

Li, J., Zhao, L., Qi, S., Zhao, W., Xue, X., Wu, L., & Huang, S. (2021b). Sublethal effects of Isoclast™ Active (50% sulfoxaflor water dispersible granules) on larval and adult worker honey bees (*Apis mellifera* L.). *Ecotoxicology and Environmental Safety, 220*, 112379. https://doi.org/10.1016/j.ecoenv.2021.112379.

Li, W., Lv, L., Wang, Y., & Zhu, Y. C. (2023). Mixture effects of thiamethoxam and seven pesticides with different modes of action on honey bees (*Aplis mellifera*). *Scientific Reports, 13*(1), 2679. https://doi.org/10.1038/s41598-023-29837-w.

Li, Z., & Fantke, P. (2023). Framework for defining pesticide maximum residue levels in feed: Applications to cattle and sheep. *Pest Management Science, 79*(2), 748–759.

Li, Z., Yu, T., Chen, Y., Heerman, M., He, J., Huang, J., Nie, H., & Su, S. (2019). Brain transcriptome of honey bees (*Apis mellifera*) exhibiting impaired olfactory learning induced by a sublethal dose of imidacloprid. *Pesticide Biochemistry and Physiology, 156*, 36–43. https://doi.org/10.1016/j.pestbp.2019.02.001.

Liang, L., Lu, Y. L., & Yang, H. (2012). Toxicology of isoproturon to the food crop wheat as affected by salicylic acid. *Environmental Science and Pollution Research, 19*, 2044–2054.

Liao, L. H., Wu, W. Y., Dad, A., & Berenbaum, M. R. (2019). Fungicide suppression of flight performance in the honeybee (*Apis mellifera*) and its amelioration by quercetin. *Proceedings of the Royal Society B, 286*(1917), 20192041. https://dx.doi.org/10.1098/rspb.2019.2041.

Liu, S., Liu, Y., He, F., Zhang, H., Li, X., & Tan, H. (2019). Enantioselective olfactory effects of the neonicotinoid dinotefuran on honey bees (*Apis mellifera* L.). *Journal of Agricultural and Food Chemistry, 67*(43), 12105–12116. https://doi.org/10.1021/acs.jafc.9b04851.

Lopes, M. P., Fernandes, K. M., Tomé, H. V. V., Gonçalves, W. G., Miranda, F. R., Serrão, J. E., & Martins, G. F. (2018). Spinosad-mediated effects on the walking ability, midgut, & Malpighian tubules of Africanized honey bee workers. *Pest Management Science, 74*(6), 1311–1318. https://doi.org/10.1002/ps.4815.

Lu, C., Hung, Y. T., & Cheng, Q. (2020). A review of sub-lethal neonicotinoid insecticides exposure and effects on pollinators. *Current Pollution Reports, 6*, 137–151. https://doi.org/10.1007/s40726-020-00142-8.

Lu, F. F., Liu, J. T., Zhang, N., Chen, Z. J., & Yang, H. (2020). OsPAL as a key salicylic acid synthetic component is a critical factor involved in mediation of isoproturon degradation in a paddy crop. *Journal of Cleaner Production, 262*, 121476.

Lu, F. F., Xu, J. Y., Ma, L. Y., Su, X. N., Wang, X. Q., & Yang, H. (2018). Isoproturon–induced salicylic acid confers *Arabidopsis* resistance to isoproturon phytotoxicity and degradation in plants. *Journal of Agricultural and Food Chemistry, 66*, 13073–13083.

Lu, Y. C., Feng, S. J., Zhang, J. J., Luo, F., Zhang, S., & Yang, H. (2016). Genome–wide identification of DNA methylation provides insights into the association of gene expression in rice exposed to pesticide atrazine. *Scientific Reports, 6*, 18985–18999.

Lu, Y. C., Zhang, S., & Yang, H. (2015). Acceleration of the herbicide isoproturon degradation in wheat by glycosyltransferases and salicylic acid. *Journal of Hazardous Materials, 283*, 806–814.

Lushchak, VI. (2012). Glutathione homeostasis and functions: Potential targets for medical interventions. *Journal of Amino Acids, 2012*, 736837.

Lv, L., Li, W., Li, X., Wang, D., Weng, H., Zhu, Y. C., & Wang, Y. (2023). Mixture toxic effects of thiacloprid and cyproconazole on honey bees (*Apis mellifera* L.). *Science of The Total Environment, 870*, 161700. https://doi.org/10.1016/j.scitotenv.2023.161700.

Ma, L. Y., Zhai, X. Y., Qiao, Y. X., Zhang, A. P., Zhang, N., Liu, J. T., & Yang, H. (2021). Identification of a novel function of a component in the jasmonate signaling pathway for intensive pesticide degradation in rice and environment through an epigenetic mechanism. *Environmental Pollution, 268*, 115802.

Ma, L. Y., Zhang, S. H., Zhang, J. J., Zhang, A. P., Li, N., Wang, X. Q., Yu, Q. Q., & Yang, H. (2018). Jasmonic acids facilitate the degradation and detoxification of herbicide isoproturon residues in wheat crops (Triticum aestivum). *Chemical Research in Toxicology, 31*, 752–761.

Macar, O., Kalefetoğlu Macar, T., Çavuşoğlu, K., Yalçın, E., & Yapar, K. (2023). Lycopene: An antioxidant product reducing dithane toxicity in *Allium cepa* L. *Scientific Reports, 13*(1), 2290.

Majeau, J. A., Brar, S. K., & Tyagi, R. D. (2010). Laccases for removal of recalcitrant and emerging pollutants. *Bioresource Technology, 101*, 2331–2350.

M'Anampiu, J. M. (2011). Organochlorine pesticides residues in fish and sediment from Lake Naivasha [Doctoral dissertation]. University of Nairobi.

Mann, R. M., Bidwell, J. R., & Tyler, M. J. (2003). Toxicity of herbicide formulations to frogs and the implications for product registration: A case study from Western Australia. *Applied Herpetology, 1*, 13–22.

Maruyama, T., Komatsu, C., Michizoe, J., Ichinose, H., & Goto, M. (2006). Laccase mediated oxidative degradation of the herbicide dymron. *Biotechnology Progress, 22*, 426–430.

Mayack, B. K. (2023). Modeling disruption of *Apis mellifera* (honey bee) odorant-binding protein function with high-affinity binders. *Journal of Molecular Recognition, 36*, e3008. https://doi.org/10.1002/jmr.3008.

Meena, R. S., Kumar, S., Datta, R., Lal, R., Vijayakumar, V., Brtnicky, M., Sharma, M. P., Yadav, G. S., Jhariya, M. K., Jangir, C. K., Pathan, S. I., Dokulilova, T., Pecina, V., & Marfo, T. D. (2020). Impact of agrochemicals on soil microbiota and management: A review. *Land, 9*(2), 34.

Meftaul, I. M., Venkateswarlu, K., Parven, A., Annamalai, P., & Megharaj, M. (2023). Human health risk assessment of pesticides in lettuce and spinach grown in urban backyard garden soils. *Journal of Food Composition and Analysis, 115*, 104977.

Mieldažys, R., Jotautienė, E., Pocius, A., & Jasinskas, A. (2016). Analysis of organic agricultural waste usage for fertilizer production. *Agronomy Research, 14*(1), 143–149.

Mohamed, I. A., Omar, E. M., Tawfik, A. I., Amro, A., & Al Naggar, Y. (2023). Sublethal effects of herbicides clethodim, haloxyfop-P-methyl, and their mixture on honey bee health. *Apidologie, 54*(1), 1–17. https://doi.org/10.1007/s13592-022-00982-x.

Moreira, S., Pereira, S. C., Seco-Rovira, V., Oliveira, P. F., Alves, M. G., & Pereira, M. D. L. (2021). Pesticides and male fertility: A dangerous crosstalk. *Metabolites, 11*(12), 799.

Murthy, K. S., Kiran, B. R., & Venkateshwarlu, M. (2013). A review on toxicity of pesticides in fish. *International Journal of Open Scientific Research, 1*(1), 15–36.

Mustard, J. A., Gott, A., Scott, J., Chavarria, N. L., & Wright, G. A. (2020). Honeybees fail to discriminate floral scents in a complex learning task after consuming a neonicotinoid pesticide. *Journal of Experimental Biology, 223*(5), jeb217174. https://doi.org/10.1242/jeb.217174.

Naiara Gomes, I., Ingred Castelan Vieira, K., Moreira Gontijo, L., & Canto Resende, H. (2020). Honeybee survival and flight capacity are compromised by insecticides used for controlling melon pests in Brazil. *Ecotoxicology (London, England), 29*, 97–107. https://doi.org/10.1007/s10646-019-02145-8.

Narasimhamurthy, R. K., Andrade, D., & Mumbrekar, K. D. (2022). Modulation of CREB and its associated upstream signaling pathways in pesticide-induced neurotoxicity. *Molecular and Cellular Biochemistry, 477*(11), 2581–2593.

Nguegang, I. N., Noutakdie, R. P., Omokolo, D. M., & Djoufack, B. M. (2022). Exposure to pesticides, fertilizers and crop preservatives during agriculture and food storage in a cameroon health district. *Journal of Biology and Today's World, 11*(3), 1–4. https://doi.org/10.35248/2322-3308.11.3.002.

Ohkawa, H., & Inui, H. (2014). Metabolism of agrochemicals and related environmental chemicals based on cytochrome P450s in mammals and plants. *Pest Management Science, 71*, 824–828.

Ottenbros, I., Lebret, E., Huber, C., Lommen, A., Antignac, J. P., Čupr, P., Mikeš, O., Szigeti, T., Középesy, S., Martinsone, I., Martinsone, Z., Akulova, L., Pardo, O., Fernández, S. F., Coscollá, C., Pedraza-Diaz, S., Krauss, M., Debrauwer, L., Wagner, K., … Vlaanderen, J. (2023). Assessment of exposure to pesticide mixtures in five European countries by a harmonized urinary suspect screening approach. *International Journal of Hygiene and Environmental Health, 248*, 114105.

Paleolog, J., Wilde, J., Gancarz, M., Wiącek, D., Nawrocka, A., & Strachecka, A. (2023). Imidacloprid pesticide causes unexpectedly severe bioelement deficiencies and imbalance in honey bees even at sublethal doses. *Animals, 13*(4), 615. https://doi.org/10.3390/ani13040615.

Palmer, M. J., Moffat, C., Saranzewa, N., Harvey, J., Wright, G. A., & Connolly, C. N. (2013). Cholinergic pesticides cause mushroom body neuronal inactivation in honeybees. *Nature Communications, 4*(1), 1634. https://doi.org/10.1038/ncomms2648.

Pervez, M., & Manzoor, F. (2023). Honey bee losses and pesticides threat: An Asian perspective. *Journal of Apicultural Research, 62*(1), 64–75. https://doi.org/10.1080/00218839.2022.2103331.

Philippe, V., Neveen, A., Marwa, A., & Basel, A. Y. A. (2021). Occurrence of pesticide residues in fruits and vegetables for the Eastern Mediterranean Region and potential impact on public health. *Food Control, 119*, 107457.

Pineaux, M., Grateau, S., Lirand, T., Aupinel, P., & Richard, F. J. (2023). Honeybee queen exposure to a widely used fungicide disrupts reproduction and colony dynamic. *Environmental Pollution, 322*, 121131. https://doi.org/10.1016/j.envpol.2023.121131.

Radosevich, S. R., Holt, J. S., & Ghersa, C. M. (2007). Ecology of weeds and invasive plants: Relationship to agriculture and natural resource management. John Wiley & Sons.

Richterova, Z., & Svobodova, Z. (2012). Pyrethroid influence on fish. *Slovenian Veterinary Research, 49*, 63–72.

Royuela, V., Suriñach, J., & Reyes, M. (2003). Measuring quality of life in small areas over different periods of time. *Social Indicators Research, 64*, 51–74.

Sabarwal, A., Kumar, K., & Singh, R. P. (2018). Hazardous effects of chemical pesticides on human health–Cancer and other associated disorders. *Environmental Toxicology and Pharmacology, 63*, 103–114.

Saladin, G., & Clément, C. (2005). Physiological side effects of pesticides on non-target plants. In Agriculture and soil pollution: New research, Nova Publishers (pp. 53–86).

Saladin, G., Magne, C., & Clement, C. (2003). Effects of fludioxonil and pyrimethanil, two fungicides used against *Botrytis cinerea*, on carbohydrate physiology in *Vitis vinifera* L. *Pest Management Science: Formerly Pesticide Science, 59*(10), 1083–1092.

Serra, R. S., Cossolin, J. F. S., de Resende, M. T. C. S., de Castro, M. A., Oliveira, A. H., Martinez, L. C., & Serrão, J. E. (2021). Spiromesifen induces histopathological and cytotoxic changes in the midgut of the honeybee *Apis mellifera* (Hymenoptera: Apidae). *Chemosphere, 270*, 129439. https://doi.org/10.1016/j.chemosphere.2020.129439.

Serra, R. S., Martínez, L. C., Cossolin, J. F. S., de Resende, M. T. C. S., Carneiro, L. S., Fiaz, M., & Serrão, J. E. (2023). The fungicide azoxystrobin causes histopathological and cytotoxic changes in the midgut of the honey bee *Apis mellifera* (Hymenoptera: Apidae). *Ecotoxicology (London, England), 32*, 234–242. https://doi.org/10.1007/s10646-023-02633-y.

Shahid, M., Zaidi, A., & Saghir Khan, M. (2021). Modulations in growth, structure, cell viability and antioxidant enzyme of a nodule bacterium *Mesorhizobium ciceri* induced by pesticides. *Environment, Development and Sustainability, 23*, 4103–4119.

Shahid, M., Zaidi, A., Ehtram, A., & Khan, M. S. (2019). In vitro investigation to explore the toxicity of different groups of pesticides for an agronomically important rhizosphere isolate *Azotobacter vinelandii*. *Pesticide Biochemistry and Physiology, 157*, 33–44.

Shakir, S. K., Kanwal, M., Murad, W., ur Rehman, Z., ur Rehman, S., Daud, M. K., & Azizullah, A. (2016). Effect of some commonly used pesticides on seed germination, biomass production and photosynthetic pigments in tomato (*Lycopersicon esculentum*). *Ecotoxicology (London, England), 25*, 329–341.

Sharma, A., Kumar, V., Kumar, R., Shahzad, B., Thukral, A.K. & Bhardwaj, R. (2018). Brassinosteroid-mediated pesticide detoxification in plants: A mini-review. *Cogent Food & Agriculture, 4*(1), p. 1436212.

Sharma, A., Kumar, V., Kohli, S. K., Kaur, R., Kaur, T., Arora, S., ... Bhardwaj, R. (2020). Pesticide metabolism in plants, insects, soil microbes and fishes: An overview. In *Pesticides in crop production: Physiological and biochemical action* (pp. 35–53). John Wiley & Sons Ltd. ISBN:9781119432241. https://doi.org/10.1002/9781119432241.

Sharma, N., Khanna, K., Jasrotia, S., Kumar, D., Bhardwaj, R., & Ohri, P. (2023). Metabolites and chemical agents in the plant roots: An overview of their use in plant-parasitic nematode management. *Nematology, 1*, 243–257. https://doi.org/10.1163/15685411-bja10220.

Sharma, N., Kour, S., Kumar, D., Kaur, R., Khajuria, A., & Ohri, P. (2022). Role of brassinosteroids (BRs) in modulating antioxidative defense mechanism in plants growing under abiotic and biotic stress conditions. In T. Aftab, & K. R. Hakeem (Eds.), *Antioxidant defense in plants* (pp. 325–367). Springer. https://doi.org/10.1007/978-981-16-7981-0_15.

Sharma, R., Jindal, R., & Faggio, C. (2021). Impact of cypermethrin in nephrocytes of freshwater fish *Catla catla*. *Environmental Toxicology and Pharmacology, 88*, 103739.

Shi, J., Yang, H., Yu, L., Liao, C., Liu, Y., Jin, M., Yan, W., & Wu, X. B. (2020). Sublethal acetamiprid doses negatively affect the lifespans and foraging behaviors of honey bee (*Apis mellifera* L.) workers. *Science of the Total Environment, 738*, 139924. https://doi.org/10.1016/j.scitotenv.2020.139924.

Singh, P., & Prasad, S. M. (2018). Antioxidant enzyme responses to the oxidative stress due to chlorpyrifos, dimethoate and dieldrin stress in palak (*Spinacia oleracea* L.) and their toxicity alleviation by soil amendments in tropical croplands. *Science of the Total Environment, 630*, 839–848.

Souto, A. L., Sylvestre, M., Tölke, E. D., Tavares, J. F., Barbosa-Filho, J. M., & Cebrián-Torrejón, G. (2021). Plant-derived pesticides as an alternative to pest management and sustainable agricultural production: Prospects, applications and challenges. *Molecules (Basel, Switzerland), 26*(16), 4835.

Stanley, D. A., Russell, A. L., Morrison, S. J., Rogers, C., & Raine, N. E (2016). Investigating the impacts of field-realistic exposure to a neonicotinoid pesticide on bumblebee foraging, homing ability and colony growth. *Journal of Applied Ecology, 53*(5), 1440–1449. https://doi.org/10.1111/1365-2664.12689.

Sule, R. O., Condon, L., & Gomes, A. V. (2022). A common feature of pesticides: Oxidative stress: The role of oxidative stress in pesticide-induced toxicity. *Oxidative Medicine and Cellular Longevity, 2022*. https://doi.org/10.1155/2022/5563759.

Tackenberg, M. C., Giannoni-Guzmán, M. A., Sanchez-Perez, E., Doll, C. A., Agosto-Rivera, J. L., Broadie, K., Moore, D., & McMahon, D. G. (2020). Neonicotinoids disrupt circadian rhythms and sleep in honey bees. *Scientific Reports, 10*(1), 17929. https://doi.org/10.1038/s41598-020-72041-3.

Teodoro, M., Briguglio, G., Fenga, C., & Costa, C. (2019). Genetic polymorphisms as determinants of pesticide toxicity: Recent advances. *Toxicology Reports, 6*, 564–570.

Thiour-Mauprivez, C., Martin-Laurent, F., Calvayrac, C., & Barthelmebs, L. (2019). Effects of herbicide on non-target microorganisms: Towards a new class of biomarkers? *Science of the Total Environment, 684*, 314–325.

Tihelka, E. (2018). Effects of synthetic and organic acaricides on honey bee health: A review. *Slovenian Veterinary Research, 55*(3), 114–140. https://doi.org/10.26873/SVR-422-2017.

Tilak, K. S., & Rao, J. (1991). Effect of pesticides mixed in different ratios to the freshwater *Lareo rohita*. *Journal of Ecotoxicology & Environmental Monitoring, 1*, 49–52.

Tiwari, P., Sangwan, R. S., & Sangwan, N. S. (2016). Plant secondary metabolism linked glycosyltransferases: An update on expanding knowledge and scopes. *Biotechnology Advances, 34*, 714–739.

Tosi, S., & Nieh, J. C. (2019). Lethal and sublethal synergistic effects of a new systemic pesticide, flupyradifurone (Sivanto®), on honeybees. *Proceedings of the Royal Society B, 286*(1900), 20190433. http://dx.doi.org/10.1098/rspb.2019.0433.

Tripathi, S., Srivastava, P., Devi, R. S., & Bhadouria, R. (2020). Influence of synthetic fertilizers and pesticides on soil health and soil microbiology. In *Agrochemicals detection, treatment and remediation* (pp. 25–54). Butterworth-Heinemann.

Ulziibayar, D., Begna, T., Ghosh, S., & Jung, C. (2021). Acute and chronic toxicity of selected pesticides used in strawberry greenhouse to honeybee (*Apis mellifera*) larvae. *Journal of Apiculture, 36*(4), 281–287. https://doi.org/10.17519/apiculture.2021.11.36.4.281.

Van der Sluijs, J. P., Simon-Delso, N., Goulson, D., Maxim, L., Bonmatin, J. M., & Belzunces, L. P. (2013). Neonicotinoids, bee disorders and the sustainability of pollinator services. *Current Opinion in Environmental Sustainability, 5*(3–4), 293–305. https://doi.org/10.1016/j.cosust.2013.05.007.

Venkaiah, K., Daveedu, T., Prathima, P., Pavani, R., Sukeerthi, S., Reddy, M. H., Pradeepkiran, J. A., & Sainath, S. B. (2022). Immobilization stress exacerbates arsenic-induced reprotoxic effects in adult rats. *Toxicology Research, 11*(3), 426–436.

Wade, A., Lin, C. H., Kurkul, C., Regan, E. R., & Johnson, R. M (2019). Combined toxicity of insecticides and fungicides applied to California almond orchards to honey bee larvae and adults. *Insects, 10*(1), 20. https://doi.org/10.3390/insects10010020.

Walsh, E. M., Sweet, S., Knap, A., Ing, N., & Rangel, J. (2020). Queen honey bee (*Apis mellifera*) pheromone and reproductive behavior are affected by pesticide exposure during development. *Behavioral Ecology and Sociobiology, 74*, 1–14. https://doi.org/10.1007/s00265-020-2810-9.

Wan, W. N., Huang, H. L., Lv, J. T., Han, R. X., & Zhang, S. Z. (2017). Uptake, translocation, and biotransformation of organophosphorus esters in wheat (*Triticum aestivum* L.). *Environmental Science & Technology, 51*, 13649–13658.

Wang, K., Sun, Y., Chen, D., Xu, Q., Li, N., Li, H., & Lu, J. (2023). Enhanced remediation of phenanthrene in water and soil by novel biochar-immobilized bacterial microspheres. *Chemical Engineering Journal, 462*, 141932.

Ward, L. T., Hladik, M. L., Guzman, A., Winsemius, S., Bautista, A., Kremen, C., & Mills, N. J. (2022). Pesticide exposure of wild bees and honey bees foraging from field border flowers in intensively managed agriculture areas. *Science of the Total Environment, 831*, 154697. https://doi.org/10.1016/j.scitotenv.2022.154697.

Williams, J. R., Swale, D. R., & Anderson, T. D (2020). Comparative effects of technical-grade and formulated chlorantraniliprole to the survivorship and locomotor activity of the honey bee, *Apis mellifera* (L.). *Pest Management Science, 76*(8), 2582–2588. https://doi.org/10.1002/ps.5832.

Wolmarans, N. J., Du Preez, L. H., Yohannes, Y. B., Ikenaka, Y., Ishizuka, M., Smit, N. J., & Wepener, V. (2018). Linking organochlorine exposure to biomarker response patterns in Anurans: A case study of Müller's clawed frog (*Xenopus muelleri*) from a tropical malaria vector control region. *Ecotoxicology (London, England), 27*, 1203–1216.

World Health Organization. (2009). Dengue guidelines for diagnosis, treatment, prevention and control: New edition. World Health Organization. https://apps.who.int/iris/handle/10665/44188.

Wu, H., Eckhardt, C. M., & Baccarelli, A. A. (2023). Molecular mechanisms of environmental exposures and human disease. *Nature Reviews Genetics, 24*, 332–344.

Wu, Y. Y., Luo, Q. H., Hou, C. S., Wang, Q., Dai, P. L., Gao, J., Liu, Y.-J., & Diao, Q. Y. (2017). Sublethal effects of imidacloprid on targeting muscle and ribosomal protein related genes in the honey bee *Apis mellifera* L. *Scientific Reports, 7*(1), 15943. https://doi.org/10.1038/s41598-017-16245-0.

Xiang, W., Wang, X., & Ren, T. (2006). Expression of a wheat cytochrome P450 monooxygenase cDNA in yeast catalyzes the metabolism of sulfonylurea herbicides. *Pesticide Biochemistry and Physiology, 85*, 1–6.

Yadav, I. C., & Devi, N. L. (2017). Pesticides classification and its impact on human and environment. *Environmental Science and Engineering, 6*, 140–158.

Yang, G., Lv, L., Di, S., Li, X., Weng, H., Wang, X., & Wang, Y. (2021). Combined toxic impacts of thiamethoxam and four pesticides on the rare minnow (*Gobiocypris rarus*). *Environmental Science and Pollution Research, 28*, 5407–5416.

Yang, Y., Wu, Y., Long, H., Ma, X., Shariati, K., Webb, J., Guo, L., Pan, Y., Ma, M., Deng, C., Cao, P., & Chen, J. (2023). Global honeybee health decline factors and potential conservation techniques. *Food Security*, 1–21. https://doi.org/10.1007/s12571-023-01346-8.

Yao, J., Zhu, Y. C., & Adamczyk, J. (2018a). Responses of honey bees to lethal and sublethal doses of formulated clothianidin alone and mixtures. *Journal of Economic Entomology, 111*(4), 1517–1525. https://doi.org/10.1093/jee/toy140.

Yao, J., Zhu, Y. C., Adamczyk, J., & Luttrell, R. (2018b). Influences of acephate and mixtures with other commonly used pesticides on honey bee (*Apis mellifera*) survival and detoxification enzyme activities. *Comparative Biochemistry and Physiology Part C: Toxicology and Pharmacology, 209*, 9–17. https://doi.org/10.1016/j.cbpc.2018.03.005.

Yen, J., Donerly, S., Levin, E. D., & Linney, E. A. (2011). Differential acetylcholinesterase inhibition of chlorpyrifos, diazinon and parathion in larval zebrafish. *Neurotoxicology and Teratology, 33*(6), 735–741.

Yin, Y. L., Zhou, Y., Zhou, Y. H., Shi, K., Zhou, J., Yu, Y., Yu, J.-Q., & Xia, X. J. (2016). Interplay between mitogen-activated protein kinase and nitric oxide in brassinosteroid-induced pesticide metabolism in *Solanum lycopersicum*. *Journal of Hazardous Materials, 316*, 221–231. https://doi.org/10.1016/j.jhazmat.2016.04.070.

Yu, G. B., Chen, R. N., Chen, Q. S., Chen, F. Q., Liu, H. L., Ren, C. Y., ... Wei, J. P. (2022). Jasmonic acid promotes glutathione assisted degradation of chlorothalonil during tomato growth. *Ecotoxicology and Environmental Safety, 233*, 113296.

Zacharia, J. T. (2011). Identity, physical and chemical properties of pesticides. In *Pesticides in the modern world: Trends in pesticides analysis* (pp. 1–18). ISBN978-953-307-437-5. https://doi.org/10.5772/701.

Zaller, J. G., & Brühl, C. A. (2019). Non-target effects of pesticides on organisms inhabiting agroecosystems. *Frontiers in Environmental Science, 7*, 75.

Zhang, C., Zhou, T., Zhu, L., Juhasz, A., Du, Z., Li, B., Wang, J., Wang, J., & Sun, Y. (2019). Response of soil microbes after direct contact with pyraclostrobin in fluvo-aquic soil. *Environmental Pollution, 255*, 113164.

Zhang, J. J., & Yang, H. (2021). Metabolism and detoxification of pesticides in plants. *Science of the Total Environment, 790*, 148034.

Zhang, J. J., Lu, Y. C., Zhang, S. H., Lu, F. F., & Yang, H. (2016). Identification of transcriptome involved in atrazine detoxification and degradation in alfalfa (*Medicago sativa*) exposed to realistic environmental contamination. *Ecotoxicology and Environmental Safety, 130*, 103–112.

Zhang, Q., Fu, L., Cang, T., Tang, T., Guo, M., Zhou, B., Zhu, G., & Zhao, M. (2021). Toxicological effect and molecular mechanism of the chiral neonicotinoid dinotefuran in honeybees. *Environmental Science and Technology, 56*(2), 1104–1112. https://doi.org/10.1021/acs.est.1c05692.

Zhang, X. D., Zhao, K. X., & Yang, Z. M. (2018). Annotation of ATP binding cassette (ABC) transporter genes and identification of Cd–responsive ABCs in rapeseed (*Brassica napus*). *Gene, 664*, 139–151.

Zhao, H., Li, G., Cui, X., Wang, H., Liu, Z., Yang, Y., & Xu, B. (2022). Review on effects of some insecticides on honey bee health. *Pesticide Biochemistry and Physiology, 105219*, 105219. https://doi.org/10.1016/j.pestbp.2022.105219.

Zhao, K. X., Zhang, M. Y., Yang, D., Zhu, R. S., Zhang, Z. F., Hu, Y. H., & Kannan, K. (2023). Screening of pesticides in serum, urine and cerebrospinal fluid collected from an urban population in China. *Journal of Hazardous Materials, 449*, 131002.

Zhou, Y. H., Xia, X. J., Yu, G. B., Wang, J. T., Wu, J. X., Wang, M. M., Yang, Y. X., Shi, K., Yu, Y. L., & Chen, Z. X. (2015). Brassinosteroids play a critical role in the regulation of pesticide metabolism in crop plants. *Scientific Reports UK, 5*, 9018.

Zhu, L., Qi, S., Xue, X., Niu, X., & Wu, L. (2020). Nitenpyram disturbs gut microbiota and influences metabolic homeostasis and immunity in honey bee (*Apis mellifera* L.). *Environmental Pollution, 258*, 113671. https://doi.org/10.1016/j.envpol.2019.113671.

Zikankuba, V. L., Mwanyika, G., Ntwenya, J. E., & James, A. (2019). Pesticide regulations and their malpractice implications on food and environment safety. *Cogent Food & Agriculture, 5*(1), 1601544.

CHAPTER 3

Pesticide biology in soil: Sorption, leaching, and accumulation

Sandeep[a,b], Sonika Sharma[b], Ashish Sharma[a], Raj Bala[a], Indu Sharma[c] and Ashutosh Sharma[b]
[a]Department of Botany & Environment Studies, DAV University, Jalandhar, Punjab, India
[b]Faculty of Agricultural Sciences, DAV University, Jalandhar, Punjab, India
[c]Department of Life Sciences, University Institute of Sciences, Sant Baba Bhag Singh University, Jalandhar, Punjab, India

3.1 Introduction

Agriculture in the 21st century is facing some problems, which were not anticipated previously. Since the last few decades, an array of agrochemicals especially pesticides are employed in agriculture to kill various target pests like insects, weeds, microorganisms, nematodes, etc. of the crops. These pesticides generally protect crops from harmful and unwanted crop pests. In order to increase production by reducing the population of competing pests has led to excessive utilization of pesticides (herbicides, insecticides, fungicides, nematicides, etc.) in agriculture. Some pesticides can retain and accumulate in agricultural soils for long periods and are referred persistent pesticides. The indiscriminate use of toxic and persistent pesticides is considered more hazardous due to their accumulation in soil. The excessive and unwarranted utilization of pesticides leads to their bio-accumulation in food chains, thereby causing a high risk to mammals and other nontarget organisms. Further, their other direct or indirect effects on such nontarget organisms may lead to ecosystem imbalances (Liu et al., 2016; Rosell et al., 2008). The pesticide residues are therefore harmful to the ecosystem as they may persist in the environment for a longer period and may have carcinogenic effects (Özkara et al., 2016).

Although pesticides have been reported to adversely affect the human-health and have several ecological disadvantages, these are required to meet the global demand for food for the ever-increasing population (Rani et al., 2021). Besides agriculture, pesticides are also applied for the management of household insects and pests, in home gardening and to control disease insect vectors. Being highly toxic, these chemical pesticides also pose a threat not only to the health of agricultural workers who are directly exposed to them while applying the pesticides but also to the neighboring communities. When pesticides are used for plant protection interventions in agriculture, the soil acts as its sink. Further, their persistence in the soil is mainly dependent on their sorption by soil

Pesticides in the Environment: Impact, Assessment, and Remediation
DOI: https://doi.org/10.1016/B978-0-323-99427-9.00013-6

components and their degradation (Narayanan et al., 2014). Upon pesticide application in soil, it has three different fates (Pérez-Lucas et al., 2019):

1. It can be adsorbed by soil particles (clay or organic matter).
2. Leaching in the soil through water.
3. Get degraded by microbial/enzymatic action.

Therefore, it is imperative to study the sorption, leaching, and degradation behavior of different pesticides in the soil on case-to-case bases. Carsel et al. (1985) provided a continuous simulation model, the Pesticide Root Zone Model (PRZM) that considers the interactions of pesticides in surface runoff, molecular diffusion, uptake by plants, sorption by soil particles, their biochemical degradation, and some other parameters. The model was successfully tested and calibrated in the United States. The pesticide leaching models are applied at large scales to assess the risk of pesticide groundwater contamination and pesticide leaching is affected by the pesticide sorption by soil and the soil organic matter is the most important soil constituent affecting this (Farenhorst, 2006). Efforts have also been made to simulate pesticide transport by SWAP-PEARL models for the prediction of pesticide leaching (Urbina et al., 2020).

After leaching from soil or run-off from agriculture fields, the soluble pesticides tend to be carried away by water molecules and reach the aquatic bodies or underground water bodies through downward percolation (Syafrudin et al., 2021). The accumulation of pesticides in water bodies leads to the deterioration of the quality of water and when consumed, these soluble pesticides may accumulate in the bodies of living organisms. Therefore, to purify water sources contaminated with pesticides, generally, methods like adsorption, coagulation-flocculation, sedimentation, and filtration are used. For the removal of pesticides from the soil, the plants have developed several biological features for the production of pesticide-metabolized enzymes, and mechanisms to detoxify the pesticide residues (Zhang & Yang, 2021). However, this herbicide-enzyme interaction is a complicated process and is discussed in detail, elsewhere (Bhardwaj et al., 2021). The concentration of pesticides above threshold levels is harmful to the plants and plants cannot survive in the soils having highly toxic and persistent pesticides, because excessive pesticides tend to hinder the bioavailability of useful minerals and nutrients in the soil.

The persistence of pesticides in soil not only affects the soil biochemistry but also the soil biology. The different types of pesticides affect soil microflora to different extents. They affect the population dynamics of soil micro-flora (Arora et al., 2019; Simon-Sylvestre & Fournier, 1980). This may also negatively affect agriculturally important microbial processes like nitrogen fixation (Mohamed et al., 2021). The reduction of nontarget soil microbial count may lead to a permanent change in the soil microbial community and affect the agricultural inputs (increased fertilizer usage). Therefore, not only the study of pesticide fate in the soil is important, but how they affect the soil microbial communities at large is also a matter of great concern. The pesticide risk assessment is an important step in studying the environmental impacts of the pesticide. However, the consequence of

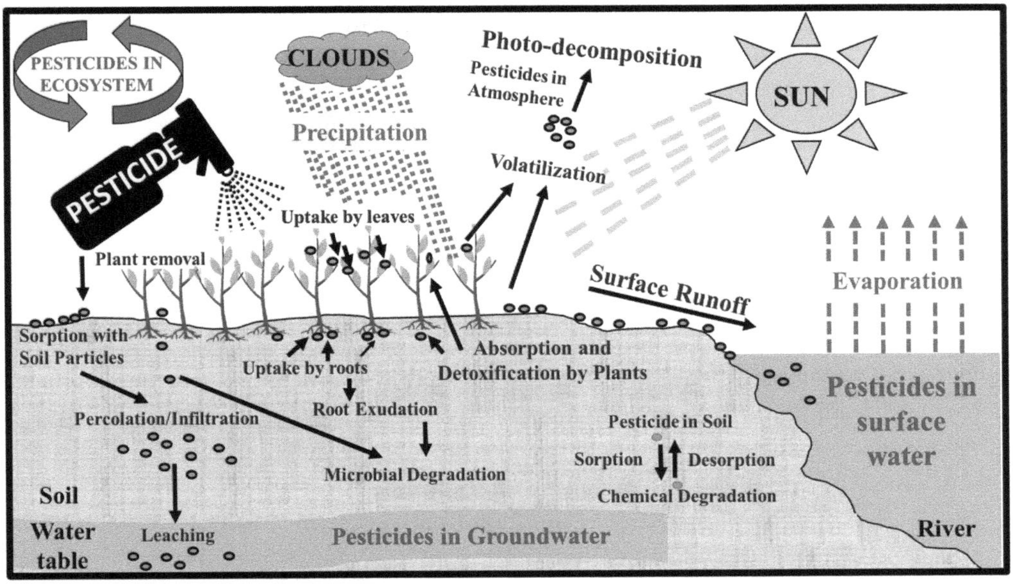

Figure 3.1 ***Mechanism of sorption, uptake, leaching, accumulation, and degradation of pesticides in the ecosystem on the basis of our current understanding.*** [The availability and degradation of pesticides in the ecosystem depend upon their type and mode of application, sorption-desorption, absorption or uptake, runoff, infiltration, leaching, soil properties, tillage practices, plant harvest and/or removal, transfer, volatilization, bio-transformations, and microbial/chemical/photo-degradation.]

an inappropriate environmental risk assessment of pesticides may lead to a decline in biodiversity (Brühl & Zaller, 2019). Although to feed the increasing human population, the use of pesticides to increase productivity by suppressing the population or the propagules of the competing organisms (pests, weeds, pathogens, etc.) has become backbone of the present-day agriculture, the use of pesticides can be made on more rational and environmentally sustainable with the cooperation of all the stakeholders (Lykogianni et al., 2021).

This chapter is, therefore, an effort to compile all the information available to date about the biology of pesticides in soil. Since pesticide sorption, leaching, and accumulation are important phenomena in their persistence in soil, an attempt has also been made in the chapter to discuss them. On the basis of our present understanding, Fig. 3.1 explains the mechanisms of pesticides' sorption, uptake, leaching, accumulation, and degradation in the ecosystem. The mobility, transformation, accumulation, persistence, degradation, and fate of pesticides in the ecosystem is dependent upon the following processes:

- In-soil or foliar applications of pesticides introduces the pesticides into the ecosystem.
- After application, pesticides usually reach the soil, which is the major sink of pesticides.

- Soil adsorption and absorption (uptake) of pesticides depends upon the organic matter, clay content, and other soil properties or tillage practices.
- Pesticides are generally absorbed (uptake) by the roots of plants or enter through leaves into plants, where they can be detoxified by the innate plant mechanisms into simpler and less toxic forms.
- Crop and/or plant removal leads to their transfer from agricultural soils and pesticide residue may retain in the crop/plant produce.
- The pesticides may undergo microbial degradation due to the activity of microorganisms in the vicinity of roots or in/near rhizosphere, or released root exudates in the soil.
- Pesticides may also undergo sorption and adsorption or hydrolysis or other chemical reactions leading to their chemical degradation in the soil.
- After sorption, pesticides may be transferred to deeper soils through percolation or infiltration on the basis of their solubility and physicochemical properties, followed by their leaching into groundwater resulting in the contamination of the water table.
- Pesticides can also runoff from the slanted soil surfaces due to irrigation, precipitation (rainfall), flood, spillage, soil-erosion, etc. that may damage nontarget plants and cause contamination of surface water bodies like ponds, lakes, rivers, etc.
- Due to strong sunlight, scorching heat, high temperatures, less relative humidity, or other factors, the volatile pesticides from leaves, above-ground plant surfaces, or unbound pesticides from the soil, can vaporize into gases and enters the atmosphere.
- Volatilized pesticides can be photo-degraded or photo-decomposed by sunlight.
- Pesticides solubilized in surface water may be evaporated and may circulate in the ecosystem through the water cycle.

The specific details of some of the above steps are discussed under the subsequent headings at the appropriate place. Further, the effect of pesticides on the soil micro-flora has also been discussed with an overall aim to provide compiled up-to-date information for the readers.

3.2 Release of pesticides in soil during agricultural activities

Pesticides are plant-protectant chemicals that are used to prevent various plant diseases, weeds, and pests and thereby enhance the quality as well as the yield of food products. Based on their target organisms, these pesticides are variously classified as herbicides, fungicides, acaricides, nematicides, insecticides, rodenticides, etc. These pesticides can either be directly applied to the soil or can be used as foliar spray/dusting treatment, besides their use as a seed treatment. These days novel techniques of pesticides like "drip chemigation" of these pesticides are also in use (Ghidiu et al., 2012), whereby their dosage can be minimized and their effect on nontarget organisms can also be minimized. Irrespective of their mode of application, ultimately pesticides go into the soil. The soil, therefore, acts as a sink for the pesticides (applied to control crop pests, plant diseases, weeds, etc.) that are frequently applied in agricultural fields.

3.3 Pesticide sorption by soil organic matter

Adsorption of pesticides by soil partials may either be physical (van der Waals forces) or chemical (electrostatic interactions) in nature (Raffa & Chiampo, 2021). Konda et al. (2002) derived a nonlinear mathematical model from the Langmuir equation to represent single-step and multistep-shaped adsorption isotherms in different pesticides. The process of pesticide adsorption was described by a single-step (Langmuir) isotherm for acetochlor and carbendazim, by a two-step curve for diazinon, isoproturon and atrazine, and by a three-step curve for imidacloprid. Further, in a study involving three different pesticides on five different soil types, it was found that the adsorption/desorption processes of these pesticides were mainly dependent on soil organic matter and octanol/water partitioning coefficient (Yu et al., 2006).

The sorption of pesticides depends on a number of soil factors like organic matter, clay percentage, size of soil partials, pH, pore space and weather pattern, etc. besides the nature of the pesticide. A strong relationship exists between total organic carbon and the sorption as well as the mobility of different pesticides in soil. In soils with low organic matter, the mobility of most of the pesticides is often related to the active components of the inorganic clay-sized fraction, therefore an increase in the clay content leads to a reduction in pesticide mobility (Spark & Swift, 2002). In order to increase crop productivity, the soil is usually supplemented with a variety of organic amendments, that lead to an increase in soil organic matter. However, the use of some of such amendments like crop residues and other organic wastes may influence the sorption of certain organic chemicals in soils. Organic materials in both raw and decomposed forms increase the sorption capacity of pesticides in soils (Jing et al., 2021). Chlorothalonil had the highest absorption in clay soil with high organic matter (Wang et al., 2021). Similar to chlorothalonil, the hydrophobic pesticide endosulfan is also adsorbed more by clay soil than sandy soils (Kumar & Philip, 2006). The soil pH also affects pesticide sorption. Higher adsorption of 2, 4-D was recorded in an acidic pH (Jamshidi et al., 2022). In a study involving three carbamate pesticides, it was found that the addition of organic matter to soil decreases pH (Bansal, 2010). Therefore, the effect of pH may also be associated indirectly with the soil organic matter also. The pesticide sorption characteristics of different types of soils in relation to their organic matter are represented in Table 3.1.

3.4 Effect of pesticide on soil process

The long-term and repeated applications of pesticides have been found to interfere with the bio-chemical balance of soil, decreasing soil fertility by affecting soil-enzymatic

Table 3.1 Pesticide sorption characteristics of different types of soils in relation to their organic matter.

S. no.	Type of soil	Organic matter (g/Kg)	Clay %	Pesticide	Sorption of pesticides (mg/g)	References
1.	Clay	36.06	15.24	Chlorothalonil	6.7158	Wang et al. (2021)
		2.29	56	Endosulfan	0.4480	Kumar and Philip (2006)
2.	Clay loam	3.62	30	2, 4-D	2.45	Jamshidi et al. (2022)
		1.4	17	Alachlor	31	Ahmad (2018)
3.	Sandy	1.07	11081	Chlorothalonil	1.2568	Wang et al. (2021)
		0.575	0.2	Endosulfan	0.1083	Kumar and Philip (2006)
4.	Loam	2.81	24	2,4-D	1.6	Jamshidi et al. (2022)
		1.1	10	Alachlor	12	Ahmad (2018)
		3.34	23.64	Pendimethalin	14.328	Đurović et al. (2009)
5.	Sandy clay loam	2.10	21	2,4-D	0.38	Jamshidi et al. (2022)
		8.69	23.44	Pendimethalin	15.231	Đurović et al. (2009)
6.	Red loam soil	1.01	10.2	Phenazine-1-carboxamide	5.57	Ou et al. (2020)
7.	Silt clay loam	1.35	35	Alachlor	19	Ahmad (2018)
8.	Loamy sand	3.17	3.44	Pendimethalin	14.502	Đurović et al. (2009)
9.	Black soil	1.98	25.7	Phenazine-1-carboxamide	12.8	Ou et al. (2020)
10.	Tidal soil	1.72	13.6	Phenazine-1-carboxamide	10.4	Ou et al. (2020)
11.	Sandy loam	0.56	14.92	Penconazole	High sorption	Jiang et al. (2016)

activities. It has also been found that pesticides also impact soil enzymes involved in nutrient recycling (like carbon, nitrogen, sulfur, and phosphorus cycles) and thereby affect the soil processes (Riah et al., 2014). Leaching of pesticides in soil may also influence soil biochemical processes driven by soil microbes and enzymes, thereby affecting soil fertility also (Hussain et al., 2009). It may be due to pesticidal effects on biological nitrogen fixers (Mohamed et al., 2021). They may also decrease the phosphate solubilizing activity of microbial phosphate solubilizes in soil (Shahid et al., 2021). Overall, a number of soil physio-bio-chemical processes as well as the agriculturally important beneficial microbes are affected by the application of pesticides in soil.

3.5 Leaching of pesticides in soil

The leaching of a pesticide is the process in which the nonabsorbed pesticide solution goes deeper from the pore space between the particles to the subsoil. This vertically downwards displacement of pesticides mainly depends upon the soil properties like soil texture, clay content, organic matter, and permeability and also on the properties of pesticide and may finally reach the groundwater and contaminate it (Navarro et al., 2007; Pérez-Lucas et al., 2019). The leaching of pesticides is not only dependent on the chemical nature of the pesticides, but also on the climatic conditions. Various agricultural operations like the amount of tillage have also been found to affect the leaching of pesticides in soil. In an attempt to study the adsorption of four different organophosphorus pesticides (*viz.*, diazinon, dimethoate, methyl parathion, and sulfate) in four different soil samples from Mexico, the sorption coefficient of organic carbon (K_{oc}) was calculated, which indicated that these pesticides had lower adsorption on the four studied soil samples, suggesting a high risk of contamination to the aquifer, thereby explaining the accumulation of these pesticides in the groundwater in considerable concentrations within the study area (Menchen et al., 2017). The organic matter of the soil is directly associated with the volume of leachate. Lower organic matter in soil may lead to a higher leaching percentage than the soil with a higher amount of organic matter (Bajeer et al., 2012). Chlorothalonil has a good leaching capacity in clay soil with high organic matter, whereas sandy soil shows a low leaching capacity (Wang et al., 2021).

Besides this, the tillage operations also affect the leaching of pesticides in soil. No-tillage cultivation improves macropore connectivity leading to an increase in pesticide leaching (Alletto et al., 2010). Further, the spray application of herbicides coincided with the rainy seasons, and the high leaching rates are recorded (Lutri et al., 2022). Glyphosate, atrazine, and aminomethylphosphonic acid were recorded from about 80% of the rainwater samples collected in Argentina grassland pampas (Alonso et al., 2018). For the in-soil mobility of pesticides, their distribution coefficients (K_d) are considered important and the pesticide leaching characteristics of the soil types can be calculated using wick lysimeters (Park et al., 2020). Some soil types have a higher rate of leaching or rate of pesticide transport than others. The runoff potential of soils was found to be an important factor affecting their transport in soils and watersheds (Lerch and Blanchard, 2003). Holten et al. (2019) suggested that the rank order in pesticide leaching of a particular soil type corresponds to its rank order of mean soil sorption coefficient (K_f) value and their sorption plays a major role in determining their leaching losses even under the frozen soil conditions. The leaching ability of different pesticides in various soil types has been presented in Table 3.2.

3.6 Impact of pesticides on the soil biology

The biological activity of soil is an important indicator of pesticide pollution in soil (Wołejko et al., 2020). Pesticides affect both macro and micro-organisms present in the soil. Earthworms are also quite susceptible to pesticides, which may lead to their immobility, biomass reduction, and disruption of growth and reproduction by disruption of various physiological activities. This may lead to a reduction in their population and biodiversity (Miglani & Bisht, 2019). Soil microbial biomass is also affected by the use of pesticides (Arora et al., 2019). In a study on the soil samples taken from agricultural fields under a rice-wheat cropping system, a negative impact of pesticide application on all soil microbes with a decrease in their average population of all major groups was reported (Gupta et al., 2000). In an experiment to study the effect of pesticides on soil microbes, an overall decrease in bacterial abundance and diversity was noticed in irrigated rice fields (Onwona-Kwakye et al., 2020). Pesticides not only decrease the microbial diversity of the soil but also interfere with other agriculturally important spoil processes like nitrogen fixation. The application of chlorpyrifos was recorded to inhibit the growth of nitrogen-fixing bacteria and decrease the biological nitrogen fixation in paddy fields. The toxic effect of chlorpyrifos on the *in vitro* nitrogenase activity of *Pseudomonas stutzeri* strain A1501 was also recorded with the down-regulation of nitrogen-fixation-related genes *viz.*, *nifA* and *nifH*. Further, the rhizosphere colonization and biofilm formation by *P. stutzeri* A1501 were also recorded to be reduced by exposure to chlorpyrifos, leading to decreased nitrogenase activity in the rhizosphere (Lu et al., 2020). However, a stimulatory effect on the *Azotobacter vinelandii* population and nitrogen fixation was also recorded earlier by the application of insecticide, diflubenzuron (Martinez-Toledo et al., 1988). Therefore, it is not easy to draw any generalization in this regard and a case-by-case evaluation is required. Some pesticides may stimulate the growth of micro-organisms, but others can have a negative effect on their growth (Lo, 2010). Therefore, the effect of different pesticides on various soil microbes has been compiled in Table 3.3.

3.7 Accumulation of pesticides in soil and its remediation

Pesticides are widely used worldwide, particularly in developing nations in Asia and Africa for pest management and enhancing crop productivity (Rasool et al., 2022).

Table 3.2 Leaching ability of different pesticides in various soil types.

S. no.	Soil type	Pesticide	Organic matter	Clay (%)	pH	Leaching	References
1.	Clay	Chlorothalonil	36.06 g/kg	15.24	8.1	0.4977 μg/g	Wang et al. (2021)
2.	Sandy		1.07 g/kg	11.81	7.71	0.0498 μg/g	
3.	Acidic loam	2-methyl-4-chlorophenoxyacetic	2.5%	19.1	5.5	24.5%	Holten et al. (2019)
4.	Acidic silt		1.2%	5.4	5.4	16.4%	
5.	Clay loam	Clothianidin	0.95%	30.4	5.06	16.09 μg/g	Singh et al. (2018)
6.	Sandy loam	Mecoprop-P	9.0 g/kg	15.9	6.2	1.36%	Fogg et al. (2004)
		Isoproturon	9.0 g/kg	15.9	6.2	0.002%	
		Penconazole	0.56 g/kg	14.92	6.20	Low leaching percentage	Jiang et al. (2016)
		Clothianidin	0.29%	10.4	8.41	13.84 μg/g	Singh et al. (2018)
7.	Silt clay	Isoproturon	36.0 g/kg	40.6	7.7	0.006%	Fogg et al. (2004)
		Mecoprop-P	36.0 g/kg	40.6	7.7	0.04%	
8.	Sandy soil (*Mollic Paleudafls*)	Alachlor	14.8 g/kg	–	4.7	6% after 5 weeks	Park et al. (2020)
		Carbofuran		–	4.7	10% after 5 weeks	
		Ethoprophos		–	4.7	4% after 4 weeks	
		Napropamide		–	4.7	Slightly after 4 weeks	
		Tebuconazole		–	4.7	0.1% after 9 weeks	
9.	Andic palehumults	Alachlor	52 g/kg	–	4.9	0.8% after 9 weeks	
		Carbofuran		–	4.9	5% after 7–9 weeks	
		Ethoprophos		–	4.9	0.9% after 7–9 weeks	
10.	Acrudoxic melanudand	Carbofuran	137 g/kg	–	6.0	Slightly after 7 weeks	

Where the symbol "–" represents that there is no information available on the clay content of a particular soil type in the research cited.

Table 3.3 Effect of various pesticides on different soil micro-organisms.

S. no.	Pesticide	Micro-organism(s)	Effect	References
1.	Butachlor (20 μg/g)	*Azospirillum* and anaerobic nitrogen fixers	Reduces the population	Jena et al. (1987)
2.	Carbofuran (2–4 μg/g)	*Azospirillum* and anaerobic nitrogen fixers	Reduces the population at 2 μg/g, but increases the number at 4 μg/g	
3.	Diflubenzuron (100–500 μg/g)	*Azotobacter vinelandii*	Increases the number in both sterile and nonsterile soil	Martinez-Toledo et al. (1988)
4.	Methylpyrimifos (100–300 μg/g)	Aerobic dinitrogen fixing bacteria	Reduces the population and the dinitrogen fixation	Martinez-Toledo et al. (1992)
5.	Chlorpyrifos (10–300 μg/g)			
6.	Metsulfuron methyl	77 strains of fluorescent pseudomonads	Toxic to a major proportion of the strains at lower concentrations	Boldt and Jacobsen (1998)
7.	Chlorsulfuron		Less toxic at lower concentrations, but toxic at higher concentrations	
8.	Thifensulfuron methyl		Toxic only to a small number of strains	
9.	Diuron	Soil bacterial community-based on 16S ribosomal ribonucleic acid (rRNA) profile	The structure of the microbial community of pesticide-treated and untreated soils was different	El fantroussi et al. (1999)
10.	Linuron			
11.	Chlorotoluron			

(continued on next page)

12.	Fenpropimorph	Soil bacteria and fungi	Inhibited the growth of fungi in the first 10 days, the total culturable bacteria were lowered at day 17 but later stimulated at day 56	Thirup et al. (2001)
13.	Iprodione (5–50 μg/g)	Soil bacterial community	Soil bacterial community increases rapidly at 30°C upon the treatment with a higher concentration	Wang et al. (2004)
14.	Methylparathion	Multiple phylotypes based on 16S ribosomal DNA (rDNA) clones	Change in phylotype diversity recorded (dominant phylotypes were replaced by others like γ-proteobacteria)	Zhang et al. (2006)
15.	Azoxystrobin	*Paraflabellula hoguae* (protozoa)	The reduced population as indicated by the absence of a band in 18S rRNA PCR-DGGE analysis	Bending et al. (2007)
16.	Chlorothalonil	*Arcuospathidium* sp. or *Bresslaua vorax* (protozoa)		
17.	Tebuconazole	*Cladosporium tenuissimum*, an ascomycete		
18.	Isoproturon	*Sphingomonas* spp.	Decreased population	Bending et al. (2003)

Therefore, adequate regulation of pesticide application in these nations is necessary to prevent pesticide-related environmental contamination on a broad scale. The organic matter content, type of soil, environmental factors, and other characteristics of the soil like moisture content and pH are the determining factors for the transit and fate of pesticides in the soil environment. Besides this, the studies have revealed that microbial diversity and activity also determine bioavailability and the rate of degradation of pesticides in the soil (Bhatt et al., 2021; Chaudhary et al., 2023; Chen et al., 2019; Rasool et al., 2022).

Whatever may be the mode of application of pesticides, ultimately, they reach the soil that acts as their sink or the natural reservoir. Eventually, they are accumulated in soil and their availability may be mediated by their sorption characteristics in the soil that may vary with soil types and the nature of pesticides. Accumulation of pesticides in the soil is a major environmental concern in achieving sustainable agriculture goals. These pesticides may persist and accumulate in the soil and therefore may contaminate the groundwater depending on their sorption kinetics (Tang & Maggi, 2021). A commonly used nitrogen-containing pesticide namely phenazine-1-carboxamide was reported to be degraded in the soil by the activities of anaerobic micro-organisms. The phenazine had high adsorption, and limited mobility and was easily degraded in soils with more amounts of organic matter and clay. Phenazine-1-carboxamide was observed to be less toxicant for groundwater systems in soils rich in both organic matter and clay as compared to soils with lowers contents of organic matter and clay content (Ou et al., 2020).

Attempts have been made to reduce the levels of accumulation or reduce the harmful effects of pesticides on the environment in a variety of ways. In an experiment to evaluate the use of biochar application to soil, it was found that it may have a long-lasting effect on the reduction of dissipation of pesticides, which are applied as a foliar application, thereby reducing the risk of environmental contamination and human exposure mediated by the transfer along the food chain. However, on the other hand, it may affect the efficacy of soil-applied herbicides (Jones et al., 2011). Therefore, we need to be careful about the use of biochar depending on the mode of application of pesticides (foliar application or soil application). Bioremediation and biodegradation are also among the eco-friendly approaches to reduce pesticide toxicity and their accumulation in soil (Rani & Dhania, 2014). Various physicochemical methods for their remediation include adsorption, oxidation-reduction, catalytic degradation, and the use of nanotechnology, etc. (Bakshi et al., 2021), that also useful besides the use of other bio-remediation strategies. Among the bio-remediation strategies, the use of microbes could be one of the promising approaches for the cleaning up of the environment.

Further, a consortium of microbial species is better than the use of a single microbe, since in the consortia the metabolic burden is divided among the microbial strains in the group, which leads to a robust, step-wise, and faster pesticide degradation (Bhatt et al., 2021). Immobilized enzyme systems are also found to be effective in the removal of pesticide residue from the soil. In an experiment, the immobilization of *Aspergillus* laccase and the redox mediator syringaldehyde on two different biomass materials *viz.*, peanut shell and wheat straw, were recorded to have a strong potential for the effective removal of pesticides from water and soil by bio-sorption coupled with degradation and degradation of nine different pesticides was recorded (Chen et al., 2019). Bacterial consortium use enhances micro-organisms' capacity to break down both organic and inorganic pesticides into less harmful forms (Chaudhary et al., 2023). The potentially beneficial microbial consortia can be employed as biocontrol agents and/or bio-stimulants. *Azotobacter, Bacillus, Pseudomonas, Rhizobium*, and *Trichoderma* are some of the most prevalent micro-organisms with biocontrol activity that are utilized in preparing microbial bio-formulations. Because of their biocontrol activity against soil-borne pathogens, the use of microorganism-based bio-formulations as bio-pesticides can be a sustainable method for plant protection. Therefore, the identification of diverse microbial consortia found in various types of soils and their degrading efficacy is warranted to develop a green alternative and ecologically safe bio-pesticides.

3.8 Conclusion and future prospects

Present-day agriculture largely relies on chemical pesticides for crop protection against a variety of bio-agent like insect pests, pathogens, weeds, etc., so that a good yield is obtained. The overreliance on these pesticides has resulted in their marked effects on ecology that include adverse effects on the nontarget organisms, and microbial diversity. Since the soil is the ultimate sink of pesticides, irrespective of their mode of application; the fate of pesticides in soil is a major concern. The fate of a pesticide in the soil is controlled by several factors like pesticide sorption by soil particles (clay or organic matter), degradation in soil, and leaching. It may enter the water bodies affecting their ecology, enter into the food chains, and may get bio-accumulated there up the tropic levels. Further, the problem of pesticide residues in food grains above the permissible limits is also a matter of concern now. However, the accumulation of pesticides in the soil varies with the nature and mode of application of pesticides, but the major phenomenon behind their accumulation remains largely the same. Despite the several efforts to predict the behavior and fate of different pesticides in soil, a careful and long-term evolution of different classes of pesticides in different soil types under varied environments is required for better prediction of pesticide behavior and device strategies for minimizing the ecological hazards associated with their judicial usage in agriculture. This chapter is an attempt to present up-to-date information regarding the sorption, leaching, and accumulation of pesticides in soil. In order to lessen the detrimental effects of pesticides on human health and the ecosystem, it is urgently necessary to identify ecologically safer and less toxic pesticides that may be used in smaller quantities and are economically feasible. Further investigations are required to develop more green and safer alternatives to chemical-based pesticides such as bio-control agents, bio-formulations, and bio-pesticides.

Acknowledgment

The authors are thankful to the administration of DAV University, Jalandhar, Punjab, India to provide all the necessary infrastructure.

Conflicts of interest

The authors declare that they do not have a conflict of competing interests.

References

Ahmad, K. S. (2018). Evaluating the adsorption potential of alachlor and its subsequent removal from soils via activated carbon. *Soil and Sediment Contamination: An International Journal, 27*(4), 249–266.

Alletto, L., Coquet, Y., Benoit, P., Heddadj, D., & Barriuso, E. (2010). Tillage management effects on pesticide fate in soils: A review. *Agronomy for Sustainable Development, 30*(2), 367–400.

Alonso, L. L., Demetrio, P. M., Etchegoyen, M. A., & Marino, D. J. (2018). Glyphosate and atrazine in rainfall and soils in agroproductive areas of the Pampas region in Argentina. *Science of the Total Environment, 645*, 89–96.

Arora, S., Arora, S., Sahni, D., Sehgal, M., Srivastava, D. S., & Singh, A. (2019). Pesticides use and its effect on soil bacteria and fungal populations, microbial biomass carbon and enzymatic activity. *Current Science, 116*(4), 643–649.

Bajeer, M. A., Nizamani, S. A., Sherazi, S. T. H., & Bhanger, M. I. (2012). Adsorption and leaching potential of imidacloprid pesticide through alluvial soil. *American Journal of Analytical Chemistry, 3*(8), 21904.

Bakshi, P., Singh, A. D., Kour, J., Jan, S., Ibrahim, M., Mir, B. A., & Bhardwaj, R. (2021). Advanced technologies for the remediation of pesticide-contaminated soils. In *Handbook of assisted and amendment: Enhanced sustainable remediation technology* (pp. 331–353). John Wiley and Sons, Inc.

Bansal, O. P. (2010). The Effects of composts on adsorption-desorption of three carbamate pesticides in different soils of Aligarh district. *Journal of Applied Sciences and Environmental Management, 14*(4), 155–158.

Bending, G. D., Lincoln, S. D., Sørensen, S. R., Morgan, J. A. W., Aamand, J., & Walker, A. (2003). In-field spatial variability in the degradation of the phenyl-urea herbicide isoproturon is the result of interactions between degradative *Sphingomonas* spp. and soil pH. *Applied and Environmental Microbiology, 69*(2), 827–834.

Bending, G. D., Rodríguez-Cruz, M. S., & Lincoln, S. D. (2007). Fungicide impacts on microbial communities in soils with contrasting management histories. *Chemosphere, 69*(1), 82–88.

Bhardwaj, L., Pandey, J., & Dubey, S.K. (2021). Effects of herbicides on soil enzymes and their regulatory factors in agroecosystem: A review. In: Dubey, S.K., Verma, S.K. (Eds.), *Plant, soil and microbes in tropical ecosystems. Rhizosphere Biology*. (pp. 71–100), Springer, Singapore.

Bhatt, P., Bhatt, K., Sharma, A., Zhang, W., Mishra, S., & Chen, S. (2021). Biotechnological basis of microbial consortia for the removal of pesticides from the environment. *Critical Reviews in Biotechnology, 41*(3), 317–338.

Boldt, T. S., & Jacobsen, C. S. (1998). Different toxic effects of the sulfonylurea herbicides metsulfuron methyl, chlorsulfuron and thifensulfuron methyl on fluorescent pseudomonads isolated from an agricultural soil. *FEMS Microbiology Letters, 161*(1), 29–35.

Brühl, C. A., & Zaller, J. G. (2019). Biodiversity decline as a consequence of an inappropriate environmental risk assessment of pesticides. *Frontiers in Environmental Science, 7*, 177.

Carsel, R. F., Mulkey, L. A., Lorber, M. N., & Baskin, L. B. (1985). The pesticide root zone model (PRZM): A procedure for evaluating pesticide leaching threats to groundwater. *Ecological Modelling, 30*(1-2), 49–69.

Chaudhary, P., Xu, M., Ahamad, L., Chaudhary, A., Kumar, G., Adeleke, B. S., Verma, K. K., Hu, D. M., Širić, I., Kumar, P., & Abou Fayssal, S. (2023). Application of synthetic consortia for improvement of soil fertility, pollution remediation, and agricultural productivity: A review. *Agronomy, 13*(3), 643.

Chen, X., Zhou, Q., Liu, F., Peng, Q., & Teng, P. (2019). Removal of nine pesticide residues from water and soil by biosorption coupled with degradation on biosorbent immobilized laccase. *Chemosphere, 233*, 49–56.

Đurović, R., Gajić-Umiljendić, J., & Đorđević, T. (2009). Effects of organic matter and clay content in soil on pesticide adsorption processes. *Pesticidi i fitomedicina, 24*(1), 51–57.

El Fantroussi, S., Verschuere, L., Verstraete, W., & Top, E. M. (1999). Effect of phenylurea herbicides on soil microbial communities estimated by analysis of 16S rRNA gene fingerprints and community-level physiological profiles. *Applied and Environmental Microbiology, 65*(3), 982–988.

Farenhorst, A. (2006). Importance of soil organic matter fractions in soil-landscape and regional assessments of pesticide sorption and leaching in soil. *Soil Science Society of America Journal, 70*(3), 1005–1012.

Fogg, P., Boxall, A. B., Walker, A., & Jukes, A. (2004). Effect of different soil textures on leaching potential and degradation of pesticides in biobeds. *Journal of Agricultural and Food Chemistry, 52*(18), 5643–5652.

Ghidiu, G., Kuhar, T., Palumbo, J., & Schuster, D. (2012). Drip chemigation of insecticides as a pest management tool in vegetable production. *Journal of Integrated Pest Management, 3*(3), E1–E5.

Gupta, R. P., Singh, J., Sultan, M. S., Hujan, R. K., Gosal, S. K., Sahota. , H., & Sharma, S. (2000). Impact of pesticides on soil biota and non-target organisms in rice wheat cropping system (abstract). In *41st annual conference of AMI, Birla Research Institute*.

Holten, R., Larsbo, M., Jarvis, N., Stenrød, M., Almvik, M., & Eklo, O. M. (2019). Leaching of five pesticides of contrasting mobility through frozen and unfrozen soil. *Vadose Zone Journal, 18*(1), 1–10.

Hussain, S., Siddique, T., Saleem, M., Arshad, M., & Khalid, A. (2009). Impact of pesticides on soil microbial diversity, enzymes, and biochemical reactions. *Advances in Agronomy, 102*, 159–200.

Jamshidi, M. H., Salehian, H., Babanezhad, E., & Rezvani, M. (2022). The adsorption and degradation of 2, 4-D affected by soil organic carbon and clay. *Bulletin of Environmental Contamination and Toxicology, 108*(1), 151–157.

Jena, P. K., Adhya, T. K., & Rajaramamohan Rao, V. (1987). Influence of carbaryl on nitrogenase activity and combinations of butachlor and carbofuran on nitrogen-fixing micro-organisms in paddy soils. *Pesticide science, 19*(3), 179–184.

Jiang, L., Lin, J. L., Jia, L. X., Liu, Y., Pan, B., Yang, Y., & Lin, Y. (2016). Effects of two different organic amendments addition to soil on sorption–desorption, leaching, bioavailability of penconazole and the growth of wheat (*Triticum aestivum* L.). *Journal of Environmental Management, 167*, 130–138.

Jing, X., Li, Q., Qiao, X., Chen, J., & Cai, X. (2021). Effects of accumulated straw residues on sorption of pesticides and antibiotics in soils with maize straw return. *Journal of Hazardous Materials, 418*, 126213.

Jones, D. L., Edwards-Jones, G., & Murphy, D. V. (2011). Biochar mediated alterations in herbicide breakdown and leaching in soil. *Soil Biology and Biochemistry, 43*(4), 804–813.

Konda, L. N., Czinkota, I., Füleky, G., & Morovján, G. (2002). Modeling of single-step and multistep adsorption isotherms of organic pesticides on soil. *Journal of Agricultural and Food Chemistry, 50*(25), 7326–7331.

Kumar, M., & Philip, L. (2006). Adsorption and desorption characteristics of hydrophobic pesticide endosulfan in four Indian soils. *Chemosphere, 62*(7), 1064–1077.

Lerch, R. N., & Blanchard, P. E. (2003). Watershed vulnerability to herbicide transport in northern Missouri and southern Iowa streams. *Environmental Science & Technology, 37*(24), 5518–5527.

Liu, Y., Li, S., Ni, Z., Qu, M., Zhong, D., Ye, C., & Tang, F. (2016). Pesticides in persimmons, jujubes and soil from China: Residue levels, risk assessment and relationship between fruits and soils. *Science of the Total Environment, 542*, 620–628.

Lo, C. C. (2010). Effect of pesticides on soil microbial community. *Journal of Environmental Science and Health Part B, 45*(5), 348–359.

Lu, C., Yang, Z., Liu, J., Liao, Q., Ling, W., Waigi, M. G., & Odinga, E. S. (2020). Chlorpyrifos inhibits nitrogen fixation in rice-vegetated soil containing *Pseudomonas stutzeri* A1501. *Chemosphere, 256*, 127098.

Lutri, V. F., Blarasin, M. T., Matteoda, E. M., Currell, M., Giacobone, D. B., Quinodóz, F. B., & Cabrera, A. E. (2022). Screening of atrazine distribution in groundwater and modeling of leaching potential to the unconfined aquifer in the Pampean Plain of Cordoba, Argentina. *Environmental Processes, 9*(2), 25.

Lykogianni, M., Bempelou, E., Karamaouna, F., & Aliferis, K. A. (2021). Do pesticides promote or hinder sustainability in agriculture? The challenge of sustainable use of pesticides in modern agriculture. *Science of the Total Environment, 795*, 148625.

Martinez-Toledo, M. V., De la Rubia, T., Moreno, J., & Gonzalez-Lopez, J. (1988). Effect of diflubenzuron on *Azotobacter* nitrogen fixation in soil. *Chemosphere, 17*(4), 829–834.

Martinez-Toledo, M. V., Salmeron, V., & Gonzalez-Lopez, J. (1992). Effect of the insecticides methylpyrimifos and chlorpyrifos on soil microflora in an agricultural loam. *Plant and Soil, 147*, 25–30.
Menchen, A., Heras, J. D. L., & Alday, J. J. G. (2017). Pesticide contamination in groundwater bodies in the Jucar River European Union pilot basin (SE Spain). *Environmental Monitoring and Assessment, 189*, 1–18.
Miglani, R., & Bisht, S. S. (2019). World of earthworms with pesticides and insecticides. *Interdisciplinary Toxicology, 12*(2), 71.
Mohamed, M., Aliyat, F. Z., Ben Messaoud, B., Simone, C., Marina, M., Filippo, G., Laila, N., & Jamal, I. (2021). Effects of pesticides use (glyphosate & paraquat) on biological nitrogen fixation. *Water Air & Soil Pollution, 232*, 1–15.
Narayanan, N., Gajbhiye, V. T., Gupta, S., & Manjaiah, K. M. (2014). Leaching behavior of chlorothalonil, chlorpyrifos and pendimethalin in soil: Effect of soil organic matter and clay. *Clay Research, 33*(1), 15–25.
Navarro, S., Vela, N., & Navarro, G. (2007). An overview on the environmental behaviour of pesticide residues in soils. *Spanish Journal of Agricultural Research, 5*(3), 357–375.
Onwona-Kwakye, M., Plants-Paris, K., Keita, K., Lee, J., Brink, P. J. V. D., Hogarh, J. N., & Darkoh, C. (2020). Pesticides decrease bacterial diversity and abundance of irrigated rice fields. *Microorganisms, 8*(3), 318.
Ou, J., Li, H., Ou, X., Yang, Z., Chen, M., Liu, K., Teng, Y., & Xing, B. (2020). Degradation, adsorption and leaching of phenazine-1-carboxamide in agricultural soils. *Ecotoxicology and Environmental Safety, 205*, 111374.
Özkara, A., Akyıl, D., & Konuk, M. (2016). Pesticides, environmental pollution, and health. In *Environmental health risk-hazardous factors to living species* (pp. 3–27). IntechOpen.
Park, W. P., Chang, K. M., Hyun, H. N., Boo, K. H., & Koo, B. J. (2020). Sorption and leaching characteristics of pesticides in volcanic ash soils of Jeju Island, Korea. *Applied Biological Chemistry, 63*, 1–13.
Pérez-Lucas, G., Vela, N., El Aatik, A., & Navarro, S. (2019). Environmental risk of groundwater pollution by pesticide leaching through the soil profile. In *Pesticides-use and misuse and their impact in the environment* (pp. 45–72). IntechOpen.
Raffa, C. M., & Chiampo, F. (2021). Bioremediation of agricultural soils polluted with pesticides: A review. *Bioengineering, 8*(7), 92.
Rani, K., & Dhania, G. (2014). Bioremediation and biodegradation of pesticide from contaminated soil and water: A noval approach. *International Journal of Current Microbiology and Applied Sciences, 3*(10), 23–33.
Rani, L., Thapa, K., Kanojia, N., Sharma, N., Singh, S., Grewal, A. S., Srivastav, A. L., & Kaushal, J. (2021). An extensive review on the consequences of chemical pesticides on human health and environment. *Journal of Cleaner Production, 283*, 124657.
Rasool, S., Rasool, T., & Gani, K. M. (2022). A review of interactions of pesticides within various interfaces of intrinsic and organic residue amended soil environment. *Chemical Engineering Journal Advances, 11*, 100301.
Riah, W., Laval, K., Laroche-Ajzenberg, E., Mougin, C., Latour, X., & Trinsoutrot-Gattin, I. (2014). Effects of pesticides on soil enzymes: A review. *Environmental Chemistry Letters, 12*, 257–273.
Rosell, G., Quero, C., Coll, J., & Guerrero, A. (2008). Biorational insecticides in pest management. *Journal of Pesticide Science, 33*(2), 103–121.
Shahid, M., Manoharadas, S., Altaf, M., & Alrefaei, A. F. (2021). Organochlorine pesticides negatively influenced the cellular growth, morphostructure, cell viability, and biofilm-formation and phosphate-solubilization activities of *Enterobacter cloacae* strain EAM 35. *ACS Omega, 6*(8), 5548–5559.
Simon-Sylvestre, G., & Fournier, J. C. (1980). Effects of pesticides on the soil microflora. *Advances in Agronomy, 31*, 1–92.
Singh, N. S., Mukherjee, I., Das, S. K., & Varghese, E. (2018). Leaching of clothianidin in two different Indian soils: Effect of organic amendment. *Bulletin of Environmental Contamination and Toxicology, 100*, 553–559.
Spark, K. M., & Swift, R. S. (2002). Effect of soil composition and dissolved organic matter on pesticide sorption. *Science of the Total Environment, 298*(1-3), 147–161.
Syafrudin, M., Kristanti, R. A., Yuniarto, A., Hadibarata, T., Rhee, J., Al-Onazi, W. A., Algarni, T. S., Almarri, A. H., & Al-Mohaimeed, A. M. (2021). Pesticides in drinking water: A review. *International Journal of Environmental Research and Public Health, 18*(2), 468.
Tang, F. H., & Maggi, F. (2021). Pesticide mixtures in soil: A global outlook. *Environmental Research Letters, 16*(4), 044051.
Thirup, L., Johnsen, K., Torsvik, V., Spliid, N. H., & Jacobsen, C. S. (2001). Effects of fenpropimorph on bacteria and fungi during decomposition of barley roots. *Soil Biology and Biochemistry, 33*(11), 1517–1524.

Urbina, C. A. F., van den Berg, F., van Dam, J. C., Tang, D. W. S., & Ritsema, C. J. (2020). Parameter sensitivity of SWAP–PEARL models for pesticide leaching in macroporous soils. *Vadose Zone Journal, 19*(1), e20075.

Wang, Y. S., Wen, C. Y., Chiu, T. C., & Yen, J. H. (2004). Effect of fungicide iprodione on soil bacterial community. *Ecotoxicology and Environmental Safety, 59*(1), 127–132.

Wang, Z., Yang, L., Cheng, P., Yu, Y., Zhang, Z., & Li, H. (2021). Adsorption, degradation and leaching migration characteristics of chlorothalonil in different soils. *European Journal of Remote Sensing, 54*(suppl. 2), 238–247.

Wołejko, E., Jabłońska-Trypuć, A., Wydro, U., Butarewicz, A., & Łozowicka, B. (2020). Soil biological activity as an indicator of soil pollution with pesticides: A review. *Applied Soil Ecology, 147*, 103356.

Yu, Y. L., Wu, X. M., Li, S. N., Fang, H., Zhan, H. Y., & Yu, J. Q. (2006). An exploration of the relationship between adsorption and bioavailability of pesticides in soil to earthworm. *Environmental Pollution, 141*(3), 428–433.

Zhang, J. J., & Yang, H. (2021). Metabolism and detoxification of pesticides in plants. *Science of the Total Environment, 790*, 148034.

Zhang, R., Jiang, J., Gu, J. D., & Li, S. (2006). Long term effect of methylparathion contamination on soil microbial community diversity estimated by 16S rRNA gene cloning. *Ecotoxicology (London, England), 15*, 523–530.

CHAPTER 4

Pesticide biology in plants: Plant uptake, translocation, and accumulation

Harleen Kaur, Ravneet Kaur, Swapnil Singh, Neetu Jagota and Ashish Sharma
Department of Botany & Environment Studies, DAV University, Jalandhar, Punjab, India

4.1 Introduction

Pests are any animals or microorganisms which deal with foodstuff, health, or console. Pesticides are defined as any chemical substances, mixtures of substances or natural agents that are used in agriculture to control or kill pests, insects, rodents, weeds, or diseases that are causing great harm to plants or can spread vector-borne disease in humans. Pesticides are broadly classified into insecticides, fungicides, herbicides, plant growth regulators, etc. The use of pesticides provides great protection to various plants in other words providing crop protection, helps in the longer preservation of foodstuff, and prevents the spread of various vector-borne diseases. The term "pesticide" is not an innovation. It is being used since ancient times; Sumerians, Greek, and Romans were applying diverse chemicals to kill insects including sulfur, mercury, copper, and plant extracts.

The widespread use of these hazardous chemical-based pesticides has a greater impact on humans and other systems. According to Environmental Protection Agency (EPA), pesticides are a group of chemical substances that are being used to control and repel the pest population. Pesticides mainly help farmers to get higher food yields. Other benefits such as improved productivity of the crops, protection damage of crops from weeds, vector disease controls, and food quality improvements pave the way for greater use of pesticides, despite having lethality (Arias-Estévez et al., 2008). Insecticides, fungicides, herbicides, rodenticides, and plant growth regulators are typical examples of pesticides. These products are also used for other purposes, such as to improve and maintain nonagricultural areas like public urban green areas and sports fields (Hoffman et al., 2000).

Being a formulation of chemicals, pesticides are persistent and somewhere poisonous in nature, the long-term presence of which have both beneficial and detrimental effects on the environment, plants and humans. Most of pesticides are nonspecific and may kill life forms that are harmless or useful (Horowitz et al., 2004). During the past three decades, efforts have been made to reduce the exposure and human risk of pesticides, especially insecticides. There is great demand for selective and safe insecticides that

Pesticides in the Environment: Impact, Assessment, and Remediation
DOI: https://doi.org/10.1016/B978-0-323-99427-9.00004-5

spare natural enemies and nontarget organisms. Some conventional pesticides have been replaced by newer bio-rational (biopesticides) or "low-risk" pesticides.

The United Nations Environment Programme (UNEP) reported that 9 of the 12 most unwanted persistent organic pollutants (POPs) are pesticides (Zhang et al., 2015). Improper use of pesticides may pose a high risk to food safety and human health (Hamza et al., 2016; Su et al., 2019; Zhang et al., 2016). Consumption of contaminated food crops is associated with many human chronic diseases, such as different kinds of cancers, diabetes, birth defects, fertility disorders, and neurodegeneration (Lushchak et al., 2018; Parween et al., 2016).

Various pesticides have been formulated having an active ingredient (Table 4.1) that function in a specific way to kill insects, weeds and to eradicate diseases, thus significantly contributing to the sustained growth of crop yield observed over the last few decades (Liu et al., 2016; Silva et al., 2019). Despite the benefits to crop yield, extensive and intensive pesticide application has led to growing public concerns about food safety, environmental contamination, biodiversity losses, and health problems (Bhandari et al., 2019; Rauh et al., 2012; Taiwo, 2019). Ingestion of edible plants (or crops) exposed to the chemicals is therefore one of the major pathways leading to threatening wildlife and human health. Growing evidence shows that the risk of prolonged exposure to toxic pesticides is associated with human multiorgan dysfunctions such as immune, neurological, cardiovascular, endocrine, and reproductive disorders leading to diverse chronic diseases like diabetes, malignant tumor, Parkinson, Alzheimer, and amyotrophic lateral sclerosis (Goodman et al., 2014; Mostafalou & Abdollahi, 2013; Pogrmic-Majkic et al., 2012).

4.2 Effect of pesticides on plants

Pesticides are applied in agricultural fields to control the activity of pests. However, one or the other way, pesticides may enter into the plants ultimately affecting plant's metabolic activity by functioning as a xenobiotic. Various toxic effects shown by pesticides in different plants species are listed in Table 4.2.

4.3 Plant uptake

Plant's capability/capacity to uptake or accumulate pesticides from soil is influenced by various characteristics of plants such as water uptake potential and root length/root architecture. Also, various physicochemical properties of the concerned pesticide, its mode of application, soil properties, environmental conditions, plant species also influences uptake of pesticides. In soil to plant transfer of pesticides, soil properties such as moisture content, organic carbon, and clay content in soil, pesticide residence time in the soil can also play a significant role (Koskinen et al., 2006).

Table 4.1 Common name, active ingredients and biological activity of some pesticides used to control pest attack.

S. no.	Common name	Active ingredients(s)	Biological activity	References
1.	Curasol	Carbofuran (5%)	Insecticide, nematicide	López-Pérez et al. (2006)
2.	Nemacur	Fenamiphos (10%)	Insecticide, nematicide	López-Pérez et al. (2006)
3.	Eclipse 70 WG	Metribuzin (70%)	Herbicide	López-Pérez et al. (2006)
4.	Keshet 2.5 EC	Deltametrin (2.5% w/v)	Insecticide	López-Pérez et al. (2006)
5.	Sponsor Comb	Metalaxyl (10%) folpet (40%)	Fungicide	López-Pérez et al. (2006)
6.	Golan S L	Actemiprid	Pesticide	Amoako et al. (2014)
7.	Deltapaz 2.5 EC	Deltamethrin	Pesticide	Amoako et al. (2014)
8.	Cypercal 50 EC	Cypermethrin	Pesticide	Amoako et al. (2014)
9.	Karate 5EC	Lambda cyhalothrin	Pesticide	Amoako et al. (2014)
10.	Pyrical 480 EC	Chloropyriphos ethyl	Pesticide	Amoako et al. (2014)
11.	Orthene 750 sp	Acephate	Pesticide	Amoako et al. (2014)
12.	Pawa 2.5 EC	Lambda cyhalothrin	Pesticide	Amoako et al. (2014)
13.	Cymethoate	Cymethoate	Pesticide	Amoako et al. (2014)
14.	Dimethoate	Dimethoate	Pesticide	Amoako et al. (2014)
15.	Sumithion	Fenitrithrin	Pesticide	Amoako et al. (2014)
16.	Dursban 4 E	Chlopyriphos	Pesticide	Amoako et al. (2014)
17.	Thionex 35 EC	Endosulphan	Pesticide	Amoako et al. (2014)
18.	Cymthox	Fenvalerate	Pesticide	Amoako et al. (2014)
19.	Thiodan	Endosulphan	Pesticide	Amoako et al. (2014)
20.	Mektin 1.5 EC	Abamectin	Pesticide	Amoako et al. (2014)
21.	Confidor 200 sl	Imidacloprid	Pesticide	Amoako et al. (2014)
22.	Diazol 50 EC	Diazion	Pesticide	Amoako et al. (2014)
23.	Wrecko 2.5 EC	Lambda cyhalothrin	Pesticide	Amoako et al. (2014)
24.	Endocel	Endosulphan	Pesticide	Amoako et al. (2014)
25.	Mancozeb	Symoxanil 72 wp		
26.	Imidachloropid	Imidachloropid 20 SL		
27.	Cocostar	Bifenthrin + pirimiphosmethyl	Pesticide	Amoako et al. (2014)
28.	Lannate	Methomyl	Insecticide	Pereira et al. (2009)
29.	Spasor	Glyphosate	Herbicide	Pereira et al. (2009)
30.	Stam Novel Flo	Propanil	Herbicide	Pereira et al. (2009)

Table 4.2 Detrimental effect of some of the pesticides on different plant species are listed below.

S. no.	Compound name	Plant species	Effect on plant	References
1.	Acifluorfen	Soybean, bean, pea, wide bean, cotton, and spinach	Chlorosis, necrosis, wilting, and shrinking of leaves	Shafeeque et al. (2020)
2.	Dimethon-S-methyl (organophosphate), dimethoate (organophosphate), pirimicarb (carbamate), dicofol (diphenylethane)	Wild tamarind plant	Affects dry matter, leaf area, and leaf number and decrease or inhibition of photosynthetic machinery	Dalzell & Mullen (2004)
3.	Fungicide	*Malus domestica* and *Cucumis sativus*	Reduction in net CO_2 absorption along with changes in stomatal conductance and intercellular CO_2 concentration	Untiedt & Blanke (2004); Xia et al. (2006)
4.	Fludioxonil, a fungicide	*Vitis vinifera*	Reduction in net CO_2 absorption and in intercellular CO_2	Shafeeque et al. (2020)
5.	Glufosinate, an herbicide	Tomato plants	Ammonium accumulation and ethylene generation as a marker of stress as well as bright senescence	Shafeeque et al. (2020)
6.	Herbicides, fungicides (phenylpyrrole aniline pyrimidine)	Grapevine	Decline in amino acid content; enhancement in protein hydrolysis, and accumulation of stress protein	Llorens et al. (2000)

(*continued on next page*)

7.	Mancozeb, imidacloprid, and sulfentrazone	*Allium cepa*	Cytotoxic and genotoxic effects by inducing different types of chromosomal abnormalities, likely sticky, disoriented, and fragmented chromosomes, abnormal DNA condensation, and chromosome coiling by spindle inactivation and thus reduced mitotic index	Bianchi et al. (2016); Fatma et al. (2018)
8.	Omethoate (OM) insecticide/organophosphorus	*Zea mays* L.	Yellowing of leaf, distortion of stomata and, loss of chloroplast, mitochondrial structure and function	Wen et al. (2017)
9.	Fenthion insecticide/organophosphate	*Vicia faba*	DNA damage, programmed cell death including chromatin condensation, cytoplasmic vacuolization, nuclear shrinkage, condensation of the protoplast, fragmentation with apoptotic-like corpse formation	Cortés-Eslava et al. (2018)
10	Imidacloprid and chlorpyrifos	Rice seedlings	Reduction in root length, shoot length, fresh weight and protein content	Sharma et al. (2018)
11.	Chlorothalonil	Tomato plants	Causes enhanced levels of H_2O_2	Zhou et al. (2015)
12.	Pesticides	Cucumber plants	Negatively affects the photosynthetic machinery of cucumber plants, resulting in reduced photosynthetic rate (Pn), stomatal conductance and intercellular CO_2	Xia et al. (2006)

(continued on next page)

Table 4.2 Detrimental effect of some of the pesticides on different plant species are listed below—cont'd

S. no.	Compound name	Plant species	Effect on plant	References
13.	Mancozeb, flusilazol, dithianon, pirimicarb	Apple tree	Inhibit the photosynthesis	Untiedt & Blanke (2004)
14.	Glyphosate and its breakdown product AMPA	Monocotyledonous plants (such as grasses) and dicotyledonous plants (most broad leaf plants).	Inhibit antioxidant enzyme activities and induce the accumulation of reactive oxygen species (ROS) that induce physiological dysfunction and cell damage	Gomes et al. (2016)
15.	Chlorpyrifos, an organophosphorus insecticide	*Vigna radiata* L.	Suppressession in the nitrogen metabolism and growth	Parween et al. (2011)
16.	Chlorpyrifos an insecticide	*Vigna radiata* L.	Responses of the antioxidative enzymatic system	Parween et al. (2012)
17.	Imazapic, an herbicide	Tobacco plants	Induction in some phytotoxic effects for tobacco plants, including carotenoids, jasmonic acid, antioxidant enzyme activity, and malondialdehyde contents	Kaya & Doganlar (2016)
18.	Organophospahtes	–	Prevent the biosynthesis of catalase, perioxidase and d-aminolevulinic acids (ALA) which are the major component of chlorophyl biosynthetic pathway by inducing Fe deficiency	Barcelos et al. (2012)

(*continued on next page*)

19.	Glyphosate	*Pisum sativum*	Reduces the soluble sugar content along with the alterations in potassium and sodium tissue concentrations	Mondal et al. (2017)
20.	Chloropyrifos	*Vigna radiata*	Reduces plant growth and nitrogen metabolism. The rate of lipid peroxidation was also found to be enhanced	Parween et al. (2011)
21.	Dimethoate	Soybean	Detrimental effect on plant growth causing inhibition of photosynthetic pigments	Murthy et al. (2005)
22.	Dimethoate	Cotton	Detrimental effect on plant growth causing inhibition of photosynthetic efficiency	Murthy et al. (2005)
23.	Imidacloprid	Rice	Reduction in seed germination and growth	Stevens et al. (2008)
24.	Dimethoate	Wheat	A hormetic response	Pandey & Gopal (2011)
25.	Imidacloprid	*Brassica juncea*	limits growth and development, reduces photosynthetic ability, generates oxidative stress, and stimulates enzymatic and nonenzymatic antioxidants resulting in deposition of IMI residues in plant parts	Jan et al. (2020)

(continued on next page)

Table 4.2 Detrimental effect of some of the pesticides on different plant species are listed below—cont'd

S. no.	Compound name	Plant species	Effect on plant	References
26.	Imidacloprid	–	Reduced photosynthetic activity by 36% as compared with control plants and this reduction was significantly associated with reduction in stomatal conductance	Xia et al. (2006)
27.	CPF	–	Adversely affects the growth and protein content of seedlings with concomitant increased in MDA level	Sharma et al. (2012)
28.	Chlorpyrifos, cypermethrin, fenvalerate	*Cenchrus setigerus* Vahl, *Pennisetumn pedicellatum* Tan	Significant reduction and delay in seed germination at higher concentrations (75 and 100 mg/kg) of Chlorpyrifos compared to cypermethrin and fenvalerate	Dubey & Fulekar (2011)
29.	Pendimethalin	*Zea mays* L.	Decrease in germination percentage by about 69% at higher concentration of pendimethalin. Also, plumule length and radicle length decreased by 77% and 90% respectively at 10 ppm	Rajashekar & Murthy (2012)
30.	Acetochlor and bensulfuronmethyl	*Oryza sativa* L.	Decline in root and shoot biomass	Huang & Xiong (2009)

The three main routes for entry of pesticides into a plant's root system are: apoplastic, symplastic and transmembrane. Apoplastic pathway refers to movement of pesticides through nonliving continuum of plant cell, i.e., through plant cell wall and exterior of plasma membrane. Few chemicals such as phenoxy acid herbicides are known to be transported in plant via apoplastic pathway (Bromilow & Chamberlain, 1995). Symplastic pathway involves movement of pesticides from one cell to another through plasmodesmata, creating a thread of cytoplasm which allows the transfer of solutes. Transmembrane transport refers to cell to cell movement of pesticides not by entering the cytoplasmic stream but only through plasma membrane.

Organic chemicals may enter into the root system of plants either via water or air. Chemicals entering by this route ultimately translocated into the transpiration stream via diffusion or passive transport. For example, organophosphate pesticides (OPPs) may enter to the plants via diffusion or in other words without the aid of any membrane transporters (Dordio & Carvalho, 2013). In aquatic plants, pesticides may opt for symplastic pathway or apoplastic pathway or may enter to root cell by passive transport (Nwoko, 2010). It has been reported that chlordecone, an organochloride compound, was absorbed by radish roots through the adsorption on the periderm, followed by diffusion into cortex, xylem, and pith (Létondor et al., 2015).

Various factors such as plant species, concentration, and the surface area below the water level influences the plant-associated pesticide sorption. Furthermore, the pesticide hydrophobicity (K_{OW}) (Lee et al., 2012; Tang et al., 2016), the barriers of the plant surface, and the thickness of subsurface wax layer can affect the level of absorption of pesticides by plants (Richins et al., 1997). For foliar uptake, most important property affecting pesticide uptake is lipophilicity. Some pesticides are very polar in nature, hence cannot easily pass through plant membranes, thus have limiting potential to enter to plant. On the other hand, very less polar compounds (lipophilic compounds) can easily cross plant membrane system, thereby sorb to the roots (Burken & Schnoor, 1996; Trapp, 2000, 2004). Many modern pesticides have intermediate lipophilicity, which favors their uptake by most plant species (Briggs et al., 1982, 1987; Trapp, 2000, 2004). Lipophilicity of a particular compound is expressed by using an octanol/water partition coefficient indicated as log o/w or log P. On the basis of value of log P, the chemicals have been classified as hydrophilic or lipophilic. Chemicals with $\log P < 0$ are considered as hydrophilic and chemicals with $\log P > 0$ lipophilic. The tendency of a pesticide to partition from water into organic phase (K_{OW}) controls its bioavailability and translocation in plant (Bromilow & Chamberlain, 1995).

Various factors are found to affect the rate of uptake of pesticides by plants. For example, pesticide accumulation capacity was found to be higher in indoor plants than outdoor plants because of the influence of air currents, soil dispersion, and photodegradation of the pesticide molecules (Hwang et al., 2015). Furthermore, pesticide uptake capacity varies from species to species. Chlordecone uptake was higher in zucchini and pumpkin compared to cucumber plants (Létondor et al., 2015).

A plant's ability to take up pesticides varies from species to species (Lunney et al., 2004) and even varies between different subspecies (Otani et al., 2007). Early reports suggested that different plant species uptake different pesticides belonging to category organochlorine: cucumber, alfa, and soyabean uptake aldrin and heptachlor (Bruce & Decker, 1966); corn and peanut uptakes dieldrin (Beetsman et al., 1969; Thompson et al., 1970); soybeans uptakes heptachlor and dieldrin (Turner et al., 1972). Bouldin et al. (2006) grown two different plants viz. *Juncus effusus* and *Ludwigia peploides* under hydroponic conditions and studied the water to plant uptake of pesticides (atrazine and lambda-cyhalothrin). It was observed that atrazine uptake was higher in *J. effuses*, while uptake of lambda-cyhalothrin was higher in *L. peploides*. In a study regarding uptake of hydrophobic organochlorine pesticides (OCPs) in vegetables conducted by Gonzalez et al. (2005), it was found that vegetables accumulated OCPs efficiently with tissue residues 4- to 45-fold greater than those in soils and OCP distribution in tissue was found to be K_{OW} dependent.

After the entry of pesticides within plant root tissue, they may be restricted to plant roots or may be translocated to other plant parts by entering to xylem stream where they can be further metabolized. The permeation from plant roots to xylem depends on hydrophobicity of the concerned compound. Slightly hydrophobic compounds having log K_{OW} 1.8 can easily permeate into xylem stream as compared to highly hydrophobic compound (Moore & Locke, 2012). Thus xylem translocation of compounds with log $K_{OW} > 3$ may be severely limited in plants (Wang et al., 2012).

4.4 Translocation of pesticides

Various physical and chemical properties of pesticides, their interaction with soil, plant microbiome, water and chemicals present in the rhizosphere determine the actions of pesticides within plant (Vryzas, 2016).

After uptake, pesticides gets translocated to different parts within the plant which occur in two distinct vascular systems: apoplastic and symplastic (Crafts & Crisp, 1971). Generally, apoplastic translocation refers to passive transport of pesticides through xylem from roots to all aerial leaves. Symplastic translocation refers to active transport of pesticides in the phloem occurring selectively from mature leaves to centers of high metabolic activity areas, associated with the young growing points of the shoot and root. Early work done by Crafts and Crisp (1971), showed that presence of an acidic group in chemical structure of some herbicides enables them to enter to the phloem by virtue of specific carrier proteins recognizing a carboxylic (acidic) group, hence, such herbicides get transported via phloem within plants (Crisp, 1971). Other classes of compounds may enter the symplast of plant cell (Peterson et al., 1978) but, nonetheless, few of these having limited permeation rates through membranes could be retained within the symplast for long-distance transport occurring through phloem, thus effectively translocated in phloem (Tyree et al., 1979).

After getting accumulated within the plant roots, xenobiotics (pesticides) gets translocated to other parts of a plant by entering to the transpiration stream (Pullagurala et al., 2018). For reaching up to xylem, a chemical needs to pass through various layers present in the interior of plant tissue including the epidermis, cortex, endodermis, and pericycle (from exterior to interior) (Collins et al., 2006). No matter whatever the mode of entry to root system was, pesticides can enter the xylem stream only via symplastic pathway and are then transported to the shoots (Miller et al., 2016; Su et al., 2010). Hydrophobic pesticides after entering to the root system either by apoplastic or symplastic pathways may not be able to cross the suberized casparian band of endodermis (Nurzhanova et al., 2013). Solute absorbed by the root hair must pass through at least one cell membrane in the endodermis via the symplastic pathway to the xylem (Foster & Miklavcic, 2016; Wang et al., 2019).

4.4.1 Factors governing translocation of pesticides

Translocation tactor (TF) is an important parameter used to evaluate the capability of a plant to relocate a chemical from its roots to shoots (Wang et al., 2011). It can be calculated as the ratio of concentration of chemical in shoots to its concentration in roots. Chemical with smaller molecular mass are generally more effective in penetrating these layers, having higher TF and subsequently transferring to the aerial part of plant through phloem and xylem (Kvesitadze et al., 2015). The reason behind excellent translocation ability is its low molecular mass enabling it to penetrate the layers more easily. On the contrary, chemicals such as avermectin have high molecular mass, hence, their penetration and translocation within different plant tissues are less efficient (Zhang et al., 2017), consequently have smallest TFs. Different pesticides have different translocation ability within plants as indicated by their concentrations within plants parts. The concentration of dinotefuran, thiamethoxam, and imidacloprid was higher in leaves than in roots indicated their higher upward translocation potential. On the other side, concentration of imazethapyr, propiconazole, and chlorpyrifos was higher in roots indicating that these pesticides had low upward translocation potential. Translocation potential of neutral pesticides is dependent on their hydrophobicity. Li et al. (2018) observed that content of dinotefuran and thiamethoxam in komatsuna shoots were higher than in roots during a 35 days incubation period in a 10 mg/L solution suggesting that neutral pesticides with high water solubility exhibited high translocation potential along with the transpiration stream.

Movement of pesticides within plants depends mainly on the rate at which substances are able to cross the membranes of various layers coming along the passage. The factor that governs the movement of substances is lipophilicity. Lipophilicity is the most important property that regulates uptake and translocation of nonionized pesticides. Lipophilicity is found to have inverse relationship with mobility of substance. In order words, substance having high polarity or high lipophilicity will have less mobility across the membranes. Pesticides having log K_{OW} values 1–3 are found to have high accumulation rate within root and also high translocation rate to shoots (Bromilow & Chamberlain, 1995).

Distribution of ionized pesticides into various plant compartments (pH ranges from 5 to 8) is also governed by pH values while ion trapping is a well-studied mechanism of accumulation of weak acids in cytoplasm (Briggs et al., 1987; Chamberlain et al., 1998).

For example, herbicides which are monobasic acids having pKa ranging from 3 to 6, and the anions having 3–4 log K_{OW} units are more polar than the corresponding undissociated acids. If two plant compartments have different pH values (e.g., vacuole having pH 5.5 and cytoplasm, having pH 7.5), then the concentration of nonionized and ionized acids in the two compartments will be different. The nonionized form of acid will be able to permeate through the membrane more easily as compared to the anion, and so nonionized molecules moving from the low-pH compartment will be substantially ionized in the higher-pH compartment and the anions so produced can only escape slowly. This process is the basis of the "ion trap" effect, whereby weak acids can be substantially accumulated in plant cells of high pH (Bromilow & Chamberlain, 1995).

4.5 Storage and metabolism of pesticides

For complete degradations of pesticides (xenobiotic) entering to the plant cell, the pesticides are subjected to detoxification process. The detoxification process of pesticides proceeds in three phases.

- Phase I refers to modification of xenobiotic (pesticide) by addition of various chemical groups.
- Phase II refers to conjugation of modified derivative formed during Phase I
- Phase III refers to compartmentalization of product formed during Phase II, i.e., conjugated product (Siminszky, 2006).

4.5.1 Phase I metabolism

Phase I metabolism occurs in both root and shoot tissues of plants. In this phase, with the aid of various enzymes, the xenobiotic that entered the plant is chemically modified into its derivatives. In other words, Phase I causes the activation of foreign molecule by creation of reactive sites via addition of one or more functional groups such as hydroxyl or carboxyl groups. The resultant molecule becomes more reactive than the parent molecule, hence, preparing the contaminant to enter to Phase II reactions (Schröder & Collins, 2002). Various reactions occurring during this phase for activation of the compound are oxidation, dealkylation, hydroxylation, epoxidation, sulfoxidation, reduction, and hydrolysis (Behringer et al., 2011). To carry out these reactions, different groups of enzymes are known to play a significant role (Eapen et al., 2007).

One of the most important enzymes participating in Phase I reaction is cytochrome P450 monooxygenases (cytochrome P450s, P450s, or CYPs, EC 1.14.14.1). It belongs to the largest family of enzymatic proteins whose activities are composed of cytochrome P450 protein and NADPH–cytochrome P450 reductase (CPR) (Aigrain et al., 2009). CYPs carry out the reaction of hydroxylation by which a more hydrophobic compound gets converted into a less hydrophobic compound (Kato et al., 2020; Khanom et al., 2019; Lu et al., 2020) ultimately causing detoxification of highly hydrophobic compound (pesticide) that have entered to the plant cell (Tan et al., 2015). One such example of hydroxylation is carbon hydroxylation of pesticide pelargonic acid mediated by CYP72A18 in rice seedlings (Imaishi & Matumoto, 2007).

Some CYPs also mediated the process of oxidation of heteroatoms such as N, S, P, I. For example, in soybean *Glycine max* L., oxygenation of pesticide fenamiphos with sulfur atoms produces sulfoxide and sulfone (Romeh & Hendawi, 2017).

Another mechanism of Phase I metabolism of pesticide in plants is dealkylation. The dealkylation reaction concerning amines, sulfides, and ethers undergoes a single electron transfer process. C–dealkylation products of isoproturon in rice and atrazine in alfalfa (*Medicago sativa*) have been identified potentially through some CYPs (Lu et al., 2020). Epoxidation likely occurs in the pesticide molecules with carbon–carbon double or triple bonds (e.g., cyclopentadiene insecticides and clethodim) to form unstable products (Guengerich, 2001).

Another enzyme involved in Phase I metabolism of pesticides is hydrolase (EC 3.x.y.z) (Cho & Kong, 2007; Zhou et al., 2015). This enzyme modifies the pesticide by adding chemical groups such as amino or amide, phosphate ester and carboxylic acid ester onto the pesticides (Wan et al., 2017). For example, enzyme organophosphate hydrolases (OPH, EC 3.1.8.1) actively hydrolyze organophosphate compounds with substrate specificities of P\\O, P\\S, P–CN, and P\\F bonds (Cho & Kong, 2007).

Certain herbicides such as atrazine and isoproturon are metabolized by the action of enzyme Laccases (EC 1.10.3.2), the encoding gene of which were found in rice viz. LOC_Os01g63180 and LOC_Os12g15680 (Huang et al., 2016). Laccases catalyze single-electron oxidation of natural metabolites or artificial substrates by the use of molecular oxygen (Majeau et al., 2010).

4.5.2 Phase II metabolism

Phase II metabolism refers to conjugation of chemically modified compound with amino acids or glutathione depending on the functional groups present in the derived compound. Phase II metabolism occurs at the expense of existing enzymatic machinery and is therefore referred as a cometabolic process (Van Eerd et al., 2003). Enzymes such as glutathione-S-transferases (Edwards et al., 2000; Pflugmacher et al., 2000) and glucosyl transferase are associated with Phase II (Sandermann, 1992, 1994). One of the enzymes involved in conjugation reaction is glutathione-S-transferases (GST, E.C.2.5.1.18). It is a dimeric multifunctional enzymatic protein causing nucleophilic attack of thiol group of GSH to various electrophilic substrates having halogen or nitro groups (Gill & Tuteja, 2010). Some members of the GST family have been shown to be also involved in the protection of the cells from the xenobiotic-imposed oxidative damages.

Another enzyme involved in conjugation reaction is glycosyltransferases (GT, E.C. 2.4.1.x) causing conjugation of OH^-, NH^{2-}, SH^-, and COOH with glucose. The most common glucose conjugates are O-glucosides because pesticide oxidation reactions form hydroxyl groups, which are suitable sites for glucose conjugation. Furthermore, enzyme uridine diphosphate–glucosyl (UDPG) transferase are involved in pesticide–glucose conjugation as well as pesticide–glucose ester conjugation reactions (Van Eerd et al., 2003).

The conjugates formed after Phase II metabolism are much less toxic than derivatives formed in Phase I as well as original xenobiotic that entered the plant cells (Hatzios & Burgos, 2004).

4.5.3 Phase III metabolism

Phase III refers to further conjugation (secondary conjugation), compartmentalization of xenobiotic (pesticide) metabolized in Phase II. Soluble pesticides are stored within vacuoles and insoluble one's are stored within apoplast (Van Eerd et al., 2003). ATP–binding cassette (ABC)–type transporters play an important role in Phase III metabolism. ABC transporters are a group of integral membrane proteins meant for allocation of a wide range of internal and external chemicals such as sugars, amino acids, drugs, xenobiotics, metals, and so forth (Zhang et al., 2018).

4.6 Conclusion

Various chemical compounds referred as pesticides are used in agriculture to reduce the damage caused by pests. However, after application in fields, these pesticides may enter to plants owing to their physiochemical properties. Different pesticides show various different detrimental effects on plants. Plants develop various adaptive strategies to deal with any foreign compound (xenobiotic). Within plants, the pesticides gets translocated and eventually gets metabolized to other less toxic compounds thus reducing the effect of pesticides within plants.

References

Aigrain, L., Pompon, D., Truan, G., & Moréra, S. (2009). Cloning, purification, crystallization and preliminary X-ray analysis of a chimeric NADPH-cytochrome P450 reductase. *Acta Crystallographica Section F: Structural Biology and Crystallization Communications, 65*(3), 210–212.

Amoako, P. K., Kumah, P., & Appiah, F. (2014). Pesticides usage in cabbage (*Brassica oleracea*) cultivation in the forest ecozone of Ghana. *Acta Horticulturae*, 401–407. doi:10.17660/ActaHortic.2014.1021.37.

Arias-Estévez, M., López-Periago, E., Martínez-Carballo, E., Simal-Gándara, J., Mejuto, J. C., & García-Río, L. (2008). The mobility and degradation of pesticides in soils and the pollution of groundwater resources. *Agriculture, Ecosystems & Environment, 123*(4), 247–260.

Barcelos, R. P., de Lima Portella, R., Lugokenski, T. H., Rosa, E. J. F., Amaral, G. P., Garcia, L. F. M., … Barbosa, N. B. V. (2012). Isatin-3-N4-benzilthiosemicarbazone, a non-toxic thiosemicarbazone derivative, protects and reactivates rat and human cholinesterases inhibited by methamidophos in vitro and in silico. *Toxicology in Vitro, 26*(6), 1030–1039.

Beetsman, G. B., Keeney, D. R., & Chesters, G. (1969). Dieldrin uptake by corn as affected by soil properties. *Agronomy Journal, 61*(2), 247–250.

Behringer, C., Bartsch, K., & Schaller, A. (2011). Safeners recruit multiple signalling pathways for the orchestrated induction of the cellular xenobiotic detoxification machinery in Arabidopsis. *Plant, Cell & Environment, 34*(11), 1970–1985.

Bhandari, G., Zomer, P., Atreya, K., Mol, H. G., Yang, X., & Geissen, V. (2019). Pesticide residues in Nepalese vegetables and potential health risks. *Environmental Research, 172*, 511–521.

Bianchi, J., Fernandes, T. C. C., & Marin-Morales, M. A. (2016). Induction of mitotic and chromosomal abnormalities on Allium cepa cells by pesticides imidacloprid and sulfentrazone and the mixture of them. *Chemosphere, 144*, 475–483.

Bouldin, J. L., Farris, J. L., Moore, M. T., Smith Jr, S., & Cooper, C. M. (2006). Hydroponic uptake of atrazine and lambda-cyhalothrin in Juncus effusus and ludwigia peploides. *Chemosphere, 65*(6), 1049–1057.

Briggs, G. G., Bromilow, R. H., & Evans, A. A. (1982). Relationships between lipophilicity and root uptake and translocation of non-ionised chemicals by barley. *Pesticide Science, 13*(5), 495–504.

Briggs, G. G., Rigitano, R. L., & Bromilow, R. H. (1987). Physico-chemical factors affecting uptake by roots and translocation to shoots of weak acids in barley. *Pesticide Science, 19*(2), 101–112.

Bromilow, R. H., & Chamberlain, K. (1995). Principles governing uptake and transport of chemicals. In S. Trapp, & J. C. McFarlane (Eds.), *Plant contamination: Modeling and simulation of organic chemical processes* (pp. 37–68). CRC Press.

Bruce, W. N., & Decker, G. C. (1966). Insecticide residues in soybeans grown in soil containing various concentrations of aldrin, dieldrin, heptachlor, and heptachlor epoxide. *Journal of Agricultural and Food Chemistry, 14*(4), 395–398.

Burken, J. G., & Schnoor, J. L. (1996). Phytoremediation: Plant uptake of atrazine and role of root exudates. *Journal of Environmental Engineering, 122*(11), 958–963.

Chamberlain, K., Patel, S., & Bromilow, R. H. (1998). Uptake by roots and translocation to shoots of two morpholine fungicides in barley. *Pesticide Science, 54*(1), 1–7.

Cho, H. Y., & Kong, K. H. (2007). Study on the biochemical characterization of herbicide detoxification enzyme, glutathione S-transferase. *Biofactors, 30*(4), 281–287.

Collins, C., Fryer, M., & Grosso, A. (2006). Plant uptake of non-ionic organic chemicals. *Environmental Science & Technology, 40*(1), 45–52.

Cortés-Eslava, J., Gómez-Arroyo, S., Risueño, M. C., & Testillano, P. S. (2018). The effects of organophosphorus insecticides and heavy metals on DNA damage and programmed cell death in two plant models. *Environmental Pollution, 240*, 77–86.

Crafts, A. S., & Crisp, C. E. (1971). *Phloem transport in plants*. WH Freeman & Co.

Crisp, C. E. (1971). The molecular design of sustemic insecticides and organic functional groups in translocation. In *International IUPAC Congress of Pesticide Chemistry, 2nd, Tel Aviv*.

Dalzell, S. A., & Mullen, B. F. (2004). Application of pesticides suppress foliar proanthocyanidin content in *Leucaena* species. *Animal Feed Science and Technology, 113*, 191–198.

Dordio, A. V., & Carvalho, A. J. P. (2013). Organic xenobiotics removal in constructed wetlands, with emphasis on the importance of the support matrix. *Journal of Hazardous Materials, 252*, 272–292.
Dubey, K. K., & Fulekar, M. H. (2011). Effect of pesticides on the seed germination of Cenchrus setigerus and Pennisetum pedicellatum as monocropping and co-cropping system: Implications for rhizospheric bioremediation. *Romanian Biotechnological Letters, 16*(1), 5908–5918.
Eapen, S., Singh, S., & D'souza, S. F (2007). Advances in development of transgenic plants for remediation of xenobiotic pollutants. *Biotechnology Advances, 25*(5), 442–451.
Edwards, R., Dixon, D. P., & Walbot, V. (2000). Plant glutathione S-transferases: Enzymes with multiple functions in sickness and in health. *Trends in Plant Science, 5*(5), 193–198.
Fatma, F., Kamal, A., & Srivastava, A. (2018). Exogenous application of salicylic acid mitigates the toxic effect of pesticides in *Vigna radiata* (L.) Wilczek. *Journal of Plant Growth Regulation, 37*(4), 1185–1194.
Foster, K. J., & Miklavcic, S. J. (2016). Modeling root zone effects on preferred pathways for the passive transport of ions and water in plant roots. *Frontiers in Plant Science, 7*, 914.
Gao, Y., & Collins, C. D. (2009). Uptake pathways of polycyclic aromatic hydrocarbons in white clover. *Environmental Science & Technology, 43*(16), 6190–6195.
Gill, S. S., & Tuteja, N. (2010). Reactive oxygen species and antioxidant machinery in abiotic stress tolerance in crop plants. *Plant Physiology and Biochemistry, 48*(12), 909–930.
Goggin, D. E., Cawthray, G. R., & Powles, S. B. (2016). 2,4-D resistance in wild radish: Reduced herbicide translocation via inhibition of cellular transport. *Journal of Experimental Botany, 67*(11), 3223–3235.
Gomes, M. P., Le Manac'h, S. G., Maccario, S., Labrecque, M., Lucotte, M., & Juneau, P. (2016). Differential effects of glyphosate and aminomethylphosphonic acid (AMPA) on photosynthesis and chlorophyll metabolism in willow plants. *Pesticide Biochemistry and Physiology, 130*, 65–70.
Gonzalez, M., Miglioranza, K. S., de Moreno, J. E. A., & Moreno, V. J. (2005). Evaluation of conventionally and organically produced vegetables for high lipophilic organochlorine pesticide (OCP) residues. *Food and Chemical Toxicology, 43*(2), 261–269.
Goodman, M., Mandel, J. S., DeSesso, J. M., & Scialli, A. R. (2014). Atrazine and pregnancy outcomes: A systematic review of epidemiologic evidence. *Birth Defects Research Part B: Developmental and Reproductive Toxicology, 101*(3), 215–236.
Guengerich, F. P. (2001). Common and uncommon cytochrome P450 reactions related to metabolism and chemical toxicity. *Chemical Research in Toxicology, 14*(6), 611–650.
Hamza, R. A., Iorhemen, O. T., & Tay, J. H. (2016). Occurance, impacts and removal of emerging substances of concern from wastewater. *Environmental Technology and Innovation, 5*, 161–175.
Hatzios, K. K., & Burgos, N. (2004). Metabolism-based herbicide resistance: Regulation by safeners. *Weed Science, 52*(3), 454–467.
Hoffman, R. S., Capel, P. D., & Larson, S. J. (2000). Comparison of pesticides in eight US urban streams. *Environmental Toxicology and Chemistry: An International Journal, 19*(9), 2249–2258.
Horowitz, A. R., & Ishaaya, I. S. A. A. C. (2004). Biorational insecticides—mechanisms, selectivity and importance in pest management. *Insect pest management* (pp. 1–28). Springer.
Huang, H., & Xiong, Z. T. (2009). Toxic effects of cadmium, acetochlor and bensulfuron-methyl on nitrogen metabolism and plant growth in rice seedlings. *Pesticide Biochemistry and Physiology, 94*(2-3), 64–67.
Huang, M. T., Lu, Y. C., Zhang, S., Luo, F., & Yang, H. (2016). Rice (*Oryza sativa*) laccases involved in modification and detoxification of herbicides atrazine and isoproturon residues in plants. *Journal of Agricultural and Food Chemistry, 64*(33), 6397–6406.
Hwang, J. I., Lee, S. E., & Kim, J. E. (2015). Plant uptake and distribution of endosulfan and its sulfate metabolite persisted in soil. *PLoS ONE, 10*(11), e0141728.
Imaishi, H., & Matumoto, S. (2007). Isolation and functional characterization in yeast of CYP72A18, a rice cytochrome P450 that catalyzes (ω-1)-hydroxylation of the herbicide pelargonic acid. *Pesticide Biochemistry and Physiology, 88*(1), 71–77.
J Edwards, R. (1998). Targeting antipeptide antibodies towards cytochrome P450 enzymes. In *Cytochrome P450 protocols* (pp. 239–250). Humana Press.
Jan, S., Singh, R., Bhardwaj, R., Ahmad, P., & Kapoor, D. (2020). Plant growth regulators: A sustainable approach to combat pesticide toxicity. *3 Biotech, 10*(11), 1–11.

Kamal, A., Ahmad, F., & Shafeeque, M. A. (2020). Toxicity of pesticides to plants and non-target organism: A comprehensive review. *Iranian Journal of Plant Physiology, 10*(4), 3299–3313.

Kato, S., Yokota, Y., Suzuki, R., Fujisawa, Y., Sayama, T., Kaga, A., ... Ishimoto, M. (2020). Identification of a cytochrome P450 hydroxylase, CYP81E22, as a causative gene for the high sensitivity of soybean to herbicide bentazon. *Theoretical and Applied Genetics, 133*(7), 2105–2115.

Kaya, A., & Doganlar, Z. B. (2016). Exogenous jasmonic acid induces stress tolerance in tobacco (*Nicotiana tabacum*) exposed to imazapic. *Ecotoxicology and Environmental Safety, 124*, 470–479.

Khanom, S., Jang, J., & Lee, O. R. (2019). Overexpression of ginseng cytochrome P450 CYP736A12 alters plant growth and confers phenylurea herbicide tolerance in Arabidopsis. *Journal of Ginseng Research, 43*(4), 645–653.

Koskinen, W. C., Calderón, M. J., Rice, P. J., & Cornejo, J. (2006). Sorption–desorption of flucarbazone and propoxycarbazone and their benzenesulfonamide and triazolinone metabolites in two soils. *Pest Management Science: Formerly Pesticide Science, 62*(7), 598–602.

Kvesitadze, G., Khatisashvili, G., Sadunishvili, T., & Kvesitadze, E. (2015). Plants for remediation: Uptake, translocation and transformation of organic pollutants. In *Plants, pollutants and remediation* (pp. 241–308). Springer.

Lee, K. Y., Strand, S. E., & Doty, S. L. (2012). Phytoremediation of chlorpyrifos by populus and salix. *International Journal of Phytoremediation, 14*(1), 48–61.

Létondor, C., Pascal-Lorber, S., & Laurent, F. (2015). Uptake and distribution of chlordecone in radish: Different contamination routes in edible roots. *Chemosphere, 118*, 20–28.

Li, Y., Long, L., Yan, H., Ge, J., Cheng, J., Ren, L., & Yu, X. (2018). Comparison of uptake, translocation and accumulation of several neonicotinoids in komatsuna (*Brassica rapa var. perviridis*) from contaminated soils. *Chemosphere, 200*, 603–611.

Liu, S., Che, Z., & Chen, G. (2016). Multiple-fungicide resistance to carbendazim, diethofencarb, procymidone, and pyrimethanil in field isolates of *Botrytis cinerea* from tomato in Henan Province, China. *Crop Protection, 84*, 56–61.

Llorens, N., Arola, L., Bladé, C., & Mas, A. (2000). Effects of copper exposure upon nitrogen metabolism in tissue cultured Vitis vinifera. *Plant Science, 160*(1), 159–163.

López-Pérez, G. C., Arias-Estévez, M., López-Periago, E., Soto-González, B., Cancho-Grande, B., & Simal-Gándara, J. (2006). Dynamics of pesticides in potato crops. *Journal of Agricultural and Food Chemistry, 54*(5), 1797–1803.

Lu, F. F., Liu, J. T., Zhang, N., Chen, Z. J., & Yang, H. (2020). OsPAL as a key salicylic acid synthetic component is a critical factor involved in mediation of isoproturon degradation in a paddy crop. *Journal of Cleaner Production, 262*, 121476.

Lunney, A. I., Zeeb, B. A., & Reimer, K. J. (2004). Uptake of weathered DDT in vascular plants: Potential for phytoremediation. *Environmental Science & Technology, 38*(22), 6147–6154.

Lushchak, V. I., Matviishyn, T. M., Husak, V. V., Storey, J. M., & Storey, K. B. (2018). Pesticide toxicity: A mechanistic approach. *EXCLI Journal, 17*, 1101.

Majeau, J. A., Brar, S. K., & Tyagi, R. D. (2010). Laccases for removal of recalcitrant and emerging pollutants. *Bioresource Technology, 101*(7), 2331–2350.

Miller, E. L., Nason, S. L., Karthikeyan, K. G., & Pedersen, J. A. (2016). Root uptake of pharmaceuticals and personal care product ingredients. *Environmental Science & Technology, 50*(2), 525–541.

Mondal, S., Kumar, M., Haque, S., & Kundu, D. (2017). Phytotoxicity of glyphosate in the germination of *Pisum sativum* and its effect on germinated seedlings. *Environmental Health and Toxicology, 32*, e2017011.

Moore, M. T., & Locke, M. A. (2012). Phytotoxicity of atrazine, S-metolachlor, and permethrin to *Typha latifolia* (Linneaus) germination and seedling growth. *Bulletin of Environmental Contamination and Toxicology, 89*(2), 292–295.

Mostafalou, S., & Abdollahi, M. (2013). Pesticides and human chronic diseases: Evidences, mechanisms, and perspectives. *Toxicology and Applied Pharmacology, 268*(2), 157–177.

Murthy, G. P., Prasad, G. M., & Sudarshana, M. S. (2005). Toxicity of different imbibition periods of dimethoate on germination, chlorophyll a/b and dry matter of *Glycine max* (L) Merrill. cv. KHSB-2, during early seedling growth. *Journal of Phytological Research, 18*(2), 199–201.

Nurzhanova, A., Kalugin, S., & Zhambakin, K. (2013). Obsolete pesticides and application of colonizing plant species for remediation of contaminated soil in Kazakhstan. *Environmental Science and Pollution Research, 20*(4), 2054–2063.

Nwoko, C. O. (2010). Trends in phytoremediation of toxic elemental and organic pollutants. *African Journal of Biotechnology, 9*(37), 6010–6016.

Otani, T., Seike, N., & Sakata, Y. (2007). Differential uptake of dieldrin and endrin from soil by several plant families and Cucurbita genera. *Soil Science and Plant Nutrition, 53*(1), 86–94.

Pandey, J. K., & Gopal, R. (2011). Laser-induced chlorophyll fluorescence: A technique for detection of dimethoate effect on chlorophyll content and photosynthetic activity of wheat plant. *Journal of Fluorescence, 21*(2), 785–791.

Parween, T. A. L. A. T., Jan, S., & Fatma, T. (2012). Evaluation of oxidative stress in *Vigna radiata* L. in response to chlorpyrifos. *International Journal of Environmental Science and Technology, 9*(4), 605–612.

Parween, T., Jan, S., & Fatma, T. (2011). Alteration in nitrogen metabolism and plant growth during different developmental stages of green gram (*Vigna radiata* L.) in response to chlorpyrifos. *Acta Physiologiae Plantarum, 33*(6), 2321–2328.

Parween, T., Jan, S., Mahmooduzzafar, S., Fatma, T., & Siddiqui, Z. H. (2016). Selective effect of pesticides on plant: A review. *Critical Reviews in Food Science and Nutrition, 56*(1), 160–179.

Pereira, J. L., Antunes, S. C., Castro, B. B., Marques, C. R., Gonçalves, A. M., Gonçalves, F., & Pereira, R. (2009). Toxicity evaluation of three pesticides on non-target aquatic and soil organisms: Commercial formulation versus active ingredient. *Ecotoxicology (London, England), 18*(4), 455–463.

Peterson, C. A., De Wildt, P. P. Q., & Edgington, L. V. (1978). A rationale for the ambimobile translocation of the nematicide oxamyl in plants. *Pesticide Biochemistry and Physiology, 8*(1), 1–9.

Pflugmacher, S., Schröder, P., & Sandermann Jr,, H. (2000). Taxonomic distribution of plant glutathione S-transferases acting on xenobiotics. *Phytochemistry, 54*(3), 267–273.

Pogrmic-Majkic, K., Kaisarevic, S., Fa, S., Dakic, V., Glisic, B., Hrubik, J., & Kovacevic, R. (2012). Atrazine effects on antioxidant status and xenobiotic metabolizing enzymes after oral administration in peripubertal male rat. *Environmental Toxicology and Pharmacology, 34*(2), 495–501.

Pullagurala, V. L. R., Rawat, S., Adisa, I. O., Hernandez-Viezcas, J. A., Peralta-Videa, J. R., & Gardea-Torresdey, J. L. (2018). Plant uptake and translocation of contaminants of emerging concern in soil. *Science of the Total Environment, 636*, 1585–1596.

Rajashekar, N., & Murthy, T. S. (2012). Seed germination and physiological behavior of maize (cv. NAC-6002) seedlings under abiotic stress (pendimethalin) condition. *Asian Journal of Crop Science, 4*(2), 80–85.

Rauh, V. A., Perera, F. P., Horton, M. K., Whyatt, R. M., Bansal, R., Hao, X., ... Peterson, B. S. (2012). Brain anomalies in children exposed prenatally to a common organophosphate pesticide. *Proceedings of the National Academy of Sciences, 109*(20), 7871–7876.

Richins, R. D., Kaneva, I., Mulchandani, A., & Chen, W. (1997). Biodegradation of organophosphorus pesticides by surface-expressed organophosphorus hydrolase. *Nature Biotechnology, 15*(10), 984–987.

Romeh, A. A., & Hendawi, M. Y. (2017). Biochemical interactions between *Glycine max* L. silicon dioxide (SiO_2) and plant growth-promoting bacteria (PGPR) for improving phytoremediation of soil contaminated with fenamiphos and its degradation products. *Pesticide Biochemistry and Physiology, 142*, 32–43.

Saladin, G., Magné, C., & Clément, C. (2003). Effects of fludioxonil and pyrimethanil, two fungicides used against *Botrytis cinerea*, on carbohydrate physiology in Vitis vinifera L. *Pest Management Science: Formerly Pesticide Science, 59*(10), 1083–1092.

Sandermann, H., Jr (1992). Plant metabolism of xenobiotics. *Trends in Biochemical Sciences, 17*(2), 82–84.

Sandermann, H., Jr (1994). Higher plant metabolism of xenobiotics: The "green liver" concept. *Pharmacogenetics, 4*(5), 225–241.

Schröder, P., & Collins, C. (2002). Conjugating enzymes involved in xenobiotic metabolism of organic xenobiotics in plants. *International Journal of Phytoremediation, 4*(4), 247–265.

Shafeeque, M. A., Ahmad, F., & Kamal, A. (2020). Toxicity of pesticides to plants and non-target organism: a comprehencive review. *Iranian Journal of Plant Physiology, 10*(4), 3299–3313.

Sharma, A., Kumar, V., Kumar, R., Shahzad, B., Thukral, A. K., & Bhardwaj, R. (2018). Brassinosteroid-mediated pesticide detoxification in plants: A mini-review. *Cogent Food & Agriculture, 4*(1), 1436212.

Sharma, I., Bhardwaj, R., & Pati, P. K. (2012). Mitigation of adverse effects of chlorpyrifos by 24-epibrassinolide and analysis of stress markers in a rice variety Pusa Basmati-1. *Ecotoxicology and Environmental Safety, 85*, 72–81.

Silva, V., Mol, H. G., Zomer, P., Tienstra, M., Ritsema, C. J., & Geissen, V. (2019). Pesticide residues in European agricultural soils: A hidden reality unfolded. *Science of the Total Environment, 653*, 1532–1545.

Siminszky, B. (2006). Plant cytochrome P450-mediated herbicide metabolism. *Phytochemistry Reviews, 5*(2), 445–458.

Stevens, M. M., Reinke, R. F., Coombes, N. E., Helliwell, S., & Mo, J. (2008). Influence of imidacloprid seed treatments on rice germination and early seedling growth. *Pest Management Science: Formerly Pesticide Science, 64*(3), 215–222.

Su, X. N., Zhang, J. J., Liu, J. T., Zhang, N., Ma, L. Y., Lu, F. F., ... Yang, H. (2019). Biodegrading two pesticide residues in paddy plants and the environment by a genetically engineered approach. *Journal of Agricultural Food Chemistry, 67*(17), 4947–4957.

Su, Y. H., Liu, T., & Liang, Y. C. (2010). Transport via xylem of trichloroethylene in wheat, corn, and tomato seedlings. *Journal of Hazardous Materials, 182*(1-3), 472–476.

Taiwo, A. M. (2019). A review of environmental and health effects of organochlorine pesticide residues in Africa. *Chemosphere, 220*, 1126–1140.

Tan, L. R., Lu, Y. C., Zhang, J. J., Luo, F., & Yang, H. (2015). A collection of cytochrome P450 monooxygenase genes involved in modification and detoxification of herbicide atrazine in rice (*Oryza sativa*) plants. *Ecotoxicology and Environmental Safety, 119*, 25–34.

Tang, X., Yang, Y., Tao, R., Chen, P., Dai, Y., Jin, C., & Feng, X. (2016). Fate of mixed pesticides in an integrated recirculating constructed wetland (IRCW). *Science of the Total Environment, 571*, 935–942.

Thompson, N. P., Wheeler, W. B., & Norden, A. J. (1970). Residues in three peanut varieties grown in dieldrin treated soil. *Journal of Agricultural and Food Chemistry, 18*(5), 862–863.

Trapp, S. (2000). Modelling uptake into roots and subsequent translocation of neutral and ionisable organic compounds. *Pest Management Science: Formerly Pesticide Science, 56*(9), 767–778.

Trapp, S. (2004). Plant uptake and transport models for neutral and ionic chemicals. *Environmental Science and Pollution Research, 11*(1), 33–39.

Turner, B. C., Taylor, A. W., & Edwards, W. M. (1972). Dieldrin and heptachlor residues in soybeans. *Agronomy Journal, 64*(2), 237–239.

Tyree, M. T., Peterson, C. A., & Edgington, L. V. (1979). A simple theory regarding ambimobility of xenobiotics with special reference to the nematicide, oxamyl. *Plant Physiology, 63*(2), 367–374.

Untiedt, R., & Blanke, M. M. (2004). Effects of fungicide and insecticide mixtures on apple tree canopy photosynthesis, dark respiration and carbon economy. *Crop Protection, 23*(10), 1001–1006.

Van Eerd, L. L., Hoagland, R. E., Zablotowicz, R. M., & Hall, J. C. (2003). Pesticide metabolism in plants and microorganisms. *Weed Science, 51*(4), 472–495.

Vryzas, Z. (2016). The plant as metaorganism and research on next-generation systemic pesticides–prospects and challenges. *Frontiers in Microbiology, 7*, 1968.

Wan, W., Huang, H., Lv, J., Han, R., & Zhang, S. (2017). Uptake, translocation, and biotransformation of organophosphorus esters in wheat (*Triticum aestivum* L). *Environmental Science & Technology, 51*(23), 13649–13658.

Wang, Q., Zhang, W., Li, C., & Xiao, B. (2012). Phytoremediation of atrazine by three emergent hydrophytes in a hydroponic system. *Water Science and Technology, 66*(6), 1282–1288.

Wang, S., Zhang, S., Huang, H., Zhao, M., & Lv, J. (2011). Uptake, translocation and metabolism of polybrominated diphenyl ethers (PBDEs) and polychlorinated biphenyls (PCBs) in maize (*Zea mays* L.). *Chemosphere, 85*(3), 379–385.

Wang, W., Wan, Q., Li, Y., Xu, W., & Yu, X. (2019). Uptake, translocation and subcellular distribution of pesticides in Chinese cabbage (*Brassica rapa var. chinensis*). *Ecotoxicology and Environmental Safety, 183*, 109488.

Wen, Y. M., Meng, L. Y., Liu, S. T., Zhu, C. M., Liu, Y. J., & Qu, A. J. (2017). Ultrastructure effects of omethoate and cypermethrin insecticides on maize seedlings. *Chemistry and Ecology, 33*(9), 883–892.

Xia, X. J., Huang, Y. Y., Wang, L., Huang, L. F., Yu, Y. L., Zhou, Y. H., & Yu, J. Q. (2006). Pesticides-induced depression of photosynthesis was alleviated by 24-epibrassinolide pretreatment in *Cucumis sativus* L. *Pesticide Biochemistry and Physiology, 86*(1), 42–48.

Zhang, C., Yao, F. E. N. G., Liu, Y. W., Chang, H. Q., Li, Z. J., & Xue, J. M. (2017). Uptake and translocation of organic pollutants in plants: A review. *Journal of Integrative Agriculture, 16*(8), 1659–1668.
Zhang, J. J., Lu, Y. C., Zhang, S. H., Lu, F. F., & Yang, H. (2016). Identification of transcriptome involved in atrazine detoxification and degradation in alfalfa (*Medicago sativa*) exposed to realistic environmental contamination. *Ecotoxicology and Environmental Safety, 130*, 103–112.
Zhang, M., Zeiss, M. R., & Geng, S. (2015). Agricultural pesticide use and food safety: California's model. *Journal of Integrative Agriculture, 14*(11), 2340–2357.
Zhang, X. D., Zhao, K. X., & Yang, Z. M. (2018). Annotation of ATP binding cassette (ABC) transporter genes and identification of Cd–responsive ABCs in rapeseed (*Brassica napus*). *Gene, 664*, 139–151.
Zhou, Y., Xia, X., Yu, G., Wang, J., Wu, J., Wang, M., … Yu, J. (2015). Brassinosteroids play a critical role in the regulation of pesticide metabolism in crop plants. *Scientific Reports, 5*, 9018.

CHAPTER 5

Impact of pesticide application on the food chain and food web

Shalini Dhiman[a], Jaspreet Kour[a], Arun Dev Singh[a], Kamini Devi[a], Raman Tikoria[c], Mohd. Ali[b], Deepak Kumar[b], Puja Ohri[b] and Renu Bhardwaj[a]
[a]Department of Botanical and Environmental Sciences, Guru Nanak Dev University, Amritsar, Punjab, India
[b]Department of Zoology, Guru Nanak Dev University, Amritsar, Punjab, India
[c]Department of Zoology, School of Bioengineering and Biosciences, Lovely Professional University, Phagwara, Punjab, India

5.1 Introduction

After the "green revolution," there has been an increase in the use of pesticides in the agriculture sector. Pesticides and agrochemicals have enhanced agricultural output, food supply, and control crop-damaging pests. Still, their residues have damaged the environment and food sources, particularly aquatic systems and fisheries, which is an essential part of the human food chain (Carvalho, 2017). Chemical pesticides are notorious for having unintended consequences that have a severe impact on biodiversity, such as polluting the air and water, contaminating the soil, and wiping out vital insects and pests (Rani et al., 2021). The application of pesticides not only impacts targeted organisms, but their usage also impacts nontargeted organisms, thus making their entry into the variety of food chains even easier. Herbicides, insecticides, and fungicides are the most widely used pesticides in agriculture, which has an adverse effect on aquatic and terrestrial organisms (de Souza et al., 2020).

According to the Agrochemicals (2018) report, only 10% of pesticides really make it to where they're supposed to be used i.e., pesticides targeted sites, with the rest ending up in the environment and reacting with nontargeted organisms that were not even intended to be affected. On the other hand, when farmers all over the globe started relying on agrochemicals, the health of the topsoil started to degrade drastically. When the health of the soil is put in jeopardy, the nutritional content of the food that is produced from that soil is also put in jeopardy. Not only do chemical pesticides strip our food of the nutritional value it formerly had, but they also pollute it. Even though pesticides are meant to kill living beings, they have been found to investigate numerous times in food (Osaili et al., 2023). Meat may have been high pesticide residues after cattle dipping, vector control, or feeding on pesticide-contaminated feedstuffs (Osaili et al., 2023). Because these pesticidal residual compounds are harmful to living things, a greater build-up of them in the food chain could jeopardize the general population's health. Several studies have connected chemical pesticides to heart disease, cancer, Alzheimer's,

Pesticides in the Environment: Impact, Assessment, and Remediation
DOI: https://doi.org/10.1016/B978-0-323-99427-9.00005-7

and Parkinson's diseases (Rajawat et al., 2023). Asthma, allergies, hypersensitivity, cancer, hormone, reproductive, and fetal development disorders have been linked to pesticide exposure (Melanda et al., 2022). Pesticides in food can cause acute or chronic liver damage, causing substantial public health implications (Nwosu & John, 2022). When it comes to using pesticides, it is important for users to have a solid understanding of the range of pesticide toxicity levels available so that they may select the product that offers the least danger while still achieving the desired results. With this approach, we must always pursue less harmful approaches to public health and the environment. This book chapter is an attempt to attain such kind of deep understanding and awareness. By knowing the impact of pesticide application on the ecosystem, food chain, and food web, we would be able to advance our understanding towards risk-free bioagrochemicals usage, developing new safe agrochemical practices for attaining the ultimate goal of organic farming and sustainable agriculture.

5.2 Pesticide types and occurrences in the environment

Pesticides are considered as world's major health issue due to their toxicity. The term "pesticide" is a hypernym for all fungicides, rodenticides, wood preservatives, insecticides, disinfectants, etc. used for killing pests. Pesticides have complex structures and compositions, and are usually organic in nature. They are grouped on the basis of toxicity, nature, and target application. They consist of plant-specific and synthetic pesticides. The data studied by the European Commission in 2019 showed 1331 forms of pesticides having various mechanisms in impeding the adverse effects of stress. Insecticides, herbicides, and fungicides contain the highest pesticidal compounds. Herbicides act as growth manager that regulates the photosynthetic efficiency of plants whereas insecticides modify the sustainability of the biome. Similarly, fungicides are used for killing the spores by inhibiting the mode of action of fungal spores inside the plant cell. The storage, disposal, and misuse of these pesticides posed a threat of contamination in the food chain and food web (de Souza et al., 2020). Some reports show a higher concentration of pesticides is presented in developing countries in surface water (Table 5.1).

5.2.1 Herbicides

Synthetic chemical compounds stimulate the regulation and death of adverse plagues (fungi, weeds, etc.). These compounds block the division of cells by inhibiting the synthesis of lipids and amino acids resulting in inhibiting photosynthesis in plants (Herrera-Herrera et al., 2016). Heap (2017) reported herbicide resistance in 70 countries that cause major human health concerns. European Union Legislation defined the allowable limits of pesticides in drinking water but still, there are pesticides detected and collected by drinking water treatment plants (DWTPs) (Climent et al., 2019). Stone et al. (2014) studied that 39 rivers in the United States have 4 insecticides, 1 fungicide, and

Table 5.1 Pesticide types and occurrences worldwide.

Pesticides types	Pesticidal compounds	Concentration (ng/L[1])	Occurrence	References
Herbicides	Acetochlor	677.51	China	Xu et al. (2019)
		72	Chile	Climent et al. (2019)
		105	Greece	Kapsi et al. (2019)
		105	China	Xie et al. (2019)
		460	USA	Glinski et al. (2018)
	Bentazone	850	China	Xu et al. (2019)
		514	Spain	Köck-Schulmeyer et al. (2019)
		761	Costa Rica	Carazo-Rojas et al. (2018)
		36	Brazil	Caldas et al. (2019)
		100	China	Sun et al. (2018)
	Des-isopropyl atrazine	0.25	China	Xu et al. (2019)
		70	China	Brauns et al. (2018)
	2,4-D	24.5	Spain	Köck-Schulmeyer et al. (2019)
		1678	Canada	Metcalfe et al. (2019)
	Alachlor	1400	United States	Glinski et al. (2018)
	Diuron	790	Brazil	Caldas et al. (2019)
		107.25	China	Peng et al. (2018)
	Metolachlor	5.5	China	Xie et al. (2019)
		77	Greece	Kapsi et al. (2019)
		44.7	United States	Battaglin et al. (2011)
	Simazine	5.3	China	Xie et al. (2019)
		24.2	Spain	Köck-Schulmeyer et al. (2019)
	Isoproturon	56.2	Spain	Köck-Schulmeyer et al. (2019)
		220	China	Sun et al. (2018)
	Terbutryn	84	Greece	Kapsi et al. (2019)
		4760	Costa Rica	Carazo-Rojas et al. (2018)
	Molinate	144	Greece	Kapsi et al. (2019)
	Terbuthylazine	59.4	Spain	Köck-Schulmeyer et al. (2019)
		87	Greece	Kapsi et al. (2019)
		10	China	Sun et al. (2018)
Insecticides	Acetamiprid	77.1	China	Yi et al. (2019)
		249	Canada	Metcalfe et al. (2019)
	Chlorpyrifos	21.95	Italy	Triassi et al. (2019)
		23.74	China	Chen et al. (2018)
		785	Ghana	Affum et al. (2018)

(continued on next page)

Table 5.1 Pesticide types and occurrences worldwide—cont'd

Pesticides types	Pesticidal compounds	Concentration (ng/L[1])	Occurrence	References
	Dimethoate	7.79	China	Xu et al. (2019)
	Heptachlor	2138	India	Khuman and Chakraborty (2019)
	Imidacloprid	11	Canada	Montiel-León et al. (2019)
		154	China	Yi et al. (2019)
Fungicides	Metconazole	12	Canada	Metcalfe et al. (2019)
	Myclobutanil	63	Greece	Kapsi et al. (2019)
	Azoxystrobin	233	Brazil	Caldas et al. (2019)
		159	United States	Battaglin et al. (2011)
	Cyprodinil	380	United States	Glinski et al. (2018)
	Propiconazole	141	United States	Battaglin et al. (2011)
	Tebuconazole	480	United States	Glinski et al. (2018)
		133	China	Peng et al. (2018)
		300	Brazil	Caldas et al. (2019)

11 herbicides concentration which is higher than the data of reference values given by the U.S. Geological Survey (USGS) and National Water-Quality Assessment (NAWQA).

5.2.2 Insecticides

These are chemical repellents, killers, or attractants that disturbed the functional structure of insects. They have an active mechanism that may vary according to the type of insecticides, i.e., organophosphates, phenylpyrazoles, carbamates, neonicotinoids, and pyrethroids (Belzunces et al., 2012). The insecticides become selective for insects by seizing and activating the receptors (nAChR) presenting minimal toxicity for animals. The carbamates and organophosphates have the same action mechanism involving inhibition of the action of acetylcholinesterase (AChE) enzyme by esters resulting in highly toxic to mammals. The pyrethroid's action mechanism is based on the voltage-sensitive sodium channel which acts on the central nervous system (CNS) causing paralysis, movement loss, and other explosive activity in insects. The phenylpyrazoles insecticidal class act by obstructing the CNS neurotransmitter and are selective and low toxicity to mammals. These are released in the environment by superficial flow, discharging, and liquidation and are still toxic to the ecological biome.

5.2.3 Fungicides

These compounds are used to prevent fungal infections in plants and seeds, therefore, protecting the fruits, tubers, and vegetables. Their action mechanism depends on the type of substance presented in fungicides including inactivation of enzymes, alteration in various specific processes of respiration, damage of the cell wall (Gupta, 2017), and also affects metabolic pathways (Atmaca et al., 2018; Correia et al., 2016). These compounds are toxic to aquatic life because of their accumulation in animal or plant tissues that trigger mutations and other alterations in animals or in the whole environment which ultimately disturbs the food chain ecology.

5.3 Pesticide transfer: from environment to food (food chain/web)

Intensive and uncontrolled applications of pesticides in agricultural fields lead to their entry into the environment. Pesticides are not only restricted to their target sites but can travel distances through water, air, and soil particles and act toxic to all the beneficial biota ranging from plants, honey bees, birds, predators, humans, and other mammals, etc. These, unchecked applications of toxic pesticides lead to an imbalance in the arrangement and working of different ecosystems (Damalas & Eleftherohorinos, 2011). The mechanistic pathway followed by the pesticides after their application in crop fields mainly undergoes absorption and uptake in plants, biosorption inside the soil particles, and photodegradation as well. Inside the soil, the pesticides undergo biological degradation as well as chemical decomposition processes. Finally, the residues generated after multiple degradation processes get up taken inside the roots xylem tissue and finally get translocated to all the aerial parts such as leaves, fruits, stem and furthermore results in deleterious impacts such as oxidative stress, ROS generation, photosynthetic blockage, chlorosis, necrosis, DNA damage, leaves twisting, and finally death of the plants (Sharma et al., 2018, 2019). Pesticide entry is, however, associated with multiple degradation and transfer processes known to take place in the environment and finally result in new chemical constituents (Marie et al., 2017). On the other hand, pesticides get relocated from their target sites and will get deposited on other nontarget ecosystems through specific processes like adsorption, spray drift, leaching, volatilization, and surface runoff (Robinson et al., 1999). These chemicals behave differently in the environment, as dichloro-diphenyl-trichloroethane (DDT) an organochlorine compound shows low acute toxicity but can accumulate inside the living tissues and result in long-term injury to the target organism. Whereas, organophosphate pesticides show high acute toxicity in mammals and are rarely persistent in the target organisms (Kim et al., 2017; Tudi et al., 2021).

However, their toxicity and collateral effects can be observed on the basis of their potential to take part in biological reactions. The sequential transfer of these pesticides takes place from the soil into the plant system and furthermore leads to their entry inside the humans and other nontarget organisms. Their uptake from the soil mainly depends on the physiochemical properties of the pesticide group such as their solubility, movement in the soil, and dissipation of pesticides. Some other factors like abiotic factors which include soil composition, temperature, moisture content, and pH as well as the physiological properties of the host plants such as plant transpiration rate, plant development, and diversities of fruits and vegetables further affect the pesticide mobility inside the plants (Hwang et al., 2017; Pullagurala et al., 2018). Thus, the solubility of pesticides in the environment is of much importance for the uptake and transfer of pesticides. As well as the cultivation techniques proved to be of importance for the bioavailability and bioconcentration of these pesticides inside the plants. For instance, the tomatoes grown in the greenhouse and in the open field conditions when provided with insecticide and fungicides are found to have more pesticidal residues in plants grown under greenhouse conditions with respect to the open plants grown under open field conditions. This concludes that the physiochemical properties of soil, and the rate of pesticide degradation and precipitation works well inside wet and open field soils (Allen et al., 2015; Bojacá et al., 2013). However, study shows that some of the plant parts are capable to accumulate more amounts of pesticidal residue with respect to other plant parts, e.g., root vegetables accumulate lower concentrations of pesticides in comparison to leafy vegetables (cabbage, spinach, lettuce, etc.) (Olatunji, 2019; Wang et al., 2015). Studies conclude that higher photosynthesis and transpiration rates in the upper leafy parts and the growth condition play crucial roles in it. Finally, the pesticide residues inside the plant tissues move to another trophic level of the food chain and biomagnification occurs. These residue particles from the lower trophic level of the food chain get an insignificant increase at the higher trophic levels (Lushchak et al., 2018). Furthermore, the bioaccumulation of persistent organic particles (POPs) fundamentally differentiates from the currently used pesticides (CUPs), as the POPs show the highest accumulation at the higher tropic levels but in the case of CUPs, the concentration is maximum in vertebrates and will show a decrease in residue concentrations at higher tropic levels (Balmer et al., 2019). Volatilization of pesticides is another very common issue found when pesticides are applied on plants through sprayers. As the herbicides when applied to the plants, it gets volatilized from the treated plants and effects severely to the nontarget plants (Straathoff, 1986). However, some of the systemic pesticides travel through the vascular tissues in plants and are found to cause contaminations in the floral as well as the extrafloral nectar parts. This contamination further leads to devastating impacts on the honey bee populations and other nectar-feeding nontarget insects (El-Wakeil et al., 2013). Whereas some other pesticides namely chlordane, endrin, dieldrin, and hexachlorobenzene thiobencarb undergoes the least environmental degradation and further gets bioaccumulated and shows 70,000 folds of its bioconcentration with respect to its original concentrations (Kim et al., 2017).

5.3.1 Pesticide dynamics in different tropical levels

Recent research has focused on the destiny and behavior of many pesticides at different trophic levels (Ginebreda et al., 2016). The prediction of preferred partitioning across environmental compartments is made possible by several physicochemical parameters, including organic carbon–water partition coefficients (K_{OC}) and octanol–water coefficients (K_{OW}), and where applicable, dissociation constants (K_A) (Kang et al., 2018).

Abiotic factors like soil composition, pH, heat, and moisture content, physiological characteristics of plants like plant water potential, development, cultivation methods, and cultivars of fruits and vegetables and physical and chemical characteristics of pesticides all affect the uptake and translocation of lethal compounds from one trophic level to the next level (Veres et al., 2020). The solubility of pesticides must be considered since this factor can greatly impact the transmission of pollutants. The bioavailability and general toxicity of pesticides can be considerably altered by the presence of liquid and particulate organic matter as well as by a compound's lipophilicity. Poor log K_{OW} pesticides, such as triazines and carbamates, have low lipophilicity and particle impact has little effect on their bioavailability across tropical levels (Knauer et al., 2017). Neonicotinoids, for instance, are easily soluble in water, ensuring both their systemic action (as the pesticide's solubility grows, the pesticide's bioavailability rises to protect plants from pests) and elimination from the environment. However, neonicotinoids are substantially more persistent in dry soils with a high level of organic matter which ultimately transfer to soil-borne biota (Chowdhary et al., 2022; Moyo et al., 2014; Tikoria et al., 2022). The toxicity of the most lipophilic pesticides, such as pyrethroids decreases when they interact with organic matter (Moyo et al., 2014). The best conditions for bioaccumulation in marine and terrestrial food chains are probably found in pesticides with moderate lipid solubility (log K_{OW} values ranging from 3 to 5) and high octanol-air partitioning (log KOA values ranging from 7 to 11) (Balmer et al., 2019). The makeup of the soil is another essential consideration. Humic acids and colloidal clay can function as adsorbents for various pollutants (Pullagurala et al., 2018). Pesticide absorption by plants may be reduced by significant pesticide sorption onto soil particulates. However, estimating bioavailability and bioaccumulation only based on K_{OW} and K_{OC} is insufficient for a precise evaluation due to the complicated connections of pesticides with particles as well as the many physicochemical features of pesticides (Knauer et al., 2017).

Neonicotinoids will be present when primary consumers ingest modified plants, notably insect herbivores, and omnivores. The roots, leaves, phloem, pollen, nectar, xylem, and extra-floral nectar of treated plants can provide fatal and sublethal dosages of neonicotinoids to primary consumers (van de Merwe et al., 2018). Pollinator exposure pathways have been discussed in depth in other places (Vickers, 2017). Neonicotinoids may transfer into untreated plants via polluted soil or water due to the same chemical properties that allow them to do so in treated plants. This poses a concern to untreated plants as well (Nerozzi et al., 2020).

When consuming plant-parasitic fungi that have ingested neonicotinoids from their target plants, even mycophagous insects may encounter deadly concentrations. When consuming senesced tissues of treated plants, primary consumers of decaying plant matter may also be exposed to fatal and sublethal dosages of neonicotinoids (Nagy et al., 2020). Particularly earthworms are able to accumulate neonicotinoids in amounts that affect their behavior and ability to operate in the environment (Langrand et al., 2020). Neonicotinoid residues in soil, sediment, and water can alter decaying processes in both terrestrial and aquatic systems (Qu et al., 2019; Socorro et al., 2016). For decomposers and detritivores in general, exposure to neonicotinoids via substrates can be just as relevant as dietary exposures (Vickers, 2017). Neonicotinoids can be administered in fatal and sublethal amounts to secondary consumers by a variety of mechanisms, including predaceous and parasitic insects. Like main consumers, predators and omnivores may eat plant tissues tainted with neonicotinoid residues, such as flowery components, nectar, or guttation droplets (Olatunji, 2019). Neonicotinoids can also be ingested at hazardous levels by predators through food webs, such as when they eat or scavenge after herbivorous invertebrates that have ingested neonicotinoids from crops. The above trophic transmission can assist pest epidemics by discharging herbivore communities from cross-face; such impacts have been demonstrated with both noninsect herbivores which have low sensitivity to neonicotinoids (such as mites and slugs) and herbivorous invertebrates that obtained nonlethal concentrations of neonicotinoids (such as scale insects, aphids, and crickets) (Defarge et al., 2018). When eating polluted honey generated by hemipteran pests that feed on the phloem, such as aphids, mealybugs, and scales, secondary consumers may also come into contact with neonicotinoids (van de Merwe et al., 2018). Neonicotinoid access can be self-reinforcing for highly omnivorous and proactive consumers like ants. When ants ingest contaminated prey, flowery assets, extrafloral nectar, honeydew, or even seeds from nontarget plants, they may be loaded with compound exposure to neonicotinoids, which can lower colony fitness and change the conformation of the ant community (Wang et al., 2018).

5.3.2 Pesticide dynamics in different food chains (terrestrial, aquatic, and other food chains)

Pesticides with greater water solubility are mostly found in the aqueous phase, whereas those with medium or low solubility can interact with sediments and suspended particles or build up in living tissues (Alegbeleye et al., 2017).

Following the Green Revolution, traditional agriculture became heavily reliant on pesticides and artificial chemical fertilizers, which had a number of detrimental repercussions on the environment and public health (Ferdous et al., 2021). A total of 4 million tons of pesticides are sprayed on crops annually to prevent pests, yet less than 1% of those chemicals actually reach their intended targets (Li, 2018). The remaining substances are absorbed by plants through soil and air, ranging from 50% to 83% depending on plant type, canopy cover, and age (Paul et al., 2016), which significantly contaminates the environment (Sarker et al., 2021), having an impact on local animals, wildlife, and aquatic life (Fantke & Juraske, 2013). Food contains traces of pesticides, and when their concentration surpasses the Maximum Residue Limits (MRL), this has a negative impact on human health. Global studies on the dynamics of pesticides following field application have taken into account intricate processes such as volatilization, leaching, sorption/desorption, degradation, plant uptake, wash-off, and runoff (Dong et al., 2019; Tikoria et al., 2022b). Pesticides may go through a number of procedures right away, as such decomposition and transformation to metabolites, or they may be transmitted to the surrounding ecosystem by a number of paths (Dromard et al., 2018). Pesticides that linger in the region where they were applied and in the surrounding environment are known as residual pesticides. Environmental elements including rainfall, wind speed, humidity, and sunshine regulate the dynamics of the applied pesticides as well as the influence of residual pesticides (Delcour et al., 2015). Several habitat-, crop-, and pesticide-specific traits were identified as crucial elements for keeping track of pesticide residues by the dynamic plant uptake model dynamiCROP, which was used to foretell the properties of leftover pesticides and the chemodynamics of applied pesticides in multiple-crop applications (Fantke et al., 2011). However, based on soil adsorption traits, soil texture, physicochemical properties of pesticides, leaching and discharge of applied pesticides, etc., pesticides sprayed on farmlands are transported to or deposited in the soils or sediments of marshes (Ali et al., 2019). Additionally, the number of soil or terrestrial organisms, their interactions with soil organic substances, and their breakdown within soil-plant transfer networks all have an impact on pesticide breakdown and residue availability (Díaz-López et al., 2019).

Based on their solubility, applied pesticides may dissolve in soil water. Pesticides in streams and rivers are highly mobile. As a result, they can be transferred downriver and through the hydrologic cycle, having a somewhat wide-ranging impact on people and the environment (Chen et al., 2019). Pesticide-containing soil water fractions can then be absorbed by plants and transmitted to various plant components, such as fruits, vegetables, and leafy parts. Aquatic habitats close to agricultural land can quickly get polluted by residual pesticide runoff because neighboring rivers act as a key sink for pesticides used on arable land. Recently, the ecohydrological model Soil and Water Assessment Tool has been used to clarify the trajectories and destinies of sprayed pesticides in the marine atmosphere (Vryzas, 2018). The trophic transmission of important persistent micropollutants and their metabolites in marine food webs is explained in a number of studies (Goutte et al., 2020). Previous research has shown that a number of parameters, including bioconcentration, biomagnification, and bioamplification, have an impact on the trophic transmission of pesticides from the application site (Wang et al., 2019). Additionally, the dynamics and degradation of pesticides can be greatly impacted by the intricate interactions between their aerosol behavior, volatility, solubility, leaching, absorption, and adsorption qualities, photodecomposition, biodegradation, chemical

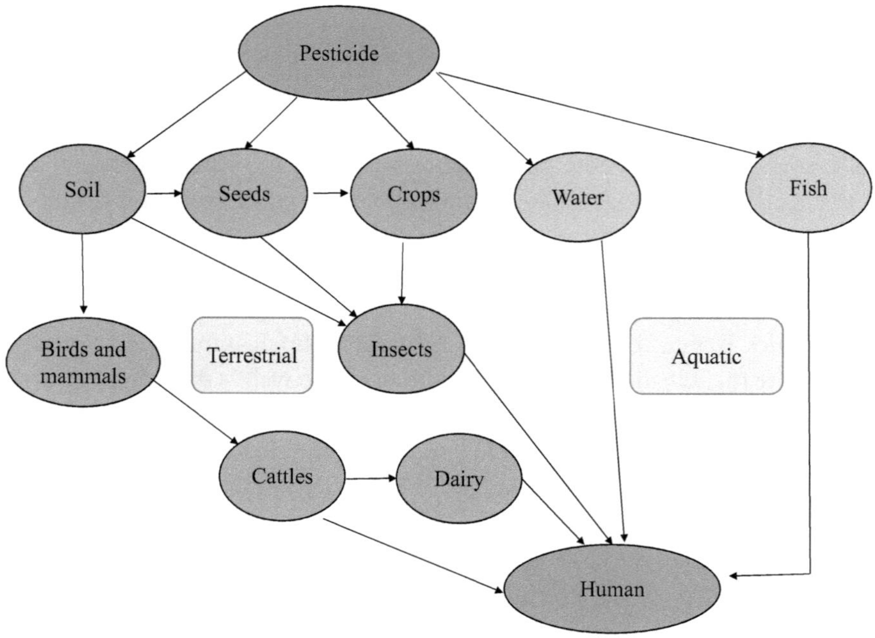

Figure 5.1 Pesticide transition in different food chains.

degradation, and detoxification (Hwang et al., 2018). Thus, it is essential to comprehend complicated mechanisms, such as multitrophic transmission and amplification, in order to set rules for the harmless use of pesticides and the wellbeing of edible products.

Pesticide residues that get into the water can also hurt large marine ecosystems like coral reefs. Australia's Great Barrier coral reef is presently monitoring for agrochemical leftovers in the seawater. Recent studies have shown that pesticides, especially herbicides, have contaminated water in a large number of places. This could hurt the symbiotic algae and kill the coral reef. (Smith et al., 2012). The deep-sea biota, which is still often regarded as a distant and pristine habitat, was also shown to have traces of persistent organic pollutants (Lyu et al., 2023). Fig. 5.1 summarizes the pesticide transition across different food chains.

5.4 Pesticide residues impact

Using pesticides to eradicate, reduce, or repel unwanted organisms has long been a crucial aspect of agriculture. It is done in order to fulfill the increasing demands of food. As a result, a huge amount of synthetic chemicals are used as insecticides, herbicides, and fungicides in agro ecosystems and natural ecosystems for protecting crops from insects, weeds, and other pathogens (Ali et al., 2021). It has been observed that about 98% of pesticides that are sprayed influence the nontarget organisms. Pesticides that are frequently used in residential gardens, agro-ecosystems, and spray drifts build up in food, soil, aquatic, and the air. Similar to how the life of beneficial microbes is impacted by the buildup of pesticide residues on soil, it has been reported that when pesticides are put on plants, their residues are responsible for the contamination of surface as well as groundwater through leaching activity. For example, due to plenty of application of insecticides and herbicides, nitrogen-fixing bacteria in soil are being disturbed. Similarly, the effects of excessive use of pesticides have also been observed on mycorrhizal fungi (Lu & Lu, 2018). Thus, it results in lowered productivity of crops. Through agricultural runoff, these pesticides can also enter into living systems directly or indirectly from the environment and then pass into various trophic levels in the food web and get accumulated in different organisms. These pesticides frequently intensify and lead to biomagnification as they pass from one trophic level to another (Katagi & Tanaka, 2016).

Studies on the nature of pesticides have revealed that these chemicals can also find their way into natural food chains before being consumed by people through the consumption of contaminated food and water (Köhler & Triebskorn, 2013). When these pesticidal residues enter the human body through contaminated food, the gastrointestinal microbiota of humans like viruses, fungi, bacteria, and protozoa also become affected. The impacts of the pesticidal residues on various ecosystems, trophic levels, food chains, and food webs have been discussed in detail in the sections below.

5.4.1 Different ecosystems (soil, water, and air)

Through the excessive use of pesticides, the agroecosystem is being directly or indirectly poisoned (Sharma et al., 2022). Pesticides can be deadly to a variety of nontarget creatures, including helpful insects, birds, fish, and nontarget floras, in addition to killing hazardous insects, pathogens, or weeds (Serrão et al., 2022; Tan et al., 2023). Herbicides can also pose risks to living things that are not their intended targets, however, insecticides are typically the most intensely poisonous set of pesticides (Ruuskanen et al., 2022). Pesticide residues disrupt agricultural systems by contaminating the soil and entering the environment through surface water and groundwater (Fig. 5.2). The presence of pesticide residues in the ecosystem is a severe problem because they constitute harm to all living things, including humans (Ali et al., 2021; Tikoria et al., 2022). Similar to this, the transformation products of pesticides are also stored in the soils and may damage the beneficial soil microbiome life. Because of high pesticide depositions in agricultural soils and their surroundings, the beneficial microfauna is diminishing. For instance, extensive soil treatments of pesticides and herbicides upset bacteria that fix nitrogen. Mycorrhizal fungi have also been observed to have similar effects (Daisley et al., 2022;

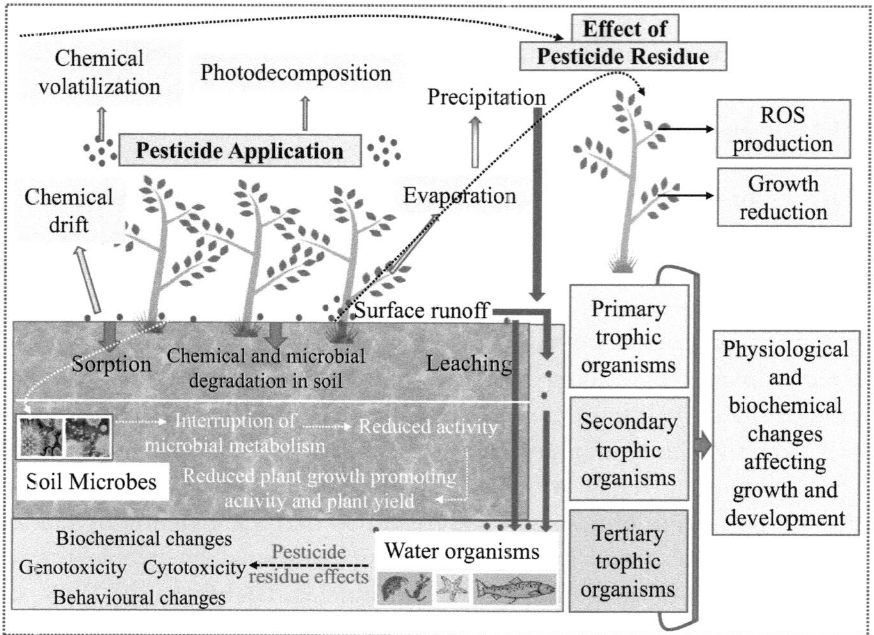

Figure 5.2 Entry of pesticide residue in ecosystem and their effect on different ecosystems and trophic levels.

Streletskii et al., 2022). Crop yield may decrease in the future as a result of declining populations of helpful bacteria brought on by prolonged pesticide exposure. For instance, Zhang et al. (2019) discovered that the fumigant 1,3-dichloropropene might alter the relative abundance of the dominant bacterial population, affect the variety of the soil bacterial community, and decrease the copy number of soil bacteria (Zhang et al., 2019).

Moreover, as pesticides are used frequently and without selection, a higher percentage is being accumulated in the soil. Afterward, through a variety of degradation processes, transformation products, and adsorption-desorption processes, soil characteristics and the soil microbiome control what happens to the sprayed pesticides and their residues (Ahmad et al., 2022; Bhatt & Shrivastav, 2022). By altering the bacterial biodiversity, biochemical, and enzyme activity in the soil, degrading pesticides have been shown to alter soil textures and microorganism activities, which may result in worrisome soil ecology and loss of soil fertility (Bokade et al., 2022). Additionally, repeated pesticide treatments alter the growth, colonization potential, and metabolic processes of beneficial root-colonizing soil microbes such as bacteria, fungi, arbuscular mycorrhiza, and algae (Mukherjee, 2022; Shahid & Khan, 2022). For instance, organochlorines can have substantial inhibitory effects on microbial growth and general metabolic activities at the community level in soil, according to various studies (Ajitha & Mathew, 2022; Arya et al., 2022). Additionally, following the application of pyraclostrobin, the numbers of culturable fungus and bacteria drastically decreased. Both the urease activity linked to the phosphorus cycle and the prevalence of the taxa (*Sphingomonas* and *Lysobacter*) were suppressed. Ammonia-oxidizing archaea and ammonia-oxidizing bacteria have abundant *amoA* genes, which indicates that this inhibits the N-cycle activity of soil microbes (Hou et al., 2022). Similarly, Goswami et al. (2013) performed tests to determine how cypermethrin affected soil MBC. After 15 days and 30 days of incubation, the bacterial biomass carbon in the samples that had been exposed to cypermethrin drastically decreased. The herbicides quizalofop-p-ethyl and imazethapyr exerted a strong harmful effect on bacterial biomass carbon at the suggested rate and double recommended rate. At 7 days and 45 days after application, a decrease of 5.9%–9.3% in soil microbial biomass was seen compared to the control (Saha et al., 2016). In another study, two samples of agricultural soil were treated with myclobutanil, a fungicide from the triazole family, at doses of 0.4 mg/kg soil (3X) and 4 mg/kg soil (10X), and their effects on soil bacterial biomass, respiration, and soil N transformations were assessed. Of these, soil bacterial biomass was found to be inhibited by a concentration of 4 mg/kg soil (10X) (Ju et al., 2016). Pesticides may have long-lasting negative or inhibitive effects on soil microbial biomass due to either a strong inhibition of pesticide-decomposing bacteria or a buildup of toxic metabolites (Zhang et al., 2014). Pesticide residues also alter specific soil microbial species or their enzyme activity, which disrupts biochemical processes like nitrogen fixation, ammonification, and nitrification (Mandal et al., 2020; Meena et al., 2020). Pesticides may have an impact on soil organic matter and mineralization, two important soil characteristics that control soil production. For instance, the use of herbicides like atrazine, glyphosate, paraquat, and prime extra greatly reduces the amount of organic matter in the soil (Sebiomo et al., 2011). It is crucial to evaluate how soil bacteria and a variety of enzymes react to pesticide depositions in order to reduce their harmful effects.

Further, surface water bodies can become contaminated by pesticide runoff from treated plants and soil, which could disrupt aquatic life over time. According to the USGS, agricultural aquatic bodies are less likely to have pesticide contamination than urban water resources. Leaching pesticides from soil surfaces build up in the groundwater system (Pérez-Lucas et al., 2019). Pesticides can enter water resources through unintentional spills, waste effluents, surface run-off, shifting from soils treated with pesticides, washing the spray equipment after use, leaching into marine bodies, and aerial spraying to destroy bugs that live in the water (Obidike et al., 2020; Okoye et al., 2022). Pesticides not only harm the fish population but also disrupt the subsequent food webs (Yamamuro et al., 2019). The main Arctic Ocean food webs were found to contain large amounts of persistent pesticides, such as polychlorinated biphenyls and organochlorines. Pesticide penetration into an organism and its solubility in water determine how long they remain active in an aquatic environment. The residual pesticide's hydrophilic or lipophobic properties reduce its availability in aquatic creatures' fatty tissues, which reduces its toxicity (Ribeiro et al., 2022). The agro-ecosystem will be greatly threatened if pesticide use is not stopped, hence it is imperative that we become fully aware of this problem and seek out environmentally appropriate solutions.

Furthermore, a significant source of air pollution that poses dangers to both public health and the health of plants and animals is pesticide contamination (Woodrow et al., 2019). Spraying pesticides for agricultural purposes always results in airborne drift, and pesticide residues in the atmosphere are mostly the result of pesticide application or volatilization from the soil or plants (Tudi et al., 2021). One way to apply pesticides is through pesticidal sprays. They are predominantly delivered out by a fan as water droplets, and after moving through the canopy in a jumbled way, they are both pulled into the earth's crust by gravity and floated by atmospheric activity like the wind. In current farming, three spraying methods—subsurface, surface, and aerial—are used. But many underdeveloped nations still practice manual spraying (Meshram et al., 2022; Stejskal et al., 2021). All methods of applying pesticides have the danger of failing to work as intended, causing pollutants in the atmosphere and exposing the general public to pesticides (Laohaudomchok et al., 2020). The pesticide remains are subject to a system of environmental recycling between the aerial and the terrestrial environment because they are volatilized, distributed, and dispersed over a long distance. Up to 25% of pesticide losses while drifting may be attributable to pesticide drift (Pan et al., 2020). This procedure not only pollutes the local surrounding but also has negative effects on the ecosystem worldwide (de Jong et al., 2008). Hexachlorocyclohexanes, toxaphene and chlordane for instance were used as pesticides in the fields of the southern United States, where they were volatilized, transported by atmospheric processes, condensed in colder climates, and then dumped from the atmosphere onto the Great Lakes in Canada (Sultana et al., 2014). It is therefore difficult to assess the level of air pollution caused by pesticides.

5.4.2 Different trophic (primary, secondary, and tertiary) levels

Chemical pesticides can infiltrate biological systems directly from the environment or indirectly through runoff from agricultural operations (Akhtar et al., 2021). Because most of the synthetic compounds have half-lives between 1 year and 4 years, they tend to accumulate in the tissues of living species at each trophic level as they go through the food chain from lowermost (algae, oysters, and fish) to topmost (eagle, bears, and people) (Tahir et al., 2021). Due to the nondecomposable nature of pesticides, which is referred to as biomagnification, pesticide residues rise in the tissues of living animals at each trophic level that follows (He et al., 2020) and results in toxic effects (Fig. 5.2). Studying numerous revealed this feature, and it was discovered that higher trophic levels have greater amounts of the poisons relative to primary exposures. For instance, Fig. 5.3 illustrates the biomagnification potential of the pesticide DDT at various trophic levels (Ali et al., 2021). Higher trophic levels are more at risk from this occurrence. Due to the biomagnification of these poisons, pesticide residues had

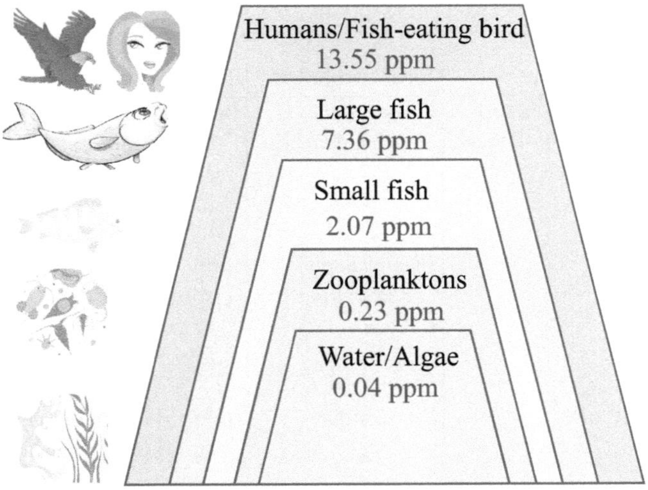

Figure 5.3 Biomagnification of dichloro-diphenyl-trichloroethane (DDT) pesticide at different trophic levels.

adverse impacts on nontarget creatures like aquatic life and people (Zaller & Brühl, 2019). Such as population losses and reproductive difficulties in numerous fish-eating birds, such as terns, gulls, and herons, were noted as a result of eggshell weakening caused by dichloro-diphenyl-dichloroethylene. It has been found that biomagnifications rise when lipophilic characteristics and pesticide residue persistence in live organisms increase. For this reason, compared to organophosphates, organochlorines are rated as compounds with significantly greater rates of biomagnification and are persistent in a wider variety of living species. They are consequently prohibited over time (Prajapati et al., 2022; Sharma et al., 2019). A study revealed that samples of honey from 27 different nations included 92 pesticide residues. According to the computed hazard indices, eating honey carries a significant danger to one's health. Additionally, it is known that contaminants in honey affect the quality of semen in both exposed people and lab animals (El-Nahhal, 2020). In another study, the mean endosalfan in fish muscle samples examined was higher than the permitted level of various international standards. However, the health risk index (HRI) varies between 0.002 and 0.1275, indicating that there is no public health risk due to Ethiopia's low per capita consumption rate of food derived from fish. Due to the likelihood of chemical residues in fish food products highlighted by this study, frequent pesticide use regulations and concentration level monitoring should be put in place to minimize hazards to human and environmental health (Mitiku & Mitiku, 2022). Further, Butralin was administered intravenously to rats for 30 days, and this caused oxidative stress, decreased antioxidant enzyme activity, increased DNA damage, and histological abnormalities in the rat kidney tissue (Refaie et al., 2021). Table 5.2 summarizes the effect

Table 5.2 Effect of pesticides and pesticide residues on various organism of different trophic levels.

Pesticide (residue)	Concentration	Target organism	Reported effect	References
Glyphosate and chlorpyrifos	0.84 nL/L and 8.4 nL/L	*Cnesterodon decemmaculatus*	Increased nuclear abnormalities and altered ultrastructure of hepatocytes	Bonifacio & Hued (2019)
Glyphosate-isopropyl-ammonium, butralin and pendimethalin	–	*Biomphalaria alexandrina*	Induced cell abnormalities, loss of viability and DNA strand breaks	Ibrahim et al. (2019)
Glyphosate	20 μM	*Lemna minor*	Induced biochemical and morphological changes	Sikorski et al. (2019)
Thiram	6.6 mM	*Solanum lycopersicum*	Induced activity of antioxidants like catalase and peroxidase and increased content of stress markers	Yüzbaşıoğlu, & Dalyan (2019)
Pyrethroids and organochlo-rine	–	*Mugil capito*	Increased malondialdehyde and glutathione level in liver tissue and altered protein profiling	Abd El Megid et al. (2020)
Imidacloprid	7.8 ppm, 15.6 ppm, and 23.4 ppm	*Cyprinus carpio*	Reduced operculum and tail beat frequency and induced hyperactivity and loss of balance	Bhardwaj & Tyor (2020)
Hexaconazole and carbendazim	3200 μg/L and 2400 μg/L	*Raphanus sativus*	Reduced plant growth, development and chlorophyll content	Khan et al. (2020)
Pyraclostrobin	146 ng/g	*Apis mellifera*	Increased DNA fragmentation and altered HSP70	Tadei et al. (2020)
Chlorpyrifos, cypermethrin and glyphosate	14.45 mg/L, 12/37 mg/L and 15.60 mg/L	*Selenastrum capricornutum*	Accumulation of lipids, disruption of thylakoids and mitochondria and formation of electrodense bodies	Fernandez et al. (2021)

(continued on next page)

Table 5.2 Effect of pesticides and pesticide residues on various organism of different trophic levels—cont'd

Pesticide (residue)	Concentration	Target organism	Reported effect	References
Chlorpyrifos	–	*Hyalella azteca*	Reduced activity of acetylcholinesterase in brain	Fuller et al. (2021)
Oryzalin	25 μM	*Caenorhabditis elegans*	Induced ultrastructural changes in meiotic oocyte nuclei along with morphological changes	Goldstein & Caliskaner (2021)
Permethrin	0.09 μg/L	*Astacus leptodactylus*	Increased haemolymph chloride and potassium and increased malondialdehyde levels in gills and muscles and induced hyperplasia	Günal et al. (2021)
Chlorpyrifos	5.50 μg/L	*Parachromis dovii*	Inhibited activity of muscle and gill cholinesterase	Jiménez et al. (2021)
Atrazine	1 nM and 100 nM	*Capra hircus*	Chromolysis in spermatids and nuclear fragmentation	Sonal et al. (2021)
Bromoxynil	825 mL/ha	*Triticum aestivum*	Reduced growth of wheat plants	Farooq et al. (2022)
Karate Zeon, pyrinex and mulan	0.7–200 μL, 0.7–8750 μL and 0.3–1000 mg	*Oreochromis mossambicus*	Induced mortality, motionlessness and loss of equilibrium	Mutshekwa et al. (2022)
Chlorpyrifos and difenoconazole	5 μg/L and 325 μg/L	*Poecilia gillii*	Induced neurotoxicity and behavioural changes	Redondo-López et al. (2022)
Malathion	100 and 200 μg/mL	Chick	Reduced embryo survival and body weight	Siddique et al. (2022)
Chlorpyrifos and cypermethrin	0.0002, 0.0003 and 0.0006 μg/L and 0.10 μg/L	*Cyprinus carpio*	Induced behavioural and biochemical changes	Yancheva et al. (2022)
Cyantraniliprole	300 g a.i./hm^2	*Procambarus clarkii*	Decreased weight gain and specific growth rate	Liao et al. (2023)

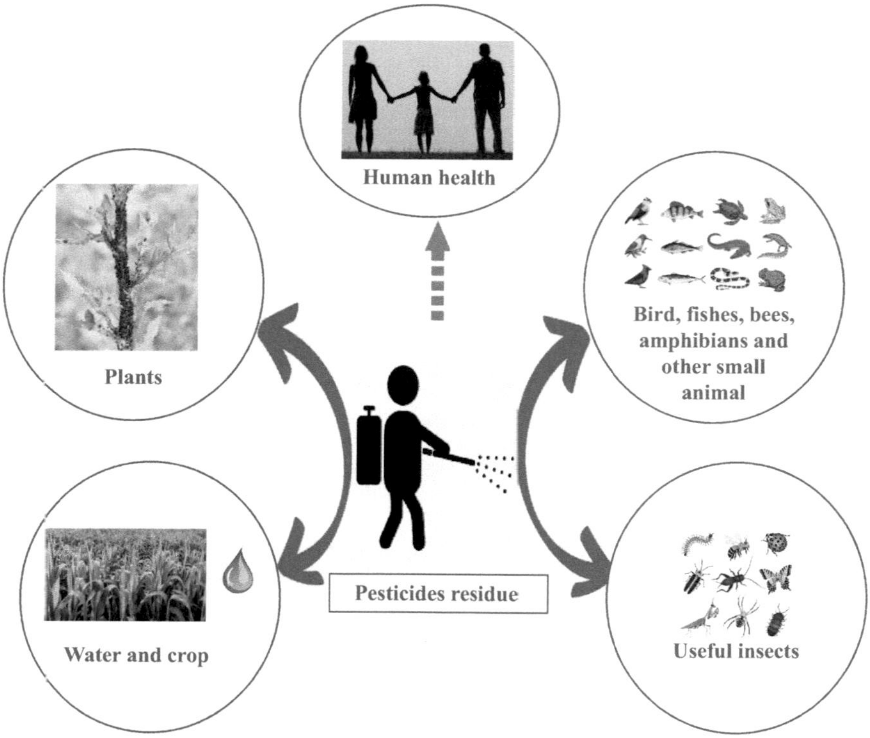

Figure 5.4 Pesticide residues impact on different food chains.

of different pesticides and pesticide residues on various organisms of different trophic level.

5.4.3 Different food chains (terrestrial, aquatic, and other)

A worldwide problem with food safety is pesticide contamination. For food consumers, the extent, and consequences of pesticide exposures through food are a major concern (Shenashen et al., 2022). Pesticides are one type of environmental contaminant that could accumulate in both terrestrial and aquatic food systems (Mali et al., 2022). Numerous studies have documented the accumulation of pesticides in soils, terrestrial ecosystems, and aquatic environments, as well as their fatal effects on human and other biota (Fig. 5.4). Innovations are used first, and then their negative effects become apparent. One of the best instances in this situation is pesticides (Ali et al., 2021). Synthetic pesticides were developed to protect crops from pests and increase agricultural yields. New pesticide chemicals are used to increase agricultural productivity due to crop yield security. Use of pesticides is still widespread, especially in tropical and southern countries (Carvalho et al., 2017). The natural human nourishment is nevertheless contaminated by pesticide deposits and late build-ups even if persistent synthetic chemicals like dichlorodiphenyl-trichloroethane, hexachlorocyclohexane, and lindane have been replaced and restricted by increasingly decomposable synthetic compounds. These pesticides are no longer allowed to be used on farms in most countries. As a result, these synthetic compound residues spread throughout the ecosystem and pollute the environment (Mishra et al., 2021; Reeves et al., 2019). The use of pesticides must be done more safely, which requires earlier testing, careful handling of risks, as well as educating farmers and customers, better insurance of biological systems, and great practices for the reasonable advancement of agribusiness, fisheries, and aquaculture (Chávez-Dulanto et al., 2021; Green et al., 2016). Agriculture must also develop excellent practices to ensure universal well-being. Pesticide chemicals are consumed by people who consume such food and water after they have made their way into the natural food chain. Companies that manufacture pesticides continuously create new synthetic chemicals due to the emergence of pesticide resistance. More than 800 synthetics are designated as pesticides in the European Union (Köhler & Triebskorn, 2013). Pesticide application frequently resulted in pesticide toxicity among farmers, employees, and their families. With the aid of a few systems, pesticide application in agriculture evolved from manual labor performed by workers walking to the truck- and plane-based application techniques. It is projected that 0.35 million people die globally each year from unintentional pesticide poisoning. Such poisonings have a close link to unneeded exposures to harmful substances and improper use of such chemicals (Carvalho et al., 2017; Sombatsawat et al., 2022). Widespread pesticide residues in the environment as well as significant losses of human and other biotas, including bees, birds, amphibians, fish, and small animals were mentioned in the literature (Bourguet & Guillemaud, 2016; Köhler & Triebskorn, 2013). Extensive research has also been done over the years to understand the behavior of these synthetic chemicals in the ecosystem in order to comprehend their decomposition cycles and destinies with regard to their deadliness to biota. It was frequently seen that the DDT used to protect crops was transferred into aquatic bodies where it was swiftly converted to dichloro-diphenyl-dichloroethylene and tend to bioaccumulated in the oceanic food chain eventually returning to final consumers (Ali et al., 2021; Spataro et al., 2021). Long before they were acknowledged as stable products, organochlorine pesticides (OCPs) accumulate in nonhuman animals and have terrible effects on population levels. They linger longer in soils and silt. They eventually travel up food chains, which has a negative impact on human health (Köhler & Triebskorn, 2013; Salihu, et al., 2022). Based on research regarding the rates of their breakdown in soil and aquatic environments, synthetic organochlorine compounds were questioned and replaced with less persistent synthetic substances such as organophosphate, carbamate, and pyrethroids (Ali et al., 2021; Leemans et al., 2019). For instance, toxaphene is no longer applied to cotton fields in Nicaragua, but the residue it left behind in agricultural soils continues to be a source of marine pollution, posing a serious threat to food production along the coast (Carvalho et al., 2017). The primary and consistent source of organochlorines is soil. The ecology is heavily contaminated by harmful organochlorines that are carried via soil erosion, surface runoff, and river forms. Large sea biological systems, such as coral reefs and other deep-sea flora fauna, are more at risk from pesticide residues that have made their way into the ocean (Jamieson et al., 2017; Razali et al., 2018). Pesticide poison left behind in water systems has the potential to wipe out marine life, reduce biodiversity, and disrupt the equilibrium of the natural ecosystem. Large-scale studies in aquatic toxicology have aided in understanding pesticide bioaccumulation pathways to define toxicity standards for a few demonstrative species (crustaceans, fish, and plants), as well as complex controls to keep pesticide pollution within acceptable bounds (Carvalho et al., 2017). Upon consumption, the pesticide remains in food and reacts with the human gut microbe, which consists of certain microbe, fungus, viruses, and protozoa. These chemicals affect the microbiome's composition, which could have fatal consequences on systemic immunity as a whole and intestinal homeostasis (Hirt, 2020; Tsiaoussis et al., 2019).

5.4.4 Pesticide residue impact on different food webs

Pesticides, often known as crop protection chemicals, contain several chemical classes that have been produced one after another, including organochlorine, organophosphate, carbamate, growth regulators, neonicotinoids, pyrethroids, and currently, biopesticides (Nicolopoulou-Stamati et al., 2016). All types of pesticides have seen an increase in sales, but herbicides had the biggest growth, followed by insecticides and fungicides. Any pesticides have the potential to interrupt a food web when vulnerable species come into direct touch with toxic residue at the time of application or come into contact with contaminated media (Tooker & Pearsons, 2021; Zhang et al., 2011) (Fig. 5.5).

According to the trophic level impacted, this ecological exposure, whether deadly or harmless, can break linkages in food webs, change competition, and free prey from top–down control. Pesticides more subtle ability to alter food chains by climbing the trophic levels. OCPs, particularly DDT, interrupt trophic cascades and, more notoriously, food chains by reducing insect abundance. As predators eat infected prey, organochlorine concentrations may rise because they bioaccumulate in the fatty tissue of animals (Kolok, 2016; Nilsen et al., 2019). This biomagnification phenomena can cause top vertebrate predators in both marine and terrestrial systems to be exposed to significantly higher hazardous doses than lower trophic levels. Like organochlorines, other synthetic insecticides such as some organophosphates and pyrethroids can biomagnify and endanger higher trophic levels (Corcellas et al., 2015; Tooker & Pearsons, 2021). However, significant bioaccumulation is uncommon because most carbamates and organophosphates degrade quickly pyrethroids are mostly metabolized and neonicotinoids are typically eliminated. When sensitive prey is in short supply and sensitive predators (both invertebrate and vertebrate) eat contaminated food, even substances that do not readily bioaccumulate can

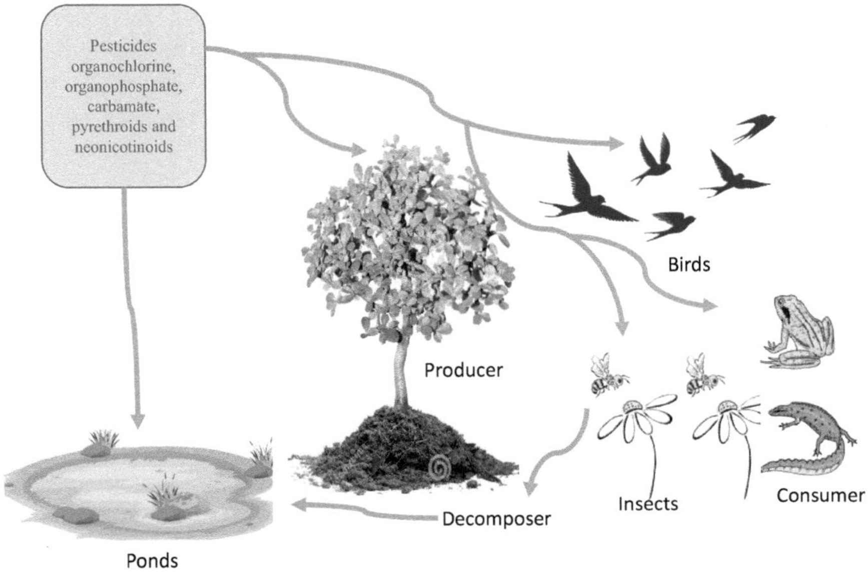

Figure 5.5 Pesticide residues impact on different food web.

disturb trophic cascades. Neonicotinoids are poised to disrupt food webs by drastically reducing insect diversity and abundance due to their widespread use and distinctive chemistry (Resende-Silva et al., 2023; Tooker & Pearsons, 2021). Indeed, the widespread use of neonicotinoids, particularly in large-scale row-crop agriculture, has transformed parts of North America and Europe into environments that are less and less hospitable even toxic for insects (Douglas et al., 2020). Additionally, the chemical properties of neonicotinoids such as their higher water solubility and soil persistence make it easier for contamination of unintended areas such as aquatic environments, natural habitats, organic crop fields. Where these toxic insecticides can adversely affect delicate tax systems. When they consume treated plants, primary consumers, particularly insect herbivores and omnivores, will meet neonicotinoids. Primary consumers who consume the roots, leaves, xylem, phloem, pollen, nectar, and extraflowery nectar of treated plants may be exposed to deadly and sublethal amounts of neonicotinoids (Hopwood et al., 2016; Tooker & Pearsons, 2021). Neonicotinoids can be ingested by secondary consumers, like predatory and parasitic insects, in lethal and sublethal amounts in a variety of ways. Predators and omnivores, like main consumers, come into contact with neonicotinoid traces when they consume contaminated plant tissues like floral resources, extra flowery nectar, or guttation droplets (Fritsch et al., 2022). Despite their distinct chemistry and toxicity, neonicotinoids have the same two effects on food webs as older pesticide classes: they directly simplify food webs and indirectly affect them by moving within them. First, neonicotinoids can affect food webs by lowering the variety and abundance of consumers directly. Members of higher trophic levels, particularly specialists, may experience prey scarcity when lower trophic levels are negatively impacted. Neonicotinoids have clear and depopulated insect populations in diverse systems (such as croplands, woodlands, and waterways), which has decreased the quantity of animals at higher trophical levels such as insects, birds, and fishes (Ngata et al., 2018; Tooker & Pearsons, 2021).

Similarly, pesticides with long-lasting residuals in the water body include OCPs, which are ubiquitous, persistent, and bioaccumulative on a global scale. The high toxicity of OCPs presents serious risks to aquatic biodiversity and human health (Bai et al., 2018). However, these pesticides are among the most hazardous substances to human health, including DDT, dieldrin, and toxaphene, among others. These pesticides are teratogens, neuroendocrine disruptors, immunological and reproductive system suppressors, dysregulation of lipid and metabolic processes. because of their persistence and severe toxicity to aquatic and terrestrial species as well as people OCPs continue to be key agrochemicals of concern (Martyniuk et al., 2020). Additionally, pyrethroids are insecticides with greater biological activity and low dose application rates. They have strong sportive qualities and poor water solubility, which lower their bioavailability in natural settings (Jaiswal et al., 2017). Aquatic insects are innately vulnerable to pyrethroids, although it is unclear exactly how these chemicals cause them to be so sensitive. Pyrethroids' severe toxicity to aquatic creatures prevents them from being used more widely in agriculture (Antwi and Reddy, 2015). Insecticide poisoning in aquatic insects can occur at even very low concentrations. Pyrethroids are only permitted to be used in locations close to aquatic ecosystems due to their severe acute toxicity to aquatic invertebrates and fish (Anderson et al., 2015; Antwi and Reddy, 2015).

5.5 Conclusion

Pesticides in the food chain can cause a variety of disturbances in organism population dynamics, unwanted adverse consequences, such as depletion of the nontargeted predator group, toxic effects on organisms that directly or indirectly eat pesticide-contaminated organisms, and causing ecosystem imbalances. Despite advancing our understanding of toxicity and environmental impact, pesticides, and other agrochemicals are still not completely under control (Carvalho, 2017). Moreover, due to the failure of modern environmental risk assessment of pesticides to capture their impact on trophic level links, additional investigation of the toxicological impacts of pesticides on various targeted or nontargeted organisms and population dynamics is needed. To ensure food safety, alternative safe and nonrisky advance bioagrochemical such as biopesticide use should be explored. Alternative approaches to the heavy use of crop protection pesticides should be investigated. Advanced techniques to safeguard public health should be created through previous testing, risk assessment analysis, licensing, and general education on farmers and users. Further, more research on food production and quality, soil and ecosystem preservation, and appropriate agriculture practices are needed to ensure ongoing food supply and safety. Novel technologies, such as nanocarriers, are being developed to lessen the danger of pesticide formulations by delaying their release, improving their precision in targeting specific pests, and decomposing harmlessly after use without any side effects on the soil, water, and air ecosystem. While progress has been made toward using safer and nonhazardous agrochemicals, organic farming and sustainable development goals are still a long way from being reached.

References

Abd El Megid, A., Abd Al Fatah, M. E., El Asely, A., El Senosi, Y., Moustafa, M. M., & Dawood, M. A. (2020). Impact of pyrethroids and organochlorine pesticides residue on IGF-1 and CYP1A genes expression and muscle protein patterns of cultured *Mugil capito*. *Ecotoxicology and Environmental Safety, 188*, 109876.

Affum, A. O., Acquaah, S. O., Osae, S. D., & Kwaansa-Ansah, E. E. (2018). Distribution and risk assessment of banned and other current-use pesticides in surface and groundwaters consumed in an agricultural catchment dominated by cocoa crops in the Ankobra Basin, Ghana. *Science of the Total Environment, 633*, 630–640.

Agrochemicals, WHO. (2018). Health and environment: Directory of resources. http://www.who.int/heli/risks/toxics/chemicalsdirectory/en/index1.html. (Accessed on 24 January 2018).

Ahmad, S., Ahmad, H. W., & Bhatt, P. (2022). Microbial adaptation and impact into the pesticide's degradation. *Archives of Microbiology, 204*(5), 288.

Ajitha, D., & Mathew, L. (2022). Impact of organochlorine pesticides on soil microflora and soil fertility. In *Pesticides bioremediation* (pp. 3–30). Springer International Publishing.

Akhtar, N., Syakir Ishak, M. I., Bhawani, S. A., & Umar, K. (2021). Various natural and anthropogenic factors responsible for water quality degradation: A review. *Water, 13*(19), 2660.

Aldas, S. S., Arias, J. L. O., Rombaldi, C., Mello, L. L., Cerqueira, M. B., Martins, A. F., & Primel, E. G. (2019). Occurrence of pesticides and PPCPs in surface and drinking water in southern Brazil: Data on 4-year monitoring. *Journal of the Brazilian Chemical Society, 30*, 71–80.

Alegbeleye, O. O., Opeolu, B. O., & Jackson, V. A. (2017). Polycyclic aromatic hydrocarbons: A critical review of environmental occurrence and bioremediation. *Environmental Management, 60*, 758–783.

Ali, N., Khan, S., Li, Y., Zheng, N., & Yao, H. (2019). Influence of biochars on the accessibility of organochlorine pesticides and microbial community in contaminated soils. *Science of the Total Environment, 647*, 551–560.

Ali, S., Ullah, M. I., Sajjad, A., Shakeel, Q.,& Hussain,A. (2021). Environmental and health effects of pesticide residues. In Inamuddin, Ahamed, M. I., & Lichtfouse, E. (Eds.), *Sustainable agriculture reviews* (Vol. 48, pp. 311-336). Springer.

Allen, G., Halsall, C. J., Ukpebor, J., Paul, N. D., Ridall, G., & Wargent, J. J. (2015). Increased occurrence of pesticide residues on crops grown in protected environments compared to crops grown in open field conditions. *Chemosphere, 119*, 1428–1435.

Anderson, J. C., Dubetz, C., & Palace, V. P. (2015). Neonicotinoids in the Canadian aquatic environment: a literature review on current use products with a focus on fate, exposure, and biological effects. *Science of the Total Environment, 505*, 409–422.

Antwi, F. B., & Reddy, G. V. (2015). Toxicological effects of pyrethroids on non-target aquatic insects. *Environmental Toxicology and Pharmacology, 40*(3), 915–923.

Arya, S., Kumar, R., Prakash, O., Rawat, A., & Pant, A. K. (2022). Impact of insecticides on soil and environment and their management strategies. In *Agrochemicals in soil and environment: Impacts and remediation* (pp. 213–230). Springer Nature Singapore.

Atmaca, N., Arikan, S., Essiz, D., Kalender, H., Simsek, O., Bilmen, F. S., & Kabakci, R. (2018). Effects of mancozeb, metalaxyl and tebuconazole on steroid production by bovine luteal cells in vitro. *Environmental Toxicology and Pharmacology, 59*, 114–118.

Bai, Y., Ruan, X., & van der Hoek, J. P (2018). Residues of organochlorine pesticides (OCPs) in aquatic environment and risk assessment along Shaying River, China. *Environmental Geochemistry and Health, 40*, 2525–2538.

Balmer, J. E., Morris, A. D., Hung, H., Jantunen, L., Vorkamp, K., Rigét, F., …, & Muir, D. C. (2019). Levels and trends of current-use pesticides (CUPs) in the arctic: An updated review, 2010–2018. *Emerging Contaminants, 5*, 70–88.

Battaglin, W. A., Sandstrom, M. W., Kuivila, K. M., Kolpin, D. W., & Meyer, M. T. (2011). Occurrence of azoxystrobin, propiconazole, and selected other fungicides in US streams, 2005–2006. *Water, Air & Soil Pollution, 218*, 307–322.

Belzunces, L. P., Tchamitchian, S., & Brunet, J. L. (2012). Neural effects of insecticides in the honey bee. *Apidologie, 43*, 348–370.

Bhardwaj, J. K., & Tyor, A. K. (2020). Imidacloprid induced alterations in behavioral and locomotory activity of fingerlings of common carp, *Cyprinus carpio*. *Toxicology International*, 158–167.

Bhatt, P., & Shrivastav, A. (2022). Removal of pesticides from water and waste water by microbes. In *Development in wastewater treatment research and processes* (pp. 371–399). Elsevier.

Bojacá, C. R., Arias, L. A., Ahumada, D. A., Casilimas, H. A., & Schrevens, E. (2013). Evaluation of pesticide residues in open field and greenhouse tomatoes from Colombia. *Food Control, 30*(2), 400–403.

Bokade, P., Gaur, V. K., Tripathi, V., Bobate, S., Manickam, N., & Bajaj, A. (2022). Bacterial remediation of pesticide polluted soils: Exploring the feasibility of site restoration. *Journal of Hazardous Materials, 441*, 129906.

Bonifacio, A. F., & Hued, A. C. (2019). Single and joint effects of chronic exposure to chlorpyrifos and glyphosate based pesticides on structural biomarkers in *Cnesterodon decemmaculatus*. *Chemosphere, 236*, 124311.

Bourguet, D., & Guillemaud, T. (2016). The hidden and external costs of pesticide use. *Sustainable Agriculture Reviews, 19*, 35–120.

Brauns, B., Jakobsen, R., Song, X., & Bjerg, P. L. (2018). Pesticide use in the wheat-maizedouble cropping systems of the North China Plain: Assessment, field study, and implications. *Science of the Total Environment, 616–617*, 1307–1316. http://dx.doi.org/10.1016/j.scitotenv.2017.10.187.

Caldas, S. S., Arias, J. L. O., Rombaldi, C., Mello, L. L., Cerqueira, M. B., Martins, A. F., & Primel, E. G. (2019). Occurrence of pesticides and PPCPs in surface and drinking water in southern Brazil: Data on 4-year monitoring. *Journal of the Brazilian Chemical Society, 30*, 71–80.

Carazo-Rojas, E., Pérez-Rojas, G., Pérez-Villanueva, M., Chinchilla-Soto, C., Chin-Pampillo, J. S., Aguilar-Mora, P., …, & Vryzas, Z. (2018). Pesticide monitoring and ecotoxicological risk assessment in surface water bodies and sediments of a tropical agro-ecosystem. *Environmental Pollution, 241*, 800–809.

Carvalho, F. P. (2017). Pesticides, environment, and food safety. *Food and Energy Security, 6*(2), 48–60.

Chávez-Dulanto, P. N., Thiry, A. A., Glorio-Paulet, P., Vögler, O., & Carvalho, F. P. (2021). Increasing the impact of science and technology to provide more people with healthier and safer food. *Food and Energy Security, 10*(1), e259.

Chen, C., Guo, W., & Ngo, H. H. (2019). Pesticides in stormwater runoff: A mini review. *Frontiers of Environmental Science & Engineering, 13*, 1–12.

Chen, Y., Yu, K., Hassan, M., Xu, C., Zhang, B., Gin, K. Y. H., & He, Y. (2018). Occurrence, distribution and risk assessment of pesticides in a river-reservoir system. *Ecotoxicology and Environmental Safety, 166*, 320–327.

Chowdhary, A. B., Singh, J., Quadar, J., Singh, S., Dutta, R., Angmo, D., & Vig, A. P. (2022). Earthworm's show tolerance and avoidance response to pesticide clothianidin: effect on antioxidant enzymes. *International Journal of Environmental Science and Technology, 20*, 1–10.

Climent, M. J., Herrero-Hernández, E., Sánchez-Martín, M. J., Rodríguez-Cruz, M. S., Pedreros, P., & Urrutia, R. (2019). Residues of pesticides and some metabolites in dissolved and particulate phase in surface stream water of Cachapoal River basin, central Chile. *Environmental Pollution, 251*, 90–101.

Corcellas, C., Eljarrat, E., & Barceló, D. (2015). First report of pyrethroid bioaccumulation in wild river fish: A case study in Iberian River basins (Spain). *Environment International, 75*, 110–116.

Correia, M., Rodrigues, M., Paíga, P., & Delerue-Matos, C. (2016). Fungicides. In *Encyclopedia of Food and Health* (pp. 169–176). USA: Elsevier, Academic Press.

Daisley, B. A., Chernyshova, A. M., Thompson, G. J., & Allen-Vercoe, E. (2022). Deteriorating microbiomes in agriculture-the unintended effects of pesticides on microbial life. *Microbiome Research Reports, 1*(1), 6.

Damalas, C. A., & Eleftherohorinos, I. G. (2011). Pesticide exposure, safety issues, and risk assessment indicators. *International Journal of Environmental Research and Public Health, 8*(5), 1402–1419.

Defarge, N., De Vendômois, J. S., & Séralini, G. E. (2018). Toxicity of formulants and heavy metals in glyphosate-based herbicides and other pesticides. *Toxicology Reports, 5*, 156–163.

de Jong, F. M., de Snoo, G. R., & van de Zande, J. C. (2008). Estimated nationwide effects of pesticide spray drift on terrestrial habitats in the Netherlands. *Journal of Environmental Management, 86*(4), 721–730.

Delcour, I., Spanoghe, P., & Uyttendaele, M. (2015). Literature review: Impact of climate change on pesticide use. *Food Research International, 68*, 7–15.

de Souza, R. M., Seibert, D., Quesada, H. B., Fagundes-Klen, M. R., & Bergamasco, R.de, J. Bassetti, F. (2020). Occurrence, impacts and general aspects of pesticides in surface water: A review. *Process Safety and Environmental Protection, 135*, 22–37.

Díaz-López, M., García, C., Garrido, I., Navarro, S., Vela, N., Nicolás, E., …, & Bastida, F. (2019). Solarisation-based pesticide degradation results in decreased activity and biomass of the soil microbial community. *Geoderma, 354*, 113893.

Dong, X., Singh, B. P., Li, G., Lin, Q., & Zhao, X. (2019). Biochar increased field soil inorganic carbon content five years after application. *Soil and Tillage Research, 186*, 36–41.

Douglas, M. R., Sponsler, D. B., Lonsdorf, E. V., & Grozinger, C. M. (2020). County-level analysis reveals a rapidly shifting landscape of insecticide hazard to honey bees (*Apis mellifera*) on US farmland. *Scientific Reports, 10*(1), 797.

Dromard, C. R., Bouchon-Navaro, Y., Cordonnier, S., Guéné, M., Harmelin-Vivien, M., & Bouchon, C. (2018). Different transfer pathways of an organochlorine pesticide across marine tropical food webs assessed with stable isotope analysis. *PLoS ONE, 13*(2), e0191335.

El-Nahhal, Y. (2020). Pesticide residues in honey and their potential reproductive toxicity. *Science of the Total Environment, 741*, 139953.

El-Wakeil, N., Gaafar, N., Sallam, A., & Volkmar, C. (2013). Side effects of insecticides on natural enemies and possibility of their integration in plant protection strategies. In *Insecticides: Development of safer and more effective technologies* (pp. 1–54). IntechOpen.

Fantke, P., & Juraske, R. (2013). Variability of pesticide dissipation half-lives in plants. *Environmental Science & Technology, 47*(8), 3548–3562.

Fantke, P., Juraske, R., Antón, A., Friedrich, R., & Jolliet, O. (2011). Dynamic multicrop model to characterise impacts of pesticides in food. *Environmental Science & Technology, 45*(20), 8842–8849.

Farooq, S., Yasmeen, T., Niaz, A., Rizwan, M., & Ali, S. (2022). Rice straw biochar in combination with farmyard manure mitigates bromoxynil toxicity in wheat (*Triticum aestivum* L.). *Chemosphere, 295*, 133854.

Ferdous, Z., Zulfiqar, F., Datta, A., Hasan, A. K., & Sarker, A. (2021). Potential and challenges of organic agriculture in Bangladesh: A review. *Journal of Crop Improvement, 35*(3), 403–426.

Fernandez, C., Asselborn, V., & Parodi, E. R. (2021). Toxic effects of chlorpyrifos, cypermethrin and glyphosate on the non-target organism *Selenastrum capricornutum* (Chlorophyta). *Anais da Academia Brasileira de Ciências, 93*, e20200233.

Fritsch, C., Appenzeller, B., Burkart, L., Coeurdassier, M., Scheifler, R., Raoul, F., …, & Pelosi, C. (2022). Pervasive exposure of wild small mammals to legacy and currently used pesticide mixtures in arable landscapes. *Scientific Reports, 12*(1), 15904.

Fuller, N., Hartz, K. E. H., Johanif, N., Magnuson, J. T., Robinson, E. K., Fulton, C. A., …, & Lydy, M. J. (2021). Enhanced trophic transfer of chlorpyrifos from resistant *Hyalella azteca* to inland silversides (*Menidia beryllina*) and effects on acetylcholinesterase activity and swimming performance at varying temperatures. *Environmental Pollution, 291*, 118217.

Ginebreda, A., Pérez, S., Rivas, D., Kuzmanovic, M., & Barceló, D. (2016). Pollutants of emerging concern in rivers of catalonia: Occurrence, fate, and risk. Experiences from surface water quality monitoring: The EU water framework directive implementation in the Catalan river basin district (Part I) (pp. 283–320). D. Barceló, A.G. Kostianoy (Eds.), The Handbook of Environmental Chemistry, Springer International Publishing, Switzerland.

Glinski, D. A., Purucker, S. T., Van Meter, R. J., Black, M. C., & Henderson, W. M. (2018). Analysis of pesticides in surface water, streamflow, and throughfall in an agricultural area in South Georgia, USA. *Chemosphere, 209*, 496–507. http://dx.doi.org/10.1016/j.chemosphere.2018.06.116.

Goldstein, P., & Caliskaner, Z. O. (2021). Exposure to the anthelmintic dinitroaniline oryzalin causes changes in meiotic prophase morphology and loss of synaptonemal complexes in the nematode *Caenorhabditis elegans*. *Experimental Results, 2*, e38.

Goswami, M. R., Pati, U. K., Chowdhury, A., & Mukhopadhyay, A. (2013). Studies on the effect of cypermethrin on soil microbial biomass and its activity in an alluvial soil. *International Journal of Agriculture and Food Science, 3*(1), 1–9.

Goutte, A., Alliot, F., Budzinski, H., Simonnet-Laprade, C., Santos, R., Lachaux, V., ..., & Labadie, P. (2020). Trophic transfer of micropollutants and their metabolites in an urban riverine food web. *Environmental Science & Technology, 54*(13), 8043–8050.

Green, H., Broun, P., Cakmak, I., Condon, L., Fedoroff, N., Gonzalez-Valero, J., ..., & Roulin, A. (2016). Planting seeds for the future of food. *Journal of the Science of Food and Agriculture, 96*(5), 1409–1414.

Günal, A. Ç., Tunca, S. K., Arslan, P., Gül, G., & Dinçel, A. S. (2021). How does sublethal permethrin effect non-target aquatic organisms? *Environmental Science and Pollution Research, 28*(37), 52405–52417.

Gupta, P. K. (2017). Herbicides and fungicides. Reproductive and developmental toxicology (pp. 503–521). Academic Press, Elsevier.

He, B., Wang, X., Yang, C., Zhu, J., Jin, Y., & Fu, Z. (2020). The regulation of autophagy in the pesticide-induced toxicity: Angel or demon? *Chemosphere, 242*, 125138.

Heap, I. (2017). The International Survey of Herbicide Resistant Weeds. www.weedscience.org. (Accessed 31 May 2017).

Herrera-Herrera, A.V., Asensio-Ramos, M., Hernández-Borges, J., & Rodriguez-Delgado, M.A. (2016). Pesticides and herbicides: Types, uses, and determination of herbicides. In *Encyclopedia of Food and Health*, Finglas, P.M., Toldrá, F. (Eds.), Academic Oxford, UK, pp. 326–332.

Hirt, H. (2020). Healthy soils for healthy plants for healthy humans: How beneficial microbes in the soil, food and gut are interconnected and how agriculture can contribute to human health. *EMBO Reports, 21*(8), e51069.

Hopwood, J., Code, A., Vaughan, M., Biddinger, D., Shepherd, M., Black, S.H., ... & Mazzacano, C. (2016). How neonicotinoids can kill bees. In *The science behind the role these insecticides play in harming bees*. Xerces Society for Invertebrate Conservation.

Hou, K., Lu, C., Shi, B., Xiao, Z., Wang, X., Zhang, J., ..., & Zhu, L. (2022). Evaluation of agricultural soil health after applying pyraclostrobin in wheat/maise rotation field based on the response of soil microbes. *Agriculture, Ecosystems & Environment, 340*, 108186.

Hwang, J. I., Lee, S. E., & Kim, J. E. (2017). Comparison of theoretical and experimental values for plant uptake of pesticide from soil. *PLoS ONE, 12*(2), e0172254.

Hwang, J. I., Zimmerman, A. R., & Kim, J. E. (2018). Bioconcentration factor-based management of soil pesticide residues: Endosulfan uptake by carrot and potato plants. *Science of the Total Environment, 627*, 514–522.

Ibrahim, A. M., Ahmed, A. K., Bakry, F. A., Rabei, I., & Abdel-Ghaffar, F. (2019). Toxicological impact of butralin, glyphosate-isopropylammonium and pendimethalin herbicides on physiological parameters of *Biomphalaria alexandrina* snails. *Molluscan Research, 39*(3), 224–233.

Jaiswal, D. K., Verma, J. P., & Yadav, J. (2017). Microbe induced degradation of pesticides in agricultural soils. In *Microbe-Induced Degradation of Pesticides* (pp. 167–189). Switzerland: Springer, Cham.

Jamieson, A. J., Malkocs, T., Piertney, S. B., Fujii, T., & Zhang, Z. (2017). Bioaccumulation of persistent organic pollutants in the deepest ocean fauna. *Nature Ecology and Evolution, 1*(3), 1–4.

Jiménez, K., Solano, K., Scholz, C., Redondo-López, S., & Mena, F. (2021). Early toxic effects in a central American native fish (*Parachromis dovii*) exposed to chlorpyrifos and difenoconazole. *Environmental Toxicology and Chemistry, 40*(7), 1938–1947.

Ju, C., Xu, J., Wu, X., Dong, F., Liu, X., & Zheng, Y. (2016). Effects of myclobutanil on soil microbial biomass, respiration, and soil nitrogen transformations. *Environmental Pollution, 208*, 811–820.

Kapsi, M., Tsoutsi, C., Paschalidou, A., & Albanis, T. (2019). Environmental monitoring and risk assessment of pesticide residues in surface waters of the Louros River (N.W. Greece). *Science of the Total Environment, 650*, 2188–2198. http://dx.doi.org/10.1016/j.scitotenv.2018.09.185.

Katagi, T., & Tanaka, H. (2016). Metabolism, bioaccumulation, and toxicity of pesticides in aquatic insect larvae. *Journal of Pesticide Science, 41*(2), 25–37.

Khan, S., Shahid, M., Khan, M. S., Syed, A., Bahkali, A. H., Elgorban, A. M., & Pichtel, J. (2020). Fungicide-tolerant plant growth-promoting rhizobacteria mitigate physiological disruption of white radish caused by fungicides used in the field cultivation. *International Journal of Environmental Research and Public Health, 17*(19), 7251.

Khuman, S. N., & Chakraborty, P. (2019). Air-water exchange of pesticidal persistent organic pollutants in the lower stretch of the transboundary river Ganga, India. *Chemosphere, 233*, 966–974.

Kim, K. H., Kabir, E., & Jahan, S. A. (2017). Exposure to pesticides and the associated human health effects. *Science of the Total Environment, 575*, 525–535.

Knauer, K., Homazava, N., Junghans, M., & Werner, I. (2017). The influence of particles on bioavailability and toxicity of pesticides in surface water. *Integrated Environmental Assessment and Management, 13*(4), 585–600.

Köck-Schulmeyer, M., Postigo, C., Farré, M., Barceló, D., & de Alda, M. L. (2019). Medium to highly polar pesticides in seawater: analysis and fate in coastal areas of Catalonia (NE Spain). *Chemosphere, 215*, 515–523.

Köhler, H. R., & Triebskorn, R. (2013). Wildlife ecotoxicology of pesticides: can we track effects to the population level and beyond? *Science, 341*(6147), 759–765.

Kang, Y., Zhang, R., Yu, K., Han, M., Pei, J., Chen, Z., & Wang, Y. (2018). Occurrence and bioaccumulation of organochlorine pesticides (Ocps) in the corals and plankton from a coast coral reef ecosystem in the South China Sea: Varied roles of coral tissue and mucus. 4111395. Available at SSRN: https://ssrn.com/abstract=4111395 or http://dx.doi.org/10.2139/ssrn.4111395.

Kolok, A. (2016). Modern poisons: A brief introduction to contemporary toxicology. Island Press.

Langrand, J., Blanc-Brisset, I., Boucaud-Maitre, D., Puskarczyk, E., Nisse, P., Garnier, R., & Pulce, C. (2020). Increased severity associated with tallow amine in acute glyphosate poisoning. *Clinical Toxicology, 58*(3), 201–203.

Laohaudomchok, W., Nankongnab, N., Siriruttanapruk, S., Klaimala, P., Lianchamroon, W., Ousap, P., …, & Woskie, S. (2020). Pesticide use in Thailand: Current situation, health risks, and gaps in research and policy. *Human and Ecological Risk Assessment: An International Journal, 27*(5), 1147–1169.

Leemans, M., Couderq, S., Demeneix, B., & Fini, J. B. (2019). Pesticides with potential thyroid hormone-disrupting effects: a review of recent data. *Frontiers in Endocrinology, 10*, 743.

Li, Z. (2018). Evaluation of regulatory variation and theoretical health risk for pesticide maximum residue limits in food. *Journal of Environmental Management, 219*, 153–167.

Liao, M., Liang, Z., Wu, R., Xiao, J., Gao, Q., & Cao, H. (2023). Residue behavior of cyantraniliprole and its ecological effects on *Procambarus clarkii* associated with the rice–crayfish integrated system. *Pest Management Science, 79*, 1868–1875. https://doi.org/10.1002/ps.7364.

Lu, X. M., & Lu, P. Z. (2018). Response of microbial communities to pesticide residues in soil restored with Azolla imbricata. *Applied Microbiology and Biotechnology, 102*, 475–484.

Lushchak, V. I., Matviishyn, T. M., Husak, V. V., Storey, J. M., & Storey, K. B. (2018). Pesticide toxicity: A mechanistic approach. *EXCLI Journal, 17*, 1101.

Lyu, L., Fang, K., Zhu, Z., Li, J., Chen, Y., Wang, L., …, & Zhang, S. (2023). Bioaccumulation of emerging persistent organic pollutants in the deep-sea cold seep ecosystems: Evidence from chlorinated paraffin. *Journal of Hazardous Materials, 445*, 130472.

Mali, H., Shah, C., Raghunandan, B. H., Prajapati, A. S., Patel, D. H., Trivedi, U., & Subramanian, R. B. (2022). Organophosphate pesticides an emerging environmental contaminant: Pollution, toxicity, bioremediation progress, and remaining challenges. *Journal of Environmental Sciences, 127*, 234–250.

Mandal, A., Sarkar, B., Mandal, S., Vithanage, M., Patra, A. K., & Manna, M. C. (2020). Impact of agrochemicals on soil health. In *Agrochemicals detection, treatment and remediation* (pp. 161–187). Butterworth-Heinemann.

Marie, L., Sylvain, P., Benoit, G., Maurice, M., & Gwenaël, I. (2017). Degradation and transport of the chiral herbicide s-metolachlor at the catchment scale: Combining observation scales and analytical approaches. *Environmental Science & Technology, 51*(22), 13231–13240.

Martyniuk, C. J., Mehinto, A. C., & Denslow, N. D (2020). Organochlorine pesticides: Agrochemicals with potent endocrine-disrupting properties in fish. *Molecular and Cellular Endocrinology, 507*, 110764.

Meena, R. S., Kumar, S., Datta, R., Lal, R., Vijayakumar, V., Brtnicky, M., …, & Marfo, T. D. (2020). Impact of agrochemicals on soil microbiota and management: A review. *Land, 9*(2), 34.

Melanda, V. S., Galiciolli, M. E. A., Lima, L. S., Figueiredo, B. C., & Oliveira, C. S. (2022). Impact of pesticides on cancer and congenital malformation: A systematic review. *Toxics, 10*(11), 676.

Meshram, A. T., Vanalkar, A. V., Kalambe, K. B., & Badar, A. M. (2022). Pesticide spraying robot for precision agriculture: A categorical literature review and future trends. *Journal of Field Robotics, 39*(2), 153–171.

Metcalfe, C. D., Helm, P., Paterson, G., Kaltenecker, G., Murray, C., Nowierski, M., & Sultana, T. (2019). Pesticides related to land use in watersheds of the Great Lakes basin. *Science of the Total Environment, 648*, 681–692.

Mishra, A. K., Arya, R., Tyagi, R., Grover, D., Mishra, J., Vimal, S. R., Mishra, S., & Sharma, S. (2021). Non-judicious use of pesticides indicating potential threat to sustainable agriculture. In *Sustainable agriculture reviews: 50* (pp. 383–400). Springer, Cham.

Mitiku, B. A., & Mitiku, M. A. (2022). Organochlorine pesticides residue affinity in fish muscle and their public health risks in North West Ethiopia. *Food Science & Nutrition, 10*, 4331–4338.

Montiel-León, J. M., Munoz, G., Duy, S. V., Do, D. T., Vaudreuil, M. A., Goeury, K., …, & Sauvé, S. (2019). Widespread occurrence and spatial distribution of glyphosate, atrazine, and neonicotinoids pesticides in the St. Lawrence and tributary rivers. *Environmental Pollution, 250*, 29–39.

Moyo, F., Tandlich, R., Wilhelmi, B. S., & Balaz, S. (2014). Sorption of hydrophobic organic compounds on natural sorbents and organoclays from aqueous and non-aqueous solutions: A mini-review. *International Journal of Environmental Research and Public Health, 11*(5), 5020–5048.

Mukherjee, S. (2022). Impact of pesticides and fertilisers on soil microbial diversity. In *Current topics in soil science: An environmental approach* (pp. 209–214). Springer International Publishing.

Mutshekwa, T., Cuthbert, R. N., Mugwedi, L., Wasserman, R. J., Dondofema, F., & Dalu, T. (2022). Behavioural responses and mortality of mozambique tilapia *Oreochromis mossambicus* to three commonly used macadamia plantation pesticides. *Water, 14*(8), 1257.

Nagy, K., Duca, R. C., Lovas, S., Creta, M., Scheepers, P. T., Godderis, L., & Ádám, B. (2020). Systematic review of comparative studies assessing the toxicity of pesticide active ingredients and their product formulations. *Environmental Research, 181*, 108926.

Nerozzi, C., Recuero, S., Galeati, G., Bucci, D., Spinaci, M., & Yeste, M. (2020). Effects of Roundup and its main component, glyphosate, upon mammalian sperm function and survival. *Scientific Reports, 10*(1), 1–9.

Ngata, T., Kim, N., Hammond, V., Dewes, A., Tapsell, P., Fraser, P., …, & Perley, C. (2018). Mountains to sea: Solving New Zealand's freshwater crisis. Bridget Williams Books.

Nicolopoulou-Stamati, P., Maipas, S., Kotampasi, C., Stamatis, P., & Hens, L. (2016). Chemical pesticides and human health: The urgent need for a new concept in agriculture. *Frontiers in Public Health, 4*, 148.

Nilsen, E., Smalling, K. L., Ahrens, L., Gros, M., Miglioranza, K. S., Picó, Y., & Schoenfuss, H. L. (2019). Critical review: grand challenges in assessing the adverse effects of contaminants of emerging concern on aquatic food webs. *Environmental Toxicology and Chemistry, 38*(1), 46–60.

Nwosu, O. K., & John, A. (2022). Chemical pesticides and food safety. In *Insecticides*. IntechOpen.

Obidike, B. M., Verla, A. W., Enyoh, C. E., Verla, E. N., & Onyekachi, O. N. (2020). Fate and distribution of pesticides residues in water sources in Nigeria: A review. *International Journal of Agrochemistry, 6*(2), 29–48.

Okoye, C. O., Okeke, E. S., Okoye, K. C., Echude, D., Andong, F. A., Chukwudozie, K. I., …, & Ezeonyeji-aku, C. D. (2022). Occurrence and fate of pharmaceuticals, personal care products (PPCPs) and pesticides in African water systems: A need for timely intervention. *Heliyon, 8*, e09143.

Olatunji, O. S. (2019). Evaluation of selected polychlorinated biphenyls (PCBs) congeners and dichlorodiphenyltrichloroethane (DDT) in fresh root and leafy vegetables using GC-MS. *Scientific Reports, 9*(1), 538.

Osaili, T. M., Al-Natour, M. Q., Al-Abboodi, A. R., Alkarasneh, A. Y., El Darra, N., Khazaal, S., & Holley, R. (2023). Detection and risk associated with organochlorine, organophosphorus, pyrethroid and carbamate pesticide residues in chicken muscle and organ meats in Jordan. *Food Control, 144*, 109355.

Pan, D., He, M., & Kong, F. (2020). Risk attitude, risk perception, and farmers' pesticide application behavior in China: A moderation and mediation model. *Journal of Cleaner Production, 276*, 124241.

Paul, M., Hossain, M. S., Rahman, M. M., Khaliq, Q. A., & Rahman, S. (2016). Chemodynamics of cypermethrin insecticide in summer country bean ecosystem in Bangladesh. *Research Journal of Environmental Toxicology, 10*(1), 50.

Peng, Y., Fang, W., Krauss, M., Brack, W., Wang, Z., Li, F., & Zhang, X. (2018). Screening hundreds of emerging organic pollutants (EOPs) in surface water from the Yangtze River Delta (YRD): occurrence, distribution, ecological risk. *Environmental Pollution, 241*, 484–493.

Pérez-Lucas, G., Vela, N., El Aatik, A., & Navarro, S. (2019). Environmental risk of groundwater pollution by pesticide leaching through the soil profile. Pesticides-use and misuse and their impact in the environment. Larramendy, M., Soloneski, S. (Eds.), IntechOpen, London, UK, pp. 1–15.

Prajapati, S., Challis, J. K., Jardine, T. D., & Brinkmann, M. (2022). Levels of pesticides and trace metals in water, sediment, and fish of a large, agriculturally-dominated river. *Chemosphere, 308*, 136236.

Pullagurala, V. L. R., Rawat, S., Adisa, I. O., Hernandez-Viezcas, J. A., Peralta-Videa, J. R., & Gardea-Torresdey, J. L. (2018). Plant uptake and translocation of contaminants of emerging concern in soil. *Science of the Total Environment, 636*, 1585–1596.

Qu, C., Albanese, S., Lima, A., Hope, D., Pond, P., Fortelli, A., ..., & De Vivo, B. (2019). The occurrence of OCPs, PCBs, and PAHs in the soil, air, and bulk deposition of the Naples metropolitan area, southern Italy: Implications for sources and environmental processes. *Environment International, 124*, 89–97.

Rajawat, N. K., Bhardwaj, K., & Mathur, N. (2023). Risk of Parkinson disease associated with pesticide exposure and protection by probiotics. *Materials Today: Proceedings, 69*, A1–A11.

Rani, L., Thapa, K., Kanojia, N., Sharma, N., Singh, S., Grewal, A. S., ..., & Kaushal, J. (2021). An extensive review on the consequences of chemical pesticides on human health and environment. *Journal of Cleaner Production, 283*, 124657.

Razali, A., Ismail, S., S. , N., Awang, S., Praveena, S. M., & Zainal Abidin, E. (2018). Land use change in highland area and its impact on river water quality: A review of case studies in Malaysia. *Ecological Processes,* 7(1), 1–17.

Redondo-López, S., León, A. C., Jiménez, K., Solano, K., Blanco-Peņa, K., & Mena, F. (2022). Transient exposure to sublethal concentrations of a pesticide mixture (chlorpyrifos–difenoconazole) caused different responses in fish species from different trophic levels of the same community. *Comparative Biochemistry and Physiology Part C: Toxicology & Pharmacology, 251*, 109208.

Reeves, W. R., McGuire, M. K., Stokes, M., & Vicini, J. L. (2019). Assessing the safety of pesticides in food: How current regulations protect human health. *Advances in Nutrition, 10*(1), 80–88.

Refaie, A. A., Shalby, A. B., Kassem, S. M., & Khalil, W. K. (2021). DNA damage and expression profile of genes associated with nephrotoxicity induced by butralin and ameliorating effect of Arabic gum in female rats. *Applied Biochemistry and Biotechnology, 193*(11), 3454–3468.

Resende-Silva, G. A., Bravim, J. N., Haro, M. M., Cutler, G. C., Silva, A. A., & Guedes, R. N. C. (2023). Imidacloprid seed treatment in soybean-associated arthropod food webs: Reason for concern, or justifiable neglect? *Journal of Pest Science, 96*(1), 129–139.

Ribeiro, Y. M., Moreira, D. P., Weber, A. A., Sales, C. F., Melo, R. M. C., Bazzoli, N., ..., & Paschoalini, A. L. (2022). Adverse effects of herbicides in freshwater Neotropical fish: A review. *Aquatic Toxicology*, 106293.

Robinson, D. E., Mansingh, A., & Dasgupta, T. P. (1999). Fate and transport of ethoprophos in the Jamaican environment. *Science of the Total Environment, 237*, 373–378.

Ruuskanen, S., Fuchs, B., Nissinen, R., Puigbò, P., Rainio, M., Saikkonen, K., & Helander, M. (2022). Ecosystem consequences of herbicides: The role of microbiome. *Trends in Ecology & Evolution, 38*, 35–43.

Saha, A., Bhaduri, D., Pipariya, A., & Jain, N. K. (2016). Influence of imazethapyr and quizalofop-p-ethyl application on microbial biomass and enzymatic activity in peanut grown soil. *Environmental Science and Pollution Research, 23*, 23758–23771.

Salihu, S. O., & Iyya, Z. (2022). Assessment of physicochemical parameters and organochlorine pesticide residues in selected vegetable farmlands soil in Zamfara state, Nigeria. *Science Progress and Research (SPR), 2*(2), 559–566. https://doi.org/10.52152/spr/2022.171.

Sarker, A., Islam, T., Rahman, S., Nandi, R., & Kim, J. E. (2021). Uncertainty of pesticides in foodstuffs, associated environmental and health risks to humans – a critical case of Bangladesh with respect to global food policy. *Environmental Science and Pollution Research, 28*(39), 54448–54465.

Sebiomo, A., Ogundero, V. W., & Bankole, S. A. (2011). Effect of four herbicides on microbial population, soil organic matter and dehydrogenase activity. *African Journal of Biotechnology, 10*(5), 770–778.

Serrão, J. E., Plata-Rueda, A., Martínez, L. C., & Zanuncio, J. C. (2022). Side-effects of pesticides on non-target insects in agriculture: A mini-review. *The Science of Nature, 109*(2), 17.

Shahid, M., & Khan, M. S. (2022). Ecotoxicological implications of residual pesticides to beneficial soil bacteria: A review. *Pesticide Biochemistry and Physiology, 188*, 105272.

Sharma, A., Kumar, V., Kumar, R., Shahzad, B., Thukral, A. K., & Bhardwaj, R. (2018). Brassinosteroid-mediated pesticide detoxification in plants: A mini-review. *Cogent Food & Agriculture, 4*(1), 1436212.

Sharma, A., Kumar, V., Shahzad, B., Tanveer, M., Sidhu, G. P. S., Handa, N., ..., & Thukral, A. K. (2019). Worldwide pesticide usage and its impacts on ecosystem. *SN Applied Sciences, 1*, 1–16.

Sharma, A., Kumar, V., Thukral, A. K., & Bhardwaj, R. (2019). Responses of plants to pesticide toxicity: An overview. *Planta Daninha, 37*, e019184291.

Sharma, N., Khanna, K., Jasrotia, S., Kumar, D., Bhardwaj, R., & Ohri, P. (2023). Metabolites and chemical agents in the plant roots: an overview of their use in plant-parasitic nematode management. *Nematology, 25*, 243–257. https://doi.org/10.1163/15685411-bja10220.

Sharma, N., Kour, S., Kumar, D., Kaur, R., Khajuria, A., & Ohri, P. (2022). Role of brassinosteroids (BRs) in modulating antioxidative defense mechanism in plants growing under abiotic and biotic stress conditions. In T. Aftab, & K. R. Hakeem (Eds.), *Antioxidant defense in plants* (pp. 325–367). Springer. https://doi.org/10.1007/978-981-16-7981-0_15.

Shenashen, M. A., Emran, M. Y., El Sabagh, A., Selim, M. M., Elmarakbi, A., & El-Safty, S. A. (2022). Progress in sensory devices of pesticides, pathogens, coronavirus, and chemical additives and hazards in food assessment: Food safety concerns. *Progress in Materials Science, 124*, 100866.

Siddique, M. H., Muzammil, S., Nadeem, H., Rasul, I., Sabri, A. N., Afzal, M., & Hayat, S. (2022). Mitigation of malathion induced toxicity in developing chicken embryo using *Zingiber officinale* and *Curcuma longa* extracts. *Pakistan Journal of Agricultural Sciences, 59*(5), 811–815.

Sikorski, Ł., Baciak, M., Bęś, A., & Adomas, B. (2019). The effects of glyphosate-based herbicide formulations on *Lemna minor*, a non-target species. *Aquatic Toxicology, 209*, 70–80.

Smith, R., Middlebrook, R., Turner, R., Huggins, R., Vardy, S., & Warne, M. (2012). Large-scale pesticide monitoring across Great Barrier Reef catchments–paddock to reef integrated monitoring, modelling and reporting program. *Marine Pollution Bulletin, 65*(4-9), 117–127.

Socorro, J., Durand, A., Temime-Roussel, B., Gligorovski, S., Wortham, H., & Quivet, E. (2016). The persistence of pesticides in atmospheric particulate phase: An emerging air quality issue. *Scientific Reports, 6*(1), 1–7.

Sombatsawat, E., Barr, D. B., Panuwet, P., Robson, M. G., & Siriwong, W. (2022). Pesticide toxicity assessment and geographic information system (GIS) application in small-scale rice farming operations, Thailand. *Scientific Reports, 12*(1), 499.

Sonal, G., Ahmad, B. R., Saoca, C., Giannetto, C., Piccione, G., & Fazio, F. (2021). Histo-morphological alterations in testis of goat (Capra hircus) induced by atrazine in vitro: Evaluation of ameliorating effect of vitamin E. *Large Animal Review, 27*(5), 279–283.

Spataro, F., Patrolecco, L., Ademollo, N., Præbel, K., Rauseo, J., Pescatore, T., & Corsolini, S. (2021). Multiple exposure of the *Boreogadus saida* from bessel fjord (NE Greenland) to legacy and emerging pollutants. *Chemosphere, 279*, 130477.

Stejskal, V., Vendl, T., Aulicky, R., & Athanassiou, C. (2021). Synthetic and natural insecticides: Gas, liquid, gel and solid formulations for stored-product and food-industry pest control. *Insects, 12*(7), 590.

Stone, W. W., Gilliom, R. J., & Ryberg, K. R. (2014). Pesticides in US streams and rivers: Occurrence and trends during 1992–2011. *Environmental Science & Technology, 48*, 11025–11030.

Straathof, H.J.M. (1986). Investigations on the phytotoxic relevance of volatilization of herbicides. *Mededelingen van de Faculteit Landbouwwetenschappen. Rijksuniversiteit Gent, 51*(2a), 433–438.

Streletskii, R., Astaykina, A., Krasnov, G., & Gorbatov, V. (2022). Changes in bacterial and fungal community of soil under treatment of pesticides. *Agronomy, 12*(1), 124.

Sultana, J., Syed, J. H., Mahmood, A., Ali, U., Rehman, M. Y. A., Malik, R. N., ..., & Zhang, G. (2014). Investigation of organochlorine pesticides from the Indus Basin, Pakistan: sources, air–soil exchange fluxes and risk assessment. *Science of the Total Environment, 497*, 113–122.

Sun, S., Chen, Y., Lin, Y., & An, D. (2018). Occurrence, spatial distribution, and seasonal variation of emerging trace organic pollutants in source water for Shanghai, China. *Science of the Total Environment, 639*, 1–7. http://dx.doi.org/10.1016/j.scitotenv.2018.05.089.

Tadei, R., Menezes-Oliveira, V. B., & Silva-Zacarin, E. C. (2020). Silent effect of the fungicide pyraclostrobin on the larval exposure of the non-target organism Africanized Apis mellifera and its interaction with the pathogen Nosema ceranae in adulthood. *Environmental Pollution, 267*, 115622.

Tahir, R., Ghaffar, A., Abbas, G., Turabi, T. H., Kausar, S., Xiaoxia, D., …, & Abdelgayed, S. S. (2021). Pesticide induced hematological, biochemical and genotoxic changes in fish: A review. *Agrobiological Records, 3*, 41–57.

Tan, H., Wu, Q., Hao, R., Wang, C., Zhai, J., Li, Q., …, & Wu, C. (2023). Occurrence, distribution, and driving factors of current-use pesticides in commonly cultivated crops and their potential risks to non-target organisms: A case study in Hainan, China. *Science of the Total Environment, 854*, 158640.

Tikoria, R., Kaur, A., & Ohri, P. (2022a). Potential of vermicompost extract in enhancing the biomass and bioactive components along with mitigation of Meloidogyne incognita-induced stress in tomato. *Environmental Science and Pollution Research, 29*(37), 56023–56036.

Tikoria, R., Kaur, A., & Ohri, P. (2022b). Modulation of various phytoconstituents in tomato seedling growth and *Meloidogyne incognita*: Induced stress alleviation by vermicompost application. *Environmental Science, 10*, 891195.

Tikoria, R., Sharma, N., Kour, S., Kumar, D., & Ohri, P. (2022). Vermicomposting: An effective alternative in integrated pest management. In A. P. Vig, S. S. Suthar, & J. Singh (Eds.), *Earthworm engineering and applications* (pp. 103–118). Nova Science Publishers.

Tooker, J. F., & Pearsons, K. A. (2021). Newer characters, same story: neonicotinoid insecticides disrupt food webs through direct and indirect effects. *Current Opinion in Insect Science, 46*, 50–56.

Triassi, M., Nardone, A., Giovinetti, M. C., De Rosa, E., Canzanella, S., Sarnacchiaro, P., & Montuori, P. (2019). Ecological risk and estimates of organophosphate pesticides loads into the Central Mediterranean Sea from Volturno River, the river of the "Land of Fires" area, Southern Italy. *Science of the Total Environment, 678*, 741–754.

Tsiaoussis, J., Antoniou, M. N., Koliarakis, I., Mesnage, R., Vardavas, C. I., Izotov, B. N., Psaroulaki, A., & Tsatsakis, A. (2019). Effects of single and combined toxic exposures on the gut microbiome: Current knowledge and future directions. *Toxicology Letters, 312*, 72–97.

Tudi, M., Daniel Ruan, H., Wang, L., Lyu, J., Sadler, R., Connell, D., …, & Phung, D. T. (2021). Agriculture development, pesticide application and its impact on the environment. *International Journal of Environmental Research and Public Health, 18*(3), 1112.

van de Merwe, J. P., Neale, P. A., Melvin, S. D., & Leusch, F. D. (2018). In vitro bioassays reveal that additives are significant contributors to the toxicity of commercial household pesticides. *Aquatic Toxicology, 199*, 263–268.

Veres, A., Wyckhuys, K. A., Kiss, J., Tóth, F., Burgio, G., Pons, X., …, & Furlan, L. (2020). An update of the worldwide integrated assessment (WIA) on systemic pesticides. Part 4: Alternatives in major cropping systems. *Environmental Science and Pollution Research, 27*, 29867–29899.

Vickers, N. J. (2017). Animal communication: When I'm calling you, will you answer too? *Current Biology, 27*(14), R713–R715.

Vryzas, Z. (2018). Pesticide fate in soil-sediment-water environment in relation to contamination preventing actions. *Current Opinion in Environmental Science & Health, 4*, 5–9.

Wang, J., Lin, K., Taylor, A., & Gan, J. (2018). In vitro assessment of pyrethroid bioaccessibility via particle ingestion. *Environment International, 119*, 125–132.

Wang, R., Yuan, Y., Yen, H., Grieneisen, M., Arnold, J., Wang, D., …, & Zhang, M. (2019). A review of pesticide fate and transport simulation at watershed level using SWAT: Current status and research concerns. *Science of the Total Environment, 669*, 512–526.

Wang, W., Xu, W., Zhou, K., & Xiong, Z. (2015). Research progressing of present contamination of Cd in soil and restoration method. *Wuhan University Journal of Natural Sciences, 20*(5), 430–444.

Woodrow, J. E., Gibson, K. A., & Seiber, J. N. (2019). Pesticides and related toxicants in the atmosphere. *Reviews of Environmental Contamination and Toxicology Volume, 247*, 147–196.

Xie, H., Wang, X., Chen, J., Li, X., Jia, G., Zou, Y., …, & Cui, Y. (2019). Occurrence, distribution and ecological risks of antibiotics and pesticides in coastal waters around Liaodong Peninsula, China. *Science of the Total Environment, 656*, 946–951.

Xu, M., Huang, H., Li, N., Li, F., Wang, D., & Luo, Q. (2019). Occurrence and ecological risk of pharmaceuticals and personal care products (PPCPs) and pesticides in typical surface watersheds, China. *Ecotoxicology and Environmental Safety, 175*, 289–298.

Yamamuro, M., Komuro, T., Kamiya, H., Kato, T., Hasegawa, H., & Kameda, Y. (2019). Neonicotinoids disrupt aquatic food webs and decrease fishery yields. *Science, 366*(6465), 620–623.

Yancheva, V., Georgieva, E., Velcheva, I., Iliev, I., Stoyanova, S., Vasileva, T., ..., & Nyeste, K. (2022). Assessment of the exposure of two pesticides on common carp (*Cyprinus carpio* Linnaeus, 1758): Are the prolonged biomarker responses adaptive or destructive? *Comparative Biochemistry and Physiology Part C: Toxicology & Pharmacology, 261*, 109446.

Yi, X., Zhang, C., Liu, H., Wu, R., Tian, D., Ruan, J., ..., & Ying, G. (2019). Occurrence and distribution of neonicotinoid insecticides in surface water and sediment of the Guangzhou section of the Pearl River, South China. *Environmental Pollution, 251*, 892–900.

Yüzbaşıoğlu, E., & Dalyan, E. (2019). Salicylic acid alleviates thiram toxicity by modulating antioxidant enzyme capacity and pesticide detoxification systems in the tomato (*Solanum lycopersicum* Mill.). *Plant Physiology and Biochemistry, 135*, 322–330.

Zaller, J. G., & Brühl, C. A. (2019). Non-target effects of pesticides on organisms inhabiting agroecosystems. *Frontiers in Environmental Science, 7*, 75.

Zhang, D., Ji, X., Meng, Z., Qi, W., & Qiao, K. (2019). Effects of fumigation with 1, 3-dichloropropene on soil enzyme activities and microbial communities in continuous-cropping soil. *Ecotoxicology and Environmental Safety, 169*, 730–736.

Zhang, W., Jiang, F., & Ou, J. (2011). Global pesticide consumption and pollution: With China as a focus. *Proceedings of the International Academy of Ecology and Environmental Sciences, 1*(2), 125.

Zhang, Y., Xu, J., Dong, F., Liu, X., Wu, X., & Zheng, Y. (2014). Response of microbial community to a new fungicide fluopyram in the silty-loam agricultural soil. *Ecotoxicology and Environmental Safety, 108*, 273–280.

CHAPTER 6

Methods for pesticide residue analysis: Recent advancements

Antul Kumar[a], Himani Gautam[b], Anuj Choudhary[a], Harmanjot Kaur[a], Sahil Mehta[c], V. Mohan Murali Achary[d], Bindu Yadav[e,f] and Om Prakash Narayan[g]

[a]Department of Botany, Punjab Agricultural University, Ludhiana, Punjab, India
[b]Department of Entomology, Dr. YS Parmar University of Horticulture and Forestry, Nauni Solan, Himachal Pradesh, India
[c]Department of Botany, Hansraj College, University of Delhi, Delhi, India
[d]International Centre for Genetic Engineering and Biotechnology, New Delhi, India
[e]School of Environmental Sciences, Jawaharlal Nehru University, New Delhi, India
[f]Department of Agricultural & Biological Engineering, University of Florida, Gainesville, FL, United States
[g]College of Liberal Arts and Sciences, University of Florida, Gainesville, FL, United States

Abbreviations

AITC	Allyl isothiocyanate
Ca^{2+}	Calcium ion
$CaCl_2$	Calcium chloride
CCD	Central composite design
DFA	Difluoroacetic acid
ELISA	Enzyme-linked immunosorbent test
GAP	Good agricultural practices
GC	Gas chromatography
HCH	Hexachlorocyclohexane
HILIC	Hydrophilic interaction liquid chromatography
IPM	Integrated pest management
LC	Liquid chromatography
LC-MS	Liquid chromatography/mass spectrometry
LLE	Low-temperature partitioning approach
LOD	Limit of detection
LOQ	Limit of quantitation
MIPs	Molecular imprinted polymers
MRLs	Maximum residue limits
MRMs	Multiresidue methods
MSN	Mesoporous silicon nanospheres
MSPD	Matrix solid–phase dispersion
MU	Measurement uncertainty
MWCNT	Multiwalled carbon nanotube
QuEChERS	Quick, easy, cheap, effective, rugged, and safe method
RSD	Relative standard deviation
SPE	Solid-phase extraction
SPME	Solid-phase microextraction
TFA	Trifluoroacetic acid
TLC	Thin-layer chromatography
TMB	Tetramethylbenzidine

Pesticides in the Environment: Impact, Assessment, and Remediation
DOI: https://doi.org/10.1016/B978-0-323-99427-9.00007-0

6.1 Introduction

Within agriculture, pesticides are being used to safeguard crops from pest infestations, enhance production, quality, and postharvest life (Narenderan & Babu, 2020). Pesticide use in agriculture has increased in recent years as a result of the rising population and rapid urbanization as per the "Food and Agriculture Organization Corporate Statistical Database of the United Nations" (FAO, 2019). Chemical pesticides, on the other hand, have damaged the environment, harmed crops, and posed potential health hazards to humans. The use of pesticides and monitoring of residues of applied pesticides in crops are an integral part of integrated pest management (IPM) (Sharma et al., 2021).

Monitoring of pesticide residues is an important tool for determining the exact dose of pesticide to be applied to the crops, as well as the time interval between sprays if they are administered more than once. It is also crucial in calculating the preharvest interval of the pesticide on a particular crop (Sharma et al., 2021). To reduce the pesticide residues in crops good agricultural practices (GAP) and principles of IPM should be adopted (Grimalt & Dehouck, 2016).

Farmers frequently use pesticides without following the GAP norms and requirements. Excessive use of pesticides has several negative consequences for the environment, but pesticide residues also constitute a health risk to individuals, leading to a variety of acute and chronic diseases. Furthermore, in many circumstances when pesticides are applied following GAP, pesticide residues may stay undetectable on food items, posing a risk to consumer safety. However, consumption of fruits and vegetables is supposed to help avoid many chronic diseases but pesticide residues in food can cause a variety of health problems in humans (Drouillet-Pinard et al., 2010). Hence, to avoid these adverse effects of pesticide residues on human beings, it is very crucial to maintain the pesticide levels as well as to detect pesticide residues in various food items.

Numerous regulating agencies, such as Codex Alimentarius Commission, European Union, and the Gulf Cooperation Council, are in charge of determining the maximum residue limit of different pesticides across 50 countries on different crops, and 23 countries, like India Food Safety and Standard Authority of India, have their own set of maximum residue limits (MRLs). These limits for various pesticides in various food commodities have been set by these regulatory authorities based on trade. Depending on the crop and pesticide used, MRLs typically ranged between 0.01 mg/kg and 5 mg/kg (Commission Regulation [EC] 2005; Ramadan et al., 2020). Highly selective, sensitive, and accurate analytical procedures and highly sensitive instruments are employed to measure these minute levels of pesticide residues. Too far, a variety of pesticide extraction and detection approaches have been developed, ranging from traditional techniques as well as advanced techniques for detection (Li et al., 2020).

Multiresidue methods (MRMs) can evaluate numerous pesticides at one time are needed since a large number of pesticides are used on a variety of food commodities.

The goal of this chapter is to discuss the conventional techniques including gas and liquid chromatography (LC), and advanced pretreatment with nanoparticles viz. polystyrene-coated magnetic nanoparticle, and detection including sensor development and nanotechnology methods employed in the assessment of pesticide residues across different systems. Validation of various analytical approaches used in pesticide residue analysis around the globe has also been highlighted.

6.2 Method validation

On a regular basis, several laboratories, such as European Union reference laboratories, provide novel analytical methodologies for residue analysis. Although there are numerous MRMs available; no universal method is there now which can meet the requirements of all the laboratories. The optimal procedure for the matrix to be analyzed, as well as the tools and facilities available, must be determined by the laboratory (Dubey et al., 2018). However, to employ a particular technique in any laboratory, the laboratory must meet all accrediting body standards, and all accreditation organizations demand adequate validation of the method before it can be used in a laboratory.

Validation of methods is very crucial for creating analytical quality control for data validity. The European Commission's directorate for health and food safety has established procedures for method validation in various edible items, which are being used in various regions of the world to instill confidence in pesticide residue analysis results. Method validation is the verification of a method's suitability for the intended goal using objective evidence. The goal of every analytical measurement is to produce data that is consistent, reliable, and accurate. The following are some of the most important parameters to consider when validating a method (Fig. 6.1).

6.3 Different parameters of method validation

6.3.1 Specificity/selectivity

The capacity which accurately detects the analyte response present in all conceivable sample components included in the sample matrix is termed specificity/selectivity. Formulation of a placebo, intermediate synthesis, excipients, degradation of products, and processing contaminants may be included in these sample components or matrix (Chandran & Singh, 2007).

6.3.2 Linearity

Linearity is defined as the ability of a method to get test results that are directly proportional to the amount of pesticide residue in the provided sample. Regardless of the stage of drug development, the method's linear range must be defined (Tiwari & Tiwari, 2010). To determine linearity, the International Council for Harmonization of technical

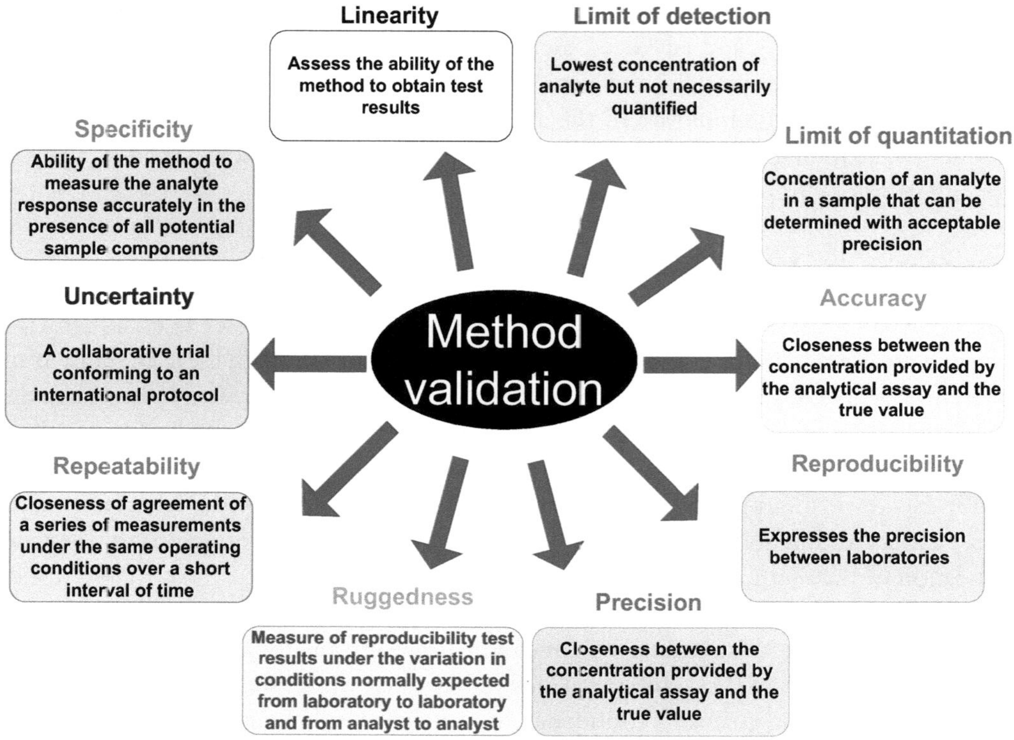

Figure 6.1 Pictorial view of various parameters of method validation during residual detection of pesticides.

requirements for pharmaceuticals for human use recommends testing a minimum of five concentrations. The upper and lower concentrations tested while studying accuracy should be bracketed by the five concentration levels (Vander et al., 2009). Response of standard solutions varies from 50% to 150% of the target concentration used to determine it. A correlation coefficient of more than 0.99 indicates that data fits the regression line nicely.

6.3.3 Method sensitivity: LOD and LOQ

6.3.3.1 Limit of detection

It is the lowest concentration of analyte which can be detected using various techniques. This detection limit method should be determined early in the method development validation process, and it should be determined again using the final procedure's unique phrasing (Chandran & Singh, 2007) (Fig. 6.2).

6.3.3.2 Limit of quantitation

Under the stipulated test conditions, it is the lowest concentration of an analyte present in a sample that can be measured with acceptable precision and accuracy. Limit of

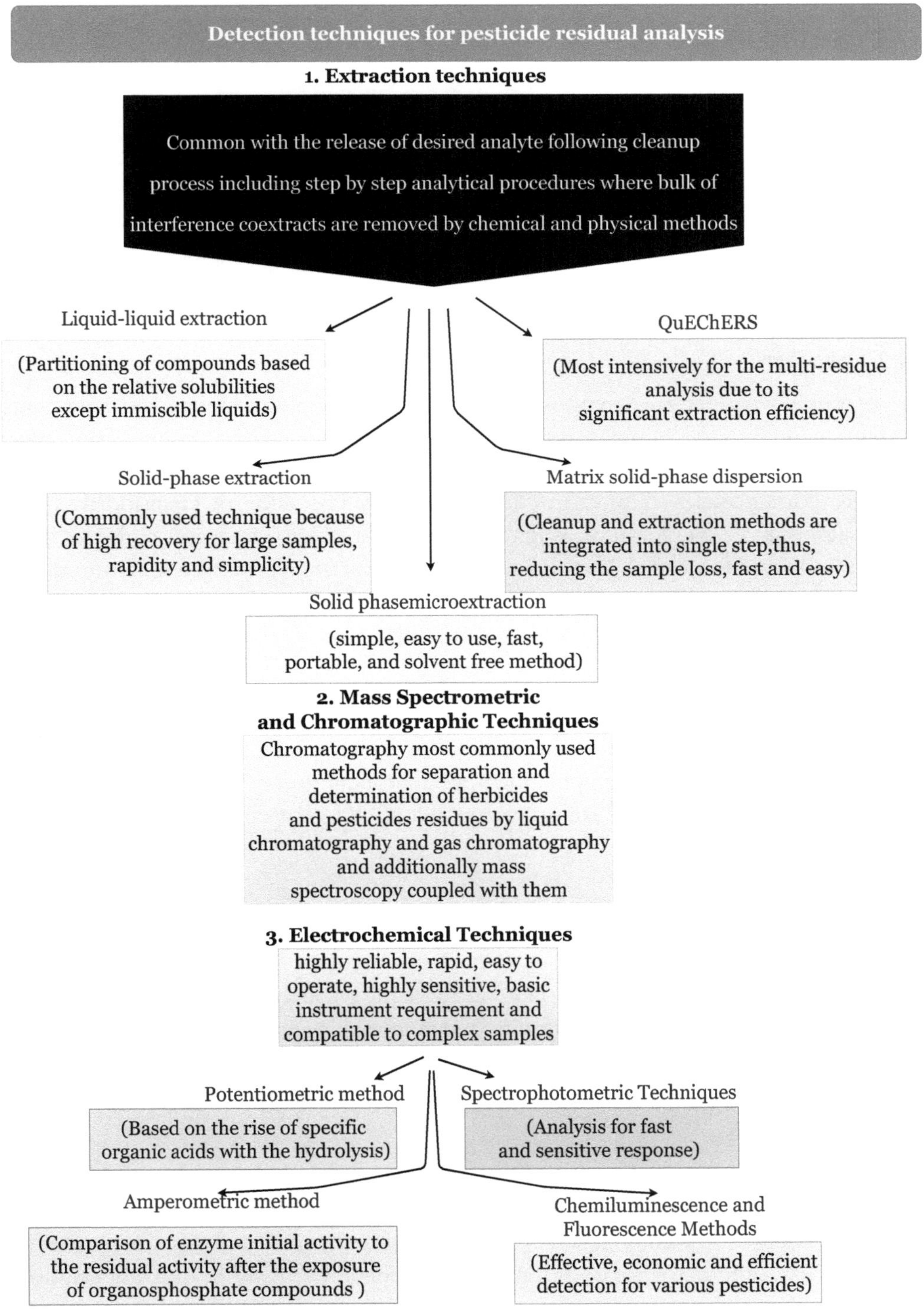

Figure 6.2 Illustration showing various detection techniques for pesticide residue analysis.

quantitation (LOQ) is determined by lowering the concentration of the analyte until it reaches a point where the method's precision becomes unacceptable (Chandran & Singh, 2007).

6.3.4 Accuracy

In the literature, the accuracy and trueness terms are used interchangeably, although having different meanings (Magnusson & Örnemark, 2015). The closeness between the concentration provided by the analytical assay and the true value is referred to as accuracy/trueness (Thompson et al., 2002).

6.3.5 Precision

The consistency of detector responses derived from multiple individual measurements of a homogeneous sample under specified conditions is defined as precision (Thompson et al., 2002). It is the agreement between the peak area values obtained from independent studies of homogeneous aliquots in chromatography. The value of precision is always positive (the sign of the deviation is ignored), and RSD near 0 indicates high precision. The method's random errors are the reasons for the variability. Although all the official guidelines only cover the first three categories of precision, in literature there are reports which cover all four types of precision (Green, 1996).

6.3.6 Repeatability

The closeness of measurements taken many times under the same conditions is referred to as repeatability. A chromatographic method's repeatability can be assessed by injecting six replicates of a single sample solution prepared at 100% test concentration (Tiwari & Tiwari, 2010).

6.3.7 Reproducibility

It is defined as the consistency of results among laboratories. An inter-laboratory trial is used to determine reproducibility. When standardizing an analytical process, for example, for inclusion in pharmacopeias, reproducibility should be taken into account. These details are not included in the dossier for marketing authorization (Lavanya et al., 2013).

6.3.8 Ruggedness

It is a measure of test results' reproducibility under the wide range of conditions that might be found from one laboratory to another and analysts to different analysts. The degree of reproducibility of test findings achieved by analyzing the same samples under different conditions, like different laboratories, reagents, temperature, analysts, instruments, time, etc. is referred to as the ruggedness of an analytical process (Lavanya et al., 2013).

6.3.9 Measurement of uncertainty

The performance features of analytical methods employed for the control of regulation limits established to foods/feeds must be examined by a joint experiment using an international protocol whenever possible and practical. Testing laboratories must use methodologies for estimating measurement uncertainty (MU) in addition to common performance parameters (accuracy, ruggedness, sensitivity, linearity, the limit of detection (LOD), and so on) (Radim et al., 2004).

For pesticide residue analysis on vine leaves, an improved SweEt/QuEChERS (quick, easy, cheap, effective, rugged, and safe) technique was developed. Deep-frozen vine leaves were cryogenically processed. After buffering with $NaHCO_3$ and adding Na_2SO_4, a 2-gram analytical fraction was then hydrated for up to a time period of 30 minutes followed by extraction with acidified ethyl acetate. Primary–secondary amine used during dispersive solid-phase clean-up. Pesticide residues were detected with the help of GC-MS/MS. The entire method was validated in fresh vine leaves for 54—59 pesticides at various concentrations (0.01, 0.02, 0.1, and 0.2 mg/kg) on *Vitis vinifera*. The RSDs for the majority of the pesticides investigated were less than 20%, while the recoveries percentage for all studied pesticides varied between 60% and 110% (Maestroni et al., 2018).

Constantinou et al. (2021) used liquid and gas chromatography (GC) mass spectrometry to validate a multiresidue approach for the simultaneous identification of a wide spectrum of pesticide residues in raisins. Ethyl acetate is used to extract pesticides from raisins, which are then centrifuged. In this study, 314 pesticides were spiked in organically grown raisins at two levels of spiking (0.025 mg/kg and 0.1 mg/kg) in GC–MS/MS and three spiking levels (0.005 mg/kg, 0.05 mg/kg, and 0.1 mg/kg) in LC-MS/MS, with six replicates at every concentration.

6.4 Detection techniques for pesticide residual analysis

6.4.1 Extraction techniques

For assessing pesticidal residues in fruits and vegetables, a variety of pretreatment and extraction procedures are used (Table 6.1) . In general, there are no approved standard methods in laboratories for extracting pesticide residues. The extraction pathway is similar, with the release of the desired analyte after a cleanup process that includes step-by-step analytical processes to remove the majority of interfering coextracts using chemical and physical methods. Initially, subsamples have been prepared from 0.5 kg to 2 kg, cleaning and follow homogenizing by the mixer. Then, a 0.5—100 g homogenized subsample was taken for further extraction. The most commonly used solvent in vegetables and fruits for analyzing residue are toluene, acetonitrile, dichloromethane, ethyl acetate, and methanol. Additionally, acetic acid and sodium hydroxide are used to neutralize the matrices for the most efficient recovery percentage (Andrade et al., 2015).

Table 6.1 Different extraction methods for the analysis of pesticide residues.

Extraction method	Techniques used	Amount of sample	Pesticides	Results	References
Solid-liquid extraction	Micro electron capture detector; Gas chromatography	5 gm	26 multiclass pesticides	Detection of α- and γ-HCH	Zhang et al. (2013)
	Mass spectrometry; high-performance liquid chromatography	2 gm	7 neonicotinoid insecticides	Detect presence of acetamiprid and imidacloprid	Yáñez et al. (2014a)
	Mass spectrometry; high-performance liquid chromatography	2 gm	Spinosyns A and spinosyns D	No pesticide residues were detected	Yáñez et al. (2014b)
	Mass spectrometry; gas chromatography	1 gm	11 multiclass pesticides	Detection of four pesticides at 4.36–89.95 μg/kg level	Li et al. (2015)
	Ultra high performance liquid and tandem mass chromatography	2 gm	3 neonicotinoid insecticides	No residues were detected	Sánchcz Hernández et al. (2016)
	Mass spectrometry; gas chromatography	2 gm	24 endocrine-disrupting chemicals	Most of the samples were detected at levels between 5.8 ng/kg and 580 ng/kg	Hejji et al. (2021)
Matrix solid-phase dispersion	Electron capture detector	4 gm	Fipronil	No residues were detected	Sánchez Brunete et al. (2008)
	Mass spectrometry; gas chromatography	0.5 gm	Organochlorine pesticides	No residues were detected	Vazquez Quintal et al. (2012)

(continued on next page)

Dispersive solid-phase extraction	High-performance liquid chromatography; tandem mass spectrometry	2 gm	80 multiclass pesticides	A total of twenty-one pesticides were detected	Wiest et al. (2011)
	Ultra-high-performance liquid chromatography	0.1 gm	20 multiclass pesticides	Detection of thirteen pesticides from 0.10–67 μg/kg levels	David et al. (2015)
	Gas chromatography; tandem mass spectrometry	2 gm	26 multiclass pesticides	Most pesticides were detect from 5.0432 μg/kg level	de Oliveira et al. (2016)
	Tandem mass spectrometry; ultra-high-performance liquid chromatography	2 gm	66 multiclass pesticides	70% pesticides residues were detected (5 μg/kg, 50 μg/kg, and 500 μg/kg)	Tong et al. (2018)

6.4.1.1 Solvent extraction and partitioning method

Except for immiscible liquids, it entails the partitioning of molecules based on their relative solubilities. It is the oldest, most versatile, compatible, and reliable with most of the instruments. Cho et al. (2013) suggested that medium polarity solvents such as ethyl acetate, hexane, and acetonitrile are often employed to remove pesticide residue from food matrices. Dinotefuran, nitenpyram, thiamethoxam, clothianidin, imidacloprid, thiacloprid, and acetamiprid insecticides were determined using the LLE method and a mixture of chloroform/dichloromethane (Wang et al., 2012). LOD values ranged from 0.002 mg/kg to 0.005 mg/kg, with an R^2 value of $\geq$ 0.99 and a recovery range of 76%–123%. Jiang et al. (2018) established a method for dipropetryn, prometon, prometryn, desmetryn, ametryn, and atraton (target contaminants of triazine herbicides). For all the analytes, the LOD value (S/N = 3) and LOQs (S/N = 10) were 0.01–0.12 ng/g and 0.04–0.35 ng/g, respectively. Limitations of this procedure include time-consuming analysis, limited polar solvent effectiveness, and the use of large hazardous solvents.

Silvaa et al. (2021) used GC to assess the residues of four pesticides viz., chlorpyrifos, bifenthrin, λ-cyhalothrin, and deltamethrin in a biological matrix using a simplified liquid–liquid extraction with a low-temperature partitioning approach (LLE/LTP). A full factorial design (2^3) with one center point was used and the approach was presented for fetal-bovine serum samples (500 μL). The amount of water injected into the sample (1000 μL), extraction solvent volume (500 μL), and the stirring duration (60 secnods) were all considered, the cleanup step was not required in this approach. Results showed that pesticide extraction from blood serum samples using the proposed miniaturized method was very effective and efficient (75%–115%). The method showed both selectivities as well as sensitiveness, and LOQ in the range from 8 μg/L to 17 μg/L. After validation, the proposed technique was tested on blood-serum samples of rural laborers in the Brazilian state of Minas Gerais. Chlorpyrifos residues were detected in two of the total analyzed samples.

6.4.1.2 Solid-phase extraction

It is the most common technique used, because of high recovery for large samples, rapidity, and simplicity. For determining the residues of neonicotinoid pesticides, a technique was developed using an SPE HLB cartridge and an Extrelut NT 20 column (Liu et al., 2010). Primary secondary amine and graphitized carbon black sorbents were employed to estimate residues in the majority of the residual estimation studies (Balinova et al., 2007; Bakırcı et al., 2014). The sorbent utilized to assess 36 pesticide residues in cauliflower and spinach was a multiwalled carbon nanotube (MWCNT) (Fan et al., 2014). LOD value was 0.1–5 μg/kg and had 57%–108% recovery rates at less than 12% relative standard deviation (RSD). Zhao et al. (2018) used dispersive solid-phase extraction (SPE) with the help of primary nanotubes and secondary nanotubes for determining the multipesticide residues in blackberries, blueberries, strawberries, and raspberries. The recovery was 71%–123%

in range and the intraday precision was less than 13%. The results of this study revealed that the proposed method is useful for analyzing residues of multiclass pesticides in natural waters.

For determining the residues of 34 multiclass chemical pesticides in natural waterways (Kapsi et al., 2020) developed an improved approach using the SPE technique. Measurement of different pesticidal residues in ambient waters was done using SPE with C18 extraction discs followed by gas and LC. This newly developed SPE method had excellent extraction efficiency and low LODs, as well as good repeatability and reproducibility. The proposed method was found effective for analyzing multiclass pesticides in natural waters.

6.4.1.3 Solid-phase microextraction

It is a very simple, fast, portable, easy-to-use, solvent-free method which is first introduced in 1990 (Arthur & Pawliszyn, 1990). SPME is based on the partitioning of analytes among matrix and a phase immobilized on the SPME fiber. A study reported the method involving headspace solid-phase microextraction (SPME) in vegetables and fruits and then analyze with a GC spectrophotometer (Abdulra'uf & Tan, 2015). Similarly, another study was conducted using the cucumber and strawberry samples to determine pesticide residue with the assistance of headspace SPME employing GC coupled with an electron capturing detector (Kin & Huat, 2010). More recently, a study has been conducted to determine the pyrethroids such as cyhalothrin, cyfluthrin, permethrin, bifenthrin, fenpropathrin, flucythrinate, deltamethrin, and fenvalerate using the SPME method with the help of GC-μECD. LODs range of this method was 0.1−0.5 ng/g and 0.3–50 ng/g of linearity range (Zhang et al., 2017).

Gondo et al. (2016) optimized HS-SPME for analyzing the organochlorine pesticides of varied polarity in some medicinal plants collected from Botswana's northern region. Analytes were isolated from the roots of five medicinal plants collected from Maun and Kasane, Botswana, using optimized SPME settings. A gas chromatograph coupled with ECD was used to determine the final analytes, and validation was performed with the help of an electron ionization mass spectrometer. *Terminalia sericea* sample, procured from Kasane, was reported to be contaminated with dieldrin only and confirmation of its presence was done with MS. Validation of the proposed method was done and recovery of analyte varied between 69.58 ± 7.20% and 113 ± 15.44%, with RSDs of 1.19%−17.97%. The method showed linearity ($R^2 > 0.9900$) in the range of 2−100 ng/g. The developed technique was found sensitive, with the values of LODs in the range from 0.48 ± 0.16 ng/g to 1.50 ± 0.50 ng/g. As a result, SPME was successfully used to analyze and extract pesticides that have varying polarities from the roots of medicinal plants.

An effective and fast method for extracting and quantifying allyl isothiocyanate (AITC) residue in cowpea beans was optimized and validated, as well as the persistence of

AITC was also studied in the grains. Headspace SPME and GC with a flame ionization detector was used. A central composite design (CCD) was used to optimize it, with the ideal circumstances for extracting AITC being attained with 15 minutes of fiber exposure at 30°C. Selectivity, linearity, LOD and LOQ, precision, and accuracy were used to evaluate the performance of the method. For AITC, the LOD and LOQ were 0.11 μg/kg and 0.33 μg/kg, respectively. (R^2) coefficient of determination was more than 0.99. With an interday coefficient of variation of lower than 9%, the relative recovery rate varied from 108.2% to 114.8%. No residues were found in the samples after 36 hours, indicating that the AITC has a low persistence and can be used safely as a bioinsecticide for grains (Ailyn et al., 2020).

6.4.1.4 Matrix solid-phase dispersion

Initially, this method was demonstrated for the extraction of semi-solid and solid samples (Barker et al., 1989). In this technique, the cleanup and extraction methods are done simultaneously in a single step, so, reducing the sample loss, fast and easy. Organophosphorus pesticides such as monocrotophos, methamidophos, methidathion, mevinphos, malathion, parathion-methyl, parathion-ethyl, ethion, and diazion are extracted from the vegetables and fruits samples (Guan et al., 2011). This method was combined with the rapid resolution LC tandem-mass spectrometry (RRLC-MS/MS) for pesticide detections. 71.2%–102.8% was recovery rates, <0.2 μg/kg of LODs value, and 2−11.8 of RSD. A method was developed using chiral LC–tandem mass spectrometry to determine the residues in soil and river sediment samples (Zhao et al., 2018). The method was based on the dispersive liquid-liquid microextraction and matrix solid-phase dispersion retreatments. The recovery rate was 87.0%−104.1% and the RSD was <10.1%. LODs and LOQs values were 0.22−1.54 ng/g and 0.91−4.00 ng/g, respectively.

A rapid, simple, and effective matrix solid-phase dispersion (MSPD) analytical technique was established and validated for determining the residues of some pesticides viz., atrazine, alachlor, bifenthrin, cypermethrin, deltamethrin, ethion, lindane, permethrin, and pirimicarb in eggs of *Gallus gallus domesticus*, requiring only minute amounts of sample and organic Florisil which were used as a dispersion agent, C18 was used as a cleaning sorbent, and acetonitrile was used as an eluant to remove pesticides from lyophilized egg matrix. GC–mass spectrometry was used to conduct the analysis (GC-MS).

Pesticide recoveries ranged from 62% to 95% at concentrations of 0.5 μg/g, 1.0 μg/g, and 1.5 μg/g, with RSDs of 0.1%−18% excluding deltamethrin having an RSD of 30%. The LOD is recorded between 0.003 and μg/g 0.015 μg/g and LOQ values were found between 0.010 and μg/g 0.050 μg/g. To ensure the interaction of insecticides with the egg matrix, lyophilization was required. This procedure was employed to analyze eight commercial egg samples procured from local sources, and bifenthrin was found in one of those samples (Michel Rubens et al., 2013).

6.4.1.5 Quick, easy, cheap, effective, rugged, and safe method

Numerous strategies have been developed and used for extracting and cleaning up different contaminants from the matrix to develop highly sensitive techniques. Although the majority of the methods were developed to achieve good recovery, these methods have certain disadvantages like a long equilibrium period, the possibility of carryover, and the matrix effect. As a result, due to its high extraction efficiency, the QuEChERS technique has been extensively investigated for multiresidue analysis (Narenderan et al., 2019; Zhang et al., 2017). This method is very easy, affordable, and compatible with other instruments like GC and LC.

It was introduced by Anastassiades and coworker (2003), the method was developed based on acetonitrile extraction and cleanup strategy using dispersive-solid phase extraction. The method has become prominent due to microscale extraction which is less time consuming, and simple. Acetonitrile is commonly used in the procedure because of its high recovery and miscibility in water (Fernández et al., 2008). Carneiro et al. (2013) have established the QuEChERS technique without involving the cleanup step. This method detected the 128 analytes with the detection limit of 5 μg/kg and the quantification limit of 10 μg/kg. This method has a 70%–120% recovery rate and <20% RSD. The acetate buffer yields a higher recovery of pesticides and depends upon pH (Chamkasem et al., 2013). The method has been modified recently by changing the solvent conditions such as ethyl acetate and methanol and facilitates better detection in gas and LC (Collimoreand Bent, 2020; Narenderan et al., 2019). With the addition of $CaCl_2$ the efficiency of the QuEChERS method and allow the sample to dehydrate by forming insoluble Ca^{2+} salts with catecholic hydroxyls. LODs ranges were 0.005−0.025 mg/kg and a recovery rate of 70%–120%.

Stringhini et al. (2021) used improved acetate QuEChERS which excluded the cleanup step for preparing samples and assessment by ultra-high-performance LC-tandem mass spectrometry to establish and validate a simple multiresidue method for the assessment of 129 pesticides and metabolites in tomato samples. The raw extract was diluted in various ratios of mobile phase, and a dilution of ten times provided appropriate results in terms of boosting analysis performance while reducing the matrix effect. Validation was done in accordance with the SANTE guidelines and yielded positive results. For the majority of the substances, LOQ was 0.01 mg/kg. For most of the chemicals and spike levels tested, recovery rates of 70%−120% were found, with a precision of ≤20%. For most of the compounds, a nonsignificant effect of the matrix was reported. Results concluded that the method was simple, robust, and effective for routine analysis.

Hrouzková and Szarka (2021) used modified QuEChERS to develop a highly effective technique for extracting 33 pesticides in textiles. After this step samples were evaluated using GC-mass spectrometry. Optimum conditions like sample amount, solvent type, and cleaning up sorbents were selected, till the validation of the technique. This method provided adequate accuracy and precision while spiked samples were injected, the value of recoveries ranged from 70% to 120% and less than 20% RSD was obtained for all pesticides evaluated. The proposed method demonstrated great separation efficiency while using little sample and sorbent.

6.4.2 Mass spectrometric and chromatographic techniques

Chromatographic techniques are the most commonly employed methods for the separation and determination of agrochemicals residues by liquid and GC. On the other hand, mass spectroscopy (MS) is coupled additionally with GC and LC to improve the detection (Yadav et al., 2022; Yadav et al., 2021). A multiresidue detection method was prepared for detecting 243 pesticides in the matrix of cardamom by using GC tandem MS (Shabeer et al., 2018). The LOQ value was 0.01 mg/kg, the recovery rate of 70%–120%, and 20% of RSD. An integrated method was established by utilizing ultra-performance LC, solid-phase extraction, and MS for detecting residues of 50 pesticides with higher sensitivity. The LOQ value was 1 ng/mL and a high linear response of 200 ng/mL (Castro et al., 2018).

Besides these, capillary electrophoresis and enzyme-linked immunosorbent assay are also being used in determining the pesticide residue in samples. The detection of organophosphate pesticides was determined with a plasmon resonance immunosensor for broad-specific monoclonal antibody development (Jiao et al., 2018). The indirect competitive ELISA results showed that monoclonal antibodies 7B2 and 2G6 had higher sensitivity to three organophosphate pesticides. CE is an effective technique because it required a lesser amount of sample and reagent with higher separation efficiency in a short time. A CE-biomimetic immunoassay coupled method to define the trichlorfon residues in the matrix of vegetables. The LOD values were 0.16–0.13 μg/L ranging with 78.8% and 103% recovery rates.

Stuke et al. (2021) prepared a new method that is based on CE-MS/MS to decrease the effect of matrix on the separation and detection of trifluoroacetic acid (TFA) and difluoroacetic acid (DFA). Before the analysis of the CE-MS/MS method, seven representative samples of fruits, grains, and vegetables were homogenized and extracted with ACN and water, proceeded by filtration. In all seven matrices examined, the CE-MS/MS method met the current legislative LOQ criteria for both TFA as well as DFA (0.01 mg/kg). For both TFA and DFA, the mean RSD determined from repeated analysis of control field samples in all the matrices was less than 10%, which met the GLP requirements.

A total of 135 beef, buffalo, and sheep meat samples were procured from Omani markets and four validated analytical methods were compared for determining the residues of organochlorines, organophosphates, anabolic steroids, and antibiotics. For the detection of pesticide residues GC was performed with various detectors, like mass spectrometer, electron capture detectors, etc. The contaminants in meat samples were also quantified with LC/mass spectrometry (LC-MS). Residues of veterinary drugs, anabolic steroids, and pesticides were evaluated using ELISA kits. Chloramphenicol, gentamicin, macrolide, sulfa-drug, and ß-lactam residues were detected in meat and meat product samples using the CHARM II instrument. To assess the amounts of veterinary medicines and anabolic steroids, a thin-layer chromatographic (TLC) approach was used. Further LC-MS was employed for crosschecking the results of TLC. These three methods allow the multiresidue analysis of anabolic steroids in meat product samples ranging from 0.06 ppm to 1.89 ppm. In red meat samples, the residues of three anabolic steroids were exceeding the MRLs. Though the amounts of organochlorine pesticides and antibiotics in meat samples were found below the MRLs, their consumption in the long run can pose health risks and will have an impact on consumers' well-being. For detecting the residues in meat and meat products, above mentioned techniques produced accurate and exact results (Al Amri et al., 2021).

Mesoporous silicon nanospheres (MSN) were used to develop a new form of enzyme-antibody conjugation that magnified the labeling signal and greatly boosted the sensitivity of an enzyme-linked immunosorbent test (ELISA) for determining residues of pesticide and veterinary drugs in the food. The conjugates were firstly obtained via layering enzyme and antibody immobilization on MSN scaffold which was used for labeling and amplification of the signal to get a sensitive colorimetric immunoassay by catalyzing the oxidation of 5, 50-Tetramethylbenzidine (TMB). When employed to quantify chloramphenicol, avermectin, tetracycline, and streptomycin in food samples, this MSN-based ELISA gave linear ranges of 0.025–25 ng/mL, 0.05—10 ng/mL, 0.025—10 ng/mL, and 0.05—25 ng/mL, respectively, with fewer LODs of 0.011 ng/mL, 0.134 ng/mL, 0.015 ng/mL and 0.106 ng/mL, respectively. In the case of avermectin, it gave a 16.7-fold reduction in the LOD compared with a typical ELISA while maintaining the specificity and accuracy of the method. This innovative immunoassay was concluded to be highly sensitive, simple, and easy which made it ideal for detecting the trace hazardous compounds in food and other high-risk products (Ji et al., 2020).

6.4.3 Electrochemical techniques

These techniques depend upon amperometric, potentiometric, and conductometric modes of measurements (Du et al., 2007; Hassani et al., 2017; Simonian et al., 2005). These methods are highly reliable, rapid, easy to operate, highly sensitive, have basic instrument requirements, and compatible with complex samples.

6.4.3.1 Potentiometric method

The potentiometric measurements are based on the rise of specific organic acids with hydrolysis (Timur & Telefoncu, 2004). A method was prepared to detect the organophosphates such as parathion-methyl, methamidophos, and malathion by inhibiting the activities of acetylcholinesterases (AChe). The enzymes were immobilized on the surface of the pH electrode with the aid of a chitosan membrane. The detection range of pesticides with efficiency was 0.1–100 mM (parathion-methyl and methamidophos) and 0.6–600 mM (Malathion). In a study, silver nano-particle coated magnetic three–dimensional graphene was employed to extract hexaconazole, fenitrothion, and chlorpyrifos (Sereshti et al., 2018).

A biosensor for diazinon pesticide analysis was established by (Mashuni et al., 2016) using a potentiometric transducer for determining the composition of glutaraldehyde cross-linked-cellulose acetate in the gold electrode, which provided the optimum performance of diazinon pesticide analysis biosensors on characterization, which includes factors like sensitivity, concentration, and detection limit (Table 6.2) . Cellulose acetate (15%) and glutaraldehyde (25%) composition achieved optimum performance in the measurement of diazinon pesticide with a working concentration range of 10−1 ppm, a sensitivity value of 20.28 mV/decade, and a detection limit ranging from 10 ppm to 6 ppm. The utilization of cellulose acetate resulted in highly sensitive devices which enable efficient residue assessment. The retention time of the electrode is roughly about 5 minutes for measuring the pesticide diazinon at concentrations ranging up to 1 ppm.

6.4.3.2 Amperometric method

The principle of this method is the comparison of the enzyme's initial activity to the residual activity after exposure to organophosphate compounds (Sun et al., 2009). A highly reproducible and super sensitive biosensor was developed consisting of electrospun chitosan/poly (vinyl alcohol). This method could bring quick detection of the pirimiphos-methyl compound found in olive oil. The detection limit is 0.2 nM (El-Moghazy et al., 2016).

6.4.4 Spectrophotometric techniques

The glyphosate can be detected by a method spectrophotometrically by flow injection analysis for a fast and sensitive response. Initially, glyphosate was converted into a photoactive compound and then measured at 405 nm. The detection limit, quantification limit and linear range were 0.20 ± 0.01 μg/mL, 0.70 ± 0.02 μg/mL, and 0.7–20 μg/mL, respectively (Jan et al., 2018). Atrazine was detected by the cloud point extraction (CPE) spectrophotometric method with the help of polysiloxane polyether (Mohd et al., 2018). LODs and LOQs were 0.09–0.59 g/L and 0.31–1.95 g/L, respectively.

6.4.5 Chemiluminescence and fluorescence methods

A method was developed based on the ZnS quantum dots (ZnSQDs) surface chemical functionalization for effective, economical, and efficient detection of various herbicides (Masteri-Farahani et al., 2018). This method wested the ZnSQDs with cysteamine hydrochloride and resulted having linear fluorescence quenching with the addition of different herbicides.

Table 6.2 Different biosensors used during pesticide residue screening.

Biosensors type	Matrix used	Analytes	Analytical method	Detection limit	References
Colorimetric	Apple	Malathion	Aptasensor employing gold nanoparticles	5.2 pM	Bala et al. (2016)
	Cabbage	Chlorpyrifosoxon, methyl paraoxon	Nanoceria coated device using an enzyme inhibition assay	0.040 mg/kg	Nouanthavong et al. (2016)
	Irrigated water	Paraoxon	Enzyme cascade along with iodine starch color reaction	10 μg/L	Guo et al. (2017)
	Water	Ethoprophos	Gold nanoparticle aggregation in combination with adenosine triphosphate	4 μM	Li et al. (2018a)
	Rice	Paraoxon	Acetylcholinesterase assay coupled to carbon dots	0.005 mg/kg	Li et al. (2018b)
	Spinach	Acetamiprid	Aptamer with DNA probe	0.022 μg/kg	Zeng et al. (2019)
	Water	Carbofuran-3-hydroxyl, carbofuran	Lateral flow immunoassay	7–10 μg/L	Lan et al. (2020)
Fluorescent	Tea	Aptasensor	Acetamiprid	0.002 mg/kg	Hu et al. (2016)
	Cabbage	Dichlorvos	Carbon dots–Cu(II) system	0.84 ng/mL	Hou et al. (2016)
	Ginger	Aldicarb	Acetylcholinesterase assay	100 μg/kg	Wang et al. (2016)
	Paraoxon	Water	BChE assay	0.25 μg/L	Wu et al. (2017)
	Cucumber	Diazinon	Aptasensor	0.13 nM	Arvand et al. (2019)
	Sweet potato	Imidacloprid	Lateral flow immunoassay	0.5 ng/g	Tan et al. (2020)

(continued on next page)

Table 6.2 Different biosensors used during pesticide residue screening—cont'd

Biosensors type	Matrix used	Analytes	Analytical method	Detection limit	References
Surface plasmon resonance biosensors	Lettuce	Chlorothalonil	Immunosensor	1 mg/kg	Hirakawa et al. (2015)
	Water	Profenofos	Fiber optic sensor	0.02 μg/L	Shrivastav et al. (2016)
	Cabbage	Parathion	Acetylcholinesterase and surface plasmon resonance	0.069 mg/L	Wu et al. (2017)
	Cucumber	Triazophos	Immunosensor	0.1 μg/kg	Guo et al. (2018)
	Corn	Mycotoxin	Surface plasmon resonance aptasensors	0.05 ng/mL	Evtugyn et al. (2021)
Surface-enhanced raman spectroscopy	Pak choi	Difenoconazole	Surface-enhanced Raman scattering	0.41 mg/kg	Huang et al. (2016)
	Apple	Paraquat	Surface-enhanced Raman scattering	100 nM	Wang et al. (2016)
	Lemon	Thiram	Surface-enhanced Raman scattering	72 ng/cm	Cui et al. (2017)
	Rice	Edifenphos	Surface-enhanced Raman scattering	0.1 mg/kg	Weng et al. (2018)
	Wheat	Prometryn	Surface-enhanced Raman scattering	20 μg/kg	Yan et al. (2019)
	Apple	Methyl parathion	Portable surface-enhanced Raman spectroscopy	0.011 μg/cm	Xie et al. (2020)
Smartphone-based method	Spinach	Chlorpyrifos	Lateral flow aptasensor	0.010 mg/kg	Cheng et al. (2018)
	Corn	Acetochlor	Lateral flow immunoassay	6.3 ng/g	Cheng et al. (2019)
	Pear	Methyl paraoxon	Nanoceria based assay	0.060 mg/kg	Wei et al. (2019)
	Cabbage	Chlorpyrifos methyl	Chemiluminescent enzyme origami paper-based biosensor	0.6 mM	Montali et al. (2020)
	Vegetable wash water	Chlorpyrifos	Lipase paper-based device	65 ng/mL	Tsagkaris et al. (2021)

Another simple method was established for the evaluation of neonicotinoid pesticides in water samples using sodium dodecyl sulfate. LOD was 0.03–0.23 μg/mL with a recovery rate of nearly 100%. The evaluated neonicotinoid pesticides include clothianidin, acetamiprid, dinotefuran, nitenpyram, imidacloprid, thiamethoxam, and thiacloprid (Ruiz-Medina et al., 2018).

The method was developed by using fluorescence polarization immunoassays in the detection of tetraconazole and thiabendazole in wheat (Boroduleva et al., 2018). This method was having fluorescence polarization equipment, fast and simple sample preparation. The LODs were 20 g/kg and 200 g/kg for thiabendazole and tetraconazole, respectively.

6.5 Instrumental detection methods

Matrix interference in the sample makes it very tough to establish a technique for pesticide residue analysis. Because of higher sensitivity, separation, and identifying abilities of gas and LC have recently become the widely used equipment for detecting and quantifying pesticides in fruits and vegetables. Furthermore, other approaches like ELISA and capillary electrophoresis have also been utilized to determine the pesticide residues in real samples (Narenderan & Babu, 2020).

6.5.1 Gas chromatography

For detecting the presence of pesticide residues GC can be used with a variety of detectors like electron capturing detectors, flame photometric detectors, nitrogen phosphorus detectors, and mass selective detectors due to their sensitivity (Bakırcı & Hışıl, 2012; Lozowicka et al., 2017). Additionally, mass detection methods coupled with analyzers like ion trap, quadrupole, triple quadrupole, and time of flight mass analyzer are used to improve method sensitivity (Cervera et al., 2012; de Oliveira et al., 2012; Tao et al., 2009; Wu, 2017). Furthermore, when selective ion monitoring (SIM) (Lima et al., 2017) or multiple reaction monitoring (Walorczyk, 2008) is employed to reduce the impact of matrix interference, the analyte mass-to-charge ratio is utilized to obtain lower LOD and LOQ with lesser interference.

Tan et al. (2021) prepared a QuEChERS-GC-MS/MS analytical technique based on the retention time of residues of 74 pesticides in the herb. This new analytical technique has two significant technical benefits: it swiftly screens pesticides in herbs without using a lot of pesticide standard solutions at first, and it aids in precisely identifying the target pesticide molecule. All chuanxiong rhizoma samples were promptly prescreened for a total of 74 chemicals. The results revealed that pesticides including chlorpyrifos, fipronil, and procymidone, were filtered out in all of the samples, and these pesticides were quantified and qualitatively assessed. Various concentrations of these pesticides studied in sample CX16 range from 0.74% to 4.15%, showing that these pesticides in the sample solutions were stable after 48 hours. These pesticides have recoveries of 95.22%, 93.03%, and 94.31%, respectively, with RSDs of less than ± 6.0%. The results of the study revealed that this novel analytical technique is very reliable and accurate. This technique is based on the use of a retention index, and the proposed technique was found very useful for analyzing multipesticide residues in herbs qualitatively and quantitatively rapidly and reliably.

Tankiewicz (2019) developed an improved QuEChERS approach for the simultaneous assessment of selected electronegative pesticides in fruits and vegetables with high water content using GC coupled with an electron capture detector. The original QuEChERS technique was improved to reduce the use of solvent in compliance with "green chemistry" principles by optimizing various parameters of extraction and cleanup. The improved method was validated for various parameters like selectivity, specificity, linearity, precision, and accuracy. For all the studied pesticides, values of LOD varied from 5.6 μg/kg to 15 μg/kg and LOQ's values were ranging from 17 μg/kg to 45 μg/kg. Results of the study revealed that the proposed method was reproducible and stable in the tested concentration range of up to 10 mg/kg, with RSDs of less than 10%. The recovery rates for spiked samples of pear at the LOQ level for each pesticide ranged from 90% to 107%, with RSDs of less than 9.6%. This developed method was evaluated on a variety of market-available fruit and vegetable samples at various seasons. This newly developed method can be used to detect and monitor pesticides not just in high-water-content fruits and vegetables, but also in samples containing considerable levels of pigments and dyes.

6.5.2 Liquid chromatography

In pesticide residues analysis, a variety of LC–based techniques are being utilized and the majority of which are equipped with diode array detectors, photodiode array, mass detectors, and ultraviolet. For multiresidue analysis, octadecyl (C18) is mostly used stationary phase for chromatographic separation, and gradient mode has been used to save runtime. Despite using LC coupled with UV, PDA, and DAD technologies, it is not easy to give structural information for the identification of pesticide residues in fruits and vegetables. As a result, utilizing tandem mass spectrometry (MS/MS), the mass detection approach has been utilized to overcome these structural interventions as well as to offer structural information from molecular masses and fragmentation patterns (Christia et al., 2015; Jallow et al., 2017).

Because of their capacity to ionize polar as well as nonpolar pesticides, ionization sources like electrospray ionization (ESI) are extensively being used in MS detection (Dzuman et al., 2015). For qualitative and quantitative evaluation, mass analyzers like the triple quadrupole and Q-Trap are utilized (Crnogorac et al., 2008; Rong et al., 2017).

Tian et al. (2016) used an improved QuEChERS approach equipped with LC-MS/MS, for detecting penflufen and one metabolite in vegetables and cereals at the same time. In addition to using LC-MS and MS/MS, a novel approach for analyzing pesticides in fruits and vegetables has been reported in recent years using ultra-performance LC due to its sensitivity and chromatographic efficiency.

Souza et al. (2021) used an improved QuEChERS method for the preparation of samples for validating a LC-tandem mass spectrometry to detect the pesticide residues in honey. Plackett Burman's design was found to help improve the recovery and lower the matrix effect during sample preparation. Extract-matched analytical curves were formed with linearity (R^2) greater than 0.99 for quantification purposes. The method's precision (RSD %) was less than 20%, and its accuracy ranged between 74% and 104%. Residues of insecticides were detected in 37.3% of the 51 honey samples tested. Imidacloprid, clothianidin, and dimethoate were the most commonly detected pesticides. Insecticide residues in honey must be monitored to ensure that consumers do not get exposed to unsafe amounts.

6.6 Advanced extraction and detection methods

Pesticide residues from fruits and vegetables have been estimated using a variety of extraction methods to date. Although the QuEChERS approach has been widely utilized for pesticide residue extraction, however, it lacks enrichment, therefore a novel improved extraction method based on polystyrene-coated magnetic nanoparticles is on the horizon that might be employed for large-scale pesticide residue analysis. This approach provides several benefits. Gas and liquid chromatographic procedures have a few drawbacks, such as being time-consuming, requiring highly skilled personnel, and requiring the use of expensive apparatus, as well as other problems. This new technology employs a sensor-based technique, which has several advantages over traditional chromatographic methods, including low cost, ease of use, operating speed, high sensitivity, and selectivity on-site detection, and reduced detection limits (Patel et al., 2019). Optical, electrochemical, piezoelectric, and molecular imprinted polymer (MIP) biosensors are commonly used detection technologies.

The use of nanomaterials in determining the residues of pesticides has also been used (Zhao et al., 2015). This method entails determining OPPs (malathion) and carbamates using graphene and gold nanoparticles (carbaryl). Detection is based on AchE immobilization by adsorption, with LODs of 4.14 pg/mL for malathion and 1.15 pg/mL for carbaryl. Another approach employing molecularly imprinted polymers (MIPs), which depends upon the function of the biological receptor, has found widespread use as a selective sensing material and in the identification of compounds with higher molecular weight (Liu et al., 2015).

Tan et al. (2015) developed a novel MIP-based lab-on-paper device that detects chemiluminescence. With the LOD value of 0.8 ng/mL, the proposed method successfully identified dichlorvos with excellent selectivity. Moreover, nanotechnology-based approaches are being used to develop a quick and simple way of detecting pesticides in food commodities (Chawla et al., 2018). Kim et al. (2018) utilized a colorimetric method using acetylcholinesterase and indoxyl acetate inhibition to detect pesticides in food. However, because of their harmful impacts, the use of nanomaterials like metal and metal oxide nanoparticles in pesticide residue analysis is a big issue (Rawtani et al., 2018). Besides the benefits offered, the main disadvantage of these detection methods was the lesser number of pesticides detected, making the method more vulnerable.

6.7 Residue analysis in various systems

6.7.1 Water

Rajput et al. (2018) performed a multiresidue evaluation of pesticides in water by using the reverse phase high-performance LC technique. They prepared and validated a quick and concurrent approach for identifying and quantifying residues of pesticides in water samples. The devised method's sensitivity and competency for identifying and quantifying pesticide residues in water samples were demonstrated by the method validation parameters. 40.02% water samples were detected with the residues of multiple pesticidal compounds and carbofuran was found in 18.18% of samples and it is reported to be the most abundant pesticide.

Lundqvist et al. (2019) investigated the toxicity of 157 water samples through in vitro bioassays in the water samples following liquid–liquid extraction and analyzed them for 129 pesticides and their transformation products. They detected a total of 51 pesticides and metabolites in water samples. The maximum net pesticide concentration in an individual sample and maximum median net concentration at a sample site was 39 μg/L and 1.1 μg/L, respectively. 31 pesticides were found in a single sample.

In another study, Hasanuzzaman et al. (2018) tested 40 water samples through HPLC along with a UV detector. In eight samples diazinon was detected at a concentration in the range of 4.11−257.91 μg/L. Malathion and chlorpyrifos residues were detected in one sample each at a concentration of 84.64 μg/L and 37.3 μg/L, respectively. Water samples collected from the fish pond were found contaminated with pesticides residues above the acceptable level.

Abdullah et al. (2017) extracted seventeen organochlorine and organophosphate pesticides from water employing an RP-C18 cartridge. They used GC equipped with an electron capture detector for analyzing all of the pesticides. The recovery experiments were performed at two concentrations of 20 μg/L and 50 μg/L and recoveries of most of the pesticides ranged from 74.2% to 116.4%. For extraction of pesticides, they used a

1000 mL sample, 10 mL of methanol and free organic water (1:1) for conditioning, and 10 mL of hexane and acetone (1:1) for extraction at the flow rate of 3 mL/min. LODs and LOQs ranged from 0.01 µg/L to 0.088 µg/L and 0.035 µg/L to 0.290 µg/L, respectively. The value of RSD % was found in the range of 1.25–6.80. With good correlation values (R^2) of 0.992–0.998, a calibration curve was obtained for the concentrations ranging from 50 g/L to 500 g/L.

6.7.2 Soil

Due to the greater affinity of organic compounds with soil, soil functions as a storage house for agrochemicals. The direct exposure of soil organisms to pesticides, as well as the increased risk to other higher organisms via nutrition, can have serious consequences on the soil ecology, water bodies, plants, and human health.

Patyal et al. (2018) concluded that the residues of spirotetramat and its metabolites in apple tree basin soil were found below the determination level (0.05 mg/kg) at harvest after two applications of spirotetramat at 20 days intervals @ 150 g (standard dose) and 300 g (double dose) g a.i./ha. Kaushik et al. (2019) experimented to evaluate the residue and dissipation behavior of tetraniliprole on tomato field soil. Tetraniliprole was sprayed at the application rate of 60 g a.i./ha and 120 g a.i./ha on tomato crop. Initial deposits of 0.092 mg/kg and 0.177 mg/kg were detected in the soil, respectively, at the application rate of 60 g a.i./ha and 120 g a.i./ha.

Riedo et al. (2021) evaluated 100 fields under organic and traditional management with an analytical procedure including 46 pesticide compounds. Pesticide residues were detected in all of the sites which included 40 organic fields. The number of pesticide residues was twice and the concentration of pesticide residues was nine times more in traditional fields in comparison to organically managed fields. With the passage of time and the duration of organic agriculture, pesticide number and concentrations significantly declined. Up to 16 various pesticide residues were present in the organic fields even after 20 years of adoption of organic agriculture.

To monitor the residues of the pesticide in Korean rice paddy soils, tandem mass spectrometry with gas chromatographic separation or ultra-performance liquid chromatographic separation was employed. Multiresidue pesticide analysis of 342 chemicals was done utilizing GC–MS/MS and LC-MS/MS, and selective MRM was also utilized. For developing an effective pretreatment method, the QuEChERS approach was combined with a dSPE clean-up. LOD and LOQ were fixed for all the selected pesticides, and method validation was carried out in the untreated soil sample for linearity and recovery at 10 mg/kg and 50 mg/kg. Except for ametoctradin and gibberellic acid, all pesticides met the acceptable recovery range of 70%–120%, with RSD values of less than 20%. Tricyclazole was the most often detected pesticide in paddy soil studies, followed by oxadiazon, endosulfan, and chlorantraniliprole.

6.7.3 Food

Negussie et al. (2021) evaluated the residues of multiclass pesticides *viz.*, fungicides, herbicides, carbamate, organophosphate, organochlorine insecticides, and pyrethroid insecticides through acetone-based salting-out assisted detection along with liquid–liquid microextraction method. The coefficient of determination achieved for all the studied pesticides was $R^2 \geq 0.9964$. LOD and LOQ were observed to be between 2.1 μg/kg and 4.5 μg/kg and 5.7 μg/kg and 12.9 μg/kg. The method's LOQ was found less than the European Union's MRLs. Recovery rates ranged from 86.8% to 109.5%. Precision repeatability and reproducibility varied from 2.9% to 8.0% and 4.9% to 9.5%, respectively. Pesticides were then quantified in tomato, pear, apple, and melon samples procured from various marketplaces using this refined and verified approach. The proposed method was concluded to be effective for the extraction of sample and pre concentration of target analytes from fruits and vegetables.

Mekonnen et al. (2021) evaluated the residues of organochlorine and organophosphate pesticides in cereal crops in Ethiopia. Samples extraction and clean–up were performed using the QuEChERS approach. Pesticide residues were then assessed by injecting the extracted samples into a GC-tandem mass spectrometer. Maximum residues of endosulfan sulfate were reported in rice, followed by dimethoate and p,p′−DDE in sorghum and common millet, respectively. The highest concentration of pesticide DDT was observed in common millet, aldrin, and dimethoate in sorghum. Moreover, aldrin residues were detected above codex and European Union maximum residual limits in rice, sorghum, and common millet samples.

Dubey et al. (2018) validated the QuEChERS process for extracting organochlorines as well as synthetic pyrethroids in different fruits and vegetables. After that samples were injected into GC equipped with an electron capture detector for further quantification of residues. Method validation was performed in line with SANCO's guidelines. Linearity, specificity, repeatability, reproducibility, and ruggedness are among the various parameters studied with the percentage of RSD less than 10% was obtained. Organochlorines and synthetic pyrethroids have LOQs of 0.01 mg/kg and 0.05 mg/kg, respectively. For all of the evaluated chemicals, MU ranged from 1% to 10%. To compensate for the matrix influence on chemical quantification, the matrix-match calibration was done. The recovery rate of the method was found in the range of 80% to 120%. The findings of this study indicated the usefulness and acceptability of the proposed method in the laboratory's routine pesticide residue estimate of these 20 pesticides in fruits and vegetables.

Kaushik et al. (2019) investigated the persistence and dissipation behavior of tetraniliprole applied at the application rate of 60 g a.i./ha and 120 g a.i./ha on fruits of tomato. Residues of tetraniliprole and its metabolite chinazolinon were determined using HPLC coupled with the RP C18 column and a PDA detector. Tetraniliprole's initial deposits were 0.865 mg/kg and 1.747 mg/kg on tomato fruits following its two applications at a respective single (60 g a.i./ha) and double dose (120 g a.i./ha). The residual half-life of tetraniliprole on tomato fruits was calculated to be 2.7 days and 3.49 days at respective single and double doses. From the standpoint of consumer safety, 15 days waiting period for tomato fruit is recommended.

Cutillas and Fernández-Alba (2021) evaluated a novel hybrid stationary phase with a mixed mode between hydrophilic interaction liquid chromatography (HILIC) and ion exchange interactions in their method. The retention modes are easily changeable and are depending on the mobile phase conditions. The performance of ten anionic chemicals was tested in hybrid stationary phase columns of various diameters. In four distinct matrices (tomato, orange, onion, and quince), the developed method was validated for sensitivity, linearity, and matrix effects with improved robustness of retention time.

Rahman et al. (2021) used QuEChERS extraction to determine the residues of 17 organochlorine, 5 pyrethroid, and 3 organophosphate pesticides in fish feed (77), fish (112), and vegetable samples (135) collected from various parts of Bangladesh. After performing QuEChERS samples were injected into GC coupled with mass spectrometry for further analysis. Residues of pyrethroids were not found in any of the fish feed samples evaluated using this approach. From all vegetable samples evaluated (country bean, green chili, tomato, eggplants, and red amaranth), 27.4% of samples contained residues of organophosphate pesticides of which 89.2% samples contained residues more than codex MRLs.

Gondo et al. (2021) validated an approach for the determination of 232 pesticide residues in fruits and vegetables based on the QuEChERS-GC-MS/MS technique. They procured 83 samples of fruit and vegetables from various marketplaces. Levels of pesticide residues varied from 0.0032 ± 0.0009 mg/kg to 70.4 ± 19.4 mg/kg. 78% of samples from the total samples analyzed were detected with pesticide residues and 13% of net samples were detected with pesticide residues above the MRLs of EU/Codex. Samples of rape vegetables surpassed MRL followed by cabbage, green pepper, and tomato. Residues of chlorpyrifos, chlorfenapyr, tebuconazole, methamidophos, and cypermethrin were detected mainly.

6.8 Conclusion and future prospective

The presence of residues of chemical pesticides can harm the fauna, flora, human health, physicochemical, and biological aspects of agricultural soil in both direct as well indirect manners (Table 6.3). Pesticides can also reduce enzyme activity and hinder the microbial populations in the soil. Due to oxidative stress, pesticides can cause chlorosis, necrosis, leaf twisting, and photosynthetic dysfunction in plants. Pesticides also suppress nitrogen metabolism, increasing and lowering the activity of specific enzymes. Furthermore, leaf coloration can change, and fruits and grains may cease to grow. Pesticides usages have

Table 6.3 Toxic effects of commercially available pesticides along with their properties.

Pesticide	Class of compound	Half-life period	Effect on various organ systems	Type of cancer	Other effects	References
Parathion	Organic phosphorus	6 months	Adverse effect on the cardiac system	Breast cancer	Nausea and headache	Garcia et al. (2003)
Mancozeb	Ethylene bisdithio-carbamate	8 weeks	Impairs thyroid system	Thyroid cancer	Headache, blurred vision, nausea and skin rashes	Adamski et al. (2011)
Endosulfan	Hexachlorocy-clopentadiene	67 days	Effect urinary system and blood chemistry	Liver cancer	Asthma, vomiting and diarrhea	Khan (2012)
Glyphosate	Organo-phosphorus	197 days	Effect reproductive development	Breast cancer	Hypersalivation, anorexia, vomiting, and diarrhea	Mesnage et al. (2015)
Aldicarb	Carbamate	2 months	Suppression of immune system and effect on reproduction and development	Colon cancer	Vomiting, diarrhea, difficulty in breathing, and muscle twitching	Ahmed et al. (2016)
Monocrotophos	Organo-phosphate	7 days	Damage central nervous system and muscular contractions	Lung cancer	Blurred vision, headache, nausea, and weakness	Kibria (2016)
Carbendazim	Carbamate	25 days	Effects functioning of respiratory system	Prostate cancer	Eye problems, skin irritation and chronic bronchitis	Singh et al. (2016)
Chlorpyrifos	Organo-phosphate	120 days	Damage nervous system damage as well as endocrine disruption	Lung cancer	Vomiting, diarrhea, dyspnea, and exothalmia	Rathod and Garg (2017)
Atrazine	Triazine	200 days	Adverse effects on the muscular system and urinary system	Colorectal cancer	Abdominal pain, vomiting, diarrhea, eye irritation, and muscle spasms	Yoon et al. (2019)
Paraquat	Ammonium compound	13 years	Accelerate oxidative stress and cause Parkinson's disease	Ovarian cancer	Muscle pain, nausea, headache, fever and temporary loss of nails	Reddy et al. (2019)
Carbofuran	Benzofurans	72 days	Deteriorate neurological system and cognitive effects	Lung cancer	Headache, nausea, chest pains, and anxiety	Ali et al. (2020)

also been linked to diseases such as soft tissue sarcoma, ovarian cancer, lung cancer, asthma, endocrine disruption, and genetic harm in humans. As a result, pesticide residue monitoring in various systems is critical, and GAP and IPM principles should be followed to reduce pesticide residues in crops.

Pesticide residues can be extracted and detected using a variety of ways in various systems. Furthermore, because of the increased consumer awareness about the importance of food safety, pesticide residue analysis has improved dramatically. Validation of the analytical method is very critical in the residue analysis of pesticides in any food product. Validation of an analytical method not only improves food safety, but also instills trust in consumers, suppliers, buyers, regulators, and specifies the quality and safety of food. So, validating the analytical method is an important step. In addition to the existing standard procedures, different pretreatment, extraction, and detection methods are being prepared, which helps in reducing time for analysis, size of the sample, and matrix interferences. However, the selection of extraction and detection systems remained the same as of the classic extraction and detection system combined with mass for quantification. But these methods take more time and are costly. Advanced methods like biosensors, molecularly imprinted polymers, and nanotechnology-based systems have recently been used as better option for pesticide detection, but these methods do have some drawbacks. Hence, it will be critical in the future to address the limitations of current advanced technologies and to prepare a pocket-friendly and environmentally safe technology that can determine a vast number of pesticide residues in a single run at lower limits.

References

Abdullah, M. P., Nabhan, K. J., Al-Qaim, F. F., Ishak, A., Othman, M. R., & Afiq, W. M. (2017). Analysis of pesticide residues in water sample: Occurrence of pesticides in paddy field. *Journal of Chemical and Pharmaceutical Sciences, 10*(3), 1159–1166.

Abdulra'uf, L. B., & Tan, G. H. (2015). Chemometric approach to the optimization of HSSPME/GC–MS for the determination of multiclass pesticide residues in fruits and vegetables. *Food Chemistry, 177*, 267–273.

Adamski, Z., Krawiec, J., Markiewicz, E., Bankiet, M., Rybska, E., Ratajczak, M., Glama, M., Scrano, L., Bufo, S., Ziemnicki, K., Erdem, M., & Büyükgüzel, E. (2011). Effect of dithiocarbamate fungicide mancozeb on development, reproduction and ultrastructure of fat body of *Agrotis segetum* Moths. *Karaelmas Science and Engineering Journal, 1*, 7–16.

Ahmed, M., Mohamed, M., Kotb, M., & Ibrahim, I. Y. (2016). Effect of aldicarb pesticide on some electrical properties of rat tissues and muscle. *Romanian Journal of Biophysics, 26*, 2016.

Ailyn, S. H., Aliannis, S. H., Díaz, G., Valdés, M. L., & Piti, F. D. C. M. M. (2020, May). PROPUESTA METODOLÓGICA PARA EL DESARROLLO DE HABILIDADES BÁSICAS EN LA ASIGNATURA SISTEMA NERVIOSO, ENDOCRINO Y REPRODUCTOR. In I *Congreso Virtual de Ciencias Básicas Biomédicas de Granma*.

Al Amiri, N., & Abu Shawali, A. (2021). Talent management strategies of a public UAE hospital in the Industry 4.0 era: A qualitative analysis. *Problems and Perspectives in Management, 19*(1), 14–27.

Ali, A., Khatoon, A., Abidin, Z., Hussain, S., Kashif, M., Abbas, R., Gul, S., & Rizvi, F. (2020). The adverse effects of carbofuran are efficiently counteracted by the supplementation of star anise (*Illicium verum*) in broiler chicks. *Toxin Reviews, 40*, 1–10.

Anastassiades, M., Lehotay, S. J., Štajnbaher, D., & Schenck, F. J. (2003). Fast and easy multiresidue method employing acetonitrile extraction/partitioning and "dispersive solid-phase extraction" for the determination of pesticide residues in produce. *Journal of AOAC International, 86*, 412–431.

Andrade, G., Monteiro, S., Francisco, J., Figueiredo, L., Botelho, R., & Tornisielo, V. (2015). Liquid chromatography–electrospray ionization tandem mass spectrometry and dynamic multiple reaction monitoring method for determining multiple pesticide residues in tomato. *Food Chemistry, 175*, 57–65.

Arthur, C. L., & Pawliszyn, J. (1990). Solid phase microextraction with thermal desorption using fused silica optical fibers. *Analytical Chemistry, 62*, 2145–2148.

Arvand, M., & Mirroshandel, A. A. (2019). An efficient fluorescence resonance energy transfer system from quantum dots to grapheme oxide nano sheets: Application in a photoluminescence aptasensing probe for the sensitive detection of diazinon. *Food Chemistry, 280*, 115–122.

Bakırcı, G. T., Acay, D. B. Y., Bakırcı, F., & Ötleş, S. (2014). Pesticide residues in fruits and vegetables from the Aegean region, Turkey. *Food Chemistry, 160*, 379–392.

Bakırcı, G. T., & Hışıl, Y. (2012). Fast and simple extraction of pesticide residues in selected fruits and vegetables using tetrafluoroethane and toluene followed by ultrahigh-performance liquid chromatography/tandem mass spectrometry. *Food chemistry, 135*(3), 1901–1913.

Bala, R., Kumar, M., Bansal, K., Sharma, R. K., & Wangoo, N. (2016). Ultrasensitive aptamer biosensor for malathion detection based on cationic polymer and gold nanoparticles. *Biosensors and Bioelectronics, 85*, 445–449.

Balinova, A., Mladenova, R., & Shtereva, D. (2007). Solid-phase extraction on sorbents of different retention mechanisms followed by determination by gas chromatography– mass spectrometric and gas chromatography–electron capture detection of pesticide residues in crops. *Journal of Chromatography A, 1150*, 136–144.

Barker, S. A., Long, A. R., & Short, C. R. (1989). Isolation of drug residues from tissues by solid phase dispersion. *Journal of Chromatography A, 475*, 353–361.

Boroduleva, A. Y., Manclus, J. J., Montoya, A., & Eremin, S. A. (2018). Fluorescence polarization immunoassay for rapid screening of the pesticides thiabendazole and tetraconazole in wheat. *Analytical and Bioanalytical Chemistry, 410*, 6923–6934.

Carneiro, H. C., Tonon, R. V., Grosso, C. R., & Hubinger, M. D. (2013). Encapsulation efficiency and oxidative stability of flaxseed oil microencapsulated by spray drying using different combinations of wall materials. *Journal of Food Engineering, 115*(4), 443–451.

Castro-Gutiérrez, V., Masís-Mora, M., Carazo-Rojas, E., Mora-López, M., & Rodríguez-Rodríguez, C. E. (2018). Impact of oxytetracycline and bacterial bioaugmentation on the efficiency and microbial community structure of a pesticide-degrading biomixture. *Environmental Science and Pollution Research, 25*, 11787–11799.

Cervera, M., Portoles, T., Pitarch, E., Beltran, J., & Hernandez, F. (2012). Application of gas chromatography time-of-flight mass spectrometry for target and non-target analysis of pesticide residues in fruits and vegetables. *Journal of Chromatography A, 1244*, 168–177.

Chamkasem, N., Ollis, L. W., Harmon, T., Lee, S., & Mercer, G. (2013). Analysis of 136 pesticides in avocado using a modified QuEChERS method with LC-MS/MS and GCMS/MS. *Journal of Agricultural and Food Chemistry, 61*, 2315–2329.

Chandran, S., & Singh, R. S. P. (2007). Comparison of various international guidelines for analytical method validation. *Die Pharmazie-An International Journal of Pharmaceutical Sciences, 62*(1), 4–14.

Chawla, P., Kaushik, R., Swaraj, V. S., & Kumar, N. (2018). Organophosphorus pesticides residues in food and their colorimetric detection. *Environmental Nanotechnology, Monitoring and Management, 10*, 292–307.

Cheng, N., Shi, Q., Zhu, C., Li, S., Lin, Y., & Du, D. (2019). Pt–Ni(OH)2 nanosheets amplified two-way lateral flow immunoassays with smartphone readout for quantification of pesticides. *Biosensors and Bioelectronics, 142*, 111498.

Cheng, N., Song, Y., Fu, Q., Du, D., Luo, Y., Wang, Y., Xu, W., & Lin, Y. (2018). Aptasensor based on fluorophore-quencher nano-pair and smartphone spectrum reader for on-site quantification of multi-pesticides. *Biosensors and Bioelectronics, 117*, 75–83.

Cho, S. K., El-Aty, A. A., Park, K. H., Park, J. H., Assayed, M., Jeong, Y. M., & Shim, J. H. (2013). Simple multiresidue extraction method for the determination of fungicides and plant growth regulator in bean sprouts using low temperature partitioning and tandem mass spectrometry. *Food Chemistry, 136*, 1414–1420.

Christia, C., Bizani, E., Christophoridis, C., & Fytianos, K. (2015). Pesticide residues in fruit samples: Comparison of different QuEChERS methods using liquid chromatography–tandem mass spectrometry. *Environmental Science and Pollution Research, 22*, 13167–13178.

Collimore, W. A., & Bent, G. A. (2020). A newly modified QuEChERS method for the analysis of organochlorine and organophosphate pesticide residues in fruits and vegetables. *Environmental Monitoring and Assessment, 192*, 128.

Commission Regulation (EC). (2005). 396/2005 on maximum residue levels of pesticides in or on foods and feed of plant and animal origin. *Journal of European Union, L70/1*, 1–16.

Constantinou, M., Louca-Christodoulou, D., & Agapiou, A. (2021). Method validation for the determination of 314 pesticide residues using tandem MS systems (GC–MS/MS and LC-MS/MS) in raisins: Focus on risk exposure assessment and respective processing factors in real samples (a pilot survey). *Food Chemistry, 360*, 129964.

Crnogorac, G., Schmauder, S., & Schwack, W. (2008). Trace analysis of dithiocarbamate fungicide residues on fruits and vegetables by hydrophilic interaction liquid chromatography/tandem mass spectrometry. *Rapid Communications in Mass Spectrometry, 22*, 2539–2546.

Cui, H., Li, S., Deng, S., Chen, H., & Wang, C. (2017). Flexible, transparent, and free-standing silicon nanowire sers platform for in situ food inspection. *ACS Sensors, 2*, 386–393.

Cutillas, V., & Fernández-Alba, A. R (2021). Analysis by LC-MS/MS of polar pesticides in fruits and vegetables using new hybrid stationary phase. *MethodsX, 8*, 101306.

David, A., Botías, C., Abdul-Sada, A., Goulson, D., & Hill, E. M. (2015). Sensitive determination of mixtures of neonicotinoid and fungicide residues in pollen and single bumblebees using a scaled down QuEChERS method for exposure assessment. *Analytical and Bioanalytical Chemistry, 407*, 8151–8162.

de Oliveira, M. L. G., Madureira, F. D., Aurelio, F., Pontelo, A. P., Silva, G., Oliveira, R., & Paes, C. (2012). A multi-residue method for the determination of pesticides in high water content matrices by gas chromatography–single quadrupole mass spectrometry with electron ionisation (EI-GC/MS). *Food Additives and Contaminants: Part A, 29*, 657–664.

de Oliveira, R. C., Queiroz, S. C., da Luz, C. F. P., Porto, R. S., & Rath, S. (2016). Bee pollen as a bioindicator of environmental pesticide contamination. *Chemosphere, 163*, 525–534.

Drouillet-Pinard, P., Boisset, M., Periquet, A., Lecerf, J. M., Casse, F., & Catteau, M. (2010). Realistic approach of pesticide residues and French consumer exposure within fruit and vegetable intake. *Journal of Environmental Science and Health, Part B, 46*, 84–91.

Du, D., Huang, X., Cai, J., & Zhang, A. (2007). Amperometric detection of triazophos pesticide using acetylcholinesterase biosensor based on multiwall carbon nanotube–chitosan matrix. *Sensors and Actuators B: Chemical, 127*, 531–535.

Dubey, J. K., Patyal, S. K., & Sharma, A. (2018). Validation of QuEChERS analytical technique for organochlorines and synthetic pyrethroids in fruits and vegetables using GC-ECD. *Environmental Monitoring and Assessment, 190*, 231.

Dzuman, Z., Zachariasova, M., Veprikova, Z., Godula, M., & Hajslova, J. (2015). Multianalyte high performance liquid chromatography coupled to high resolution tandem mass spectrometry method for control of pesticide residues, mycotoxins, and pyrrolizidine alkaloids. *Analytica Chimica Acta, 863*, 29–40.

El-Moghazy, A. Y., Soliman, E. A., Ibrahim, H. Z., Marty, J. L., Istamboulie, G., & Noguer, T. (2016). Biosensor based on electrospun blended chitosan-poly (vinyl alcohol) nanofibrous enzymatically sensitized membranes for pirimiphos-methyl detection in olive oil. *Talanta, 155*, 258–264.

Evtugyn, G., Porfireva, A., Kulikova, T., & Hianik, T. (2021). Recent achievements in electrochemical and surface plasmon resonance aptasensors for mycotoxins detection. *Chemosensors, 9*, 180.

Fan, S., Zhao, P., Yu, C., Pan, C., & Li, X. (2014). Simultaneous determination of 36 pesticide residues in spinach and cauliflower by LC-MS/MS using multi-walled carbon nanotubes-based dispersive solid-phase clean-up. *Food Additives and Contaminants: Part A, 31*, 73–82.

FAO, (2019). Food and agriculture data for all FAO regional groupings. FAOSTAT. https://www.fao.org/faostat/en/#data.

Fernández Moreno, J. L., Garrido Frenich, A., Plaza Bolaños, P., & Martínez Vidal, J. L. (2008). Multiresidue method for the analysis of more than 140 pesticide residues in fruits and vegetables by gas chromatography coupled to triple quadrupole mass spectrometry. *Journal of Mass Spectrometry, 43*, 1235–1254.

García, S., Abu Qare, A., Meeker, O'C., Winifred, Borton, A., & Mohamed, B. (2003). Methyl parathion: A review of health effects. *Journal of Toxicology and Environmental Health. Part B, Critical Reviews., 6*, 185–210.

Garcia-Barros, M., Paris, F., Cordon-Cardo, C., Lyden, D., Rafii, S., Haimovitz-Friedman, A., … Kolesnick, R. (2003). Tumor response to radiotherapy regulated by endothelial cell apoptosis. *Science, 300*(5622), 1155–1159.

Gondo, T. F., Kamakamaa, M., Oatametsea, B., Samua, T., Bogopaa, J., & Keikotlhailea, B. M. (2021). Pesticide residues in fruits and vegetables from the southern part of Botswana. *Food Additives and Contaminants: Part B, 14*, 271–280. https://doi.org/10.1080/19393210.2021.1950845.

Gondo, T. T., Obuseng, V. C., Mmualefe, L. C., & Okatch, H. (2016). Employing solid phase microextraction as extraction tool for pesticide residues in traditional medicinal plants. *Journal of Analytical Methods in Chemistry, 2016*, 2890219.

Green, J. M. (1996). A practical guide to analytical method validation. *Analytical Chemistry, 68*, 305A–309A.

Grimalt, S., & Dehouck, P. (2016). Review of analytical methods for the determination of pesticide residues in grapes. *Journal of Chromatography A, 1433*, 1–23.

Guan, S. X., Yu, Z. G., Yu, H. N., Song, C. H., Song, Z. Q., & Qin, Z. (2011). Multi-walled carbon nanotubes as matrix solid-phase dispersion extraction adsorbent for simultaneous analysis of residues of nine organophosphorus pesticides in fruit and vegetables by rapid resolution LC–MS–MS. *Chromatographia, 73*, 33–41.

Guo, L., Li, Z., Chen, H., Wu, Y., Chen, L., Song, Z., & Lin, T. (2017). Colorimetric biosensor for the assay of paraoxon in environmental water samples based on the iodine-starch color reaction. *Analytica Chimica Acta, 967*, 59–63.

Guo, Y., Liu, R., Liu, Y., Xiang, D., Liu, Y., Gui, W., Li, M., & Zhu, G. (2018). A non-competitive surface plasmon resonance immunosensor for rapid detection of triazophos residue in environmental and agricultural samples. *Science of the Total Environment, 613–614*, 783–791.

Hasanuzzaman, M., Rahman, M. A., Islam, M. S., Salam, M. A., & Nabi, M. R. (2018). Pesticide residues analysis in water samples of Nagarpur and Saturia Upazila, Bangladesh. *Applied Water Science, 8*, 8.

Hassani, S., Momtaz, S., Vakhshiteh, F., Maghsoudi, A. S., Ganjali, M. R., & Norouzi, P. (2017). Biosensors and their applications in detection of organophosphoruspesticides in the environment. *Archives of Toxicology, 91*, 109–130.

Hejji, L., Azzouz, A., Colón, L. P., Souhail, B., & Ballesteros, E. (2021). A multi-residue method for determining twenty-four endocrine disrupting chemicals in vegetables and fruits using ultrasound-assisted solid-liquid extraction and continuous solid-phase extraction. *Chemosphere, 263*, 128158.

Hirakawa, Y., Yamasaki, T., Watanabe, E., Okazaki, F., Murakami-Yamaguchi, Y., Oda, M., Iwasa, S., Narita, H., & Miyake, S. (2015). Development of an immunosensor for determination of the fungicide chlorothalonil in vegetables, using surface plasmon resonance. *Journal of Agriculture Food and Chemistry, 63*, 6325–6330.

Hou, J., Dong, G., Tian, Z., Lu, J., Wang, Q., Ai, S., & Wang, M. (2016). A sensitive fluorescent sensor for selective determination of dichlorvos based on the recovered fluorescence of carbon dots-Cu(II) system. *Food Chemistry, 202*, 81–87.

Hrouzková, S., & Szarka, A. (2021). Development of a modified QuEChERS procedure for the isolation of pesticide residues from textile samples, followed by GC–MS determination. *Separations, 8*, 106.

Hu, W., Chen, Q., Li, H., Ouyang, Q., & Zhao, J. (2016). Fabricating a novel label-free aptasensor for acetamiprid by fluorescence resonance energy transfer between NH_2-$NaYF_4$: Yb, Ho@SiO_2 and Au nanoparticles. *Biosensors and Bioelectronics, 80*, 398–404.

Huang, S., Yan, W., Liu, M., & Hu, J. (2016). Detection of difenoconazole pesticides in pak choi by surface-enhanced Raman scattering spectroscopy coupled with gold nanoparticles. *Analytical Methods, 8*, 4755–4761.

Jallow, M. F., Awadh, D. G., Albaho, M. S., Devi, V. Y., & Ahmad, N. (2017). Monitoring of pesticide residues in commonly used fruits and vegetables in Kuwait. *International Journal of Environmental Research and Public Health, 14*, 833.

Jan, M. R., Shah, J., Muhammad, M., & Ara, B. (2018). Flow injection spectrophotometric determination of glyphosate herbicide in wheat grains via condensation reaction with p-dimethylaminobenzaldehyde. *South African Journal of Chemistry, 71*, 39–45.

Ji, C., Song, Q., Chen, Y., Zhou, Z., Wang, P., Liu, J., … Zhao, M. (2020). The potential endocrine disruption of pesticide transformation products (TPs): The blind spot of pesticide risk assessment. *Environment International, 137*, 105490.

Jiang, Y. X., Ma, P. Y., Li, X. P., Piao, H. L., Li, D., Sun, Y., & Song, D. Q. (2018). Application of metal-organic framework MIL-101(Cr) to microextraction in packed syringe for determination of triazine herbicides in corn samples by liquid chromatography tandem mass spectrometry. *Journal of Chromatography A, 1574*, 36–41.

Jiao, S. S., Liu, P. Y., Liu, Y., Zou, R. B., Zhao, Y., Liu, Y. H., & Guo, Y. R. (2018). Binding properties of broad-specific monoclonal antibodies against three organophosphorus pesticides by a direct surface plasmon resonance immunosensor. *Analytical and Bioanalytical Chemistry, 410*, 7263–7273.

Kapsi, M., Tsoutsi, C., & Albanis, T. (2020). Simple analytical methodology based on solid phase extraction for monitoring pesticide residues in natural waters. *MethodsX, 7*, 101011.

Kaushik, E., Dubey, J. K., Patyal, S. K., Katna, S., Chauhan, A., & Devi, N. (2019). Persistence of tetraniliprole and reduction in its residues by various culinary practices in tomato in India. *Environmental Science and Pollution Research, 26*, 22464–22471. https://doi.org/10.1007/s11356-019-04738-6.

Khan, K. H. (2012). Impact of endosulfan on living beings. *International Journal of Biosciences, 2*, 9–17.

Kibria, G. (2016). Pesticides and its impact on environment, biodiversity and human health-A short review. 5p. https://doi.org/10.13140/RG.2.1.4487.4965/1.

Kim, H. J., Kim, Y., Park, S. J., Kwon, C., & Noh, H. (2018). Development of colorimetric paper sensor for pesticide detection using competitive-inhibiting reaction. *BioChip Journal, 12*, 326–331.

Kin, C. M., & Huat, T. G. (2010). Headspace solid-phase microextraction for the evaluation of pesticide residue contents in cucumber and strawberry after washing treatment. *Food Chemistry, 123*, 760–764.

Lan, J., Sun, W., Chen, L., Zhou, H., Fan, Y., Diao, X., Wang, B., & Zhao, H. (2020). Simultaneous and rapid detection of carbofuran and 3-hydroxy-carbofuran in water samples and pesticide preparations using lateral-flow immunochromatographic assay. *Food and Agricultural Immunology, 31*, 165–175.

Lavanya, G., Sunil, M., Eswarudu, M. M., Eswaraiah, M. C., Harisudha, K., & Spandana, B. N. (2013). Analytical method validation: An updated review. *International Journal of Pharmaceutical Sciences and Research, 4*, 1280–1286.

Li, C., Begum, A., & Xue, J. (2020). Analytical methods to analyze pesticides and herbicides. *Water Environment Research, 92*, 1770–1785.

Li, H., Yan, X., Lu, G., & Su, X. (2018b). Carbon dot-based bioplatform for dual colorimetric and fluorometric sensing of organophosphate pesticides. *Sensors Actuators B Chemistry, 260*, 563–570.

Li, X., Cui, H., & Zeng, Z. (2018a). A simple colorimetric and fluorescent sensor to detect organophosphate pesticides based on adenosine triphosphate-modified gold nanoparticles. *Sensors, 18*, 4302.

Li, Y., Kelley, R. A., Anderson, T. D., & Lydy, M. J. (2015). Development and comparison of two multiresidue methods for the analysis of select pesticides in honey bees, pollen, and wax by gas chromatography–quadrupole mass spectrometry. *Talanta, 140*, 81–87.

Lima, V. G., Campos, V. P., Santana, T. C., Santana, F. O., & Costa, T. A. (2017). Determination of agrochemical multi-residues in grapes. Identification and confirmation by gas chromatography-mass spectrometry. *Analytical Methods, 9*, 5880–5889.

Liu, S., Zheng, Z., Wei, F., Ren, Y., Gui, W., Wu, H., & Zhu, G. (2010). Simultaneous determination of seven neonicotinoid pesticide residues in food by ultraperformance liquid chromatography tandem mass spectrometry. *Journal of Agricultural and Food Chemistry, 58*(6), 3271–3278.

Liu, W., Guo, Y., Luo, J., Kou, J., Zheng, H., Li, B., & Zhang, Z. (2015). A molecularly imprinted polymer based a lab-on-paper chemiluminescence device for the detection of dichlorvos. *Spectrochimica Acta Part A: Molecular and Biomolecular Spectroscopy, 141*, 51–57.

Łozowicka, B., Rutkowska, E., & Jankowska, M. (2017). Influence of QuEChERS modifications on recovery and matrix effect during the multi-residue pesticide analysis in soil by GC/MS/MS and GC/ECD/NPD. *Environmental Science and Pollution Research, 24*(8), 7124–7138.

Lundqvist, J., Brömssen, C. V., Rosenmai, A. K., Ohlsson, A., Godec, T. L., Jonsson, O., Kreuger, J., & Oskarsson, A. (2019). Assessment of pesticides in surface water samples from Swedish agricultural areas by integrated bioanalysis and chemical analysis. *Environmental Sciences Europe, 31*, 53.

Maestroni, B., Alnaser, A. A., Ghanem, I., Islam, M. D., Cesio, V., Heinzen, H., Kelly, S., & Cannavan, A. (2018). Validation of an Analytical Method for the Determination of Pesticide Residues in Vine Leaves by GC-MS/MS. *Journal of Agricultural and Food Chemistry, 66*(25), 6421–6430. https://doi.org/10.1021/acs.jafc.8b00453.

Magnusson, B., and Örnemark, U. (Eds.). (2015). EURACHE'M Guide: The fitness for purpose of analytical methods: a laboratory guide to method validation and related topics. EURACHEM, Europe. 62p.

Masteri-Farahani, M., Mahdavi, S., & Khanmohammadi, H. (2018). Chemically functionalized ZnS quantum dots as new optical nanosensor of herbicides. *Materials Research Express, 5*, 035055.

Mashuni, Ramadhan, L. O. A. N., Jahiding, M., & Herniati. (2016, January). Analysis of diazinon pesticide using potentiometric biosensor based on enzyme immobilized cellulose acetate membrane in gold electrode. In *IOP Conference Series: Materials Science and Engineering* (Vol. 107, No. 1, p. 012013). IOP Publishing.

Mekonnen, B., Siraj, J., & Negash, S. (2021). Determination of pesticide residues in food premises using QuECHERS method in Bench-Sheko Zone, Southwest Ethiopia. *BioMed Research International Volume, 2021*, 6612096.

Mesnage, R., Defarge, N., Vendômois, J., & Séralini, G. E. (2015). Potential toxic effects of glyphosate and its commercial formulations below regulatory limits. *Food and Chemical Toxicology, 84*, 133–153. https://doi.org/10.1016/j.fct.2015.08.012.

Michel, R., Reis, S., Cybelle, O., M. , T., Lima, G., Adriano, A., & Haroldo, S. D. (2013). Validation of a matrix solid phase dispersion (MSPD) technique for determination of pesticides in lyophilized eggs of the chicken *Gallus gallus domesticus*. *Microchemical Journal, 110*, 395–401.

Mohd, N. I., Raoov, M., Mohamad, S., & Zain, N. N. M. (2018). Performance evaluation of non-ionic silicone surfactants OFX 0309 and DC 193C as a new approach in cloud point extraction–spectrophotometry for determination of atrazine in water samples. *RSC Advances, 8*(24), 13556–13566.

Montali, L., Calabretta, M. M., Lopreside, A., D'Elia, M., Guardigli, M., & Michelini, E. (2020). Multienzyme chemiluminescent foldable biosensor for on-site detection of acetylcholinesterase inhibitors. *Biosensors and Bioelectronics, 162*, 112232.

Narenderan, S. T., & Babu, M. B. (2020). Review of pesticide residue analysis in fruits and vegetables. Pre-treatment, extraction and detection techniques. *Food Research International, 133*, 109141.

Narenderan, S., Meyyanathan, S., Karri, V. V. S. R., Babu, B., & Chintamaneni, P. (2019). Multivariate response surface methodology assisted modified QuEChERS extraction method for the evaluation of organophosphate pesticides in fruits and vegetables cultivated in Nilgiris, South India. *Food Chemistry, 300*, 125188.

Negussie, B. T., Dube, S., & Nindi, M. M. (2021). Multiclass pesticide residue analysis in fruit and vegetable samples by combining acetone-based salting-out assisted extraction with dispersive liquid-liquid microextraction. *Journal of Chemistry, 2021*, 6417093.

Nouanthavong, S., Nacapricha, D., Henry, C. S., & Sameenoi, Y. (2016). Pesticide analysis using nanoceria-coated paper-based devices as a detection platform. *Analyst, 141*, 1837–1846.

Patel, H., Rawtani, D., & Agrawal, Y. (2019). A newly emerging trend of chitosan-based sensing platform for the organophosphate pesticide detection using Acetylcholinesterase: A review. *Trends in Food Science and Technology, 5*, 1–5.

Patyal, S. K., Dubey, J. K., Katna, S., Chauhan, A., Devi, N., & Wani, A. A. (2018). Residue and risk assessment of spirotetramat in/on apple. *Pesticide Research Journal, 30*, 51–58.

Radim, S., Jana, H., Vladimirr, K., & Jana, T. (2004). Uncertainties of gas chromatographic measurement of troublesome pesticide residues in apples employing conventional and mass spectrometric detectors. *Analytica Chimica Acta, 520*, 245–255.

Rahman, M., Hoque, M. S., Bhowmik, S., Ferdousi, S., Kabiraz, M. P., & Brakel, M. L. V. (2021). Monitoring of pesticide residues from fish feed, fish and vegetables in Bangladesh by GC-MS using the QuEChERS method. *Heliyon, 7*, e06390.

Ramadan, M. F. A., Abdel-Hamid, M. M. A., Altorgoman, M. M. F., AlGaramah, H. A., Alawi, M. A., Shati, A. A., & Awwad, N. S. (2020). Evaluation of pesticide residues in vegetables from the asir region, Saudi Arabia. *Molecules (Basel, Switzerland), 25*, 205.

Rajput, S., Kumari, A., Arora, S., & Kaur, R. (2018). Multi-residue pesticides analysis in water samples using reverse phase high performance liquid chromatography (RP-HPLC). *MethodsX, 5*, 744–751.

Rathod, A., & Garg, R. (2017). Chlorpyrifos poisoning and its implications in human fatal cases: A forensic perspective with reference to Indian scenario. *Journal of Forensic and Legal Medicine, 47*, 29–34. https://doi.org/10.1016/j.jflm.2017.02.003.

Rawtani, D., Khatri, N., Tyagi, S., & Pandey, G. (2018). Nanotechnology-based recent approaches for sensing and remediation of pesticides. *Journal of Environmental Management, 206*, 749–762.

Reddy, K. B., Jeevanalatha, M., Lakshman, M., & Rani, M. (2019). The toxic effects of paraquat (PQ) on body weights and haematological parameters in male albino wistar rats and its amelioration with vitamin C. *International Journal of Current Microbiology and Applied Sciences, 8*, 314–320.

Riedo, J., Wettstein, F. E., Rösch, A., Herzog, C., Banerjee, S., Büchi, L., Charles, R., Wächter, D., Martin-Laurent, F., Bucheli, T. D., Walder, F., & Heijden, M. G. A. (2021). Widespread occurrence of pesticides in organically managed agricultural soils-the ghost of a conventional agricultural past? *Environmental Science and Technology, 55*, 2919–2928.

Rong, L., Wu, X., Xu, J., Dong, F., Liu, X., Pan, X., & Zheng, Y. (2017). Simultaneous determination of three pesticides and their metabolites in unprocessed foods using ultraperformance liquid chromatography-tandem mass spectrometry. *Food Additives and Contaminants: Part A, 5*, 273–281.

Ruiz-Medina, A., Soler-Gallardo, M. I., & Llorent-Martinez, E. J. (2018). Enhanced quenching effect of neonicotinoid pesticides on time-resolved terbium luminescence in presence of surfactants. *Journal of Chemistry, 2018*, 5378736. https://doi.org/10.1155/2018/5378736.

Sánchez-Brunete, C., Miguel, E., Alberto, B., & Tadeo, J. L. (2008). Determination of fipronil residues inhoney and pollen by gas chromatography. *Spanish Journal of Agricultural Research, 6*, 7–14.

Sánchez-Hernández, L., Hernández-Domínguez, D., Martín, M. T., Nozal, M. J., Higes, M., & Bernal Yagüe, J. L. (2016). Residues of neonicotinoids and their metabolites in honey and pollen from sunflower and maize seed dressing crops. *Journal of Chromatography A, 1428*, 220–227.

Sereshti, H., Afsharsaveh, Z., Gaikani, H., & Nodeh, H. R. (2018). Electroless-coated magnetic three-dimensional graphene with silver nanoparticles used for the determination of pesticides in fruit samples. *Journal of Separation Science, 41*, 1567–1575.

Shabeer, T. A., Girame, R., Utture, S., Oulkar, D., Banerjee, K., Ajay, D., ... Menon, K. R. K. (2018). Optimization of multi-residue method for targeted screening and quantitation of 243 pesticide residues in cardamom (Elettaria cardamomum) by gas chromatography tandem mass spectrometry (GC-MS/MS) analysis. *Chemosphere, 193*, 447–453.

Sharma, A., Dubey, J. K., Katna, S., Shandil, D., Brar, G. S., & Singh, S. (2021). Validation of analytical methods used for pesticide residue detection in fruits and vegetables. *Food Analytical Methods, 14*, 1919–1926. https://doi.org/10.1007/s12161-021-02027-y.

Shrivastav, A. M., Usha, S. P., & Gupta, B. D. (2016). Fiber optic profenofos sensor based on surface plasmon resonance technique and molecular imprinting. *Biosensors and Bioelectronics, 79*, 150–157.

Silvaa, T. L. R., de Queiroza, M. E. L. R., Nevesa, A. A., Vieirab, P. A. F., de Oliveiraa, A. F., & de A. Oliveirac, M. G. (2021). Miniaturized liquid-liquid extraction with low-temperature partitioning technique (LLE/LTP) for pesticide analysis in a biological matrix by gas chromatography. *Quimica Nova, 44*, 804–808.

Simonian, A. L., Good, T. A., Wang, S. S., & Wild, J. R. (2005). Nanoparticle-based optical biosensors for the direct detection of organophosphate chemical warfare agents and pesticides. *Analytica Chimica Acta, 534*, 69–77.

Singh, S., Singh, N., Kumar, V., Datta, S., Wani, A., Singh, D., Singh, K., & Singh, J. (2016). Toxicity, monitoring and biodegradation of the fungicide carbendazim. *Environmental Chemistry Letters, 14*, 317–329. https://doi.org/10.1007/s10311-016-0566-2.

Souza, A. P. F., Petrarca, M. H., Braga, P. A., Rodrigues, N. R., & Reyes, F. G. R. (2021). Analysis of insecticide residues in honey by liquid chromatography tandem mass spectrometry using QuEChERS optimized by the Plackett Burman design. *CyTA - Journal of Food, 19*, 326–332.

Stringhini, F. M., Ribeiro, L. C., Rocha, G. I., Kuntz, J. D., Zanella, R., Prestes, O. D., & Adaime, M. B. (2021). Dilution of QuEChERS extracts without cleanup improves results in the UHPLC-MS/MS multiresidue analysis of pesticides in tomato. *Food Analytical Methods, 14*, 1511–1523.

Stuke, H., Kress, E., Weilnhammer, V. A., Sterzer, P., & Schmack, K. (2021). Overly strong priors for socially meaningful visual signals are linked to psychosis proneness in healthy individuals. *Frontiers in Psychology, 12*, 583637.

Sun, X., Wang, X., & Jia, C. (2009). Study on carrier material of immobilization acetylcholinesterase for biosensor in detection of organophosphorus pesticide residues. In D. Li, & Z. Chunjiang (Eds.), *IFIP international federation for information processing, volume 295, computer and computing technologies in agriculture II: Vol. 3* (pp. 2283–2292). Springer.

Tan, P., Xu, L., Wei, X. C., Huang, H. Z., Zhang, D. K., Zeng, C. J., Geng, F. N., Bao, X. M., Hua, H., & Zhao, J. N. (2021). rapid screening and quantitative analysis of 74 pesticide residues in herb by retention index combined with GC-QQQ-MS/MS. *Journal of Analytical Methods in Chemistry Volume, 2021*, 8816854.

Tan, G., Zhao, Y., Wang, M., Chen, X., Wang, B., & Li, Q. X. (2020). Ultrasensitive quantitation of imidacloprid in vegetables by colloidal gold and time-resolved fluorescent nanobead traced lateral flow immunoassays. *Food Chemistry, 311*, 126055.

Tan, X., Hu, Q., Wu, J., Li, X., Li, P., Yu, H., & Lei, F. (2015). Electrochemical sensor based on molecularly imprinted polymer reduced graphene oxide and gold nanoparticles modified electrode for detection of carbofuran. *Sensors and Actuators B: Chemical, 220*, 216–221.

Tankiewicz, M. (2019). Determination of selected priority pesticides in high water fruits and vegetables by modified QuEChERS and GC-ECD with GC-MS/MS confirmation. *Molecules (Basel, Switzerland), 24*, 417.

Tao, C. J., Hu, J. Y., Li, J. Z., Zheng, S. S., Liu, W., & Li, C. J. (2009). Multi-residue determination of pesticides in vegetables by gas chromatography/ion trap mass spectrometry. *Bulletin of Environmental Contamination and Toxicology, 82*, 111–115.

Thompson, M., Ellison, S. L. R., & Wood, R. (2002). Guidelines for single laboratory validation of methods of analysis. *Pure and Applied Chemistry, 74*, 835–855.

Tian, F., Liu, X., Wu, Y., Xu, J., Dong, F., Wu, X., & Zheng, Y. (2016). Simultaneous determination of penflufen and one metabolite in vegetables and cereals using a modified quick, easy, cheap, effective, rugged, and safe method and liquid chromatography coupled to tandem mass spectrometry. *Food Chemistry, 213*, 410–416.

Timur, S., & Telefoncu, A. (2004). Acetylcholinesterase (AChE) electrodes based on gelatin and chitosan matrices for the pesticide detection. *Artificial Cells, Blood Substitutes, and Biotechnology, 32*, 427–442.

Tiwari, G., & Tiwari, R. (2010). Bioanalytical method validation: an updated review. *Pharmaceutical Methods, 1*, 25–38.

Tong, Z., Duan, J., Wu, Y., Liu, Q., He, Q., Shi, Y., … Cao, H. (2018). A survey of multiple pesticide residues in pollen and beebread collected in China. *Science of the Total Environment, 640–641*, 1578–1586.

Tsagkaris, A. S., Pulkrabova, J., & Hajslova, J. (2021). Optical screening methods for pesticide residue detection in food matrices. *Advances and Emerging Analytical Trends. Food, 10*, 88.

Vander, H. Y., Nijhuis, A., Verbeke, J. S., Vandeginste, B. G., & Massart, D. L. (2009). Guidance for rubustness/ruggedness test in method validation. *Journal of Pharmaceutical and Biomedical Analysis, 24*, 723–753.

Vazquez-Quintal, P. E., Muñoz-Rodríguez, D., Medina-Peralta, S., & Moguel-Ordóñez, Y. B. (2012). Extraction of organochlorine pesticides from bee pollen by matrix solid-phase dispersion: recovery evaluation by GC–MS and method validation. *Chromatographia, 75*, 923–930.

Walorczyk, S. (2008). Application of gas chromatography/tandem quadrupole mass spectrometry to the multi-residue analysis of pesticides in green leafy vegetables. *Rapid Communications in Mass Spectrometry, 22*(23), 3791–3801.

Wang, C., Wu, X., Dong, P., Chen, J., & Xiao, R. (2016). Hotspots engineering by grafting Au@Ag core-shell nanoparticles on the Au film over slightly etched nanoparticles substrate for on-site paraquat sensing. *Biosensors and Bioelectronics, 86*, 944–950.

Wang, P., Yang, X., Wang, J., Cui, J., Dong, A., Zhao, H., & Li, W. (2012). Multi-residue method for determination of seven neonicotinoid insecticides in grains using dispersive solid-phase extraction and dispersive liquid–liquid micro-extraction by high performance liquid chromatography. *Food Chemistry, 134*, 1691–1698.

Wang, X., Hou, T., Dong, S., Liu, X., & Li, F. (2016). Fluorescence biosensing strategy based on mercury ion-mediated DNA conformational switch and nicking enzyme-assisted cycling amplification for highly sensitive detection of carbamate pesticide. *Biosensors and Bioelectronics, 77*, 644–649.

Wei, J., Yang, L., Luo, M., Wang, Y., & Li, P. (2019). Nanozyme-assisted technique for dual mode detection of organophosphorus pesticide. *Ecotoxicology and Environmental Safety, 179*, 17–23.

Weng, S., Wang, F., Dong, R., Qiu, M., Zhao, J., Huang, L., & Zhang, D. (2018). Fast and quantitative analysis of ediphenphos residue in rice using surface-enhanced raman spectroscopy. *Journal of Food Science, 83*, 1179–1185.

Wiest, L., Buleté, A., Giroud, B., Fratta, C., Amic, S., Lambert, O., ... Arnaudguilhem, C. (2011). Multi-residue analysis of 80 environmental contaminants in honeys, honeybees and pollens by one extraction procedure followed by liquid and gas chromatography coupled with mass spectrometric detection. *Journal of Chromatography A, 1218*, 5743–5756.

Wu, C. C. (2017). Multiresidue method for the determination of pesticides in Oolong tea using QuEChERS by gas chromatography-triple quadrupole tandem mass spectrometry. *Food Chemistry, 229*, 580–587.

Wu, S., Li, D., Gao, Z., & Wang, J. (2017). Controlled etching of gold nanorods by the Au(III)-CTAB complex, and its application to semi-quantitative visual determination of organophosphorus pesticides. *Microchimica Acta, 184*, 4383–4391.

Wu, X., Song, Y., Yan, X., Zhu, C., Ma, Y., Du, D., & Lin, Y. (2017). Carbon quantum dots as fluorescence resonance energy transfer sensors for organophosphate pesticides determination. *Biosensors and Bioelectronics, 94*, 292–297.

Xie, J., Li, L., Khan, I. M., Wang, Z., & Ma, X. (2020). Flexible paper-based SERS substrate strategy for rapid detection of methyl parathion on the surface of fruit. *Spectrochimica Acta, Part A: Molecular and Biomolecular Spectroscopy, 231*, 118104.

Yadav, B., Kaur, V., Narayan, O. P., Yadav, S. K., Kumar, A., & Wankhede, D. P. (2022). Integrated omics approaches for flax improvement under abiotic and biotic stress: Current status and future prospects. *Frontiers in Plant Science, 13*, 931275. https://doi.org/10.3389/fpls.2022.931275.

Yadav, B., Dubey, R., Gnanasekaran, P., & Narayan, O. P. (2021). OMICS approaches towards understanding plant's responses to counterattack heavy metal stress: an insight into molecular mechanisms of plant defense. *Plant Gene, 28*, 100333. https://doi.org/10.1016/j.plgene.2021.100333.

Yan, M., She, Y., Cao, X., Ma, J., Chen, G., Hong, S., Shao, Y., Abd El-Aty, A. M., Wang, M., & Wang, J. (2019). A molecularly imprinted polymer with integrated gold nanoparticles for surface enhanced Raman scattering based detection of the triazine herbicides, prometryn and simetryn. *Microchimica Acta, 186*, 1–9.

Yáñez, K. P., Martín, M. T., Bernal, J. L., Nozal, M. J., & Bernal, J. (2014a). Trace analysis of seven neonicotinoid insecticides in bee pollen by solid–liquid extraction and liquid chromatography 88 M. Á. González-Curbelo et al. coupled to electrospray ionization mass spectrometry. Food Analytical Methods 7, 490–499.

Yáñez, K. P., Martín, M. T., Bernal, J. L., Nozal, M. J., & Bernal, J. (2014b). Determination of spinosad at trace levels in bee pollen and beeswax with solid–liquid extraction and LC–ESI-MS. *Journal of Separation Science, 37*, 204–210.

Yoon, D. S., Park, J. C., Lee, J. S., & Han, J. (2019). Effects of atrazine on life parameters, oxidative stress, and ecdysteroid biosynthetic pathway in the marine copepod *Tigriopus japonicus*. *Aquatic Toxicology, 213*, 105213. https://doi.org/10.1016/j.aquatox.2019.05.015.

Zeng, L., Zhou, D., Wu, J., Liu, C., & Chen, J. (2019). A target-induced and equipment-free biosensor for amplified visual detection of pesticide acetamiprid with high sensitivity and selectivity. *Analytic Methods, 11*, 1168–1173.

Zhang, H., Wang, J., Li, L., & Wang, Y. (2017). Determination of 103 pesticides and their main metabolites in animal origin food by QuEChERS and liquid chromatography–tandem mass spectrometry. *Food Analytical Methods, 10*, 1826–1843.

Zhang, L., Wang, Y., Sun, C., Yang, S., & He, H. (2013). Simultaneous determination of organochlorine, organophosphorus, and pyrethroid pesticides in bee pollens by solid-phase extraction cleanup followed by gas chromatography using electron-capture detector. *Food Analytical Methods, 6*, 1508–1514.

Zhang, S., Yang, Q., Yang, X., Wang, W., Li, Z., Zhang, L., & Wang, Z. (2017). A zeoliticimidazolate framework based nanoporous carbon as a novel fiber coating for solid phase microextraction of pyrethroid pesticides. *Talanta, 166*, 46–53.

Zhao, H., Ji, X., Wang, B., Wang, N., Li, X., Ni, R., & Ren, J. (2015). An ultra-sensitive acetylcholinesterase biosensor based on reduced graphene oxide-Au nanoparticles-β-cyclodextrin/Prussian blue-chitosan nanocomposites for organophosphorus pesticides detection. *Biosensors and Bioelectronics, 65*, 23–30.

Zhao, P. F., Zhao, J., Lei, S., Guo, X. J., & Zhao, L. S. (2018). Simultaneous enantiomeric analysis of eight pesticides in soils and river sediments by chiral liquid chromatography tandem mass spectrometry. *Chemosphere, 204*, 210–219.

CHAPTER 7

Impact of pesticide application: Positive and negative side

Shveta Saroop[a] and Sonam Tamchos[b]
[a]Department of Botany, Government Degree College, Kathua, Jammu and Kashmir, India
[b]Regional Ayurveda Research Institute, Jammu (CCRAS, Ministry of AYUSH, Government of India), Jammu and Kashmir, India

7.1 Introduction

In order to sustain the hunger as well as nutritional security and economics of the rising global population, significant usage of fertilizers, insecticides, rodenticides, nematicides, herbicides, fungicides, and other agrochemicals is on record rise in the agricultural system. They increase agricultural production, product quality, and minimize losses of agricultural products caused by insects during storage. These agents or chemical compounds primarily are employed in public health prevention initiatives for protection against noxious weeds, rodents, insects, fungi, or infections, and to safeguard living beings from vector-borne ailments. Significant benefits in the field of forestry and livestock have been achieved due to pesticide application (Nkongolo & Narendrula-Kotha, 2020; Parween et al., 2018; Sharma et al., 2019).

As effective mitigation agents, they are applied in the form of sprays and powder to mitigate mosquitoes, fleas, cockroaches, termites, bugs, etc. Apart from these more well-known applications, these chemicals also find utility in lesser-known products like building materials, pet shampoos, and boat docks to eliminate or restrict the presence of undesirable species (Aktar et al., 2009; Delaplane, 2000; Nicolopoulou-Stamati et al., 2016; Ross, 2005).

As per the Food and Agriculture Organization of the United Nations (FAO), a pesticide is "any substance, or mixture of substances of chemical or biological ingredients intended for repelling, destroying or controlling any pest, or regulating plant growth".

Copper sulfate, ferrous sulfate, lime, copper, and sulfur are categories of inorganic insecticides. On the other hand, organic insecticides have more complexes and include metaldehyde D group vitamin-based rodenticides, metal phosphide rodenticides, molluscicides, carbamate insecticides, organophosphorus insecticides, carbamate herbicides, metabolite and hormone analog herbicides, synthetic pyrethroid herbicides, synthetic triazine herbicides, urea herbicides, and benzimidazole. Of all, organochlorine pesticides have the longest half-lives, significant persistence, greater accumulation index, slow degradation rates, and are stable (Akashe et al., 2018). Owing to this a large

Pesticides in the Environment: Impact, Assessment, and Remediation
DOI: https://doi.org/10.1016/B978-0-323-99427-9.00006-9

number of organochlorine pesticides are banned. Chemical pesticides can be replaced by biopesticides, plant-incorporated protectants (PIPs), microbial-generated pesticides, and biochemical agrochemicals (Alewu & Nosiri, 2011; Brühl et al., 2013; Nkongolo & Narendrula-Kotha, 2020; Parween et al., 2018; Sharma et al., 2019). Pesticide's active ingredients are coupled with other substances like solvents, mineral clays, adhesives, wetting agents, or other adjuvant components to improve their handling and storage. This mixture is known as a "pesticide formulation."

Herbicides: employed in agriculture fields to prevent the growth of weeds.

Fungicides: pesticides that either slow or stop the growth of fungi.

Insecticides: aid in the control of insects such as flies, bugs, and mosquitoes. By destroying their nervous system, insecticides quickly stop their growth.

Algicides antifoulants: destroying or repelling microorganisms in water bodies to control or stop their growth if desired.

Antimicrobials: kill microorganisms such as bacteria and viruses.

Biopesticides: pesticides of biological origin controlling or mitigating the pest, weeds, or insects.

Biocides: any chemical substance, mixture, or organism intended to control any hazardous microorganisms rather than just physically or mechanically.

Disinfectants and sanitizers: on material objects, disease-causing microorganisms are destroyed or rendered inert.

Fumigants: to destroy pests, for instance in soil or structures, create a gas or mist.

Insect growth regulators: interfere with the molting, pupal to adult development, or other life stages of an insect.

Miticides (also called acaricides): eliminate animal and plant-eating mites.

Microbial pesticides (Biopesticides): microbes or pests can be removed, hindered, or outcompeted by other microorganisms.

Molluscicides: kill slugs and snails.

Nematicides: destroy nematodes.

Plant-incorporated protectants: alter the projected pace of plant growth, blooming, or reproduction (does not include fertilizers).

Rodenticides: mice and other rodents under control.

Due to difficulties concerning their safe management and disposal, pesticides have emerged into an environmental pollutant (Chourasiya et al., 2015; Nicolopoulou-Stamati et al., 2016; Pimentel et al., 2014; Zheng et al., 2016). Using pesticides regularly pollutes the soil and water, persists in the crops plants and subsequently enters the food chain and thus gets biomagnified resulting in severe consequences (Chourasiya et al., 2015; Goulson, 2014; Nicolopoulou-Stamati et al., 2016). Commercial pesticide use in agriculture results in the formation of pesticide vapors that have the potential to contaminate the air. The amount of pesticides that are released into the air mostly relies on the physico-chemical attributes of the active substance, the application process, and shifting environmental parameters (Nicolopoulou-Stamati et al., 2016; Sharma et al., 2019).

The insecticides are transported and dispersed throughout from one region to another as deteriorated products. The synthetic pesticides being used in agriculture are absorbed into the soil through surface runoff from treated plants. Pesticides assimilate in the soil in accordance with the type of organic compound present, farming techniques, irrigation practices, and climatic conditions. Direct application of pesticides for residential and agricultural use leads to the accumulation of pesticides in the soil; alternatively, airborne pollutants that have previously come from other locations or sites can also contribute to this accumulation (Chourasiya et al., 2015; Pereira et al., 2009; Godheja et al., 2016; Tran et al., 2021). Pesticide accumulation in soil affects soil micro-complex directly thereby raising the risk for other higher hierarchy organisms through nutrition, and poses many detrimental effects on the soil ecosystem, vegetation, and ultimately human health (Chourasiya et al., 2015; Cooper, 2007; Kim et al., 2017; Nicolopoulou-Stamati et al., 2016).

Dichlorodiphenyltrichloroethane (DDT), aldrin, chlordane, dieldrin, endrin, heptachlor, mirex, and hexachlorobenzene are a few such pesticides that have adversely affected both human health and the environment. Some commonly used pesticides like DDT, malathion, chlordane, aldrin, dieldrin, endrin, mirex, glyphosate, acephate, diazinon, heptachlor, and hexachlorobenzene impart deleterious effects on the health of human and environment (Fang et al., 2017; Raffa et al., 2021; Randika et al., 2022).

Pesticide residues have been detected in a wide variety of commonly consumed beverages and foods, including cooked meals, wine, fruit drinks, confectionery, and livestock feed. It is also crucial to understand that rinsing and peeling does not completely remove pesticide byproducts. Despite being too little, the significant ratios might have serious effects if two or more chemicals were exposed in succession. This is owing to the synergistic effects of the same. Even minute amounts of pesticides are now being found in human breast milk, which raises questions regarding prenatal exposure and the potential health effects on the unborn child (Tran et al., 2021).

Nevertheless, producers, food suppliers, investment firms, and authorities must be devoted to harmonizing the risks and benefits of periodic losses created by the reckless usage of agrochemicals in environments, ecosystems, consumers, as well as every economic sector either directly or indirectly engaged in the agrarian chain. The present review puts forth the advantages and disadvantages of pesticide application to manage agricultural pests, environmental dangers, and effects on the health of humans and animals.

7.2 Positive side of pesticide application

7.2.1 Food and nutritional security

By 2030, the UN's Sustainable Development Goal (SDG) aims to eradicate "all types of malnutrition," end hunger, and guarantee access to safe, nutritious, and sufficient food all year round. However, the FAO states that hunger has increased in many areas of the world during 2015, with the most severe increases coming in sub-Saharan Africa and the Middle East approximately 23.0% of the population is categorically under hunger. Conflict-related hunger now affects 12.4% of the Middle Eastern population. Moreover, an expanding population worldwide and limited resources have further generated concern for more food. In light of this, pesticide companies assert the role of pesticides in ensuring food and nutritional security for the expanding global population. Estimates of postharvest losses range from 14% globally to 50% in some developing nations. However, in countries with rapid population growth, in addition to food security, the re-emergence of diseases and pests manifests the biggest yield loss (Pandya, 2018; Zhang, 2018). Pesticides have evolved as a crucial instrument for crop production enhancement and crop protection with the evolution of agriculture. A wide variety of pesticides must be used against effective pests, weeds, and toxic microbial compounds in order to boost agricultural productivity, which accounts for about 45% of the yearly food production loss. As a cost-effective technique, it enhances food security for the world's population that is constantly expanding (Fig. 7.1). Pesticides are used in about one-third of agricultural products. Fruit production would decrease by 78%, vegetable production by 54%, and cereal production by 32% without the usage of pesticides. Therefore, pesticides are essential for controlling pests and enhancing crop yields (Pandya, 2018; Zhang, 2018). Pesticides indirectly improve household income and nourishment significantly and these two elements together ensure food security (US Environmental Protection Agency, "Regulation of Pesticide Residues on Food").

Currently, 2 million tons of pesticides are being used worldwide, among which herbicides make up 47.5%, insecticides 29.5%, fungicides 17.5%, and the rest 5.5% by other pesticides. Argentina, Brazil, China, Canada, France, United States, Thailand, Italy, India, and Japan are the top ten pesticide-using nations around the globe. (Atwood & Paisley-Jones, 2017; Zhang 2018).

7.2.2 Increased productivity and low food cost

Growing population demands year-round access to high-quality, economical, and safe food. Crops must be protected in the field in order to ensure continuous quantities of high-quality products. By influencing agro-productivity pesticides minimizes the postharvest and storage cost. By ensuring the cost of food the sustainable development goal of food for all can be achieved. As a result, consumers can buy food at reasonable costs while simultaneously benefiting from the viability of farms and farmers' livelihoods (Arora & Mishra, 2019).

Pesticides, particularly weedicides assist farmers to increase their crop production, which directly adds to the affordable and easily available nature of food. Pesticides contribute to a bountiful harvest by ensuring that crops are protected against pests and are healthy to be harvested. The risk of many human diseases can be decreased by eating

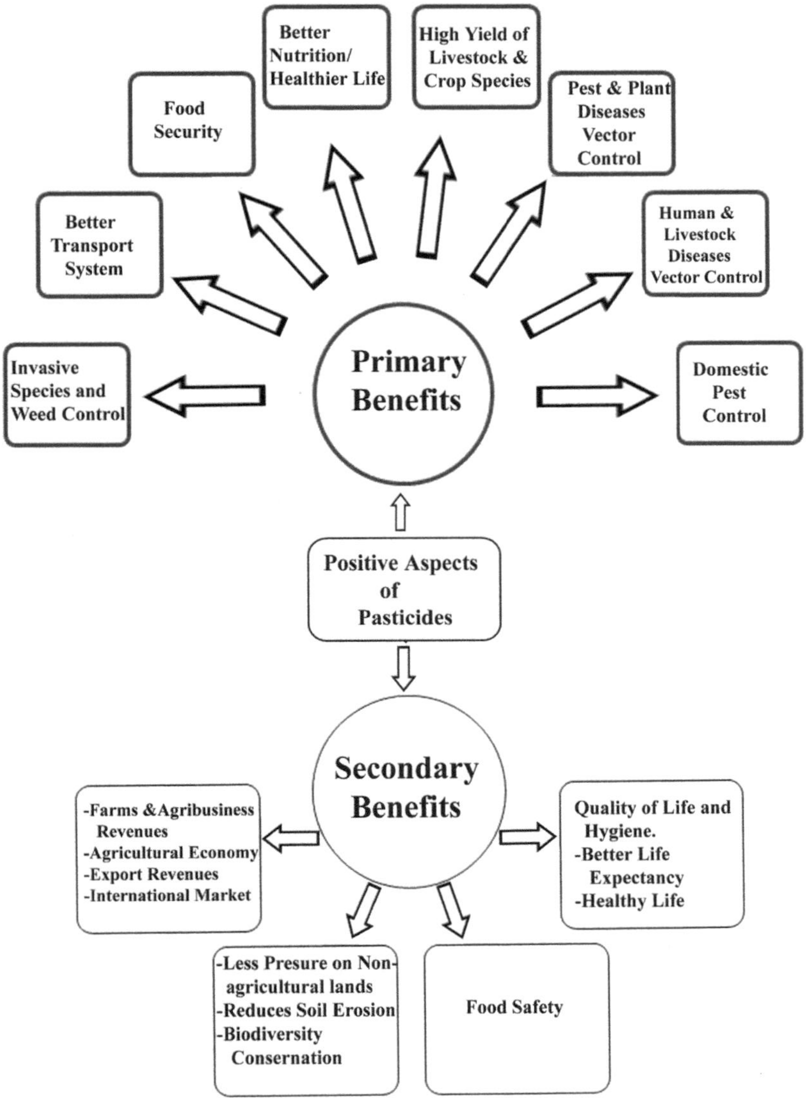

Figure 7.1 Positive aspects of pesticide application.

healthy food (both fruits and vegetables). Thus, protecting crops against pests or pathogens indirectly improve human health (Weaver, 2017).

Use of herbicides enhance weed management to the doubling of wild blueberry yield and consequent increase in consumption. As a health promoter, the importance of wholesome, affordable food extends life expectancy (Sarkar et al., 2021; Tudi et al., 2021). Australian sheep producers boosted the wool yield by US$ 50/ha by controlling red-legged earth mites in clover with a single, precise, and timely insecticidal spray that costs US$ 10/ha. This reveals how significant gains in livestock productivity can be made by reducing pasture pests.

7.2.3 Weed and invasive species control

Pesticides control weed proliferation as they compete with crop plants for light, water, and nutritional resources thus putting pressure on the survival and sustenance of economically important crop plants. Significant declines in agricultural productivity by 50%–90% are due to weeds. Because of this reason farmers mitigate the weeds via hoeing or hand-pulling and then by ploughing. Catchweed (*Galium aparine*), melde (*Chenopodium album*), and cursed thistle (*Cirsium arvensis*) are a few significant and challenging broad-leaved weeds. Other weeds are grasses, such as black grass, couch grass (*Elymus repens*), or ryegrass (Kniss, 2018; Strassemeyer et al., 2017; Bajwa et al., 2017).

Chemical pesticides have been increasingly utilized in recent years to control agricultural weeds. For most crops in the United States, weedicide is applied to 85% of the planted area. Canola (99% of its cultivated area), blueberry (95%), citrus (95%), carrot (98%), maize (98%), dry beans (99%), rice (98%), green bean (96%), sugar beet (98%), peanut (97%), soybean (96%), tomato (96%), cotton (95%), and potato (93%) are the crops that receive the most herbicide. One of the main reasons for the loss of biodiversity worldwide is invasive alien species. Today, pests, weeds, diseases, and non-native animals and plants are spreading more quickly via land, sea, and air. They are detrimental to human health and well-being, farming, and both. According to estimates, invasive species cause US$ 1.4 trillion in damage annually, the majority of which is related to agriculture, particularly crop productivity (Pimentel et al., 2014). Thus, pesticides efficiently remove weeds and invasive flora thereby allowing the crop plants to grow without any competition from the more aggressive, competitive, and powerful weeds (Fig. 7.1).

7.2.4 Pesticides have transformed developing countries into food producers

A nation's economy is heavily reliant on the expansion of food production, and pesticides assist farmers in raising a healthy harvest, which stimulates economic growth. Crop protection products have significantly aided farmers in poor nations in growing two or three crops annually, enabling these nations to serve as "economic powerhouses" for the rest of the globe. Weeds diminish the productivity of dry land crops by 37%–79%. A 40% output drop can ultimately be attributed to severe weed infestation, especially in the early stages of crop establishment. Herbicides had positive effects on labor and the economy. Profit-driven agricultural chemical manufacturers and associated trade organizations are growing worldwide. Two-thirds of the $14 billion US pesticide market are used in agriculture. Due to demand for pesticides for agro-products and changing regional sectors significant pressure is being generated on the agro-chemical market. Tons of pesticides are being marketed via import export processes therefore contributing share to gross domestic product and hence economy of any country (Atwood & Paisley-Jones, 2017; Sarkar et al., 2021; Tudi et al., 2021).

7.2.5 Pesticides help reduce waterborne and insect transmitted diseases

Natural poisons like mycotoxins, which are produced by fungi, can be harmful to both humans and animals. One of the most significant mycotoxins in food production, for instance, is aflatoxin, which is generated by the *Aspergillus fungus*. Liver necrosis, nausea, and vomiting are symptoms of acute toxic doses. Zearalenone is a significant poison that disrupts human hormone homeostasis. Depending on the climatic conditions, several fungicides have been proven to lower these natural toxins by up to 50% under field conditions (Doan et al., 2021; Hladik et al., 2018).

Fungicide is applied to prevent pathogenic fungi like late blight *Phytophthora infestans* (of potatoes), *Venturia inequalis* (of apple scab), *Sphaerotheca pannosa* (of peaches), and *Pythium* spp. (seed rot and damping-off of numerous crop species), *Claviceps purpurea* (ergot of cereal crops), *Fusarium*, etc. Fungicide aids in preventing the growth of fungus like *Aspergillus flavus*, which can deteriorate stored legumes, grains, and nuts by producing fatal aflatoxins that render them dangerous for humans and animal consumption (Doan et al., 2021; Hladik et al., 2018).

Pesticides aid in the prevention of insects and waterborne diseases like malaria (caused by *Plasmodium* and disseminated by *Anopheles* mosquitoes, yellow fever, encephalitis and West Nile virus, typhoid fever, typhus, and Rocky Mountain spotted fever. Sleeping sickness by the protozoan *Trypanosoma* and spread by the tsetse fly, plague or black death by the bacterium *Yersinia pestis* and transmitted by the rat are important human diseases transmitted by vectors. Insecticides contribute to improving human health and minimizing the risk of these diseases by inhibiting the vectors of the said diseases (Fig. 7.1). Pesticides safeguard human health by controlling disease outbreaks by controlling rodent and insect pest populations at domestic spaces. Through the use of insecticides to eradicate disease vectors, roughly seven million deaths have been avoided globally. The most notable example is malaria eradication, which protected an average of 5000 deaths daily (Hladik et al., 2018; Ross, 2005). Pesticides, especially in tropical nations, have reduced the prevalence of vectors and alternative hosts over the past 50 years, protecting millions of people from the severe or fatal consequences owing to many diseases.

Crop productivity would reduce if crops are not shielded from pests' destructive effects. A reduction in agricultural yield would result in a food shortage that elevates the cost of foodstuff. Since many agricultural products are susceptible to aflatoxins Aflatoxin is a carcinogen that can harm children's growth and development and result in liver or any other types of human cancer. It also reduces the body's normal immunological behavior. Controlling insect-mediated aflatoxin contamination involves the use of crop protection agents. As a result, pesticides indirectly contribute to the regulation of food costs and at the same time minimize the disease transfer from plant to human (Sarkar et al., 2021; Tudi et al., 2021).

7.2.6 Pesticides help conserve the environment

Protecting agricultural and farming sectors involves safeguarding all life forms. Invasive plant species, alien insects, and other pests are prevented from flourishing by pesticides in any ecosystem. All these restore the native biodiversity and thus maintain the ecological system of any ecosystem.

Modern crop protection techniques encourage higher farmland productivity, relieving strain on uncultivated/marginal lands that function as significant habitats for species of birds, bees, and other insects. Improved agronomic productivity allows farmers to increase agricultural production per unit area while requiring less tillage. By this they help to prevent deforestation, preserve natural resources, and stop soil erosion (Fig. 7.1). Farmers can minimize conventional cultivation practices like ploughing for weed control, which reduces emissions of greenhouse gases and thus improves soil quality and function. Moreover, creating a pollen- and seed-rich environment that is more alluring to pollinators, seed dispersal agents, and other beneficial insects. The availability of beneficial insects on field margins not only benefits wildlife, but it also contributes to improved pest control in neighboring crops, reducing the need to use insecticides subsequently thereafter (Stoytcheva, 2011; Strassemeyer et al., 2017; Jepson et al., 2020).

7.2.7 Improving nonagricultural surroundings like transport, sports, and building complex

By reducing the number of bugs, insecticides help enhance the sanitary conditions in homes (Delaplane, 2000). Additionally, pesticides prevent extensive damage linked to termite infestations and protect the beauty of public parks by controlling weeds. Herbicides and insecticides are widely used in sports fields, golf courses and soccer fields (Aktar et al., 2009). Herbicides in particular are largely used in the transportation sector. Buildings and wooden structures are shielded from termites and other wood-boring insects using insecticides. Many undesirable insects, like cockroaches and mosquitoes, are repelled or controlled by pesticides. Rodents or insects are expelled from eateries, hospitals, parks, and shopping centers by the use of insecticides. By this way a safe, healthy, and hygienic environment is being maintained (Fig. 7.1).

7.3 Negative side of pesticide application

7.3.1 Eco-environ complexes

Although pesticides have an important role in crop yields and the production of economical and high-quality food by killing the targeted pest, however, pesticides negatively

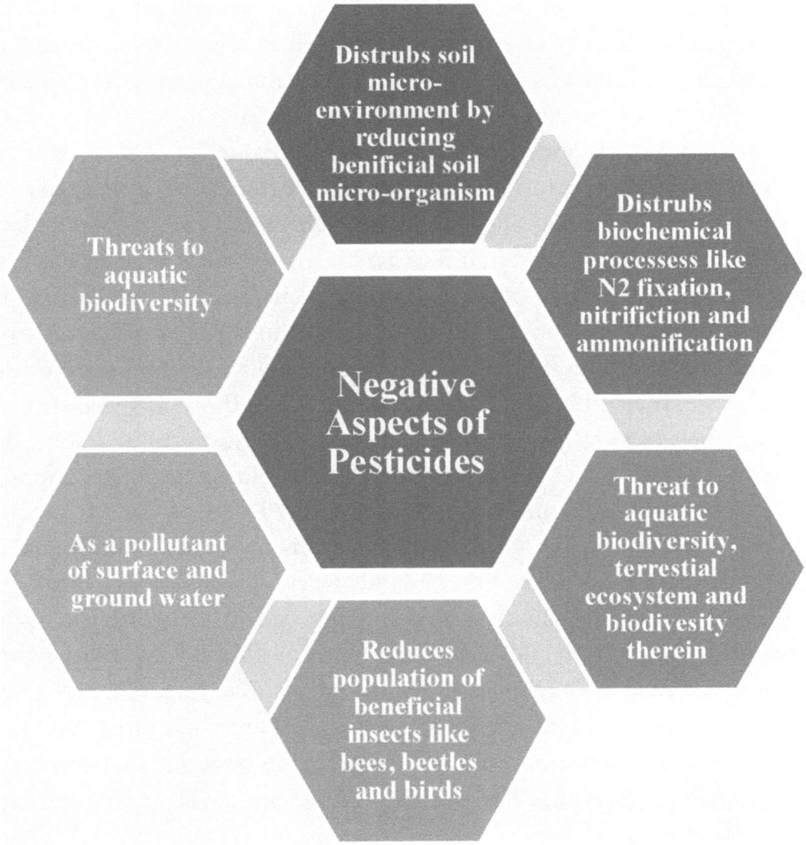

Figure 7.2 Negative aspects of pesticide application.

influence the environment, including air, water, soil, and crops (Fig. 7.2) (Gouin et al., 2008; Kawahara et al., 2005; Kim et al., 2017; Skrbic et al., 2007; Tuncel et al., 2008) and human health as well (Miraglia et al., 2009). It also affects the nontarget organisms (Wanwimolruk et al., 2016) including the wildlife, birds, honeybees, beneficial insects and aquatic ecosystems pests (Fig. 7.2) (Aktar el al., 2009; Elgueta et al., 2017; Mingo et al., 2017). The population growth and climate change are the key factors driving an increase in pesticide use. It is predicted that future pesticide output will also increase globally (Tudi et al., 2021).

When pesticides are used, only a fraction of it exhibits the role to fight against diseases and larger portions indirectly enter into the environment through transfer and degradation processes (Singh, 2012). As pesticides break down in the environment, new chemicals are produced (Marie et al., 2017). The pesticides or their degraded secondary toxins chemicals mobilize from the target site to other environmental complexes via a variety of mechanisms, including volatilization, microbiological breakdown, photo-degradation breakdown, spray drift, leaching, adsorption, absorption, chemical breakdown, and runoff.

The persistence, bioaccumulation and biomagnification of pesticides modulate its toxicity potential. Biomagnification refers to the spike in pesticide concentrations induced by their persistent and nonbiodegradable nature in the tissues of organisms at successive levels of the food chain. Biomagnifications grow more significant as a result of a pesticide's increasing persistence and lipophilic (fat-loving) characteristics. Organochlorines are known to have a higher rate of biomagnification than organophosphates because they are more persistent in a wider range of species. Some pesticides, such as Aldrin, chlordane, DDT, dieldrin, heptachlor, and hexa-chlorobenzene, include persistent organic pollutants (POPs) (Kim et al., 2017; Mingo et al., 2017). The species at the top of the food chain suffer more damage as a result of this phenomenon than those at the bottom. One of the classical examples of hazardous chemical biomagnification is DDE-induced eggshell thinning that causes fish-eating avian populations to decline due to poor reproduction (such as gulls, terns, herons, etc). The highest residual levels of organochlorines were found in tertiary consumers like green-backed herons (*Butorides striatus*) and snakes as compared to secondary consumers like bluegills (*Lepomis macrochirus*), blacktail shiners (*Notopis venustus*), and others. Catfish (*Clarias gariepinus*), the primary consumer fish in Lake Ziway, showed much higher levels of dichlorodiphenyltrichloroethane (4,4′-DDE) than did Nile tilapia (*Oreochromis niloticus*), tilapia (*Tilapia zillii*), and goldfish, the lowest consumers (*Carassius auratus*). Some of the elements that affect the persistence of pesticides in food, soil, water, and the environment are pesticide dissolution, volatility, surface adsorption, hydrolysis, redox potential, UV irradiation, and chemical modification (Assey, 2002; Tudi et al., 2021).

7.3.2 Disturbs the soil microenvironment and complexes therein

The fate of applied pesticides is evaluated by various parameters, like soil characteristics and soil microflora, as a result of which it experiences a range of adsorption/desorption, transport, and degradation processes. The interaction between the soil's native microorganisms and the decomposed pesticides alters the microbial diversity, biochemical and enzymatic processes of the soil.

Pesticides that penetrate the soil can change soil microfauna. Pesticide application-related changes in the behavior of soil microorganisms eventually cause disturbances in the soil profile and reduce soil fertility (Fig. 7.2). Excessive pesticide application has resulted into the deterioration of soil micro-environment and system. Pesticide overuse can reduce beneficial soil microorganisms. Numerous soil bacteria are involved in the transformation of atmospheric nitrogen into nitrates. There is evidence that the fungicides chlorothalonil and dinitrophenyl interfere with the activities of nitrifying and denitrifying bacteria (Lang & Cai, 2009). The pesticide triclopyr prevents soil bacteria from converting ammonia to nitrite (Pell et al., 1998). Glyphosate, a quasi herbicide, also lowers the proliferation and activity of nitrogen-fixing bacteria in soil (Santos & Flores, 1995), whereas 2,4-D inhibits the soil bacteria's transformation of ammonia into nitrates (Frankenberger et al., 1991). It has been reported that the sulfonylurea herbicides chlorsulfuron, metsulfuron methyl, and thifensulfuron methyl inhibited the proliferation and development of fluorescent *Pseudomonas* strains (Boldt & Jacobson, 1998; Tudi et al., 2021 and references). *Pseudomonas* sp. are known to have a great ecological importance in the soil ecology, their absence imposes a negative impact on soil fertility. Similar reports indicate that benomyl, Captan, and chlorothalonil can reduce soil respiration by 30% to 50% in untreated soils, which is a measure of microbial biomass (Boldt & Jacobsen, 1998).

Some microbial groups may potentially use applied pesticides as a source of energy, which could accelerate their proliferation which ultimately affects soil ecology. Bacterial strains recovered from wastewater-irrigated agricultural soil, for example, have been reported to use chlorpyriphos as a carbon source.

Herbicides also inflict significant effects on fungal populations present in soil. Insecticides like oryzalin and trifluralin are reported to restrict the proliferation of symbiotic mycorrhizal fungi that facilitate nutrient uptake by plants (Kelley & South, 1978). Ticlopyr is poisonous to some species of mycorrhizal fungi, while oxadiazon has been shown to decline the production of fungal spores (Chakravarty & Sidhu, 1987).

By serving as indicators of soil toxicity and assessment, earthworms contribute significantly to the soil ecosystem. They significantly contribute to increasing soil fertility by turning organic matter into humus. By forming channels in the soil that allow for soil aeration and drainage, they significantly contribute to enhancing and preserving soil structure. Pesticides have been reported to affect these natural soil engineers. Applications of pesticides may result in a drop in earthworm populations. For instance, several organophosphates have been found to lower the number of earthworms, and carbamate insecticides are extremely poisonous to them. They are mostly exposed to the pesticides through polluted soil pore water. The neurotoxic effects of insecticides and/or fungicides on earthworms lead to physiological decline after extended exposure (Schreck et al., 2008). DNA damage is caused by glyphosate and chlorpyrifos' harmful effects on earthworms at the cellular level. Glyphosates have an impact on earthworm feeding activity and survivability (Casabé et al., 2007; Liu et al., 2015). Neonicotinoids have a propensity to build up in the soil, they can harm earthworm species like *Eisenia foetida* (Goulson, 2013). Earthworms are negatively impacted by glyphosate and chlorpyrifos, which cause DNA damage at the cellular level. According to Casabé et al. (2007), glyphosates have an impact on earthworm feeding behavior and survivability.

The control of pest populations is greatly aided by predators. Predators are a crucial element of the "biological control" strategy, which forms the part of IPM. Cypermethrin and imidacloprid spraying produced greater mortality rate in braconid wasps, coccinellids and predatory spiders in *Solanum melongena* L. The life history characteristics of predators, like proliferation rate, maturation time, and other reproductive processes, can also be impacted by pesticides (Fig. 7.2). Glyphosate-based herbicides influence spider and ground beetle behavior and survival, in addition to influencing arthropod population dynamics (Brussaard, 1997; Evans et al., 2010). The body dimension, hemoglobin level and morphometric parameters of carabid beetles (*Pterostichus melas* italicus) were also revealed to be drastically reduced by dimethoate.

The micro metabolic or the soil enzymatic activity may both be affected by pesticides that make it to the soil. Soil serves as a reservoir of enzymes full of free enzymes, immobilized extracellular enzymes secreted by microorganisms that are indicators of biological stability including soil fertility (Hussain et al., 2009; Mayanglambam & Singh 2005). This enzymatic pool serves as a biocatalyst for the degradation of pesticides and organic compounds. By activating/deactivating soil microbes and/or enzymes, pesticides adversely affect the soil's important biochemical processes, such as nitrogen fixation, nitrification, and ammonification. Increase or decrease in soil biochemical processes is ultimately controlled by the additive interactions among pesticides, microorganisms, and soil characteristics. For instance, *Azospirillum* sp. populations and ammonification rate were reported to rise in both laterite and vertisol soils planted with *Arachis hypogaea* L. at a specific pesticide dose (2.5–5.0 kg/ha). However, at greater doses, the said pesticides exhibit antagonistic effects on the population of *Azospirillum* sp. population and ammonification (7.5–10.0 kg/ha) (Đurišíc-Mladenovíc et al 2010).

7.3.3 Threats to terrestrial ecosystem and biodiversity therein

An attribute or function of soil is its ability to filter, break down, and detoxify pesticides. Pesticide breakdown produces residues that remain in terrestrial environments over time, posing a harm to the environment. In fact, pesticide contamination of soil and sediment has been a major issue in terrestrial areas, having a negative effect on food quality and the agriculture viability (Fig. 7.2). Additionally, soil serves as the primary repository for environmental pesticides in terrestrial areas due to its high capacity to retain poisons in its structures through adsorption. Secondary sources involve re-emission of older organic pollutants into the atmosphere, groundwater, and living organisms. This makes soil a significant factor in pesticide pollution. It is assumed that pesticides that are hydrophobic, persistent, and bio-accumulable will build up and remain in soils (Aktar et al., 2009; Langenbach, et al., 2017; Hussain et al., 2009). Because of their persistent nature, some pesticides, notably the organochlorine DDT, endrin, endosulfan, heptachlor, and lindane, are banned in some countries. Other pesticides, such as carbamates, organophosphorus and fungicides insecticides, do not persistent in soil and get mobilized through distinct runoff and leak into various other environmental media. In light of this, pesticide-contaminated soil poses a serious hazard to the water supply and the food chain (Langenbach et al., 2017).

As a result of pesticide toxicity, plants produce less protein and have lower chlorophyll levels, which affects their photosynthesis. Additionally, pesticide toxicity results in stunting, burns, necrosis, chlorosis, and twisted leaves in plants (Sharma et al., 2018). Both active and passive absorption of pesticides through the root system are possible which in turn disturbs the normal nutrient uptake by roots. The herbicide glyphosate lowers seed quality and makes plants susceptible to disease. Pesticides that are absorbed/adsorbed are either digested by plants or build up in plants, which causes biomagnification in the ecosystem (Brammall & Higgins 1988).

Exposure to pesticides can have sub-lethal effects to the terrestrial ecosystem. Phenoxy herbicides harm surrounding trees and shrubs through drifting or volatilization (Dreistadt et al., 1994) thereby posing threat to the plant community and ecological networking over there. Herbicides, sulphonamides, sulfonylureas, and imidazolinones show devastating effect on nontarget crops, natural plant communities, and wildlife, even at low doses (Fletcher et al., 1993). which has had a severe impact on many raptor species, including the imperial eagle and *Gypaestus barbatus* (bearded vulture) (Margalida et al., 2008). Biological pest management, soil fertility, and adequate crop pollination (Gill & Garg, 2014). are further significant ecosystem components that are adversely impacted by pesticide toxicity (Tscharntke et al., 2005).

According to experts birds are among the most sensitive species to pesticide exposure. Globally around 87 percent of the bird species are threatened by pesticides. Numerous farmland bird species have experienced dramatic decreases in population size and geographic range during the past 40 years. According to an estimate, 672 million birds are directly toxicated with pesticides on agri-lands per year. 10% of these birds die off due to acute effects of pesticides (Sharma et al., 2019). Birds are a crucial biotic factor because they serve to maintain the natural balance by predator of bugs, feeding on a variety of insects, including mosquitoes and many other species that are thought to be among the most dangerous pests to agriculture and human health. Without birds, bug pest population will become more in number, eventually leading to higher pesticide use. Pesticides kill birds in a variety of ways, including direct ingestion of pesticide granules and via treated crops, treated seeds, direct spray exposure, contaminated polluted, or feeding on infected prey. Pesticides in sublethal doses can damage the neurological system and alter behavior.

By changing their reproduction and feeding habits, several synthetic pesticides like carbamates, organophosphates, and organochlorines result in reduction of raptorial bird populations. Rice fields being treated with a variety of organochlorines, cholinesterase-inhibiting insecticides such monocrotophos, carbofuran, phorate, fenthion, diazinon, phosphamidon, and azinphos-methyl as well as fungicides, herbicides, and molluscicides are extremely toxic to bird which used to feed on this crop during harvest season, affecting their reproductive efficiencies and sometimes even killing them. Insecticides affect the breeding behavior of yellowhammer (*Emberiza citrinella*) and corn bunting (*Miliaria calandra*). DDT exposure and its metabolites plays a significant role in the reducing bald eagle populations in the United States (Liroff, 2000). By eliminating the earthworms that birds and other mammals depend on for food, fungicides can indirectly lower animal and bird populations.

The use of broad-spectrum pesticides like pyrethroids, carbamates, and organophosphates significantly reduce populations of important insects like bees and beetles. In the pollination process, pollinators, which are biotic agents, are crucial. Several kinds of bees, honeybees (*Apis* sp.), bumble bees (*Bombus* sp.), beetles, fruit flies, birds, etc. Since their behavior is influenced by environmental stress imposed on by parasites, predators, pesticides, habitat modifications and competition. Pesticide usage, on the other hand, causes direct loss of insect pollinators and indirect loss of crops due to a lack of adequate pollinator populations (Fishel, 2011). Honeybees are negatively impacted by the synergistic actions of pyrethroids and triazole or imidazole fungicides (Pilling & Jepson, 2006). Bees are poisoned by neonicotinoids, which are present in insecticides like clothianidin and imidacloprid. Even at low dosages, imidacloprid impairs foraging behavior of bees (Decourtye et al., 2003; Yang et al., 2008).

Application of pesticides also has an impact on the efficiency of pollen collection, foraging behavior, and colony mortality in pollinator species. Since bees make up 80% of the pollinator population, bee studies have contributed significantly to our current understanding of how pesticides might alter pollinator behavior. Neonicotinoid insecticides (acetamiprid, imidacloprid, thiamethoxam, clothianidin, dinotefuran, thiacloprid and nitenpyram) have both lethal and sub-lethal effects on bees' foraging behavior as well as their capacity for learning and memory (Blacquière et al., 2012). Moreover, thiamethoxam, a neonicotinoid insecticide, leads to significant mortality from homing inability to a level that increases the likelihood of bee colony collapse (Biesmeijer et al., 2006). Imidacloprid, the most widely used pesticide globally, has an impact on honeybee lifetime and foraging of *A. mellifera* and *Nosema ceranae.* Microsporidial infections in the guts of bees from imidacloprid infected hives increased significantly. It has been hypothesized that interactions between diseases and the pesticide imidacloprid may be the primary cause of CCD (colony collapse disorder) and other cases of honeybee colony mortality globally. A decrease in bumblebee (*B. terrestris*) fecundity has also been linked to imidacloprid induced lower brood production. Agricultural field study conducted revealed lower species richness of bumble bee and butterfly in response to pesticide application. Honeybee populations have decreased by 29%–36% annually since 2006, on average (Yang et al., 2008).

7.3.4 As a pollutant/contaminant of surface and ground water and threats to biodiversity therein

According to Singh (2012) many chemicals used in the certain pesticides have been traced out in both surface and groundwater. Application of pesticides to the targeted field and disposal areas pollute the water bodies such as lakes, streams, rivers, etc. highly risk accumulation chemicals used in the pesticides (Ansara-Ross et al., 2012; Ben Salem et al., 2016). Surface water bodies such as lakes, rivers, streams, and estuaries are particularly susceptible to pesticides (Salem et al., 2016). Because of this they act as modest captive sinks of these potentially toxic byproducts. Runoff from agricultural fields, urban use, and industrial effluent contribute to the presence of pesticides in the water body. Point-source pesticide contamination is a result of pesticide leaching processes that occur during the production of pesticides as well as at disposal sites and in wastewater effluents. Despite the soil matrix acting as a pesticide sink due to the high affinity of agrochemicals with soil, surface water resources are extremely sensitive to pesticide contamination due to the intimate relationship between soil and water systems (Ansara-Ross et al., 2012). The hydrologic cycle binds the systems of surface water to those of groundwater and that of atmosphere. Once groundwater becomes polluted it takes a long time for the pollution to evaporate or be cleared up (Waskom, 1994; O'Neil & Raucher, 1998; US EPA, 2001). Pesticides leach into groundwater and contaminate drinking water, which is one of the key concerns for environmentalists (Rasmussen et al., 2015). Additionally, via seepage through soil, pesticides find their pathway to surface water and ultimately to groundwater and via vaporization to atmospheric water.

The less quantity of pesticides that have accumulated in water gets amplified via the food chain and penetrates aquatic complexes (Ansara-Ross et al., 2012). Pesticides lower the amount of dissolved oxygen in the water. Aquatic organisms are exposed to pesticides via direct skin absorption, gill uptake via breathing. In aquatic habitats, algae serve as an essential component of the food chain. Atrazine and alachlor affect algae and diatoms even at relatively less concentrations. These cause cell damage, obstruct photosynthesis, and restrict growth and proliferation (Sharma et al., 2019).

As a result of their tight interactions with the physical, biological, and chemical environments, fish constitute an essential component of the marine ecosystem. The pesticide pentachlorophenol (NaPCP) is reportedly responsible for significant fish death. Atrazine and alachlor alter fish population's physiology and behavior. Ability to mate, find food, avoid predators, and homing, among other behavior responses become altered (Kumari et al., 2012). Pesticides not only affect fish but the food chains that are connected to them. The main Arctic Ocean food webs have already been shown to contain persistent pesticides (polychlorinated biphenyls and organochlorine pesticides). The vitellogenesis process of catfish (*Heteropneustes fossilis* Bloch.) may be altered by the organophosphate insecticide Abate, posing a significant negative impact on catfish farming (Gill & Garg, 2014). High glyphosate concentrations were simultaneously exposed to freshwater fish (*Galaxias anomalus*), trematode parasite (*Telogaster opisthorchis*), and snails, which drastically affected their development and ability to survive. Herbicide spraying adjacent to weedy fish nurseries diminishes the reproductive capacity of aquatic life forms and eventually lessens the amount of shelter that juvenile fish require to hide from predators (Helfrich et al., 2009).

Amphibians are tetrapod ectothermic animals of the Amphibia. They live in a large range of habitats, with the majority of species being found in terrestrial, freshwater aquatic, fossorial, and arboreal ecosystems. An environmental concern on a global scale is the amphibian population reduction. The most widely used herbicide, Roundup (Glyphosate), has adverse effects on nontarget frogs, including high death of larval tadpoles and mortality of *Rana temporaria*, which has been found to range from 100% past one hour to 40% past seven days at the recommended quantities of this pesticide. While glyphosate, a herbicide, is known to cause substantial tadpole and juvenile frog mortality, carbaryl has been discovered to be harmful for a number of amphibian species (Relyea, 2005). Malathion has been demonstrated to alter plankton and periphyton populations in small concentrations, which in turns affect frog tadpoles' development and survival (Relyea & Hoverman, 2008). Additionally, it was discovered that repeated low dose administrations of malathion had greater effects on an aquatic system than a single high dose application. Herbicides (glyphosate, 2,4-D) and insecticides (carbaryl, malathion), affect aquatic populations. species richness with a reduction value of 15% with carbaryl, 22% with roundup and 30% with malathion. Additionally, endosulfan and chlorpyrifos both significantly damage amphibians (Evans et al., 2010; Sparling & Feller, 2009).

7.3.5 Threat to air

Airborne pesticide contamination is a significant source of pollution that poses danger to both human health and the health of plants and animals (Lee et al., 2002; Langenbach et al., 2017). Pesticide applications or volatilization from the soil or plants are the principal sources of pesticide residues in the air, and the same is true for all pesticides employed in agricultural operations (Baker et al., 1996). Pesticide sprays are typically propelled by a fan in the form of water droplets, and after passing through the canopy in a turbulence-filled manner, gravity and atmospheric activity like wind force the droplets to drift. (Đurišíc-Mladenovi et al., 2010). Three significant spraying techniques that are employed extensively in modern agriculture are surface application, subsurface application, and aerial spraying. Because the pesticide residues are volatilized, diffused, and mobilized over great distance, they are exposed to a recycling process involving air and the terrestrial environment (Doan et al., 2021; Langenbach et al. 2017; Pan et al., 2020). As a result, air pollution pesticides are challenging.

7.3.6 Human health risk

By reducing the spread of vector-borne diseases, pesticides have raised human health standards; but long-term indiscriminate pesticide use has had detrimental impacts on human health. Because of the generic character and indiscriminate application of pesticides, humans, especially infants and children, are extremely susceptible to their adverse effects. Pesticide has invaded almost every environmental complex, such as soils, sediments, groundwater, and surface water as discussed above. Either via the food chain pathway or through direct exposure, humans are being continuously exposed to chemical pesticides (Benka-Coker et al., 2020; Doan et al., 2021; Year et al., 2020). Pesticide residues in foodstuffs, direct exposure during application in field or in nonfield areas, and via contaminated drinking water or through air finds direct entry into human bodies (Machemer, 1986). The most preferred pathways by which people are exposed to pesticides are by inhalation, skin absorption, oral absorption, and pesticide residue exposure. Pesticide exposure has two types of effects on humans: acute and chronic.

Shortly after touch or exposure to the pesticide, acute sickness typically manifests. Pesticide flow from agricultural fields, pesticide poisoning during spraying, and any other intentional or unintentional exposure can contribute to acute sickness in humans. Direct consequences of pesticides include reduced eyesight, skin itching, rashes and blisters on the skin, headaches, salivation, diarrhea, vomiting, nausea, wheezing. Paralysis, lack of coordination, coma, and even death are among the most frequent acute health effects of pesticides. These effects depend on the pesticide's toxicity. Moderate pesticide poisoning can mimic the symptoms of underlying gastroenteritis, bronchitis, and asthma. The toxicity and dosage of the chemicals employed, their mode of action, how they are administered, how long and frequently individuals come into contact with pesticides, and how they are exposed are typically linked to how severe these risks are. When organochlorines are consumed, they produce hypersensitivity to light, touch and sound as well as tremors, dizziness, seizures, nausea, vomiting and anxiousness. The symptoms of elevated acetylcholine levels are caused by exposure to carbamates and organophosphates. Due to the pesticides' interference with normal nerve signal transduction, exposure to these can result in headaches, nausea, vomiting, chest pain, disorientation, and headache-like symptoms. Pyrethroids, in addition to inducing tremors and convulsions, cause skin allergy, aggression, hyperexcitation, reproductive or developmental effects (Fang et al., 2017; Raffa et al., 2021; Randika et al., 2022; Tago et al., 2014).

It takes years for the fatal consequences of pesticides to manifest chronically. Multiple body organs are damaged over time by these long-term impacts. Neurological consequences such as Parkinson's disease, loss of coordination, reduced motor signaling, memory abnormalities are examples of chronic health issues. Long-term negative impacts on human health include damages to the immune system, hypersensitivity, asthma and allergy, diabetes, respiratory illnesses, multiple sclerosis, and genetic disorders (Kim et al., 2017). Effects on reproductive health include changes in sexual behavior, pregnancy, and infertility as well as neurological symptoms like weakness in the arms, legs, and hands, and concentration problems, as well as kidney, liver, blood, or digestive tract disorders. Pesticides are linked to Leukemia, cancers of the breast, prostate, ovaries, testicles, lymphoma, brain, and lymph nodes.

Long-term, low-dose exposure to certain environmental chemicals like pesticides referred to as endocrine disruptors, is significantly related to human risk like immune suppression, decreased intelligence, hormone disruption, reproductive abnormalities, and cancer (Brouwer et al., 1999; Crisp et al., 1998; Hurley et al., 1998).

The high-risk occupational categories exposed to pesticides involve agricultural workers, factory employees, pesticide formulators, spray men and pesticide mixers and loaders. As a result of the procedures involved not being risk-free, there may be more dangers during manufacture and formulation. Because they handle a variety of dangerous toxics, including insecticides, toxic solvents, and inert carriers, workers in industrial environments are more at risk. The National Institute of Occupational Health (NIOH) evaluated the level of toxicity risk associated with the field application of the carbamate pesticide methomyl (Saiyed et al., 1992). The spray men's ECG, cholinesterase (ChE) and serum LDH levels all showed significant changes that pointed to methomyl's cardiotoxic effects. In addition to neurological, psychological, gastrointestinal, and cardiorespiratory symptoms health surveillance in male formulators producing liquid and dust formulations of different pesticides (malathion, DDT, methyl parathion and lindane) in industrial sector revealed greater incidence of generalized symptoms like nausea, headache, fatigue, vomiting, eye, and skin irritation (Pan et al., 2020).

7.4 Conclusion and future prospects

Despite the benefits, pesticides have sparked concerns about the structure, equilibrium, function, and sustainability of the ecosystems. They have imposed a serious threat to the biological diversity and synergism existing along trophic levels in a food chain in marine, aquatic, and soil ecosystems. Apparently, pesticides have made their way into the air. The risks connected with pesticide application cannot be completely eradicated, but they can be mitigated in certain ways.

Better, safer, and more environmentally friendly pesticide formulations could decrease the negative impacts of pesticide application. Pesticide toxicity can be minimized if the pesticides are utilized in the appropriate proportions. Akin to this, the pesticide toxicity can be mitigated by employing a less hazardous version or a small dose of a hazardous product.

Further, the farmers sometimes are not aware of the potential toxicity of pesticides. These individuals lack the knowledge pertaining to different pesticide types, poisoning potential, risks, and safety protocols to be implemented before their application. Hazardous and ecologically pervasive pesticides often are applied, which raises the possibility of intentional, unintentional, or professional exposure. The health of individuals is influenced by these chemicals over time. Thus, it becomes important to educate and sensitize these producers/farmers about the necessity of employing minimal chemical pesticides.

By employing alternate cropping techniques or well-maintained spraying equipment, it is possible to minimize pesticide toxicity and the negative impacts that result from it. Additional strategies include using incentives to stimulate healthier and sustainable market trends, modifying crop management techniques, integrating ecosystem-based methodologies, employing economic resources to ensure environmentally sustainable products and strategies, encouraging the use of project payments to sustain sustainable farming, and toughening norms and introducing organizational policies for regulating supply chains that are self-sustaining.

Future pest and insect control methods may combine the use of chemical pesticides with organic cures for longer-lasting pest and insect control. In addition to promising environmental sustainability, this combination has a variety of applications for eliminating invasive species and pests in urban areas.

Integrating regulatory frameworks for value fertilizer regulation; strengthening international regulations for the safe and responsible use of fertilizer application; enhancing the training with all key stakeholders in fertilizer management; and guaranteeing that appropriate and affordable fertilizers availability are all measures that can be taken to enhance fertilizer and nutrient management.

A thorough understanding of the interactions between plants and nanomaterials opens new possibilities for enhancing farming practices through higher disease resistance, crop output, and nutrient uptake. To minimize the harmful effects of pesticides in the environment, community engagement and other extended projects that inform and motivate farmers to embrace the cutting-edge IPM tactics are important. Integrated pest management (IPM) to reduce pesticide use have had limited success due to low farmer participation for a variety of reasons. An all-India survey revealed that 34% of respondents had no knowledge of IPM and only 5% were using complete IPM technology. Furthermore, the IPM package may potentially lead to increased pesticide use in multi-cropped complex agroecosystems. Alternative policy designs are needed in this instance to regulate the use and handling of pesticides in agriculture.

Integrated pest management encompasses a range of strategies used by farms commencing with detection of pests before spraying, cultivation of pest-resistant crop varieties, implementing mechanical pest traps, introducing predatory insects to target plant-eating pests, and destroying pest nesting sites. IPM farmers also entail a tailored strategy in case of extreme control regulation, switching to broadcast spraying in only extreme situations. Planting cover crops, timely weed removal and tilling, and early or late sowing of crops to outcompete weeds are some methods sustainable farms use to lessen the need for pesticides. The on-farm environment must be carefully observed in order to make any of these options, and many of them include more work than simply applying a pesticide.

References

Akashe, M. M., Pawade, U. V., & Nikam, A. V. (2018). Classification of pesticides: Review. *International Journal of Research in Ayurveda and Pharmacy, 9*(4), 144–150.

Aktar, M. W., Paramasivam, M., SenGupta, D., Purkait, S., Ganguly, M., & Banerjee, S. (2009). Impact assessment of pesticide residues in fish of Ganga river around Kolkata in West Bengal. *Environmental Monitoring and Assessment, 157*(1–4), 97–104. https://doi.org/10.1007/s10661-008-0518-9.

Aktar, M. W., SenGupta, D., & Chowdhury, A. (2009). Impact of pesticides use in agriculture: Their benefits and hazards. *Interdisciplinary Toxicology, 2*(1), 1–12. https://doi.org/10.2478/v10102-009-0001-7.

Alewu, B., & Nosiri, C. (2011). Pesticides and human health. *Pesticides in the modern world: Effects of pesticides exposure* (pp. 231–250). InTech Open.

Ansara-Ross, T., Wepener, V., van den Brink, P. V. D., & Ross, M. (2012). Pesticides in South African fresh waters. *African Journal of Aquatic Science, 37*(1), 1–16. https://doi.org/10.2989/16085914.2012.666336.

Arora, N. K., & Mishra, I. (2019). United Nations Sustainable Development Goals 2030 and environmental sustainability: Race against time. *Environmental Sustainability, 2*(4), 339–342.

Assey, G. E. (2002). Evaluation of the state of the art analytical technologies for the study of pesticides and other persistent organic pollutants. *Chemosphere, 305*, 135342.

Atwood, D., & Paisley-Jones, C. (2017). *Pesticides industry sales and usage: 2008–2012 market estimates*. US Environmental Protection Agency.

Bajwa, A. A., Walsh, M., & Chauhan, B. S. (2017). Weed management using crop competition in Australia. *Crop Protection, 95*, 8–13.

Baker, L. W., Fitzell, D. L., Seiber, J. N., Parker, T. R., Shibamoto, T., Poore, M. W., Longley, K. E., Tomlin, R. P., Propper, R., & Duncan, D. W. (1996). Ambient air concentrations of pesticides in California. *Environmental Science and Technology, 30*(4), 1365–1368. https://doi.org/10.1021/es950608l.

Benka-Coker, W., Hoskovec, L., Severson, R., Balmes, J., Wilson, A., & Magzamen, S. (2020). The joint effect of ambient air pollution and agricultural pesticide exposures on lung function among children with asthma. *Environmental Research, 190*, 109903. https://doi.org/10.1016/j.envres.2020.109903.

Biesmeijer, J. C., Roberts, S. P., Reemer, M., Ohlemüller, R., Edwards, M., Peeters, T., & Kunin, W. E. (2006). Parallel declines in pollinators and insect pollinated plants in Britain and the Netherlands. *Science (New York, NY), 313*(5785), 351–354. https://doi.org/10.1126/science.1127863.

Blacquière, T., Smagghe, G., Cornelis, A. M., & van Gestel, M. V. (2012). Neonicotinoids in bees: a review on concentrations, side-effects and risk assessment. *Ecotoxicology, 21*(21), 973–992.

Boldt, T. S., & Jacobsen, C. S. (1998). Different toxic effects of the sulfonylurea herbicides metsulfuron methyl, chlorsulfuron and thifensulfuron methyl on fluorescent pseudomonads isolated from agricultural soil. *FEMS Microbiology Letters, 161*(1), 29–35. https://doi.org/10.1111/j.1574-6968.1998.tb12925.x.

Brammall, R. A., & Higgins, V. J. (1988). The effect of glyphosate on resistance of tomato to Fusarium crown and root rot disease and on the formation of host structural defensive barriers. *Canadian Journal of Botany, 66*(8), 1547–1555. https://doi.org/10.1139/b88-213.

Brouwer, A., Longnecker, M. P., Birnbaum, L. S., Cogliano, J., Kostyniak, P., Moore, J., Schantz, S., & Winneke, G. (1999). Characterization of potential endocrine related health effects at low dose levels of exposure to PCbs. *Environmental Health Perspectives, 107*(Suppl. 4), 639–649. https://doi.org/10.1289/ehp.99107s4639.

Brühl, C. A., Schmidt, T., Pieper, S., & Alscher, A. (2013). Terrestrial pesticide exposure of amphibians: An underestimated cause of global decline? *Scientific Reports, 3*(1), 1–4.

Brussaard, L. (1997). Biodiversity and ecosystem functioning in soil. *Ambio – A Journal of the Human Environment, 26*, 563–570.

Casabé, N., Piola, L., Fuchs, J., Oneto, M. L., Pamparato, L., Basack, S., Giménez, R., Massaro, R., Papa, J. C., & Kesten, E. (2007). Ecotoxicological assessment of the effects of glyphosate and chlorpyrifos in an Argentine soya field. *Journal of Soils and Sediments, 7*(4), 232–239. https://doi.org/10.1065/jss2007.04.224.

Chakravarty, P., & Sidhu, S. S. (1987). Effects of glyphosate, hexazinone and triclopyr on in vitro growth of five species of ectomycorrhizal fungi. *Forest Pathology, 17*(4–5), 204–210. https://doi.org/10.1111/j.1439-0329.1987.tb01017.x.

Chourasiya, S., Khillare, P. S., & Jyethi, D. S. (2015). Health risk assessment of organochlorine pesticide exposure through dietary intake of vegetables grown in the periurban sites of Delhi, India. *Environmental Science and Pollution Research, 22*(8), 5793–5806.

Cooper, J., & Dobson, H. (2007). The benefits of pesticides to mankind and the environment. *Crop Protection, 26*(9), 1337–1348.

Crisp, T. M., Clegg, E. D., Cooper, R. L., Wood, W. P., Anderson, D. G., Baetcke, K. P., Hoffmann, J. L., Morrow, M. S., Rodier, D. J., Schaeffer, J. E., Touart, L. W., Zeeman, M. G., & Patel, Y. M. (1998). Environmental endocrine disruption: An effects assessment and analysis. *Environmental Health Perspectives, 106*(Suppl. 1), 11–56. https://doi.org/10.1289/ehp.98106s111.

Decourtye, A., Lacassie, E., & Pham-Delègue, M. H. (2003). Learning performances of honeybees (*Apis mellifera* L.) are differentially affected by Imidacloprid according to the season. *Pest Management Science, 59*(3), 269–278. https://doi.org/10.1002/ps.631.

Delaplane, K. S. (2000). *Pesticide usage in the United States: History, benefits, risks, and trends*. Cooperative Extension Service.

Doan, N. H., Duong, H. T., Trinh, H. T., Tanaka, Y., & Kadokami, K. (2021). Comprehensive study of insecticides in atmospheric particulate matter in Hanoi, Vietnam: Occurrences and human risk assessment. *Chemosphere, 262*, 128028. https://doi.org/10.1016/j.chemosphere.2020.128028.

Dreistadt, S. H., Clark, J. K., & Flint, M. L. (1994). *Pests of landscape trees and shrubs: An integrated pest management guide*. University of California Division of Agriculture and Natural Resources Publication.

Đurišíc-Mladenovíc, N., Škrbíc, B., & Cvejanov, J. (2010). Organochlorine pesticides in soil and sediment from an urban zone of Novi Sad, Serbia. In *Survival and sustainability: Environmental concerns in the 21st century* (pp. 1469–1481). Springer.

Elgueta, S., Moyano, S., Sepúlveda, P., Quiroz, C., & Correa, A. (2017). Pesticide residues in leafy vegetables and human health risk assessment in North Central agricultural areas of Chile. *Food Additives and Contaminants. Part B, Surveillance, 10*(2), 105–112. https://doi.org/10.1080/19393210.2017.1280540.

Evans, S. C., Shaw, E. M., & Rypstra, A. L. (2010). Exposure to a glyphosate-based herbicide affects agro bio predatory arthropod behavior and long-term survival. *Ecotoxicology, 19*(7), 1249–1257. https://doi.org/10.1007/s10646-010-0509-9.

Fang, Y., Nie, Z., Die, Q., Tian, Y., Liu, F., He, J., & Huang, Q. (2017). Organochlorine pesticides in soil, air, and vegetation at and around a contaminated site in southwestern China: concentration, transmission, and risk evaluation. *Chemosphere, 178*, 340–349.

Fishel, F. M. (2011). *Pesticides effects on nontarget organisms. PI-85*. University of Florida Institute of Food and Agricultural Sciences. http://edis.ifas.ufl.edu/pi122.

Fletcher, J. S., Pfleeger, T. G., & Ratsch, H. C. (1993). Potential environmental risks associated with the new sulfonylurea herbicides. *Environmental Science and Technology, 27*(10), 2250–2252. https://doi.org/10.1021/es00047a037.

Frankenberger, W. T., & Tabatabai, M. A. (1991). Factors affecting l-asparaginase activity in soils. *Biology and Fertility of Soils, 11*(1), 1–5. https://doi.org/10.1007/BF00335825.

Gill, H. K., & Garg, H. (2014). Pesticides: Environmental impacts and management strategies. In *Pesticide*. IntechOpen. https://www.intechopen.com/chapters/46083. https://doi.org/10.5772/57399.

Godheja, J., Shekhar, S. K., Siddiqui, S. A., & Modi, D. R. (2016). Xenobiotic compounds present in soil and water: a review on remediation strategies. *Journal of Environmental & Analytical Toxicology, 6*(392) 2161-0525.

Gouin, T., Shoeib, M., & Harner, T. (2008). Atmospheric concentrations of current-use pesticides across south-central Ontario using monthly-resolved passive air samplers. *Atmospheric Environment, 42*(34), 8096–8104. https://doi.org/10.1016/j.atmosenv.2008.05.070.

Goulson, D. (2013). An overview of the environmental risks posed by neonicotinoid insecticides. *Journal of Applied Ecology, 50*(4), 977–987. https://doi.org/10.1111/1365-2664.12111.

Goulson, D. (2014). Pesticides linked to bird declines. *Nature, 511*(7509), 295–296.

Helfrich, L. A., Weigmann, D. L., Hipkins, P. A., & Stinson, E. R. (2009). *Pesticides and aquatic animals: A guide to reducing impacts on aquatic systems*. Virginia Polytechnic Institute and State University.

Hladik, M. L., Main, A. R., & Goulson, D. (2018). Environmental risks and challenges associated with neonicotinoid insecticides. https://www.beyondpesticides.org/programs/wildlife

Hurley, P. M., Hill, R. N., & Whiting, R. J. (1998). Mode of carcinogenic action of pesticides inducing thyroid follicular cell tumors in rodents. *Environmental Health Perspectives, 106*(8), 437–445. https://doi.org/10.1289/ehp.98106437.

Hussain, S., Siddique, T., Saleem, M., Arshad, M., & Khalid, A. (2009). Impact of pesticides on soil microbial diversity, enzymes, and biochemical reactions. *Advances in Agronomy, 102*, 159–200. https://doi.org/10.1016/S0065-2113(09)01005-0.

Jepson, P. C., Murray, K., Bach, O., Bonilla, M. A., & Neumeister, L. (2020). Selection of pesticides to reduce human and environmental health risks: a global guideline and minimum pesticides list. *The Lancet Planetary Health, 4*(2), e56–e63.

Kawahara, J., Horikoshi, R., Yamaguchi, T., Kumagai, K., & Yanagisawa, Y. (2005). Air pollution and young children's inhalation exposure to organophosphorus pesticide in an agricultural community in Japan. *Environment International, 31*(8), 1123–1132. https://doi.org/10.1016/j.envint.2005.04.001.

Kelley, W. D., & South, D. B. (1978). In vitro effects of selected herbicides on growth and mycorrhizal fungi. In *Weed Science Society of America Meeting Abstracts* (p. 38). Auburn University.

Kim, K. H., Kabir, E., & Jahan, S. A. (2017). Exposure to pesticides and the associated human health effects. *Science of the Total Environment, 575*, 525–535.

Kniss, A. R. (2018). Genetically engineered herbicide-resistant crops and herbicide-resistant weed evolution in the United States. *Weed Science, 66*(2), 260–273.

Kumari, M. (2012). Effects of organophosphate pesticide Abate on the ovary of the catfish, *Heteropneustes fossilis* (Bloch.). *Bangladesh Journal of Zoology, 40*(2), 207–212. https://doi.org/10.3329/bjz.v40i2.14314.

Lang, M., & Cai, Z. (2009). Effects of chlorothalonil and carbendazim on nitrification and denitrification in soils. *Journal of Environmental Sciences, 21*(4), 458–467. https://doi.org/10.1016/S1001-0742(08)62292-5.

Langenbach, T., Mano, D., Campos, M. M., Cunha, A. L. M. C., & De Campos, T. M. P. (2017). Pesticide dispersion by spraying under tropical conditions. *Journal of Environmental Science and Health Part B, Pesticides, Food Contaminants, and Agricultural Wastes, 52*(12), 843–849. https://doi.org/10.1080/03601234.2017.1359040.

Lee, S., McLaughlin, R., Harnly, M., Gunier, R. B., & Kreutzer, R. (2002). Community exposures to airborne agricultural pesticides in California: Ranking of inhalation risks. *Environmental Health Perspectives, 110*(12), 1175–1184. https://doi.org/10.1289/ehp.021101175.

Liroff, R. A. (2000). Balancing risks of DDT and malaria in the global POPs treaty. *Pesticide Safety News, 4*, 3.

Liu, Y., Mo, R., Tang, F., Fu, Y., & Guo, Y. (2015). Influence of different formulations on chlorpyrifos behavior and risk assessment in bamboo forest of China. *Environmental Science and Pollution Research International, 22*(24), 20245–20254. https://doi.org/10.1007/s11356-015-5272-2.

Machemer, L. H. (1986). Hazards for man and the environment from pesticides. *Pflanzenschutz – Nachrielite Bayer, 39*, 1–19.

Margalida, A., Heredia, R., Razin, M., & Hernández, M. (2008). Sources of variation in mortality of the Bearded vulture *Gypaetus barbatus* in Europe. *Bird Conservation International, 18*(1), 1–10. https://doi.org/10.1017/S0959270908000026.

Marie, L., Sylvain, P., Benoit, G., Maurice, M., & Gwenaël, I. (2017). Degradation and transport of the chiral herbicide S-metolachlor at the catchment scale: Combining observation scales and analytical approaches. *Environmental Science and Technology, 51*(22), 13231–13240. https://doi.org/10.1021/acs.est.7b02297.

Mayanglambam, T., Vig, K., & Singh, D. K. (2005). Quinalphos persistence and leaching under field conditions and effects of residues on dehydrogenase and alkaline phosphomonoesterase activities in soil. *Bulletin of Environmental Contamination and Toxicology, 75*(6), 1067–1076. https://doi.org/10.1007/s00128-005-0858-x.

Mingo, V., Lötters, S., & Wagner, N. (2017). The impact of land use intensity and associated pesticide applications on fitness and enzymatic activity in reptiles: A field study. *Science of the Total Environment, 590–591*, 114–124. https://doi.org/10.1016/j.scitotenv.2017.02.178.

Miraglia, M., Marvin, H. J., Kleter, G. A., Battilani, P., Brera, C., Coni, E., Cubadda, F., Croci, L., De Santis, B., Dekkers, S., Filippi, L., Hutjes, R. W., Noordam, M. Y., Pisante, M., Piva, G., Prandini, A., Toti, L., van den Born, G. J., & Vespermann, A. (2009). Climate change and food safety: An emerging issue with special focus on Europe. *Food and Chemical Toxicology, 47*(5), 1009–1021. https://doi.org/10.1016/j.fct.2009.02.005.

Nicolopoulou-Stamati, P., Maipas, S., Kotampasi, C., Stamatis, P., & Hens, L. (2016). Chemical pesticides and human health: the urgent need for a new concept in agriculture. *Frontiers in Public Health, 4*, 148.

Nkongolo, K. K., & Narendrula-Kotha, R. (2020). Advances in monitoring soil microbial community dynamic and function. *Journal of Applied Genetics, 61*(2), 249–263.

O'Neil, W., & Raucher, R. (1998, August). *Ground water public policy leaflet series no. 4: The costs of groundwater contamination*. Groundwater Policy Education Project. http://www.dnr.state.wi.us/org/water/dwg/gw/costofgw.htm.

Pan, D., He, M., & Kong, F. (2020). Risk attitude, risk perception, and farmers' pesticide application behavior in China: A moderation and mediation model. *Journal of Cleaner Production, 276*, 124241. https://doi.org/10.1016/j.jclepro.2020.124241.

Pandya, I. Y. (2018). Pesticides and their applications in agriculture. *Asian Journal of Applied Science and Technology, 2*(2), 894–900.

Parween, T., Bhandari, P., Sharma, R., Jan, S., Siddiqui, Z. H., & Patanjali, P. K. (2018). Bioremediation: a sustainable tool to prevent pesticide pollution. In *Modern age environmental problems and their remediation* (pp. 215–227). Springer.

Pell, M., Stenberg, B., & Torstensson, L. (1998). Potential denitrification and nitrification tests for evaluation of pesticide effects in soil. *Ambio – A Journal of the Human Environment, 27*, 24–28.

Pereira, J. L., Antunes, S. C., Castro, B. B., Marques, C. R., Gonçalves, A. M., Gonçalves, F., & Pereira, R. (2009). Toxicity evaluation of three pesticides on non-target aquatic and soil organisms: Commercial formulation versus active ingredient. *Ecotoxicology, 18*(4), 455–463.

Pilling, E. D., & Jepson, P. C. (1993). Synergism between EBI fungicides and a pyrethroid insecticide in the honeybee (*Apis mellifera*). *Pesticide Science, 39*(4), 293–297. https://doi.org/10.1002/ps.2780390407.

Pimentel, D., & Burgess, M. (2014). Environmental and economic costs of the application of pesticides primarily in the United States. In *Integrated pest management* (pp. 47–71). Springer.

Raffa, C. M., & Chiampo, F. (2021). Bioremediation of agricultural soils polluted with pesticides: A review. *Bioengineering, 8*(7), 92.

Randika, J. L. P. C., Bandara, P. K. G. S. S., Soysa, H. S. M., Ruwandeepika, H. A. D., & Gunatilake, S. K. (2022). Bioremediation of pesticide-contaminated soil: a review on indispensable role of soil bacteria. *Journal of Agricultural Sciences–Sri Lanka, 17*(1).

Rasmussen, J. J., Wiberg-Larsen, P., Baattrup-Pedersen, A., Cedergreen, N., McKnight, U. S., Kreuger, J., Jacobsen, D., Kristensen, E. A., & Friberg, N. (2015). The legacy of pesticide pollution: An overlooked factor in current risk assessments of freshwater systems. *Water Research, 84*, 25–32. https://doi.org/10.1016/j.watres.2015.07.021.

Relyea, R. A. (2005). The lethal impact of roundup on aquatic and terrestrial amphibians. *Ecological Applications, 15*(4), 1118–1124. https://doi.org/10.1890/04-1291.

Relyea, R. A., & Hoverman, J. T. (2008). Interactive effects of predators and a pesticide on aquatic communities. *Oikos, 117*(11), 1647–1658. https://doi.org/10.1111/j.1600-0706.2008.16933.x.

Ross, G. (2005). Risks and benefits of DDT. *The Lancet, 366*(9499), 1771–1772.

Saiyed, H. N., Sadhu, H. G., Bhatnagar, V. K., Dewan, A., Venkaiah, K., & Kashyap, S. K. (1992). Cardiac toxicity following short term exposure to methomyl in spraymen and rabbits. *Human and Experimental Toxicology, 11*(2), 93–97. https://doi.org/10.1177/096032719201100206.

Salem, F. B., Said, O. B., Aissa, P., Mahmoudi, E., Monperrus, M., Grunberger, O., & Duran, R. (2016). Pesticides in Ichkeul Lake–Bizerta Lagoon Watershed in Tunisia: Use, occurrence, and effects on bacteria and free-living marine nematodes. *Environmental Science and Pollution Research, 23*(1), 36–48. https://doi.org/10.1007/s11356-015-4991-8.

Santos, A., & Flores, M. (1995). Effects of glyphosate on nitrogen fixation of free-living heterotrophic bacteria. *Letters in Applied Microbiology, 20*(6), 349–352. https://doi.org/10.1111/j.1472-765X.1995.tb01318.x.

Sarkar, S., Gil, J. D. B., Keeley, J., & Jansen, K. (2021). *The use of pesticides in developing countries and their impact on health and the right to food*. European Union.

Schreck, E., Geret, F., Gontier, L., & Treilhou, M. (2008). Neurotoxic effect and metabolic responses induced by a mixture of six pesticides on the earthworm *Aporrectodea caliginosa nocturna*. *Chemosphere, 71*(10), 1832–1839. https://doi.org/10.1016/j.chemosphere.2008.02.003.

SDG Knowledge Hub. UNEP report identifies top actions to minimize adverse impacts of pesticides. https://sdg.iisd.org/news/unep-report-identifies-top-actions-to-minimize-adverse-impacts-of-pesticides-fertilizers/.

Sharma, A., Kumar, V., Kumar, R., Shahzad, B., Thukral, A. K., & N Bhardwaj, R. (2018). Brassinosteroid-mediated pesticide detoxification in plants: A mini-review. *Cogent Food and Agriculture, 4*(1). https://doi.org/10.1080/23311932.2018.1436212.

Sharma, A., Kumar, V., Shahzad, B., Tanveer, M., Sidhu, G. P. S., Handa, N., … Thukral, A. K. (2019). Worldwide pesticide usage and its impacts on ecosystems. *SN Applied Sciences, 1*(11), 1446. https://doi.org/10.1007/s42452-019-1485-1.

Singh, D. K. (2012). Pesticides and environment. *Pesticide Chemistry and Toxicology, 1*, 114–122.

Škrbic, B., Cvejanov, J., & Djurisic-Mladenovic, N. (2007). Organochlorine pesticides and polychlorinated biphenyls in surface soils of Novi Sad and bank sediment of the Danube River. *Journal of Environmental Science and Health. Part B, 42*(3), 311–319. https://doi.org/10.1080/03601230701229312.

Sparling, D. W., & Fellers, G. M. (2009). Toxicity of two insecticides to California, USA, anurans and its relevance to declining amphibian populations. *Environmental Toxicology and Chemistry, 28*(8), 1696–1703. https://doi.org/10.1897/08-336.1.

Stoytcheva, M. (2011). *Pesticides in the modern world: Effects of pesticides exposure*. Books on Demand.

Strassemeyer, J., Daehmlow, D., Dominic, A. R., Lorenz, S., & Golla, B. (2017). SYNOPS-WEB, an online tool for environmental risk assessment to evaluate pesticide strategies on field level. *Crop Protection, 97*, 28–44.

Tago, D., Andersson, H., & Treich, N. (2014). Pesticides and health: A review of evidence on health effects valuation of risks and benefit cost analysis. *Advances in Health Economics and Health Services Research, 24*, 203–295. https://doi.org/10.1108/S0731-219920140000024006.

Tran, H. T., Lin, C., Bui, X. T., Ngo, H. H., Cheruiyot, N. K., Hoang, H. G., & Vu, C. T. (2021). Aerobic composting remediation of petroleum hydrocarbon-contaminated soil. Current and future perspectives. *Science of the Total Environment, 753*, 142250.

Tscharntke, T., Klein, A. M., Kruess, A., Steffan-Dewenter, I., & Thies, C. (2005). Landscape perspectives on agricultural intensification and biodiversity–ecosystem service management. *Ecology Letters, 8*(8), 857–874. https://doi.org/10.1111/j.1461-0248.2005.00782.x.

Tudi, M., Daniel Ruan, H., Wang, L., Lyu, J., Sadler, R., Connell, D., & Phung, D. T. (2021). Agriculture development, pesticide application and its impact on the environment. *International Journal of Environmental Research and Public Health, 18*(3), 1112. https://doi.org/10.3390/ijerph18031112.

Tuncel, S. G., Oztas, N. B., & Erduran, M. S. (2008). Air and groundwater pollution in an agricultural region of the Turkish Mediterranean coast. *Journal of the Air and Waste Management Association, 58*(9), 1240–1249. https://doi.org/10.3155/1047-3289.58.9.1240.

US Environmental Protection Agency. *Regulation of pesticide residues on food*. https://www.epa.gov/pesticide-tolerances#:~:text=We%20set%20tolerances%2C%20which%20are,maximum%20residue%20limits%20(MRLs).

US Environmental Protection Agency. (2001, July). *Source water protection practices bulletin: Managing small scale application of pesticides to prevent contamination of drinking water* (p. 816-F-01-031). Office of Water.

Wanwimon, S., Phopin, K., Boonpangrak, S., & Prachayasittikul, V. (2016). Food safety in Thailand 4: Comparison of pesticide residues found in three commonly consumed vegetables purchased from local markets and supermarkets in Thailand. *PeerJ, 4*, e2432. https://doi.org/10.7717/peerj.2432.

Waskom, R. (1994, August). *Best management practices for private well protection*. Colorado State University Cooperative Extension. http://hermes.ecn.purdue.edu:8001/cgi/convert?7488.

Weaver, E. R. (2017). *Sustainable development through urban agriculture*. University of South Florida.

Yang, E. C., Chuang, Y. C., Chen, Y. L., & Chang, L. H. (2008). Abnormal foraging behavior induced by sublethal dosage of Imidacloprid in the honey bee (Hymenoptera: Apidae). *Journal of Economic Entomology, 101*(6), 1743–1748. https://doi.org/10.1603/0022-0493-101.6.1743.

Yera, A. M. B., Nascimento, M. M., da Rocha, G. O., de Andrade, J. B., & Vasconcellos, P. C. (2020). Occurrence of pesticides associated with atmospheric aerosols: Hazard and cancer risk assessments. *Journal of the Brazilian Chemical Society, 31*, 1317–1326. https://doi.org/10.21577/0103-5053.20200017.

Zhang, W. (2018). Global pesticide use: Profile, trend, cost/benefit and more. *Proceedings of the International Academy of Ecology and Environmental Sciences, 8*(1), 1.

Zheng, S., Chen, B., Qiu, X., Chen, M., Ma, Z., & Yu, X. (2016). Distribution and risk assessment of 82 pesticides in Jiulong River and estuary in South China. *Chemosphere, 144*, 1177–1192.

CHAPTER 8

Pesticides-mediated ROS generation in plants

Ashutosh Sharma[a], Sanjay Kumar Sharma[b], Narayan Singh[a,c], Vasudha Maurya[a,c], Sukhwinder Kaur[b], Rahul Kumar[a] and Indu Sharma[b]
[a]Faculty of Agricultural Sciences, DAV University, Jalandhar, Punjab, India
[b]Department of Life Sciences, University Institute of Sciences, Sant Baba Bhag Singh University, Jalandhar, Punjab, India
[c]Department of Biotechnology, DAV University, Jalandhar, Punjab, India

8.1 Introduction

The agrochemicals such as fertilizers and pesticides are being used for enhancing crop productivity and protecting the plants from competing organisms (pathogens, pests and weeds, etc.) respectively to ensure global food security for the increasing human population. One of the broad categories of these agrochemicals employed during agronomic practices, to kill, destroy, attract, repel, or control any harmful crop pests, or prevent the plants from any targetted pests, are referred to as pesticides. Pesticides are an important attribute of contemporary agriculture that helps manage various biotic stresses in crops, promoting both the quality and quantity of the yield. The safest pesticides are nonpersistent in nature and can be easily transformed, broken, or converted into simpler forms in the presence of sunlight (Abubakar et al., 2020). Whereas, the nonpersistent pesticides are considered highly toxic and hazardous. These pesticides are categorized based on their composition, mode of action, source of their origin, toxicity induced, function, and targetted pest. The generally used pesticides are agrochemicals that include insecticides, herbicides, and fungicides. Among the farmers or crop producers, these pesticides are preferred-choice over other alternatives for crop protection because of their ready availability and a broad range of applications.

The usage of pesticides varies as per their mode of action and function. According to an estimate, globally, among all the pesticides, herbicides (47.55%) are the most commonly used pesticides followed by insecticides (29.5%), fungicides (17.5%), and other pesticides (5.5%) (Sharma et al., 2019). Among all the countries, China, United States, Argentina, Thailand, Brazil, Italy, France, Canada, Japan, and India are the topmost pesticide users (Alengebawy et al., 2021). Nearly 85% of total pesticides are used in agricultural fields for the enhancement of crop production. However, the matter of concern is that pesticides are used profusely beyond their actual requirement to enhance crop productivity (Intisar et al., 2022). Out of all pesticides that are used, only about 1% of them actually reach their target, whereas the rest 99% either degrade or drift off. When applied in agricultural soils

Pesticides in the Environment: Impact, Assessment, and Remediation
DOI: https://doi.org/10.1016/B978-0-323-99427-9.00001-X

or to plants, these pesticides stay for a long time in the environment (Alengebawy et al., 2021). Due to their long-time persistence, most of these pesticides have become one of the most potent environmental toxicants, threatening the entire mankind and also the entire ecosystem.

Even though these pesticides are associated with plant pest protection but their unwarranted use has posed a major environmental concern (Hatamleh et al., 2022). They not only affect the targeted pests or organisms but also negatively influence plants that are not their designated targets. Various studies have reported that the usage of pesticides affects several physiological processes like growth and development, photosynthesis, and also the crop yields (Bakshi et al., 2022; Hatamleh et al., 2022; Peng et al., 2023; Shakir et al., 2018; Sharma et al., 2019; Xi et al., 2022). Pesticides also induce some phytotoxic effects in plants such as decreased rate of seed germination (Hatamleh et al., 2022), photo-transformation on the leaf surface (Xi et al., 2022), reduced levels of photosynthetic pigments (Hatamleh et al., 2022), decreased contents of proteins and chlorophylls (Sharma et al., 2019), enhanced generation of reactive oxygen species (ROS) leading to oxidative burst (Peng et al., 2023), altered activities of antioxidant enzymes (Sharma et al., 2019), and oxidative stress (Shakir et al., 2018) in affected plants. The herbicides derived from glyphosate were reported to amend the photosynthetic apparatus and affect the photochemistry of the plants thereby affecting photosynthesis, the shikimate pathway, and antioxidant defense mechanisms (Zaller & Brühl, 2019).

In paddy fields, the application of an organophosphorus-derived chemical pesticide named malathion at 30 μg/mL dose was found to negatively affect the cell growth, the physiological and biochemical indices in *Anabaena sphaerica,* an important nitrogen-fixing cyanobacterium present in paddy fields (Chakraborty et al., 2017). It was also reported that malathion enhances the generation of ROS and results in a decrease of total pigments, total proteins, and carbohydrates, thereby reduces photosynthetic efficiency. Also, the activities of superoxide dismutase, ascorbate peroxidase, and catalase were enhanced as a result of the activation of ROS scavenging mechanisms. For ROS detoxification, the levels of antioxidants namely phenolics, flavonoids, and carotenoids were also stimulated in *A. sphaerica.* The pesticides not only affect the growth and development of standing crops or cyanobacteria by causing oxidative stress; but being persistent in nature, they may get accumulated in plants and have long-term effects on living organisms.

The persistent and hazardous pesticides are accumulated and detected as pesticide residue in the products derived from plants (pesticide-stressed). In single crude herbal drugs originating from four medicinal plants viz., *Emblica officinalis, Terminalia chebula, Terminalia belerica*, and *Withania somnifera*, the pesticide residues of organochlorine-derived, organophosphorus-derived, and pyrethroid pesticides were detected (Rao et al., 2011). The consumption of such affected herbal formulations by humans poses a severe threat to mankind. To overcome the side effects of synthetic and hazardous pesticides, some alternatives may be developed at a large scale for enhancing crop productivity.

Moreover, various plant growth regulators may be applied to combat pesticide-induced stress in plants (Jan et al., 2020). The present chapter is an attempt to discuss the classification of pesticides, their adverse impact, pesticide-mediated ROS production, oxidative stress, and mechanisms to overcome pesticide-induced toxicities in plants.

8.2 Various classes of pesticides and their comparison in terms of ROS generation

The pesticides are chemical or biological substances or compounds that differ from each other on the basis of their structure, composition, biochemical properties (Yadav & Devi, 2017) and are used to manage the crop pests. On the basis of their mode of entry, the pesticide may be chiefly categorized into five types as mentioned below.

1. **Systemic pesticides**: These pesticides are absorbed, transported and transferred by the plants to the targeted site, e.g., 2, 4-Dichlorophenoxyacetic acid (2, 4-D), neonicotinoids, phenyl-pyrazole fipronil, glyphosate, etc.
2. **Nonsystemic pesticides**: Those pesticides which only act when coming in physical contact with the target pest and thus, are also called contact pesticides, e.g., diquat dibromide, paraquat, pyrethroids, acephate, carbaryl, liquid fipronil, etc.
3. **Fumigant pesticides**: These pesticides kill the pest through vapors or fumes produced by the pesticide which form attacks the respiratory system leading to poisoning and death of targeted pest, e.g., nicotine, hydrogen cyanide, naphthalene, methyl bromide, dibromochloropropane, chloropicrin, etc.
4. **Stomach Toxicant or Poisoning pesticides**: Those pesticides which enter through the mouth into the digestive tract of the pest and then cause its death due to stomach poisoning, e.g., boric acid, borates, malathion, *Bacillus thuringiensis* (Bt), etc.
5. **Repellent pesticides**: Those volatile pesticides which are not used to kill, but only to repel the pests and discourage the pest to reach the crop, e.g., picaridin, permethrin, neem oil, etc.

As per other criteria for classification, the pesticides may be classified on the type of targeted pest and function of the pesticide (Kim et al., 2017; Yadav & Devi, 2017). Fig. 8.1 explains the classification of pesticides based on different criteria. Bactericide such as azoxystrobin targets pest namely bacteria, blights, mildews, molds, rusts, etc. Insecticides such as aldicarb target pests like insects and other arthropods. The algicide (or algaecide), avicide, fungicide, herbicide, miticide, molluscicide, nematicide, piscicide, rodenticide, larvicide, virucide, desiccant, lampricide, silvicides, and termiticides targets algae, birds, fungi, weeds, mites, snails and slugs, nematodes, fish, rodents, larvae, viruses, plants (dries plant tissues), larvae of lampreys, woody vegetation, and termites, respectively. On the basis of chemical compositions, the pesticides may be derived from chemicals referred to as synthetic, and derived from natural plant products or mineral oils are called natural or bio-pesticides (Yadav & Devi, 2017). Further, on the basis of their chemical structure,

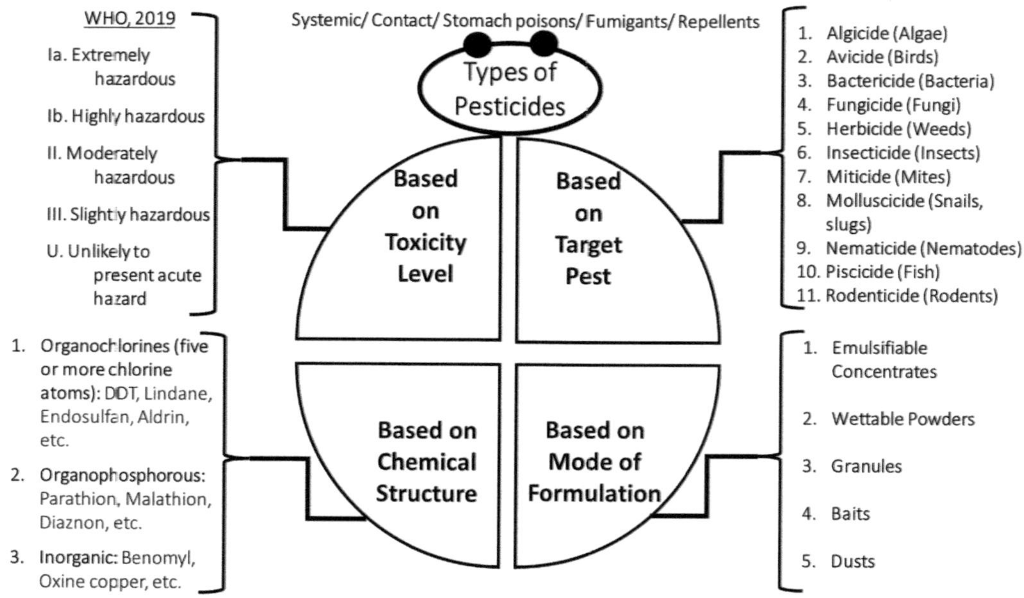

Figure 8.1 Classification of pesticides based on four different criteria.

the pesticides may be categorized as organic (organochlorines: five or more chlorine atoms, organophosphorus, carbamates, and pyrethroids), and inorganic pesticides (Kim et al., 2017).

Another type of classification is based on the mode of action of pesticides (Yadav & Devi, 2017). Accordingly, pesticides may be categorized as (1) physical poison (kills pest through physical effect); (2) protoplasmic poison (acts through protein precipitation-induced poisoning); (3) respiratory poison (acts through inactivation of respiratory enzymes); (4) nerve poison (inhibition of impulse conduction); and (5) chitin inhibition (inhibits the synthesis of chitin in pests). The pesticides may also be classified on the basis of types of formulations, such as active ingredients (control target pests) and inert ingredients (added to active ingredients to prepare safer, and more effective pesticides). Further, on the basis of the mode of formulations, these may be classified as liquids, ultra-low volume concentrates, powders, dust, granules, and baits (Kim et al., 2017).

Further, on the basis of their toxicity, World Health Organisation (WHO) has classified pesticides as per hazard and guidelines for classification (WHO, 2019). WHO has prescribed criteria for the classification of pesticides that was revised in 2019. It is the most recent recommended criteria for classification under "The Globally Harmonized System of Classification and Labelling of Chemicals" (GHS) by WHO developed in consultation with participating countries, international agencies and regional bodies (Table 8.1).

Table 8.1 Recommended criteria for the classification of pesticides by WHO (2019).

Criteria for classification of pesticides		LD_{50} for the rat (mg/kg body weight)	
Class	**Levels of toxicity**	**Oral**	**Dermal**
Ia	Extremely hazardous	<5	<50
Ib	Highly hazardous	5–50	50–200
II	Moderately hazardous	50–2000	200–2000
III	Slightly hazardous	Over 2000	Over 2000
U	Unlikely to present acute hazard	5000 or higher	5000 or higher

Different types of pesticides act as xenobiotics when used excessively and they cause the stimulation of ROS production and related pathways in the affected living organisms. In a biological unicellular model alga named *Chlamydomonas reinhardtii*, the effect of fluroxypyr (herbicide) applied at 0.05–1.00 mg/L concentrations was observed (Zhang et al., 2011). Fluroxypyr was found to be significantly accumulated by the alga and it was further degraded in the cells of *C. reinhardtii*. Further, 0.05–1.00 mg/L fluroxypyr was observed to enhance ROS production in the alga, which was associated with increased levels of peroxides as well as the degradation of DNA. In cucumber plants, imidacloprid and dichlorvos were used to manage the cucumber whiteflies by Homayoonzadeh et al. (2020). The pesticide application resulted in the stimulation of ROS scavenging antioxidant defense system.

8.3 The pesticide-induced oxidative burst

Pesticides are applied to plants especially economically important crops to enhance their productivity by reducing the adverse impact of harmful crop pests. The fate of pesticides and their mode of action may depend on the following factors:

- Type of Crop or plant where the pesticide is to be applied;
- Agronomic or cultivation practices of particular crop/plant;
- Agro-meteorological factors or environmental conditions during crop period;
- Type of pest and its severity on affected crops;
- Selection of particular pesticide;
- Dose and Mode of Application of pesticide;
- Mechanism of absorption and translocation of pesticide by the crop;
- Mode of action of pesticide;
- Effect of pesticide on target pest;
- Potential effect of pesticide on nontarget organisms and ecosystem;
- Effect of pesticide on regular physiological processes and/or metabolism of the crop.

Different pesticides invariably affect the physiology of plants and cause toxicity at their higher doses. Many of these effects of pesticide toxicity are linked to the enhanced production of reactive oxygen species (ROS) like in the case of other abiotic stresses in plants. The pesticide-induced physiological changes were reported in *Solanum lycopersicum*, when it was exposed to three different pesticides namely diazinon, imidacloprid, and mancozeb. A progressive decline in various physiological and biometric parameters of tomatoes was reported in a dose-dependent manner and it differed with the type of pesticide used (Hatamleh et al., 2022). Mancozeb at 200 μg/mL concentration was found to be most most harmful concentration for the germination, roots, contents of sugars, proteins, lycopene, chlorophylls and carotenoids, and fruits. These pesticides also negatively affected the flowering, fruiting, fruit attributes, root cell viability, activities of antioxidant enzymes, and cellular membranes. Also, cell death and oxidative stress were induced in tomato plants that also affected the activities of antioxidant enzymes (ascorbate peroxidase, catalase, superoxide dismutase, peroxidase, and guaiacol peroxidase). The higher concentration of diazinon, imidacloprid, and mancozeb enhanced the production of ROS, altered the membrane potential of mitochondria, induced root cell death, and caused severe oxidative damage to tomatoes. Hence, pesticides significantly affect the biological processes at the cellular, tissue, and molecular level. In both plants and humans, during various normal cellular metabolic processes, ROS are generated as byproducts that are essential to produce energy (Snezhkina et al., 2019). These reactive free radicals also act as physiological regulators of diverse intracellular signaling cascades (Sule et al., 2022). Although ROS are required at a threshold level for regular signaling pathways in plants, their accumulation higher than the desired amounts results in an irreparable loss in plant metabolism. Fig. 8.2 highlights the ROS equilibrium during regular cellular processes under normal conditions in living organisms.

However, during pesticide stress (pesticide-induced oxidative stress), the delicate balance between ROS production and removal is disrupted (Sule et al., 2022). Due to the imbalance between ROS generation and its subsequent scavenging by the antioxidant defense system, the excessively produced ROS are not neutralized in the affected cells. It is also reported that besides ROS, reactive nitrogen species (RNS) are also involved in the pesticide-induced oxidative burst (Sharma et al., 2021). Fig. 8.3 is an attempt to explain the pesticide-induced ROS-mediated oxidative stress in plants. The ROS is a collective term used to define oxygen-containing reactive species such as hydroxyl ion free radical (·OH), hydrogen peroxide (H_2O_2), singlet oxygen ($^1O_2$), superoxide ($O_2·^-$), alkoxyl (RO·), peroxyl (ROO·), hypochlorous acid (HOCl) free radicals, ozone (O_3) and lipid hydroperoxide (LOOH) (Sule et al., 2022). Whereas, RNS is a term used to define nitrogen-containing reactive species namely, nitric oxide ion (·NO), nitroxide anion (NO^-), nitroxyl anion or azanone (HNO), nitrosonium cation (NO^+), higher oxides of nitrogen such as nitrogen dioxide ($·NO_2$) and dinitrogen trioxide (N_2O_3), peroxynitrite ($ONOO^-$), S-nitrosothiols, nitrosoperoxycarbonate ($ONOOCO_2^-$), and dinitrosyl iron complexes, etc. Both ROS and RNS are sometimes also referred to as Reactive oxygen and nitrogen species (RONS) (Weidinger & Kozlov, 2015). The levels

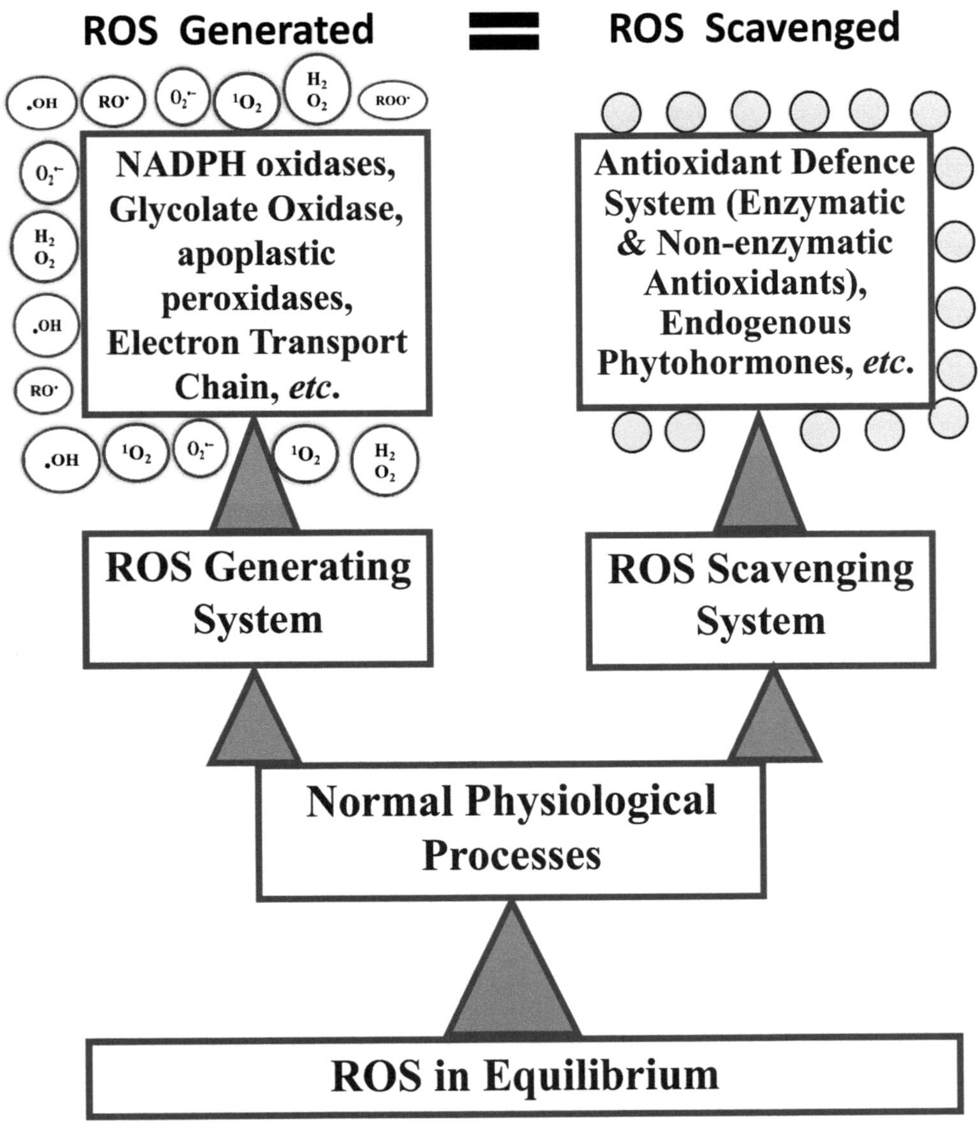

Figure 8.2 ROS equilibrium: ROS generation (*red circles*) and subsequent scavenging (*green circles*) under normal conditions (*no stress*) in plants. *(Color version of figure is available online).*

of ROS namely H_2O_2, $O_2^{-}\cdot$, and ^{-}OH ($\cdot OH$) were enhanced which led to the increased rate of lipid peroxidation in rice subject to quinclorac toxicity (Wang et al., 2016).

Recently, a study reported that three carbamate-derived pesticides such as carbofuran, carbamoyl, and methomyl were reported to affect the rate of germination, plant length, tolerance index, survival of the plant, and membrane permeability in *Vigna mungo*

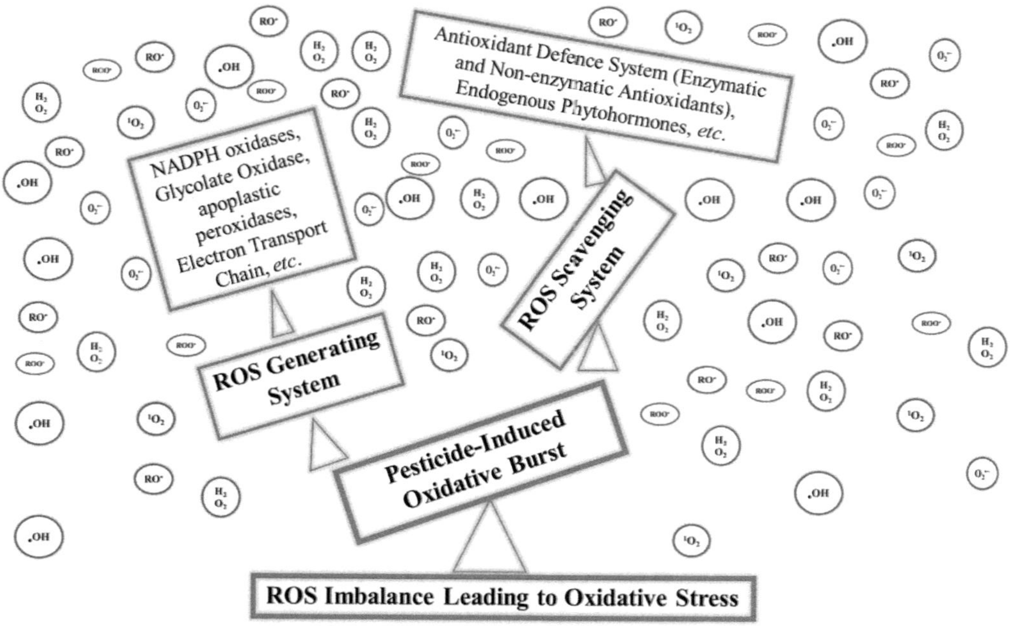

Figure 8.3 Pesticide-induced oxidative burst in plants caused by the imbalance between ROS generating and scavenging systems.

(Shahid et al., 2021). The roots treated with pesticides were found to have enhanced levels of ROS and more prominent symptoms of oxidative stress. In cucumber, two fungicides namely Ridomil and Bavistin, and Agrothoate, an insecticide were found to affect the mycorrhizal colonization and upregulate the activity of phosphatase enzymes (Rabab & Reda, 2019). These also lowered the levels of potassium ions (K^+), whereas contents of malondialdehyde (MDA) were enhanced. Besides this, the contents of several important antioxidants like flavonoids and phenolic and flavonoids were also enhanced by the applications of pesticides. A dimethyl dithiocarbamate-based broad-spectrum fungicide called tetramethyl thiuram disulfide or thiram was found to enhance the activity of ROS scavenging antioxidant and pesticide detoxification enzymes and influenced the stress parameters in tomato (Yüzbaşıoğlu & Dalyan, 2019). Table 8.2 highlights the negative impact of different pesticides on normal plant growth and development.

8.4 Effect of pesticides on ROS-generating systems

The abiotic or biotic stress-induced production of ROS is associated with the altered activity of the antioxidant defense system, which is mediated by the stimulation of the ROS-generating system (Sharma et al., 2021). Among ROS-generating mechanisms, the electron transport chain, glycolate oxidase, and NADPH oxidase are the major players. ROS are usually produced as a by-product of the common biological processes through

Table 8.2 Effect of different pesticides on physiological and biochemical attributes of different plant species.

Pesticide	Dose of pesticide	Plant	Negative impact of pesticides on plants	References
Agrothoate 40% EC	1.25 ml/L and 2.50 ml/L H_2O	*Cucumis sativus* (Cucumber)	• Decreased the percentage of mycorrhizal colonization • Enhanced activity of phosphatase enzymes • Reduced levels of potassium (K^+) ions • Increased malondialdehyde (MDA) contents	Rabab & Reda (2019)
Alpha-cypermethrin	Recommended application dose (X), 1/4X, 1/2X, 2X and 4X	*Solanum lycopersicum* (Tomato)	• Affected cell viability • Induced cell injury • Altered the contents of proline, total soluble sugar, and total soluble proteins • Modulated the activities of antioxidant enzymes (catalases, superoxide dismutase, peroxidases, glutathione reductase, and ascorbate peroxidase) • Enhanced ROS levels • Enhanced lipid peroxidation • Damaged the membrane integrity	Shakir et al. (2018)
Bavistin	500 and 1000 mg/L	*Cucumis sativus* (Cucumber)	• Increased the percentage of mycorrhizal colonization • Stimulated the activity of phosphatase enzymes • Enhanced levels of MDA • Decreased the contents of K^+	Rabab & Reda (2019)

(*continued on next page*)

Table 8.2 Effect of different pesticides on physiological and biochemical attributes of different plant species—cont'd

Pesticide	Dose of pesticide	Plant	Negative impact of pesticides on plants	References
Carbendazim	10.4 mM	*Solanum lycopersicum* (Tomato)	• Induced peroxidase activity • Increased transcript levels of glutathione reductase (*GR*) and glutathione peroxidase (*GPX*)	Wang et al. (2010)
Carbofuran	100 μg/mL	Vigna mungo (Mung bean)	• 50% decline in germination efficiency • 75% reduction in root length • 65% of shoot length was decreased • Plant survival reduced to 70% • 66% tolerance index of mung plants • More generations of ROS • Altered membrane permeability • Enhanced oxidative stress	Shahid et al. (2021)
Chlorothalonil	2.8 mM, 5.6 mM, 11.2 mM, and 22.4 mM	*Solanum lycopersicum* (Tomato)	• Enhanced contents of glutathione • Increased activities of glutathione S-transferases and glutathione reductase • Increased the transcriptional levels of three glutathione S-transferase genes namely *GST1, GST2,* and *GST3*, glutathione reductase gene (*GR*), glutathione synthetase gene (*GSH*), and glutathione peroxidase gene (*GPX*)	Wang et al. (2010)

(continued on next page)

Chlorpyrifos	50 μg/dm^3, 100 μg/dm^3, and 150 μg/dm^3	Three macrophytes namely Canadian waterweed (*Elodea canadensis*), water mint (*Mentha aquatica*), and needle spike rush (*Eleocharis acicularis*)	• Contents of flavonoids were maximum at 150 μg/dm^3 concentration of chlorpyrifos • Decreased levels of chlorophyll a, chlorophyll b, anthocyanins, and carotenoids • Concentrations of main assimilating plant pigments were significantly reduced in the highest concentration of chlorpyrifos (150 μg/dm^3) • Levels of polyphenols were significantly increased in the highest concentration of chlorpyrifos (150 μg/dm^3)	Sobiecka et al. (2022)
Diazinon	200 μg/mL	*Solanum lycopersicum* (Tomato)	• 54% decrease in root cell viability • Enhanced electrolyte leakage in leaf to 79% • 64% decline in the content of total soluble sugar • 67% reduction in contents of total soluble protein (TSP)	Hatamleh et al. (2022)

(*continued on next page*)

Table 8.2 Effect of different pesticides on physiological and biochemical attributes of different plant species—cont'd

Pesticide	Dose of pesticide	Plant	Negative impact of pesticides on plants	References
Emamectin benzoate	Recommended application dose (X), 1/4X, 1/2X, 2X and 4X	*Solanum lycopersicum* (Tomato)	• Hydrogen peroxide (H_2O_2) levels were enhanced indicating higher levels of ROS • Levels of thiobarbituric acid reactive substances (TBARS) were increased which revealed increased lipid peroxidation • Activities of antioxidant enzymes were altered • Contents of nonenzymatic antioxidants were altered • Adversely affected cellular integrity and its membranes	Shakir et al. (2018)
Imidacloprid	200 μg/mL	*Solanum lycopersicum* (Tomato)	• Decreased the rate of germination to 58% • Reduced root length to 46% • Affected fruit set and fruit attributes • Enhanced electrolyte leakage in leaf to 71%	Hatamleh et al. (2022)
Imidacloprid	Recommended application dose (X), 1/4X, 1/2X, 2X and 4X	*Solanum lycopersicum* (Tomato)	• Higher pesticide concentrations increased ROS levels • An oxidative burst was induced • The cell membrane was damaged • Enhanced the cell injury • Lowered the cell viability • Reduction in the levels of sugars and proteins	Shakir et al. (2018)

(continued on next page)

Mancozeb	200 μg/mL	*Solanum lycopersicum* (Tomato)	• Decreased rate of germination • Reduced biomass of roots • Lowered contents of Chlorophyll a, chlorophyll b, total chlorophyll, and Carotenoids • Reduced content of lycopene in fruits • Reduced flowering • Lowered the flower number per branch, fruit number per branch, size, fresh weight, and biomass of fruit • Enhanced electrolyte leakage in leaf to 80% • Increased activities of antioxidant enzymes	Hatamleh et al. (2022)
Quinclorac	0.1 g/L and 0.5 g/L	*Oryza sativa* (Rice)	• Levels of H_2O_2, $O_2^{-}\bullet$, and ^{-}OH ($\bullet OH$) were enhanced • Increased lipid peroxidation • Modulated activities of antioxidant enzymes • Affected the gene expression of abscisic acid genes	Wang et al. (2016)

(continued on next page)

Table 8.2 Effect of different pesticides on physiological and biochemical attributes of different plant species—cont'd

Pesticide	Dose of pesticide	Plant	Negative impact of pesticides on plants	References
Ridomil	250 mg/L and 500 mg/L	*Cucumis sativus* (Cucumber)	• Enhanced mycorrhizal colonization • Stimulated the activity of phosphatase enzymes • Increased MDA contents in cucumber plants • Lowered levels of K^+ ions	Rabab & Reda (2019)
Thiram	6.6 mM	*Solanum lycopersicum* (Tomato)	• Caused oxidative stress in a time-dependent manner • Enhanced levels of hydrogen peroxide (H_2O_2) and MDA • Reduced contents of total chlorophyll and carotenoids • Enhanced activities of catalases, glutathione reductase, ascorbate peroxidase, peroxidase, and glutathione-S-transferase • Up-regulated the expression of glutathione-S-transferase encoding genes namely *GST1, GST2*, and *GST3* • Up-regulated the expression of cytochrome P-450 monooxygenases gene	Yüzbaşıoğlu & Dalyan (2019)

various redox reactions. The molecular oxygen (O_2) is the initial molecule from where ROS generation starts during normal metabolic processes (Birben et al., 2012). The produced ROS may be categorized as either free radicals or nonradicals. Those molecules which have either one or more than one unpaired electrons that result in their high reactivity are referred to as free radical ROS. For example, superoxide anion ($O_2^{-}\cdot$) and hydroxyl radical ($\cdot OH$) are free radical ROS. However, when unpaired electrons are shared by any two free radicals, then no unpaired electrons are left which makes nonradical ROS. For example, hydrogen peroxide (H_2O_2) and ozone (O_3) are nonradical ROS. When one electron is added to molecular oxygen (O_2), then a superoxide anion ($O_2^{-}\cdot$) is formed. Generally, ROS production takes place in mitochondria.

The first step of ROS may production i.e., superoxide anion ($O_2^{-}\cdot$) is mediated by or mitochondrial electron transport chain, xanthine oxidase, nicotine adenine dinucleotide phosphate [NAD(P)H] oxidase or other specialized oxidases or enzymes other than NADPH oxidases in living organisms (Birben et al., 2012; Kaur et al., 2014). Besides NADPH oxidases, nitric oxide synthase and myeloperoxidase are also involved in the production of RONS free radicals in mammalian cells (Weidinger & Kozlov, 2015). Both apoplastic peroxidases and NADPH oxidases are associated with the accumulation of ROS as a response to pathogen invasion and elicitation in plants (Sharma et al., 2013). NADPH oxidases are one of the main ROS-producing enzymes which help in the catalysis of reaction that produces the superoxide (O_2^{-}) free radical in the cell (Kaur et al., 2014). Being homologs of the catalytic subunit of mammalian phagocyte $gp91^{phox}$, these enzymes are also referred to as respiratory burst oxidase homologs (Rbohs).

The pesticide-induced NADPH-mediated ROS and RNS production can be identified by fluorescent probe (e.g., hydroethidine and coumarin boronate) and radiolabeling or positron emission tomography (PET) by using PET-labeled hydroethidine analogs (Kalyanaraman et al., 2016). The organochlorine pesticides such as aldrin, β-endosulfan, α-HCH, and p,p′-DDE were detected in humans and found to cause various diseases like atherosclerosis and cardiovascular disease (Ghosh et al., 2017). In the Human Umbilical Vein Endothelial Cells, the effect of aldrin, β-endosulfan, α-HCH, and p,p′-DDE was studied. These four pesticides enhanced intracellular ROS accumulation, increased endothelial Nitric-oxide production and induced oxidative stress. The associated expression of NADPH oxidase, mRNA expression of NAPDH oxidase (p47phox), and endothelial Nitric Oxide Synthase (eNOS) was increased under pesticides stress.

8.5 Effect of pesticides on ROS scavenging systems

The pesticides affect both enzymatic and nonenzymatic antioxidant defense system in plants (Homayoonzadeh et al., 2020; Sobiecka et al., 2022; Wang et al., 2016; Yüzbaşıoğlu & Dalyan, 2019). In the next sub-sections, the effects of pesticides on both enzymatic and nonenzymatic antioxidant defense systems are discussed.

8.5.1 Nonenzymatic antioxidants

A recent investigation on three popular macrophytes viz., needle spikerush (*Eleocharis acicularis*), Canadian waterweed (*Elodea canadensis*), and water mint (*Mentha aquatica*) revealed that macrophytes protect themselves from chlorpyrifos-induced oxidative stress by stimulating the antioxidant defense system (Sobiecka et al., 2022). As a response to chlorpyrifos-induced toxicity, the levels of antioxidants namely polyphenols and flavonoids were increased, whereas the levels of assimilating pigments viz., chlorophyll a, chlorophyll b, anthocyanins, and carotenoids were decreased. The study revealed the crucial role of nonenzymatic antioxidants and low-molecular-weight compounds to manage pesticide-induced oxidative stress in macrophyte plant cells. The levels of gluatathione were affected by quinclorac in rice (Wang et al., 2016). In cucumber, the contents of nonenzymatic antioxidants namely proline, protein, total phenolics and total soluble carbohydrates were enhanced by both imidacloprid and dichlorvos (Homayoonzadeh et al., 2020).

8.5.2 Enzymatic antioxidants

Thiram, a broad-spectrum fungicide stimulated the activities of three ROS-scavenging antioxidant enzymes i.e., glutathione reductase (GR; EC 1.6.4.2), ascorbate peroxidase (APX; EC 1.11.1.11), and catalase (CAT; EC 1.11.1.6) in tomatao (Yüzbaşıoğlu & Dalyan, 2019). This fungicide also enhanced the activities of two pesticide detoxification enzymes namely, glutathione S-transferase (GST; EC.2.5.1.18), and peroxidase (POX; EC 1.11.1.7). Thiram application was able to up-regulate the gene expression of key pesticide detoxification enzymes viz., GST1, GST2, GST3, and P450s encodes for cytochrome P-450 monooxygenases enzyme in tomato. In rice, the application of a commonly used herbicide named quinclorac was found to modulate the activity of antioxidant enzymes and affect the gene expression of two important abscisic acid (ABA) genes namely OsABA8oxs, and OsNCEDs (Wang et al., 2016). The activity of antioxidant enzymes namely catalase, superoxide dismutase, guaiacol peroxide, phenylalanine ammonia-lyase, and ascorbate peroxidase, were adversely affected by the application of both imidacloprid and dichlorvos in cucumber (Homayoonzadeh et al., 2020). However, the activity of polyphenol oxidase was decreased under pesticides stress.

8.6 Various methods to reduce pesticide-induced oxidative stress in plants

The application of pesticides has been found to not only affect the targeted but nontargeted organisms like plants which subsequently pose a serious threat to humans and other animals (Hatamleh et al., 2022). The pesticides have been found to negatively impact plant growth and development in a concentration-dependent manner and are also subject to the type of pesticide used. Pesticides are detoxified in plants by a three-phase or four-phase enzymatic mechanism to biotransform the pesticides to alleviate their harmful adverse impact on the crops (Sun et al., 2018). The three-phase biotransformation strategy includes the following:

- Phase-I: Conversion of more-toxic and harmful pesticides into a less-toxic and relatively safer form: The pesticides are transformed by the action of two main detoxifying enzymes namely, cytochrome P450 monooxygenases and peroxidases. These enzymes catalyze hydrolysis, oxidation, and reduction reactions to transform the pesticide.
- Phase-II: Conjugation of pesticide metabolites: The glutathione-S-transferase and uridine diphosphate-glycosyltransferase are two main enzymes involved in the conjugation of pesticide metabolites. These are conjugated by the enzyme action into less toxic and more water-soluble compounds such as amino acid, glutathione, or sugar.
- Phase-III: Transport of less-toxic pesticide metabolites: The transformed less-toxic pesticides metabolites are then transported from cytosol to those cellular compartments (such as vacuoles and apoplasts), where they are accumulated.

The above-mentioned innate strategies of the plants indicate that plants' self-detoxification mechanism majorly focuses on the degradation or transformation of pesticides from more toxic to less toxic products or relatively unstable forms that may easily disintegrate. Moreover, it has been recently reported that several pesticides are also known to be unstable in sunlight on the leaf surface of crops in field conditions (Xi et al., 2022). The rate of photodegradation of pesticides is dependent on agro-meteorological conditions such as light, temperature, humidity, etc., the characteristics and growth conditions of plants, and the properties, and concentration of pesticides. Further, the application of adjuvants having photosensitizing potential enhances the phototransformation of pesticides. The phototransformation of pesticides is a crucial step in their disintegration or degradation in the presence of sunlight which usually occurs on leaves. The process of phototransformation of pesticides is also associated with the production of ROS on leaves. The levels of ROS are also enhanced under pesticide stress in plants (Shahzad et al., 2018). Plants have an innate system to overcome pesticide-induced stress, and for pesticide detoxification, an antioxidant defense system including nonenzymatic antioxidants and enzymatic antioxidant nzymes is stimulated (Sharma et al., 2019). Various strategies for pesticide-induced oxidative stress detoxification are presented in Fig. 8.4.

The pesticide-metabolized enzymes namely, methyltransferases, glycosyltransferases, and laccases; and phytohormones such as jasmonic acid, salicylic acid, and brassinosteroids also assist the pesticide detoxification (Bhardwaj et al., 2009; Zhang & Yang, 2021).

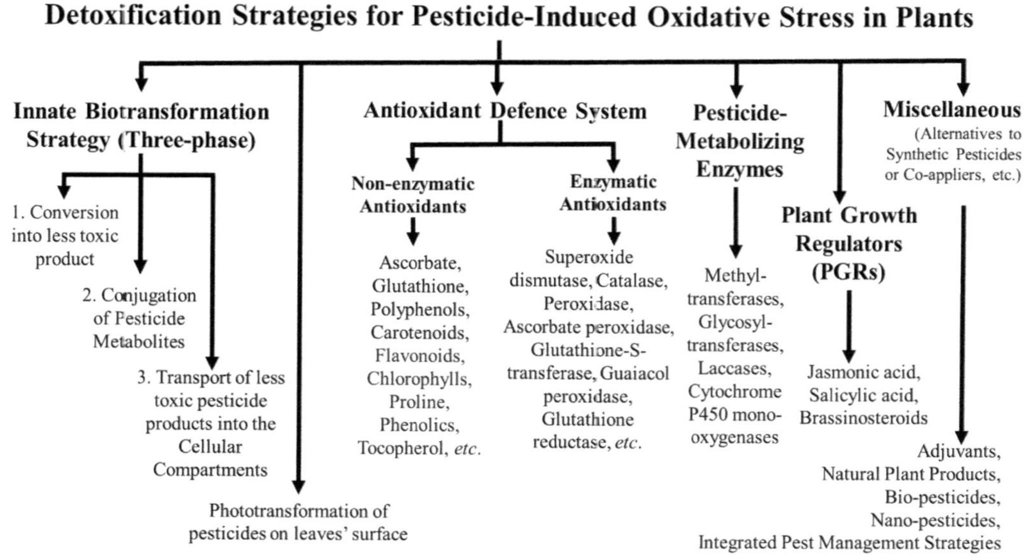

Figure 8.4 Various strategies employed for the detoxification of pesticide-induced oxidative stress in plants.

Salicylic acid was reported to ameliorate the detrimental effects of a fungicide named thiram through the modulation of antioxidant and pesticide detoxification enzymes in the tomato (Yüzbaşıoğlu & Dalyan, 2019). It could reciprocate the levels of lipid peroxidation and hydrogen peroxide in pesticide-stressed tomatoes. The contents of photosynthetic pigments such as chlorophyll *a*, chlorophyll *b*, and carotenoids were stimulated by salicylic acid application. In rice, salicylic acid was found to stimulate the antioxidant defense system in order to reduce quinclorac-induced oxidative stress (Wang et al., 2016). It reduced ROS levels and regulated the gene expression analysis of ABA genes namely OsABA8oxs, and OsNCEDs for pesticide-stress amelioration. Apart from salicylic acid, other plant growth regulators such as brassinosteroids have also been recognized to have immense potential to overcome abiotic stress-induced phytotoxicities (Bakshi et al. 2022; Bhardwaj et al., 2009; Shahzad et al., 2018; Sharma et al., 2018a, 2018b). Various mechanisms involved in the detoxification of pesticides in plants by using brassinosteroids were discussed by Bakshi et al. (2022). These plant steroidal hormones can be employed to ameliorate pesticide-induced oxidative stress by stimulating the enzymatic and nonenzymatic defense system by scavenging the excessively generated ROS.

A commercially used brassinosteroid named 24-epibrassinolide was found to be effective in scavenging pesticide-induced ROS (Shahzad et al., 2018). Epibrassinolides maintained the osmotic potential, enhanced the contents of proteins and prolines, and stimulated the activities of superoxide dismutase, catalases, and ascorbate peroxidases to reduce ROS. The epibrassinolide or homobrassinolide is also reported to interact either synergistically or antagonistically with other plant growth regulators such as ethylene, jasmonic acids, auxins, ABA, gibberellins, and cytokinins to manage the environmental stresses induced oxidative bursts (Sharma et al., 2018a). Moreover, brassinosteroids are also reported to up-regulate the gene expression of antioxidant enzymes to manage excessive ROS accumulation induced by abiotic stress in radish plants (Sharma et al., 2018b). The antioxidant enzymes are the major enzymatic ROS scavengers. Along with these, nonenzymatic antioxidants and three major ROS sensing systems also assist in the amelioration of ROS (Sharma et al., 2013). The direct inhibition of phosphatases, unidentified receptor proteins, and heat shock proteins or redox-sensitive transcription factors are the major ROS sensing mechanisms in the plant defense system. Based on the type of stress, the stress signal is identified and perceived by the sensor molecule that further generates a response by stimulating the series of kinase-mediated signaling reactions to overcome the stress (Dhiman et al., 2022).

8.7 Conclusion and future prospects

The pesticide usage varies for the different pests associated with the crops and the agro-climatic zone of crop production. A secure and systematic process be utilized before the selection of pesticides for improving the production of agronomically significant commercial crops grown in diverse agro-climatic zones. Various cellular- and molecular-level approaches need to be devised to know the cellular and molecular mechanisms involved in pesticide-induced toxicities in nontarget plants. The detrimental effects of pesticides on plants might be mitigated with a better understanding of pesticide-induced phytotoxicities. Moreover, alternative, cost-friendly, ecologically safe, plant-based approaches, or bio-pesticides may be employed for sustainable agriculture.

Burgeoning studies have emphasized that natural plant-based pesticides may be employed as potentially safe alternatives to chemical pesticides (Souto et al., 2021). Various plant-derived products such as alkaloids, flavonoids amino acids, amides, cyanogenic glucosides, polyphenols, polyketides, terpenes, aldehydes, quinones, thiophenes, saccharides, etc., show insecticidal properties. Natural plant products can also act as an attractant, repellents, fungicides, insecticides, nematicides, antifeedants, allelopathic agents, insect growth regulators, etc. The natural products or living organisms that can act as pesticides to control or manage plant pests are called Biopesticides (Samada & Tambunan, 2020). Based on their origin, the biopesticides are plant-incorporated protectants, biochemicals, and microbial pesticides. *Bacillus thuringiensis* (Bt)-based pesticides, *Baculoviruses*, Neem derived biochemicals, *Trichoderma*, and *Trichogramma* are the major commercially available biopesticides that can easily replace hazardous chemical pesticides. However, large-scale production of these biopesticides is a global challenge. The partnership between public and private sectors and research centres can help in the development of innovative and more effective biopesticide formulations, their rapid manufacturing, and commercialization at a large scale. Further, the integrated use of chemical and biological pesticides can be a sustainable, futuristic integrated pest management (IPM) strategy.

The application of both IPM techniques and eco-friendly pesticide alternatives such as EcoSMART might be helpful in managing the pesticides' impacts on human and plant health (Kim et al., 2017). The co-application of arbuscular mycorrhizal fungi along with pesticides may help either partially or completely to ameliorate the toxic effects of pesticides (Rabab & Reda, 2019). Researchers also highlighted the use of novel nanotechnology in the form of nano-enabled sensors, nanopesticides, nanofertilizers, etc., might be a turning point in the current agronomic practices (Adisa et al., 2019). The application of such nano-engineered pesticides might suppress pesticide-mediated excessive generation of ROS and/or control pests and replace conventionally used chemical-based pesticides. Hence, nanotechnology-based fertilizers might be a potential alternative that may play a crucial role in ensuring sustainable agriculture and ensuing global food security.

Conflicts of interest

The authors declare no conflicts of interest.

Acknowledgment

Authors AS and RK acknowledge the financial grant received from the Council of Scientific and Industrial Research (CSIR-HRDG), New Delhi, India, through an extramural research grant with project number 38(1477)/19/EMR-II, to work on salt stress. Author IS acknowledges the Sant Baba Bhag University for providing financial support (Seed Money Minor Project) and infrastructure research facility. Author AS, NS, VM, and RK acknowledge the DAV University for providing all the infrastructure facilities to carry out the present work.

References

Abubakar, Y., Tijjani, H., Egbuna, C., Adetunji, C. O., Kala, S., Kryeziu, T. L., Ifemeje, J. C., & Patrick-Iwuanyanwu, K. C. (2020). Pesticides, history, and classification. In *Natural remedies for pest, disease and weed control* (pp. 29–42). Academic Press.

Adisa, I. O., Pullagurala, V. L. R., Peralta-Videa, J. R., Dimkpa, C. O., Elmer, W. H., Gardea-Torresdey, J. L., & White, J. C. (2019). Recent advances in nano-enabled fertilizers and pesticides: A critical review of mechanisms of action. *Environmental Science: Nano, 6*(7), 2002–2030.

Alengebawy, A., Abdelkhalek, S. T., Qureshi, S. R., & Wang, M. Q. (2021). Heavy metals and pesticides toxicity in agricultural soil and plants: Ecological risks and human health implications. *Toxics, 9*(3), 42.

Bakshi, P., Bali, S., Sharma, P., Ibrahim, M., Devi, K., Sharma, N., Sharma, A., Singh, A. P., Mir, B. A., & Bhardwaj, R. (2022). Mechanism associated with brassinosteroids-mediated detoxification of pesticides in plants. In *Brassinosteroids signalling: Intervention with phytohormones and their relationship in plant adaptation to abiotic stresses* (pp. 203–221). Singapore: Springer Singapore.

Bhardwaj, R., Sharma, I., Arora, N., Uppal, P., Kanwar, M. K., Sharma, N., & Hundal, G. S. (2009). Regulation of oxidative stress by BRs in plants. In *Oxidative stress: Role of antioxidants in plants* (pp. 215–231). New Delhi, India: Studium Press Pvt. Ltd.

Birben, E., Sahiner, U. M., Sackesen, C., Erzurum, S., & Kalayci, O. (2012). Oxidative stress and antioxidant defense. *World Allergy Organization Journal, 5*, 9–19.

Chakraborty, S., Tiwari, B., Singh, S. S., Srivastava, A. K., & Mishra, A. K. (2017). Differential physiological, oxidative and antioxidative responses of cyanobacterium *Anabaena sphaerica* to attenuate malathion pesticide toxicity. *Biocatalysis and Agricultural Biotechnology, 11*, 56–63.

Dhiman, S., Handa, N., Kohli, S. K., Ibrahim, M., Bhardwaj, T., Kapoor, D., Sharma, I., Sareen, S., Bhardwaj, R., & Mir, B. A. (2022). Kinase-mediated signaling cascades in plant abiotic stress physiology. In *Antioxidant defense in plants: Molecular basis of regulation* (pp. 109–134). Singapore: Springer Nature Singapore.

Ghosh, R., Siddharth, M., Singh, N., Kare, P. K., Banerjee, B. D., Wadhwa, N., & Tripathi, A. K. (2017). Organochlorine pesticide-mediated induction of NADPH oxidase and nitric-oxide synthase in endothelial cell. *Journal of Clinical and Diagnostic Research: JCDR, 11*(1), BC09.

Hatamleh, A. A., Danish, M., Al-Dosary, M. A., El-Zaidy, M., & Ali, S. (2022). Physiological and oxidative stress responses of *Solanum lycopersicum* (L.) (tomato) when exposed to different chemical pesticides. *RSC advances, 12*(12), 7237–7252.

Homayoonzadeh, M., Moeini, P., Talebi, K., Roessner, U., & Hosseininaveh, V. (2020). Antioxidant system status of cucumber plants under pesticides treatment. *Acta Physiologiae Plantarum, 42*, 1–11.

Intisar, A., Ramzan, A., Sawaira, T., Kareem, A. T., Hussain, N., Din, M. I., Bilal, M., & Iqbal, H. M. (2022). Occurrence, toxic effects, and mitigation of pesticides as emerging environmental pollutants using robust nanomaterials: A review. *Chemosphere, 293*, 133538.

Jan, S., Singh, R., Bhardwaj, R., Ahmad, P., & Kapoor, D. (2020). Plant growth regulators: A sustainable approach to combat pesticide toxicity. *3 Biotech, 10*, 1–11.

Kalyanaraman, B., Hardy, M., & Zielonka, J. (2016). A critical review of methodologies to detect reactive oxygen and nitrogen species stimulated by NADPH oxidase enzymes: Implications in pesticide toxicity. *Current Pharmacology Reports, 2*, 193–201.

Kaur, G., Sharma, A., Guruprasad, K., & Pati, P. K. (2014). Versatile roles of plant NADPH oxidases and emerging concepts. *Biotechnology Advances, 32*(3), 551–563.

Kim, K. H., Kabir, E., & Jahan, S. A. (2017). Exposure to pesticides and the associated human health effects. *Science of the Total Environment, 575*, 525–535.

Peng, X., Wang, N., Sun, S., Geng, L., Guo, N., Liu, A., Chen, S., & Ahammed, G. J. (2023). Reactive oxygen species signaling is involved in melatonin-induced reduction of chlorothalonil residue in tomato leaves. *Journal of Hazardous Materials, 443*, 130212.

Rabab, A. M., & Reda, E. A. (2019). Impact of Ridomil, Bavistin and Agrothoate on arbuscular mycorrhizal fungal colonization, biochemical changes and potassium content of cucumber plants. *Ecotoxicology, 28*, 487–498.

Rao, M. M., KumarMeena, A., & Galib (2011). Detection of toxic heavy metals and pesticide residue in herbal plants which are commonly used in the herbal formulations. *Environmental Monitoring and Assessment, 181*, 267–271.

Samada, L. H., & Tambunan, U. S. F. (2020). Biopesticides as promising alternatives to chemical pesticides: A review of their current and future status. *OnLine Journal of Biological Sciences, 20*, 66–76.

Shahid, M., Manoharadas, S., Chakdar, H., Alrefaei, A. F., Albeshr, M. F., & Almutairi, M. H. (2021). Biological toxicity assessment of carbamate pesticides using bacterial and plant bioassays: An in-vitro approach. *Chemosphere, 278*, 130372.

Shahzad, B., Tanveer, M., Che, Z., Rehman, A., Cheema, S. A., Sharma, A., Song, H., ur Rehman, S., & Zhaorong, D. (2018). Role of 24-epibrassinolide (EBL) in mediating heavy metal and pesticide induced oxidative stress in plants: A review. *Ecotoxicology and Environmental Safety, 147*, 935–944.

Shakir, S. K., Irfan, S., Akhtar, B., Rehman, S. U., Daud, M. K., Taimur, N., & Azizullah, A. (2018). Pesticide-induced oxidative stress and antioxidant responses in tomato (*Solanum lycopersicum*) seedlings. *Ecotoxicology, 27*, 919–935.

Sharma, A., Kumar, V., Thukral, A. K., & Bhardwaj, R. (2019). Responses of plants to pesticide toxicity: An overview. *Planta Daninha, 37*, e019184291.

Sharma, A., Nagar, P. K., & Pati, P. K. (2013). ROS modulated responses during pathogen infestation in plants. In *Physiology of nutrition and environmental stresses on crop productivity, advances in plant physiol: An international treatise series14* (pp. 331–347). Jodhpur: Scientific Publishers (India).

Sharma, A., Sharma, P., Kumar, R., Sharma, V., Bhardwaj, R., & Sharma, I. (2021). Role of reactive oxygen species in the regulation of abiotic stress tolerance in legumes. In *Abiotic stress and legumes* (pp. 217–243). Academic Press.

Sharma, I., Bhardwaj, R., Gautam, V., Kaur, R., & Sharma, A. (2018a). Brassinosteroids: Occurrence, structure and stress protective activities. *Frontiers in Natural Product Chemistry, 4*, 204–239.

Sharma, I., Sharma, A., Pati, P., & Bhardwaj, R. (2018b). Brassinosteroids reciprocates heavy metals induced oxidative stress in radish by regulating the expression of key antioxidant enzyme genes. *Brazilian Archives of Biology and Technology,, 61*, e18160679.

Snezhkina, A. V., Kudryavtseva, A. V., Kardymon, O. L., Savvateeva, M. V., Melnikova, N. V., Krasnov, G. S., & Dmitriev, A. A. (2019). ROS generation and antioxidant defense systems in normal and malignant cells. *Oxidative Medicine and Cellular Longevity, 2019*, Article ID: 6175804.

Sobiecka, E., Mroczkowska, M., & Olejnik, T. P. (2022). The influence of chlorpyrifos on the nonenzymatic antioxidants content in macrophytes leaves. *Antioxidants, 11*(4), 684.

Souto, A. L., Sylvestre, M., Tölke, E. D., Tavares, J. F., Barbosa-Filho, J. M., & Cebrián-Torrejón, G. (2021). Plant-derived pesticides as an alternative to pest management and sustainable agricultural production: Prospects, applications and challenges. *Molecules, 26*(16), 4835.

Sule, R. O., Condon, L., & Gomes, A. V. (2022). A common feature of pesticides: Oxidative stress: The role of oxidative stress in pesticide-induced toxicity. *Oxidative Medicine and Cellular Longevity, 2022*, Article ID: 5563759. https://doi.org/10.1155/2022/5563759.

Sun, L., Xu, H., Su, W., Xue, F., An, S., Lu, C., & Wu, R. (2018). The expression of detoxification genes in two maize cultivars by interaction of isoxadifen-ethyl and nicosulfuron. *Plant Physiology and Biochemistry, 129*, 101–108.

Wang, J., Jiang, Y., Chen, S., Xia, X., Shi, K., Zhou, Y., Yu, Y., & Yu, J. (2010). The different responses of glutathione-dependent detoxification pathway to fungicide chlorothalonil and carbendazim in tomato leaves. *Chemosphere, 79*(9), 958–965.

Wang, J., Lv, M., Islam, F., Gill, R. A., Yang, C., Ali, B., Yan, G., & Zhou, W. (2016). Salicylic acid mediates antioxidant defense system and ABA pathway related gene expression in *Oryza sativa* against quinclorac toxicity. *Ecotoxicology and Environmental Safety, 133*, 146–156.

Weidinger, A., & Kozlov, A. V. (2015). Biological activities of reactive oxygen and nitrogen species: Oxidative stress versus signal transduction. *Biomolecules, 5*(2), 472–484.

WHO. (2019). *The WHO Recommended Classification of Pesticides by Hazard and guidelines to classification, 2019 edition*. World Health Organization. ISBN 978-92-4-000566-2. https://www.who.int/publications/i/item/9789240005662.

Xi, N., Li, Y., & Xia, X. (2022). A review of pesticide phototransformation on the leaf surface: Models, mechanism, and influencing factors. *Chemosphere, 308*, 136260. https://doi.org/10.1016/j.chemosphere.2022.136260.

Yadav, I. C., & Devi, N. L. (2017). Pesticides classification and its impact on human and environment. *Environmental Science and Engineering, 6*, 140–158.

Yüzbaşıoğlu, E., & Dalyan, E. (2019). Salicylic acid alleviates thiram toxicity by modulating antioxidant enzyme capacity and pesticide detoxification systems in the tomato (*Solanum lycopersicum* Mill.). *Plant Physiology and Biochemistry, 135*, 322–330.

Zaller, J. G., & Brühl, C. A. (2019). Non-target effects of pesticides on organisms inhabiting agroecosystems. *Frontiers in Environmental Science, 7*, 75.

Zhang, J. J., & Yang, H. (2021). Metabolism and detoxification of pesticides in plants. *Science of the Total Environment, 790*, 148034.

Zhang, S., Qiu, C. B., Zhou, Y., Jin, Z. P., & Yang, H. (2011). Bioaccumulation and degradation of pesticide fluroxypyr are associated with toxic tolerance in green alga *Chlamydomonas reinhardtii*. *Ecotoxicology, 20*, 337–347.

CHAPTER 9

Pesticide impacts on humans

Raman Jasrotia[a], Seema Langer[a,b] and Menakshi Dhar[a]
[a]Department of Zoology, University of Jammu, Jammu, Jammu and Kashmir, India
[b]Institute of Human Genetics, University of Jammu, Jammu, Jammu and Kashmir, India

9.1 Introduction

A pesticide can be defined as any substance or a combination of substances that are purposive for obstructing, destroying, or suppressing any pest including the vectors that transmit the diseases in animals and humans or any undesirable plant, or animal species that interfere with the plant and animal production (FAO, 1989). According to World Health Organisation (WHO), every year there are three million cases of poisoning due to pesticides predominantly in developing countries (Lah, 2011). Reactive oxygen species are formed due to the pesticides that tend to decrease the levels of antioxidants and their defense mechanisms against damage due to oxidation in the cellular system (Kaur et al., 2019). UNEP (1990) defined the term pesticide as chemicals formulated to hinder various pest and vector attacks on agricultural products, domesticated animals, and humans. A wide range of compounds such as fungicides, insecticides, rodenticides, herbicides, and plant growth regulators is collectively referred to as pesticides (Aktar et al., 2009). In India, pesticide production began in 1952 with the production of benzene hexachloride (BHC) near Kolkata. At present, China is the leading producer of pesticides in the Asian continent, followed by India. Mathur (1999) reported that on the global level, India ranks 12th in terms of pesticide production. Approximately 5000 metric tons of pesticides were produced in 1958 which increased to 102,240 metric tons in 1998. In India, the pattern of usage of pesticides is different from that of other countries. Across the globe, more than 1000 pesticides are utilized all of which have unique properties. About 56 billion pounds of pesticides are applied annually all over the world and only 0.1% of these pesticides come in contact with the target pests (Singh et al., 2021).

Pesticides in the Environment: Impact, Assessment, and Remediation
DOI: https://doi.org/10.1016/B978-0-323-99427-9.00008-2

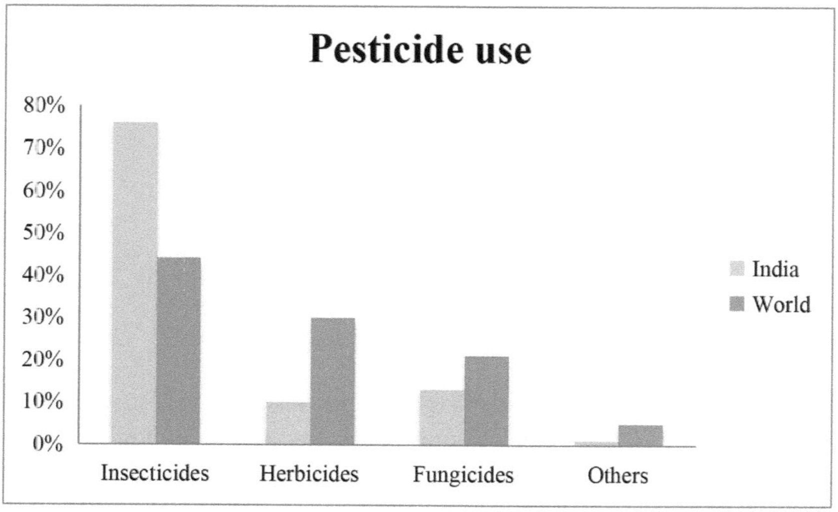

9.2 History of pesticides

The lack of herbicide use leads to the destruct destruction of the coffee industry at the end of the 19th century in Sri Lanka. Similarly, the lack of fungicide affected when *Phytophthora infestans* affected the tomato and potato production in Ireland and that became the cause of famine resulting in the death of approximately millions of individuals. The absence of insecticides had a great impact on cotton plantations in the United States due to the arrival of *Anthonomus grandis* from Mexico. Before the herbicide development, huge labor power was required to remove the weeds from the fields. Even today, about 10%–90% of the crops are lost due to pests and diseases. The insecticides were used for the first time by Sumerians about 4500 years ago. They used compounds of sulfur to control insects and mites. Literature from the Rome and Greece suggested that chemical methods were used for the control of insect pests, weeds, and animal diseases. Due to the absence of large-scale chemical industries, smoke was used against blight and mildew disease in plants. To trap the crawling insects on trees, tar was used. Pyrethrum obtained from the flowers of *Chrysanthemum cinerariaefolium* was used to control insect pests for over 2000 years. Since ancient times, several inorganic chemicals have been used as insecticides (Smith & Secoy, 1976), even today the mixture of lime and copper sulfate called the bordeaux mixture is used against fungal diseases. Salt or seawater was used as a weedicide (Smith & Secoy, 1975). Until the 1940s, various inorganic substances such as sulfuric acid and sodium chlorate were used as a pesticide. Some pesticides used were the by-products of various industrial processes. Organic compounds such as naphthalene, nitrophenol, petroleum oil, and chlorophenol were used against the fungal and insect attack. But due to their high rate of application, phytotoxicity and lack of selectivity accelerated the growth of synthetic pesticides in the 1940s which began with the discovery of 2,4-D, endrin, DDT, dieldrin, BHC, chlordane, and parathion. Among these, DDT became the most popular due to its broad-spectrum activity (Delaplane, 2000). Dr. Paul Miller, a Swiss chemist received Nobel Prize in medicine in the year 1948 for recognizing the insecticidal properties of DDT. But, in 1946 it was reported that the houseflies have developed a resistance to DDT and lindane (Toppozada et al., 1967) and it also affects the nontarget organisms with residual particles. Similar resistance to DDT was observed among *Culex*, bedbugs, cockroaches, and fleas (Livadas & Georgopoulos, 1953).

Throughout the 1950s people were least concerned about the ill effects of pesticides. New pesticides were considered safe when compared to the arsenic that lead to the death of people in the 1920s. However, problems that arise due to the haphazard use of pesticides were highlighted by Rachel Carson in 21,962 in the book Silent Spring. Research continued and in the 1980s new pesticides such as glyphosate, pyrethroids, benzoylurea, triazole, imidazole, dicarboxamide, and pyrimidine were introduced. In the 1990s research became more focussed on the reproduction of pesticides having a good environmental profile and more selectivity. New pesticide chemistry has allowed improved selectivity and better resistance management.

In India, the use of pesticides began in 1948 with the import of DDT and BHC for controlling malaria and locust, respectively. In India, the manufacturing plant for the production of DDT and BHC was set up in the year 1952 and India produced 5000 metric tons of pesticides in 1958. At present, about 145 pesticides are registered for the use which has to an increase in crop production by 85,000 metric tons. In 1958, about 1000 people died in Kerala after the consumption of wheat that was contaminated with an organophosphate, the parathion. In 1984, approximately, 40 tons of methylisocyanate gas leaked from the pesticide manufacturing unit in Bhopal due to which about 3800 people died. Because of the nonjudicious use of pesticides, it has been reported that food commodities get contaminated with pesticide residues. Approximately, 51% of food commodities are contaminated with the residues of pesticides in India, and exposure to such chemicals is linked to human health effects (Gupta, 2004).

9.3 Classification of pesticides

According to WHO, pesticides are generally classified into five main types:

1. Unlikely to cause a human hazard
2. Slightly dangerous
3. Moderately dangerous
4. Highly dangerous
5. Extremely dangerous

Based on the mode of action, pesticides are categorized as nonsystemic pesticides and systemic pesticides.

a. Systemic pesticides: Pesticides that are taken up by the plant issues and pass through its xylem and phloem soon to get the desired effects are referred to as systemic pesticides. 2,4-D and glyphosate are some of the systemic pesticides.

b. Nonsystemic pesticides: Such pesticides that do not penetrate the plant tissues and as a result, they do not move through the vascular system of plants. They are also known as contact pesticides as they produce the desired effect only when they come across the targeted pest. Diquat bromide and paraquat are some examples of nonsystemic pesticides.

Based on target pest species, pesticides are classified as:

- **Insecticides**: Used to kill or disrupt the growth and development of insects.
- **Miticides**: Control the mites that feed on animals and plant tissues.
- **Fungicides**: Controls the growth of fungi such as molds, rust, mildew, and blights.
- **Herbicides**: Inhibits the growth of unwanted plants.
- **Algicides**: Used for killing and slowing down algal growth.
- **Rodenticides**: Controls rodents such as mice and rats.
- **Molluscicides**: Controls molluscs such as snails and slugs
- **Antimicrobials**: Inhibit the growth of microbes such as viruses and bacteria.
- **Ovicides**: Controls eggs of mites and insects.
- **Repellents**: Used to repel unwanted pest species.
- **Avicides**: Controls bird pests.
- **Virucides**: Inhibits viral growth.
- **Biocides**: Kills micro-organisms.
- **Antifeedants**: Chemical substances present in some plants that adversely affect the insects and other animals that feed on them.
- **Nematicides**: Kills the microscopic worm-like organisms that feed on the roots of plants.
- **Antifoulants**: Kills micro-organisms such as barnacles that remain attached to underwater surfaces.
- **Larvicides**: Pesticides that inhibit larval growth.
- **Insect growth regulator**: Chemical that disrupts the growth and development of insects
- **Acaricides**: Chemicals that disrupt the growth and development of mites and ticks.
- **Bactericides**: A substance used to kill bacteria.
- **Chemosterillant**: Chemicals that control disease-causing pests by causing reproductive sterility in them.
- **Insect attractant**: Chemicals that attract the insects to trap and thus help to remove them from livestock and crops.
- **Silvicides**: Chemicals used to kill unwanted forest species.
- **Ovicides**: Substances that are effective at the egg stage of any pest.

Based on chemical composition, pesticides are classified as:

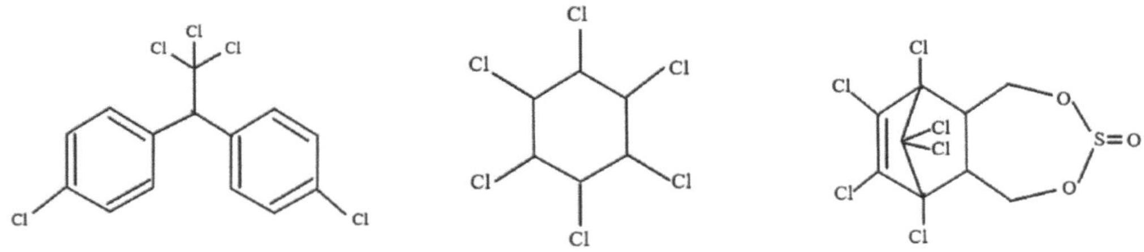

(A) (B) (C)

Figure 9.1 Chemical structure of various organochlorines (A) DDT, (B) lindane, and (C) endosulfan.

- Organochlorines
- Organophosphates
- Carbamates
- Pyrethroids

This system of classification provides an idea about the physicochemical properties of pesticides (Singh et al., 2021).

Organochlorines: These are the organic compounds containing chlorine atoms bonded by the covalent attachment. These compounds have both aromatic rings and aliphatic chains. Aromatic organic compounds are used as pesticides such as endosulfan, aldrin, endrin, DDT, chlordane, and dieldrin (Shukla et al., 2014). Due to their toxic nature, they are considered persistent organic pollutants. Copious amounts of organochlorines like polychlorinated biphenyls exist in an aquatic and terrestrial environment (Salvo et al., 2012). In most countries, the use of various organochlorines has been restricted for several years, but these are the stable molecules that have been accumulated within the organism's body and biomagnified through food chains. DDT was extensively used during the 1950s as an insecticide and was effectual against various pest species. Exposure to DDT or DDE produces estrogenic and anti-androgenic effects and thus acts as an endocrine-disrupting chemical. Organochlorines were the first synthetic pesticides used in agricultural fields. These organic compounds trigger the central nervous system which leads to nausea, tremors, and mental confusion. DDT, chlordane, lindane, dieldrin, and endosulfan are some examples of organochlorines (Fig. 9.1).

Organophosphates: All the insecticides that contain phosphorus are referred to as organophosphates (Fig. 9.2). Of all the pesticides, they are the most toxic for all vertebrate species (Jepson, 2001). Organophosphates are classified into three main groups:

1. *Aliphatic organophosphates*: They persist in the environment for a short period and are toxic to mammals, e.g., acephate, malathion, dimethoate, dichlorvos, and phorate.
2. *Phenyl organophosphates*: They are highly toxic for mammals and persist in the environment for long periods, e.g., methyl parathion, temephos, and fenitrothion.

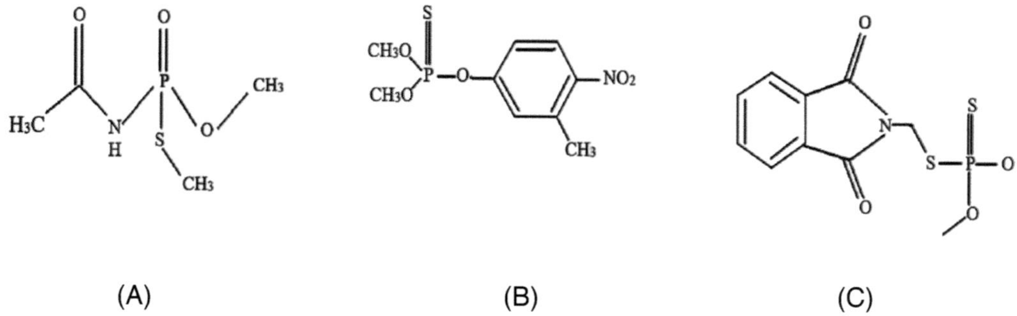

Figure 9.2 Chemical structure of various organophosphates (A) acephate, (B) fenitrothion, and (C) phosmet.

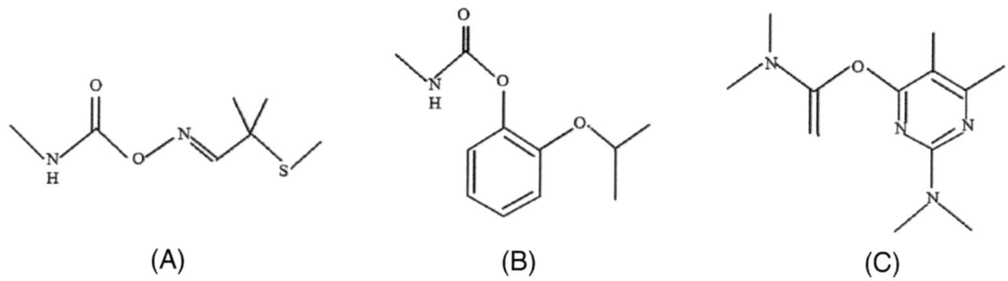

Figure 9.3 Chemical structure of various carbamates (A) aldicarb, (B) propoxur, and (C) pirimicarb.

3. *Heterocyclic organophosphates*: Such compounds are more persistent and active in the soil, e.g., phosmet, chlorpyriphos, and pyrazophos.

Such organic compounds are the esters of phosphoric acid, soluble in water and so they do not get accumulated in the tissues of the living organism.

Carbamates: Derivatives of carbamic acid are called carbamates. These esters of carbamic acid get hydrolyzed easily and as a result, they persist in the environment for a short period (Horsak et al.,1964). The carbamate pesticides inhibit acetylcholine esterase due to which acetylcholine accumulates in the central nervous system. The structure and mechanism of carbamates are similar to organophosphates (Silberman & Taylor, 2021). Examples of carbamates are: aldicarb, carbofuran, carbaryl, pirimicarb, propoxur, oxamyl, and fenobucarb (Fig. 9.3).

There are three different classes of carbamate pesticides:

1. *Carbamate ester derivatives*: They are used as insecticides and nematicides and they are highly stable in the environment.
2. *Carbamate fungicides*: They consist of the benzimidazole group.
3. *Carbamate herbicides*: These are of low mammalian toxicity such as thiocarbamic acid and dithiocarbamic acid.

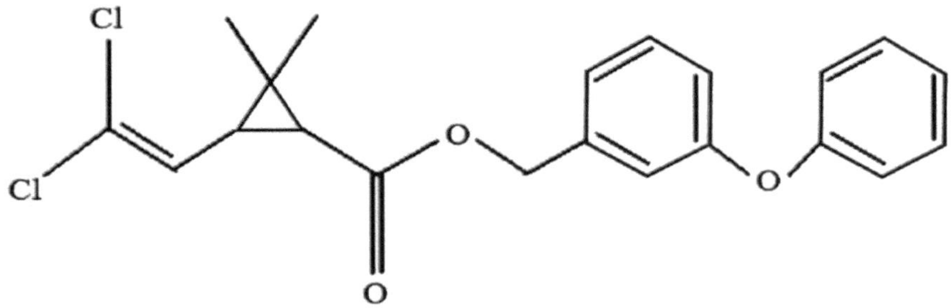

(A)

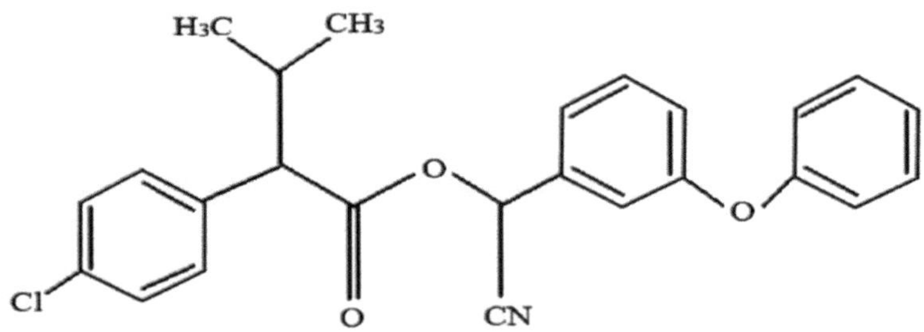

(B)

Figure 9.4 Chemical structure of various pyrethroids (A) permethrin and (B) fenvalerate.

9.4 Pyrethroids

These are the insecticides that are manufactured synthetically and are utilized for the control of insect pests in agricultural fields and animal health. These chemicals are synthetic analogs of pyrethrins, which are the insect-killing substances in the pyrethrum, derived from chrysanthemum flowers (Urkude et al., 2019). Some examples of pyrethroids are bifenthrin, cypermethrin, esfenvalerate, permethrin, tetramethrin, tefluthrin, and prallethrin. Oral intake of high doses of pyrethroids affects the central nervous system. They act by binding with sodium channels that lead to the paralysis of insects and thenceforth death (Davies et al., 2007). Pyrethroids can be classified into two main groups; based on their chemical structure, symptoms of poisoning, and effects on nerve preparations (Dong, 2007; Vais et al., 2001). Type 1 pyrethroids such as bifenthrin, telfuthrin, permethrin, cismethrin, and tetramethrin lack an alpha-cyano group (Fig. 9.4A). Type II pyrethroids such as acrinathrine, fenvalerate, cypermethrin, fenpropathrin, deltamethrin have an alpha-cyano group at the phenoxybenzyl alcohol moiety (Fig. 9.4B). These chemicals are more effective at low concentrations. Due to the small body size and low body temperature of insects, these pyrethroids are 2250 times more toxic for them as compared to mammals (Bradberry et al., 2005). Allethrin, a type I pyrethroid was the first pyrethroid pesticide that was identified in 1949. These chemicals are degraded by the chemical effects of light due to which they do not last in the environment for a long time. Pyrethroids are strong toxins that act on the nervous system (Singh et al., 2021).

According to formulation, pesticides are classified as:

- *Bait*: Invertebrate pest control, pesticides are used in the bait form that leads to the death of an animal such as a rabbit or mouse following its ingestion. Formulation of baits involves the use of food items that are consumed by the pest species along with an adequate quantity of toxicants, e.g., Klerat and Nibam.
- *Eartags*: Eartags impregnated with insecticides are used for cattle to control horn flies and face flies. Insecticide ear tags contain organophosphate and pyrethroid insecticides, e.g., Dominator.
- *Granules*: In granular formulation, the pesticide is bound to coarse particles such as clay walnut shells and corncobs. Such formulations are used for the control of ants, weeds, and nematodes living in the soil. These formulations are also used to control mosquito larvae, e.g., Merit 0.5 G and Durban G.
- *Dust*: To monitor and control rodents and insects, special dust known as tracking powders is used. Rodents move through the dust combined with the poison, which sticks to their feet and fur and they consume it when they clean themselves, e.g., Ficam D and Drione.
- *Dry flowable*: These are also known as water-dispersible granules. Such pesticides are blended with water and then used in the form of a spray.
- *Wettable powder*: Dust-like, dry, finely formulations are mixed with water and used in the form of a spray. By weight, 5%–95% of active ingredient is present in wettable powders. On mixing with water, fine particles do not dissolve and form a suspension. It is one of the most commonly used pesticide formulations, e.g., Demon WP and Captam 50WP.
- *Soluble powder*: Soluble powders are similar to wettable powders except that they form a true solution when mixed with water. In the soluble powders, the active ingredient varies from 15% to 95% by weight. Most commonly sprayed for the control of insects and weeds, e.g., Cyper WSP.
- *Aerosols*: These formulations that are used to kill insects contain s active pesticides and inert ingredients. There are two types of aerosol formulations (1) ready-to-use type as seen in household sprays and (2) fog generators that involve the use of electric generators that release the formulation in the form of fog. The use of aerosols is one of the quick ways to kill unwanted insects, e.g., Pyrid aerosol and FiPro foaming aerosol.

- *Emulsifiable concentrate*: It is a liquid formulation in which a certain amount of pesticide is dissolved in organic solvents such as xylene and benzene. It appears milky-colored. Per gallon, about 25%–75% of active ingredient is present in an emulsifiable concentrate. ECs have the most versatile formulation and are used to control pests in agriculture, forestry, food processing, animals, and public health, e.g., Buctril 4EC and Chlorpyrifos EC.
- *Microencapsulation*: Pesticide particles either in liquid or dry form and are covered by plastic coating to form microencapsulated pesticides. These pesticides are then mixed with water and applied in the form of a spray. After spraying, the plastic coating collapses and the active ingredient gets slowly released. Microencapsulation ensures the timely release of pesticides, e.g., Cyzmic CS and Optimate CS.
- *Low concentrate solution*: In this type of formulation, 1% or less than 1% of an active ingredient is dissolved in an organic solvent. Such solutions are ready to use and do not require dilution before application. They are used for structural and institutional pests, e.g., Round-up.
- *Fumigants*: These are the gaseous pesticides that are toxic to microbes, plants, and animals, e.g., Vikane and Phostoxin.

9.5 Importance of pesticides

9.5.1 Economic benefits

Improving crop production: framers are greatly benefitted from the use of pesticides as they protect the yield and quality of their agricultural products, it also reduces the cost of labor and fuel for them. There will be a decrease in food production and an increase in food prices if pesticides are not used in agricultural fields. Due to a decrease in production and an increase in food prices, the farmers will not be able to compete in global markets (Damalas, 2009). High crop production leads to a favorable export market and thus benefits the trade balance. In 1948–49, the food grain production was merely 50 million tons and it reached to 198 million tons by 1996–97. This increase in food production has been attributed to the utilization of high-yielding seed varieties, advanced irrigation facilities, and agricultural chemicals (Chandigarh Labour Bureau, 1994). Pesticides have also played an important role in this process by decreasing losses that occurred due to the growth of unwanted plants, diseases, and insect pests, according to the studies done by Webster et al. (1999); it has been found that without the use of pesticides, there would have been considerable economic losses.

9.5.2 Protection of crop losses

Studies done by Gianessi & Reigner (2006) and Gianessi (2009) revealed that every year, 9.2 billion US dollars have been used on pesticides and their crops. This use of pesticides saved 60 billion US dollars that otherwise would have been lost due to destruction by the pests. It has now been found that crop productivity gets affected by weeds because of the competition for inorganic nutrients present in the soil (Boote et al., 1983). To prevent and control the loss of crops, due to pest attacks during the field and storage conditions, crop protection methods have been developed (Popp et al., 2021). Since the early 1960s, there has been a dramatic increase in pesticide use and this has doubled the yield of rice, maize, and wheat, the major source of nutrition source for humans.

9.5.3 Control of vector-borne diseases

Vector-borne diseases such as malaria, yellow fever, leishmaniasis, chikungunya, onchocerciasis, Chagas disease, and tick-borne diseases are responsible for 17% of the burden due to communicable diseases across the globe. Onchocerciasis, commonly known as river blindness, has been controlled by the use of insecticides at the breeding sites of blackflies (James, 2004). According to this study, the weekly aerial spray was done against the blackflies and after 4 years of vector control, the endemicity level showed a significant decline. Acaricides are used during the rainy season to ravage the ticks and decrease their rate of multiplication (Basu et al., 2017). Every year, approximately 4000–5000 metric tons of DDT are used across the globe for the control of vector-borne diseases, and among these the largest consumer is India. The use of pesticides in the storage containers containing the water is helpful in decreasing the larvae of mosquitoes as a result the risk of various diseases like dengue and chikungunya lessens (WHO, 2011). On the global scale, during 2006–07, approximately 299 tons of pyrethroids, 16 tons of carbamates, 795 tons of organophosphates, and 3962 tons of organochlorines are used for controlling vector-borne diseases. About 40,000 tons of DDT was used during 2006–07 for the control of vectors (Himeidan et al., 2012). Zain and Jambulingam (2004) reported that among all the organochlorines used at the global level, only DDT is used for vector control. Recently, the WHO initiated the Global Vector Response, 2017–30 to promote the cause of integrated vector management (Alonso et al., 201; WHO, 2012). In some countries, pyrethroids were used in insecticide-treated bed nets and indoor residual spraying for eliminating visceral leishmaniasis and malaria. Control of vector-borne diseases relies on the use of pesticides applied at different places inhabited by the pest species (van den Berg et al., 2012).

9.6 Pesticides and humans

The use of pesticides has increased the production of food in the agricultural sector; but they have a considerable influence on human health and the environment (Jasrotia et al., 2021). Chemicals infiltrate into the human body through inhalation, ingestion, or contact with the skin and may produce acute or chronic poisoning effects. Exposure to the pesticide over long periods leads to chronic poisoning whereas acute is caused due to accidental exposure to the pesticides (Alewu & Nosiri, 2011). Once the pesticides enter the human or animal body, they may undergo breakdown or may get stored and amassed in the body fat (Pirsaheb et al., 2015). So, chemical pesticides have serious negative health effects. Damgaard et al. (2006) have reported the presence of pesticide residues in the samples of human breast milk which has raised concern for serious health issues in children. Pesticides contain chemicals that disrupt the functioning of hormones and thus alter the endocrine and homeostatic system of an organism (Dimanti-Kandarakis et al., 2009). Pesticides utilized for the growth of crops in the agricultural sector mimic the sex steroid hormones and thus have a significant impact on the homeostasis, reproduction, and immune system of humans (Pironti et al., 2021).

9.6.1 Effect of organochlorines

It has been reported that the areas where pesticides are used extensively in the agricultural fields showed an increased incidence of orchidopexy in male populations (García-Rodríguez et al., 1996). Patients who were operated on for cryptorchidism revealed high levels of pesticides in the samples of fat tissue than the other patients (Hosie et al., 2000). The study conducted by Damgaard et al. (2006) has also revealed a significant link between organochlorine pesticides accumulated in breast, milk, and congenital cryptorchidism. Aneck-Hahn et al. (2007) reported that humans are exposed to high levels of DDT due to indoor spraying. The uncontrolled use of an insecticide, dichloro diphenyl trichloroethane, DDT has raised various issues of human health. Exposure to high levels of p,p'-DDT, and p,p'-DDE are of serious concern as they have a great impact on reproductive health, p,p'-dichlorodiphenyldichloroethylene (p,p'-DDE) acts as potent antiandrogens. Most of the attention has been gained by organochlorine pesticides as they persist in the environment for a long time and get accumulated in the food web or the adipose tissue of humans and animals (Snedeker, 2001). Chlordane used for the control of termites and insects in agricultural fields poses a serious risk to human health as it acts as a toxin for the liver and nervous system (Koshlukova & Reed, 2014). This chlorinated cyclodiene persists in the environment for a long time and acts as a possible human carcinogen. Humans are exposed to chlordane through direct inhalation and it gets accumulated in the fat tissues of humans (Thalur & Pathania, 2020). Chlordane and heptachlor are known to cause diabetes, lymphoma, obesity, prostate, breast, and testes cancer. Hexachlorobenzene has been categorized as a possible human carcinogen (Zamora et al., 2014).

Exposure to organochlorine pesticides increases the risk of breast cancer, endometriosis, and impaired ovary functioning in human females (Ingber et al., 2013; Porpora et al., 2009). Humans are exposed to hexachlorobenzene through the skin, breast milk, inhalation, and contaminated food. Besides cancer, hexachlorobenzene impairs the functioning of the liver, skin, and thyroid gland and damages the immune system, nervous system, and endocrine system (Reed et al., 2007). Organochlorine chemicals such as 2,3,7,8-tetrachlorodibenzo-p-dioxin (TCDD) are known to play an important role in the pathogenesis of endometriosis (Matta et al., 2021). Lindane, an organochlorine pesticide, commonly known as gammaxene is a toxin for the nervous system and impedes the function of GABA, a neurotransmitter as it interacts with the binding site of the GABA receptor-chloride channel complex (Anand et al., 1998). Another organochlorine pesticide, i.e., endosulfan used to inhibit the growth of arthropods and insect pests on crops that enter to the human body through the lungs, stomach, and skin which leads to acute and chronic toxicity. The main target of endosulfan toxicity is the central nervous system. The metabolites of endosulfan get accumulated in the body fat (Cerrillo et al., 2005) and cause mental disturbances, congenital disorders, and even death. The toxic effects of pesticides on the male reproductive system have become a matter of great concern (Murray et al., 2001; Sharpe, 2001).

9.6.2 Effect of organophosphates

Exposure to acephate causes inhibition of cholinesterase, which leads to overstimulation of the central nervous system (Karanth, 2005). Organophosphates are the chemical substances formed by the esterification between alcohol and phosphoric acid. These pesticides hamper the action of cholinergic enzyme viz., acetylcholinesterase (AChE) which leads to accumulation of acetylcholine esterase (AChE) which leads to accumulation of acetylcholine at the neuronal junction (Costa, 2013). Such pesticides alter the functioning of the reproductive system by decreasing the acetylcholinesterase activity in the brain and thus affect the release of gonadotrophin from the pituitary and thus cause infertility (Perry et al., 2007). Numerous studies conducted by researchers disclosed that men exposed to organophosphate pesticides are at a high risk of fertility issues such as decreased sperm count, sperm concentration, and sperm motility (Jungwirth et al., 2015). Dhanushka and Pieris (2017) suggested that the function of sperm gets impaired on exposure to acephate, thus deteriorating the quality of human semen which triggers male infertility. Malathion inhibits the activity of acetylcholinesterase and thus interferes with nerve impulse transmission which causes serious health issues such as headache, nausea, dizziness, bradycardia, and vomiting (Sweeney & Lyon, 1999). A possible mechanism for the toxicity of malathion in humans is oxidative stress due to the production of reactive oxygen species (Moore et al., 2010). DNA damage has been seen in individuals exposed to malathion (Blasiak et al., 1999). Malathion is also called maldison, carbophos, and mercaptothion. Human exposure to malathion occurs through skin contact, ingestion, or direct inhalation. Chemical sprayers and farmworkers are at great risk of malathion toxicity. It leads to inflammation, oxidative stress, hepatic toxicity, and metabolic disorders in humans (Alp et al., 2012; Kalender et al., 2010; Lasram et al., 2009).

Dimethoate poisoning reduces systolic blood pressure, which can even become a cause of death for an individual (Davies et al., 2008). Field pesticide applicators are known to suffer from methyl parathion intoxication that leads to cholinergic overstimulation. The health concern raised due to methyl parathion has led to a decline in its use in crops in the United States (Garcia et al., 2003). Health effects related to the use of methyl parathion include diarrhea, mental disturbance, headache, nausea, difficulty in breathing, and dyspraxia (Edwards & Tchounwou, 2005). Organophosphate pesticides are known to inhibit the activity of acetylcholinesterase. Chlorpyrifos, a chlorinated organophosphate pesticide induces neurological disorders and cardiovascular diseases, immunotoxicity, endocrine disruption, and organo-toxicity in humans. It does not persist in the human body for a long time but its metabolites viz., chlorpyrifos-oxon and 3,5,6-trichloro-2-pyridinol are persistent in the body and can cause serious health issues (Rahman et al., 2021). Chlorpyrifos is the most widely used pesticide for vegetables and fruits. Fenitrothion is an organophosphate pesticide used for the control of molluscs and insects in various agricultural fields. This pesticide affects the transmission of nerve impulses by inhibiting the acetylcholinesterase in the peripheral and central nervous system (Stoytcheva et al., 2016).

9.6.3 Effect of carbamates

In developing countries, highly toxic pesticides are used freely due to a lack of control and this leads to the risk of severe toxicity after acts of self–harm. Aldicarb causes acute toxicity due to the repression of acetylcholinesterase. This leads to the accumulation of acetylcholine and causes central and peripheral toxicities. Symptoms include nausea, sweating, blurred vision, pain in the abdominal region, and even death. Recent studies have indicated that due to oxidative stress, aldicarb can damage the tissue (Blacker et al., 2010; Moser, 2014). Exposure to aldicarb leads to the accumulation of acetylcholine in the synaptic cleft of the neuromuscular junction which causes acute paralysis (Caldwell & Caldwell, 2015). Carbofuran, a carbamate pesticide used to kill insects, mites, and nematodes have low persistence as compared to organochlorine pesticides. Carbofuran poisoning causes vomiting, difficulty in breathing, sweating, abdominal cramps increase blood pressure, and blurred vision. High doses of carbofuran may even lead to death due to failure of the respiratory system (Kuswandi et al., 2017). Primicarb has a skin-sensitizing property and proves to be toxic for mammals. It enters the human body through the oral route or by direct inhalation, reduces the activity of cholinesterase, and affects different hematological parameters (EFSA Scientific Report, 2005).

Methomyl, a carbamate pesticide is used for ornamental plants and various crops. It is not able to persistent in the environment for long and has less potential to bioaccumulate in adipose tissue. Severe headaches accompanied by vomiting, pain in the abdominal region, and nausea are the main symptoms of methomyl poisoning

(Pope, 2005). Carbamates are the derivatives of carbamic acid and they inhibit the acetylcholinesterase reversibly by carbamylting the enzyme. Propoxur, another important carbamate is used to control household insects and other insects which are harmful to plant tissues and domestic animals. The half-life of propoxur is 14–15 days. It inhibits the action of acetylcholinesterase as a result of which accumulates in the synaptic cleft. This creates a cholinergic crisis that is marked by urination, salivation, muscle fasciculation, and diarrhea (Costa, 2014; Harp, 2005). Despite being relatively safe for mammals, carbaryl causes damage to liver cells. It also inhibits the acetylcholinesterase enzyme in humans leading to several problems related to the neural system. Recent studies revealed that carbaryl can bind the melatonin receptors in humans which may cause serious health issues (Sahana & Aggarwal, 2018). Meeker et al. (2004) have reported that exposure to carbaryl damages the DNA of human sperm. Carbaryl causes carbamoylation of a serine hydroxyl group at the active site and inhibits acetylcholinesterase. Generally, carbamate decomposes in the ecosystem within a few weeks or months (Goel & Aggarwal, 2007). Carbamate pesticides can increase the risk of dementia, neurobehavioural effects, and non-Hodgkin's lymphoma (Lin et al., 2015; Wesseling et al., 2002; Zheng et al., 2001).

9.6.4 Effect of pyrethroids

Various insecticide formulations having considerable outdoor and indoor applications such as pest control in agricultural fields, public, and residential areas contain synthetic pyrethroids. In recent years, a continuous increase has been seen in pyrethroid use. Humans exposed to this class of pesticides suffer from various health hazards (Saillenfait et al., 2015). They are found in shampoos and medicines for pets for the treatment of scabies and topical louse. The primary target of the pyrethroid in mammals is the nervous system, as they directly act on the voltage-sensitive sodium channels of nerve cell axons that incite hyperexcitation (Macan et al., 2006). Pyrethroids get absorbed into the human body through the skin and cause paraesthesia, due to the increased activity of nerve fibers of the skin (Bradberry et al., 2005). Widespread exposure of the US adult population to pyrethroid pesticides was found to be associated with an increased risk of cardiovascular diseases in them as reported by Bao et al. (2020). Chronic exposure of humans to mosquito repellants containing allethrin and prallethrin may lead to serious health effects. Narendra et al. (2008) evaluated a significant rise in lipid peroxides, nitrites, nitrates, phospholipids, plasma glucose levels, and a decline in plasma cholesterol levels.

Persons exposed only to allethrin showed a significant increase in the activity of serum glutamic pyruvic transaminase. Cypermethrin has wide use as a pesticide in agriculture, veterinary, and household for insect control. Due to the increase in the use of cypermethrin in daily activities, humans are getting more exposed to it. It causes various health hazards such as neurotoxicity, molecular toxicity, and reproductive toxicity (Sharma et al., 2018). It also causes deformities in sperm morphology. This acts on the nerve cell membrane and blocks the closure of sodium channel ion gates during repolarization. As a result, there is a strong disruption in nerve impulse transmission that leads to the depolarization of the membranes spontaneously (Nasuti et al., 2007). Permethrin has a detrimental effect on immunity, fertility, cardiovascular, and hepatic metabolism. Deltamethrin induces a toxic effect on nephrons and hepatocytes and has a serious impact on the activity of antioxidant enzymes in the tissues. Alpha-cypermethrin impairs immunity and increases the levels of glucose and lipid in the blood (Chrustek et al., 2018). Sumithrin, permethrin, and fen valerate used for crops and public health practices can act as endocrine-disrupting chemicals (Kolaczinski & Curtis, 2004; Ray & Fry, 2006).

9.7 Conclusion

The data relating to the health risks associated with pesticide use may be considered as an aid towards understanding the problem in a better way. Very scanty data related to the ill effects of pesticides is available from developing countries. Workers exposed to these chemicals should be well-trained to ensure the safe use of pesticides. Although pesticides are a quick and easy solution for controlling pests, they have posed a great problem in about every part of the earth. Residues of pesticides are found everywhere in the air, water, and soil. In addition to surface water, even the groundwater has also become contaminated with pesticides. They even pose a serious risk to nontarget organisms. Various steps are required to regulate the use of pesticides:

- Biopesticides such as canola oil and baking soda can be utilized.
- Insect traps can be used to control insects.
- Beneficial insects should be added to gardens.
- Genetically modified crops can be adopted.
- Garlic spray, tomato leaf spray, or vegetable oil spray should be done in fields.
- Pesticides obtained from natural plant sources like neem extract should be used.
- Integrated pest management methods can be used to control pests.
- Crop rotation and organic farming should be preferred in fields.

Acknowledgment

The authors thank the DST-PURSE, FIST, and RUSA programs of the Government of India to provide the necessary facilities for this research.

References

Aktar, M. W., Sengupta, D., & Chowdhury, A. (2009). Impact of pesticides use in agriculture: Their benefits and hazards. *Interdisciplinary Toxicology, 2*(1), 1–12. https://doi.org/10.2478/v10102-009-0001-7.

Alewu, B., & Nosiri, C. (2011). Pesticides and human health. In M. Stoytcheva (Ed.), *Pesticides in the modern world: Effects of pesticides exposure.* (pp. 231–250) InTech.

Alonso, P., Engels, D., & Reeder, J. (2017). Renewed push to strengthen vector control globally. *Lancet, 389*(10086), 2270–2271.

Alp, H., Aytekin, I., Hatipoglu, N. K., Alp, A., & Ogun, M. (2012). Effects of sulforophane and curcumin on oxidative stress created by acute malathion toxicity in rats. *European Review for Medical and Pharmacological Sciences, 16*(Suppl 3), 144–148.

Anand, M., Agrawal, A. K., Rehmani, B. N., Gupta, G. S., Rana, M. D., & Seth, P. K. (1998). Role of GABA receptor complex in low dose lindane (HCH) induced neurotoxicity: Neurobehavioural, neurochemical and electrophysiological studies. *Drug and Chemical Toxicology, 21*(1), 35–46. https://doi.org/10.3109/01480549809017849.

Aneck-Hahn, N. H., Schulenburg, G. W., Bornman, M. S., Farias, P., & de Jager, C. (2007). Impaired semen quality associated with environmental DDT exposure in young men living in a malaria area in the Limpopo Province, South Africa. *Journal of Andrology, 28*(3), 23–34. https://doi.org/10.2164/jandrol.106.001701.

Bao, W., Liu, B., Simonsen, D. W., & Lehmler, H. J. (2020). Association between exposure to pyrethroid insecticides and risk of all-cause and cause-specific mortality in the general US adult population. *JAMA Internal Medicine, 180*(3), 367–374. https://doi.org/10.1001/jamainternmed.2019.6019.

Basu, A. K., & Charles, R. A. (2017). A general account of ticks. In A. K. Basu, & R. A. Charles (Eds.), *Ticks of Trinidad and Tobago: An overview* (pp. 1–33). Academic Press. https://doi.org/10.1016/B978-0-12-809744-1.00001-3.

Blacker, A. M., Kelly, I. D., Lantz, J. L., Mihlan, G. J., Jones, R. L., & Young, B. M. (2010). Aldicarb: Toxicity, exposure and risks to humans. In R. Krieger (Ed.), *Hayes' handbook of pesticide toxicology* (3rd ed.) (pp. 1619–1632). Academic Press. https://doi.org/10.1016/B978-0-12-374367-1.00075-6.

Błasiak, J., Jałoszynski, P., Trzeciak, A., & Szyfter, K. (1999). In vitro studies on the genotoxicity of the organophosphorus insecticide malathion and its two analogues. *Mutation Research, 445*(2), 275–283. https://doi.org/10.1016/s1383-5718(99)00132-1.

Boote, K. J., Jones, J. W., Mishoe, J. W., & Berger, R. D. (1983). Coupling pests to crop growth simulators to predict yield reductions. *Phytopathology, 73*, 1581–1587.

Bradberry, S. M., Cage, S. A., Proudfoot, A. T., & Vale, J. A. (2005). Poisoning due to pyrethroids. *Toxicological Reviews, 24*(2), 93–106. https://doi.org/10.2165/00139709-200524020-00003.

Caldwell, G. A., & Caldwell, K. A. (2015). Use of *Caenorhabditis elegans* to model human *movement disorders*. In M. S. LeDoux (Ed.), Movement disorders (2nd ed., pp. 97–116). Academic Press. https://doi.org/10.1016/B978-0-12-405195-9.00006-8.

Cerrillo, I., Granada, A., Lopez-Espinosa, M. J., Olmos, B., Jimenez, M., Cano, A., Olea, N., & Olea-Serrano, M. F. (2005). Endosulfan and its metabolites in fertile women, placenta, cord blood, and human milk. *Environmental Research, 98*, 233–239.

Chandigarh Labour Bureau, Ministry of Labour. (1994). *Employment Information*. Indian Labour Statistics.

Chrustek, A., Hołyńska-Iwan, I., Dziembowska, I., Bogusiewicz, J., Wróblewski, M., Cwynar, A., & Olszewska-Słonina, D. (2018). Current research on the safety of pyrethroids used as insecticides. *Medicina, 54*(4), 61. https://doi.org/10.3390/medicina54040061.

Costa, L. G. (2013). Toxic effects of pesticides. In C. D. Klaassen (Ed.), *Toxicology: The basic science of poisons* (8th ed., pp. 763–810). McGraw-Hill.

Costa, L. G. (2014). Propoxur. In P. Wexler (Ed.), *Encyclopedia of toxicology* (3rd ed., pp. 1111–1112). Academic Press. https://doi.org/10.1016/B978-0-12-386454-3.00190-1.

Damalas, C. A. (2009). Understanding benefits and risks of pesticide use. *Scientific Research and Essay, 4*(10), 945–949.

Damgaard, I. N., Skakkebaek, N. E., Toppari, J., Virtanen, H. E., Shen, H., Schramm, K. W., Petersen, J. H., Jensen, T. K., & Main, K. M., Nordic Cryptorchidism Study Group. (2006). Persistent pesticides in human breast milk and cryptorchidism. *Environmental Health Perspectives, 114*(7), 1133–1138. https://doi.org/10.1289/ehp.8741.

Davies, J., Roberts, D., Eyer, P., Buckley, N., & Eddleston, M. (2008). Hypotension in severe dimethoate self-poisoning. *Clinical Toxicology, 46*(9), 880–884. https://doi.org/10.1080/15563650802172063.

Davies, T. G., Field, L. M., Usherwood, P. N., & Williamson, M. S. (2007). DDT, pyrethrins, pyrethroids and insect sodium channels. *Iubmb Life, 59*(3), 151–162. https://doi.org/10.1080/15216540701352042.

Delaplane, K. S. (2000). *Pesticide usage in the United States: History, benefits, risks, and trends*. Cooperative Extension Service, The University of Georgia College of Agricultural and Environmental Sciences.

Dhanushka, M. A. T., & Peiris, L. D. C. (2017). Cytotoxic and genotoxic effects of acephate on human sperm. *Journal of Toxicology, 2017*, 1–6.

Diamanti-Kandarakis, E., Bourguignon, J. P., Giudice, L. C., Hauser, R., Prins, G. S., Soto, A. M., Zoeller, R. T., & Gore, A. C. (2009). Endocrine-disrupting chemicals: An Endocrine Society scientific statement. *Endocrine Reviews, 30*(4), 293–342. https://doi.org/10.1210/er.2009-0002.

Dong, K. (2007). Insect sodium channels and insecticide resistance. *Invertebrate Neuroscience,* 7(1), 17–30. https://doi.org/10.1007/s10158-006-0036-9.

Edwards, F. L., & Tchounwou, P. B. (2005). Environmental toxicology and health effects associated with methyl parathion exposure: A scientific review. *International Journal of Environmental Research and Public Health, 2*(3-4), 430–441. https://doi.org/10.3390/ijerph2005030007.

EFSA Scientific Report., (2005). Conclusion regarding the peer review of the pesticide risk assessment of the active substance Pirimicarb. *EFSA Journal, 43*, 1–76.

FAO. (1989). *International code of conduct on the distribution and use of pesticides*. FAO Council.

Garcia, S. J., Abu-Qare, A. W., Meeker-O'Connell, W. A., Borton, A. J., & Abou-Donia, M. B. (2003). Methyl parathion: A review of health effects. *Journal of Toxicology and Environmental Health. Part B, Critical Reviews, 6*(2), 185–210. https://doi.org/10.1080/10937400306471.

García-Rodríguez, J., García-Martín, M., Nogueras-Ocaña, M., de Dios Luna-del-Castillo, J., Espigares García, M., Olea, N., & Lardelli-Claret, P. (1996). Exposure to pesticides and cryptorchidism: Geographical evidence of a possible association. *Environmental Health Perspectives, 104*, 1090–1095.

Gianessi, L. P. (2009). *The value of insecticides in U.S. crop production*. Croplife Foundation, Crop Protection Research Institute (CPRI).

Gianessi, L. P., & Reigner, N. (2006). *The value of herbicides in U.S. crop production*. Update Croplife Foundation, Crop Protection Research Institute (CPRI).

Goel, A., & Aggarwal, P. (2007). Pesticide poisoning. *National Medical Journal of India, 20*(4), 182–191.

Gupta, P. K. (2004). Pesticide exposure—Indian scene. *Toxicology, 198*(1-3), 83–90. https://doi.org/10.1016/j.tox.2004.01.021.

Harp, P. R. (2005). Propoxur. In P. Wexler (Ed.), *Encyclopedia of toxicology* (2nd ed., pp. 541–542). Elsevier. https://doi.org/10.1016/B0-12-369400-0/00804-8.

Himeidan, Y.E., Temu, E.A., Eliningaya, J.K. (2012). Insecticides for vector-borne diseases: Current use, benefits, hazard and resistance insecticides. In *Advances in integrated pest management*, Intech. (pp. 683–707).

Horsak, R. D., Bedient, P. B., Hamilton, M. C., & Thomas, F. B. (1964). Pesticides. In R. D. Morrison, & B. L. Murphy (Eds.), *Environmental forensics* (pp. 143–165). Academic Press. https://doi.org/10.1016/B978-012507751-4/50030-6.

Hosie, S., Loff, S., Witt, K., Niessen, K., & Waag, K. L. (2000). Is there a correlation between organochlorine compounds and undescended testes? *European Journal of Pediatric Surgery, 10*(5), 304–309.

Ingber, S. Z., Buser, M. C., Pohl, H. R., Abadin, H. G., Murray, H. E., & Scinicariello, F. (2013). DDT/DDE and breast cancer: A meta-analysis. *Regulatory Toxicology and Pharmacology: Rtp, 67*, 421–433.

James, E. R. (2004). Onchocerciasis control by insecticides and chemotherapy stimulates agricultural development in Central West Africa. *Journal of Agromedicine, 9*(2), 7–20.

Jasrotia, R., Langer, S., & Dhar, M. (2021). Endocrine disrupting chemicals in aquatic ecosystem: An emerging threat to wildlife and human health. *Proceedings of the Zoological Society, 74*(4), 634–647.

Jepson, P. C. (2001). Pesticides, uses and effects of. In S. A. Levin (Ed.), *Encyclopedia of biodiversity* (pp. 509–522). Elsevier. https://doi.org/10.1016/B0-12-226865-2/00220-0.

Jungwirth, A., Giwercman, A., Tournaye, H., Diemer, T., Kopa, Z., Dohle, G., & Krausz, C. (2015). *Guild lines on male infertility*. European Association of Urology.

Kalender, S., Uzun, F. G., Durak, D., Demir, F., & Kalender, Y. (2010). Malathion-induced hepatotoxicity in rats: The effects of vitamins C and E. *Food and Chemical Toxicology, 48*(2), 633–638. https://doi.org/10.1016/j.fct.2009.11.044.

Karanth, S. (2005). Acephate. In P. Wexler (Ed.), *Encyclopedia of toxicology* (2nd ed., pp. 13–15). Elsevier. https://doi.org/10.1016/B0-12-369400-0/00008-9.

Kaur, R., Mavi, G. K., & Raghav, S. (2019). Pesticides classification and its impact on Environment. *International Journal of Current Microbiology and Applied Science, 8*(3), 1889–1897. https://doi.org/10.20546/ijcmas.2019.803.224.

Kolaczinski, J. H., & Curtis, C. F. (2004). Chronic illness as a result of low-level exposure to synthetic pyrethroid insecticides: A review of the debate. *Food and Chemical Toxicology, 42*, 697–706. https://doi.org/10.1016/j.fct.2003.12.008.

Koshlukova, S. E., & Reed, N. R. (2014). Chlordane. In P. Wexler (Ed.), *Encyclopedia of toxicology* (3rd ed., pp. 841–845). Academic Press. https://doi.org/10.1016/B978-0-12-386454-3.00111-1.

Kuswandi, B., Futra, D., & Heng, L. Y. (2017). Nanosensors for the Detection of Food Contaminants. In A. E. Oprea, & A. M. Grumezescu (Eds.), *Nanotechnology applications in food* (pp. 307–333). Academic Press. https://doi.org/10.1016/B978-0-12-811942-6.00015-7.

Lah, K. (2011). *Effects of pesticides on human health*. Toxipedia.

Lasram, M. M., Annabi, A. B., Elj, N. E., Selmi, S., Kamoun, A., El-Fazaa, S., & Gharbi, N. (2009). Metabolic disorders of acute exposure to malathion in adult Wistar rats. *Journal of Hazardous Materials, 163*(2-3), 1052–1055. https://doi.org/10.1016/j.jhazmat.2008.07.059.

Lin, J. N., Lin, C. L., Lin, M. C., Lai, C. H., Lin, H. H., Yang, C. H., et al. (2015). Increased risk of dementia in patients with acute organophosphate and carbamate poisioning: A nationwide population-based cohort study. *Medicine, 94*, e1187. https://doi.org/10.1097/MD.0000000000001187.

Livadas, G. A., & Georgopoulos, G. (1953). Development of resistance to DDT by Anopheles sacharovi in Greece. *Bulletin of the World Health Organization, 8*(4), 497–511.

Macan, J., Varnai, V. M., & Turk, R. (2006). Zdravstveni ucinci piretrina i piretroida (Health effects of pyrethrins and pyrethroids). *Arhiv Za Higijenu Rada I Toksikologiju, 57*(2), 237–243.

Mathur, S. C. (1999). Future of Indian pesticides industry in next millennium. *Pesticide Information, 24*(4), 9–23.

Matta, K., Koual, M., Ploteau, S., Coumoul, X., Audouze, K., Bizec, B. L., Antignac, J. P., & Cano-Sancho, G. (2021). Associations between exposure to organochlorine chemicals and endometriosis: A systematic review of experimental studies and integration of epidemiological evidence. *Environmental Health Perspectives, 129*(7), 76003.

Meeker, J. D., Ryan, L., Barr, D. B., Herrick, R. F., Bennett, D. H., Bravo, R., & Hauser, R. (2004). The relationship of urinary metabolites of carbaryl/naphthalene and chlorpyrifos with human semen quality. *Environmental Health Perspectives, 112*, 1665–1670.

Moore, P. D., Yedjou, C. G., & Tchounwou, P. B. (2010). Malathion-induced oxidative stress, cytotoxicity, and genotoxicity in human liver carcinoma ($HepG_2$) cells. *Environmental Toxicology, 25*(3), 221–226. https://doi.org/10.1002/tox.20492.

Moser, V. C. (2014). Aldicarb. In P. Wexler (Ed.), *Encyclopedia of toxicology* (3rd ed., pp. 123–125). Academic Press. https://doi.org/10.1016/B978-0-12-386454-3.00093-2.

Murray, T. J., Lea, R. G., Abramovich, D. R., Haites, N. E., & Fowler, P. A. (2001). Endocrine disrupting chemicals: Effects on human male reproductive health. *Early Pregnancy, 5*, 80–112.

Narendra, M., Kavitha, G., Kiranmai, A. H., Rao, N. R., & Varadacharyulu, N. C. (2008). Chronic exposure to pyrethroid-based allethrin and prallethrin mosquito repellents alters plasma biochemical profile. *Chemosphere, 73*(3), 360–364. https://doi.org/10.1016/j.chemosphere.2008.05.070.

Nasuti, C., Gabbianelli, R., Falcioni, M. L., Di Stefano, A., Sozio, P., & Cantalamessa, F. (2007). Dopaminergic system modulation, behavioral changes, and oxidative stress after neonatal administration of pyrethroids. *Toxicology, 229*(3), 194–205. https://doi.org/10.1016/j.tox.2006.10.015.

Perry, M. J., Venners, S. A., Barr, D. B., & Xu, X. (2007). Environmental pyrethroid and organophosphorus insecticide exposures and sperm concentration. *Reproductive Toxicology, 23*(1), 113–118.

Pironti, C., Ricciardi, M., Proto, A., Bianco, P. M., Montano, L., & Motta, O. (2021). Endocrine-disrupting compounds: An overview on their occurrence in the aquatic environment and human exposure. *Water, 13*, 1347. https://doi.org/10.3390/w13101347.

Pirsaheb, M., Limoee, M., Namdari, F., & Khamutian, R. (2015). Organochlorine pesticides residue in breast milk: A systematic review. *Medical Journal of the Islamic Republic of Iran, 29*, 228.

Pope, C. N. (2005). Methomyl. In P.Wexler (Ed.), *Encyclopedia of toxicology* (2nd ed., pp. 57–58). Elsevier. https://doi.org/10.1016/B0-12-369400-0/00605-0.

Popp, J., Pető, K., & Nagy, J. (2013). Pesticide productivity and food security: A review. *Agronomy for Sustainable Development, 33*, 243–255. https://doi.org/10.1007/s13593-012-0105-x.

Porpora, M. G., Medda, E., Abballe, A., Bolli, S., De Angelis, I., di Domenico, A., Ferro, A., Ingelido, A. M., Maggi, A., Panici, P. B., & De Felip, E. (2009). Endometriosis and organochlorinated environmental pollutants: A case-control study on Italian women of reproductive age. *Environmental Health Perspectives, 117*, 1070–1075.

Rahman, H. U., Asghar, W., Nazir, W., Sandhu, M. A., Ahmed, A., & Khalid, N. (2021). A comprehensive review on chlorpyrifos toxicity with special reference to endocrine disruption: Evidence of mechanisms, exposures and mitigation strategies. *Science of the Total Environment, 755*(2), 142649. https://doi.org/10.1016/j.scitotenv.2020.142649.

Ray, D. E., & Fry, J. R. (2006). A reassessment of the neurotoxicity of pyrethroid insecticides. *Pharmacology & Therapeutics, 111*, 174–193. https://doi.org/10.1016/j.pharmthera.2005.10.003.

Reed, L., Buchner, V., & Tchounwou, P. B. (2007). Environmental toxicology and health effects associated with hexachlorobenzene exposure. *Reviews on Environmental Health, 22*(3), 213–243. https://doi.org/10.1515/reveh.2007.22.3.213.

Sahana, A., & Agarwal, S. (2018). Carbaryl and human health. *A Review Journal of Science, 2*(5), 12–22.

Saillenfait, A. M., Ndiaye, D., & Sabaté, J. P. (2015). Pyrethroids: Exposure and health effects: An update. *International Journal of Hygiene and Environmental Health, 218*(3), 281–292. https://doi.org/10.1016/j.ijheh.2015.01.002.

Salvo, L. M., Bainy, A. C., Ventura, E. C., Marques, M. R., Silva, J. R., Klemz, C., & Silva de Assis, H. C. (2012). Assessment of the sublethal toxicity of organochlorine pesticide endosulfan in juvenile common carp (*Cyprinus carpio*). *Journal of Environmental Science and Health, Part A: Toxic/Hazardous Substances and Environmental Engineering, 47*(11), 1652–1658. https://doi.org/10.1080/10934529.2012.687236.

Sharma, A., Yadav, B., Rohatgi, S., & Yadav, B. (2018). Cypermethrin toxicity: A review. *Journal of Forensic Sciences & Criminal Investigation, 9*(4), 555767. https://doi.org/10.19080/JFSCI.2018.09.555767.

Sharpe, R. M. (2001). Hormones and testis development and the possible adverse effects of environmental chemicals. *Toxicology Letters, 120*, 221–232.

Shukla, S. K., Mangwani, N., Rao, T. S., & Das, S. (2014). Biofilm-mediated bioremediation of polycyclic aromatic hydrocarbons. *Microbial biodegradation and bioremediation* (pp. 203–232). Elsevier. https://doi.org/10.1016/B978-0-12-800021-2.00008-X.

Silberman, J., & Taylor, A. (2021). Carbamate toxicity. *StatPearls [Internet]*. StatPearls Publishing https://www.ncbi.nlm.nih.gov/books/NBK482183/.

Singh, S., Sharma, S., Sarma, S. J., Misra, K., & Brar, S. K. (2021). Pesticides in water. In S. Ahuja (Ed.), *Handbook of water purity and quality* (2nd ed.) (pp. 231–253). Academic Press. https://doi.org/10.1016/B978-0-12-821057-4.00004-5.

Smith, A. E., & Secoy, D. M. (1975). Forerunners of pesticides in classical Greece and Rome. *Journal of Agricultural and Food Chemistry, 23*(6), 1050.

Smith, A. E., & Secoy, D. M. (1976). A compendium of inorganic substances used in European pest control before 1850. *Journal of Agricultural and Food Chemistry, 24*(6), 1180.

Snedeker, S. M. (2001). Pesticides and breast cancer risk: A review of DDT, DDE, and dieldrin. *Environmental Health Perspectives, 109*(Suppl 1), 35–47. https://doi.org/10.1289/ehp.01109s135.

Stoytcheva, M., Zlatev, R., Montero, G, Torres, R., Toscano, L., 2016. Fenitrothion: Structural features, applications, toxicity mechanisms and effects of exposure on human health: A review In N. Wilkinson (Ed.). *Organophosphorus pesticides.* Publisher: Nova Science Publishers, Inc.

Sweeney, M. I., & Lyon, M. E. (1999). Selective effect of malathion on blood coagulation versus locomotor activity. *Journal of Environmental Pathology, Toxicology and Oncology, 18*(3), 203–211.

Thakur, M., & Pathania, D. (2020). Environmental fate of organic pollutants and effect on human health. In P. Singh, A. Kumar, & A. Borthakur (Eds.), *Abatement of environmental pollutants* (pp. 245–262). Elsevier. https://doi.org/10.1016/B978-0-12-818095-2.00012-6.

Toppozada, A., Mohamed, M. A., & Eldefrawi, M. E. (1967). Resistance of the housefly (*Musca domestica* L.) to insecticides in Egypt. *Bulletin of the World Health Organization, 36*, 937–948.

Urkude, R., Dhurvey, V., & Kochhar, S. (2019). Pesticide residues in beverages. In A. Mihai Grumezescu, & A. Maria Holban (Eds.), *Quality control in the beverage industry* (pp. 529–560). Academic Press. https://doi.org/10.1016/B978-0-12-816681-9.00015-1.

Vais, H., Atkinson, S., Pluteanu, F., Goodson, S. J., Devonshire, A. L., Williamson, M. S., & Usherwood, P. N. (2003). Mutations of the para sodium channel of *Drosophila melanogaster* identify putative binding sites for pyrethroids. *Molecular Pharmacology, 64*(4), 914–922. https://doi.org/10.1124/mol.64.4.914.

van den Berg, H., Zaim, M., Yadav, R. S., Soares, A., Ameneshewa, B., Mnzava, A., et al. (2012). Global trends in the use of insecticides to control vector-borne diseases. *Environmental Health Perspectives, 120*(4), 577–582.

Webster, J. P. G., Bowles, R. G., & Williams, N. T. (1999). Estimating the economic benefits of alternative pesticide usage scenarios: Wheat production in the United Kingdom. *Crop Production, 18*, 83.

Wesseling, C., Keifer, M., Ahlbom, A., McConnell, R., Moon, J.-D., Rosenstock, L., & Hogstedt, C. (2002). Long-term neurobehavioral effects of mild poisonings with organophosphate and n-methyl carbamate pesticides among banana workers. *International Journal of Occupational and Environmental Health, 8*, 27–34. https://doi.org/10.1179/oeh.2002.8.1.27.

WHO/UNEP. (1990). *Public health impact of pesticides use in agriculture*. Geneva, Switzerland.

World Health Organization. (2011). *Guidelines for drinking-water quality*, (4th ed.) (pp. 1-564). Geneva.

WHO. (2012). *Handbook for integrated vector management*. WHO/HTM/NTD/VEM/2012.3. http://whqlibdoc.who.int/publications/2012/9789241502801_eng.pdf.

Zaim, M., & Jambulingam, P. (2004). *Global insecticide use for vector-borne disease control. World Health Organization Pesticide Evaluation Scheme (WHOPES)*. World Health Organization.

Zamora, E. M. U. (2014). Hexachlorobenzene. In P. Wexler (Ed.), *Encyclopedia of toxicology* (3rd ed., pp. 869–871). Academic Press. https://doi.org/10.1016/B978-0-12-386454-3.00150-0.

Zheng, T., Zahm, S. H., Cantor, K. P., Weisenburger, D. D., Zhang, Y., & Blair, A. (2001). Agricultural exposure to carbamate pesticides and risk of non-hodgkin lymphoma. *Journal of Occupational and Environmental Medicine, 43*, 641–649. https://doi.org/10.1097/00043764-200107000-00012.

CHAPTER 10

Role of microbes in pesticide bioremediation: Recent advances and biotechnological implications

Roohi Sharma[a] and Shveta Saroop[b]
[a]Department of Botany, University of Jammu, Jammu, Jammu and Kashmir, India
[b]Department of Botany, Government Degree College, Kathua, Jammu and Kashmir, India

10.1 Pesticides as an environmental toxic and bio-remediation strategy thereof

Pesticides are the organic chemicals that are purposely used in agricultural systems for increasing soil productivity, agriculture productivity, food quality, and storage (Nkongolo & Narendrula-Kotha, 2020; Parween et al., 2018; Sharma et al., 2019). With the increasing population, pesticides have been extensively used in agriculture practices to meet the rising demand for food and fulfill higher yield requirements. Due to this the use of pesticides in agricultural production has become a necessity, and as a result numerous environmental issues have risen including the threat to living organisms. Pesticides have an impact on various nontarget organisms as well as the pest that they are intended to control, such as earthworms, birds, amphibians, fish, small mammals, natural predators, and pollinators. The hazardous effects of pesticides affect not just terrestrial species but also the majority of aquatic creatures, plankton, birds, and microbes (Pereira et al., 2009). The equilibrium of the natural ecosystem may be negatively impacted by this. Most prior research has shown that although pesticides are administered to a limited area, they can travel long distances until they reach the water table at detectable concentration and spread over a large region. Further, only a small portion of them reach the intended target organism. The remainder is squandered and either builds up in the soil or ends up in the environment. Pesticides can have a variety of fates, including staying in the air, penetrating the soil, and dissolving in water. residues of pesticides in surface water bodies are accessible to the soil through surface discharge or other potential ground contamination percolation to get water (Kumar et al., 2018; Raffa & Chiampo, 2021). Only a very small portion (0.3%) of the total amount of pesticides applied to land reaches the target pest, while the remainder moves somewhere else (Prabha et al., 2017; Puckett et al., 2020; Rajmohan et al., 2020).

Pesticides not only contaminate the ecosystem to a large extent but also cause chronic threats to human life through long-term exposure to pesticides. By invading

Pesticides in the Environment: Impact, Assessment, and Remediation
DOI: https://doi.org/10.1016/B978-0-323-99427-9.00011-2

the ecosystem these pesticides bioaccumulate and then biomagnify across the food chain (Mishra et al., 2018). Prolonged exposure to pesticides causes hormonal disruption, reproductive abnormalities, respiratory, immunological, cancer, and neurological issues, and diminished intelligence in humans (Sharma & Singhvi, 2017; Sharma et al., 2018, 2019). Pesticide use has skyrocketed in both agricultural and nonagricultural applications. Due to these threats, the degradation of persistent pesticides is very essential for the environment's health and equilibrium (Raffa et al., 2021; Randika et al., 2022). Earlier techniques to remove pesticides from the environment including landfills, recycling, pyrolysis, etc. have opposing effects on the environment by forming various toxic intermediates (Gilden et al., 2010). All these methods of pesticide remediation possess many limitations thus becoming less suitable for efficient removal.

Sustainable development necessitates the creation and promotion of environmental management, as well as the continuous search for green technology to treat a wide range of aquatic and terrestrial environments damaged by rising anthropogenic activity (Bharagava et al., 2017; Roy et al., 2022). Bioremediation is an extremely desirable alternative to traditional methods of removing pesticide compounds since it allows pollutants to be degraded by natural microbial activity mediated by various consortiums of microbial strains. The said alternative thus detoxifies the environmental contaminants (Abatenh et al., 2017; Iqbal, 2019; Pande et al., 2020; Paul et al., 2005). The use of living organisms (bacteria and fungi or plants) to detoxify environmental toxins into less toxic states is known as bioremediation. Microbial remediation of xenobiotic substances is considered a quick, efficient, and economically beneficial method that involves a diversity of microbes using organic pollutants as carbon or nitrogen supplements to support their developmental stages (Abatenh et al., 2017; Arora, 2020; Arora et al., 2018; Bhatt et al., 2019; Chen et al., 2019; Mahmoud, 2016). Countless microbial communities can be found thriving in both naturally occurring and highly stressful environments like acid mine runoff, hydrothermal vents, human gut, and oil reservoirs (Arora, 2020; Birolli et al., 2016; Chaudhary, 2016; Delegan et al., 2019; Shekhar et al., 2020). Because of their genetic diversity and functioning, microbial communities have the capacity to clean up any polluted environment (Bhatt et al., 2021; Chen et al., 2015; Dangi et al., 2019). As a result, understanding the precise microbial traits that increase degradation rates can be gained from research on the microbial community that exists in contaminated areas.

But for microbial remediation solutions to be implemented successfully, advanced technological approaches are required, which give a thorough into microbial dynamics and survival in stressed environments (Lima-Morales et al., 2016; Mao et al., 2019; Marco & Abram, 2019). Gene-level methods of bioremediation have been developed as a result of advances in molecular, biotechnological, and bioinformatics techniques in relation to bioremedial issues (Aora et al., 2018; Jaiswal & Shukla, 2020; Nkongolo & Kotha, 2020; Shekha et al., 2020). In addition, the direct examination of microbes in a polluted environment opened a new field for scientific inquiry into the uncultivated microbial world (Panigrahi et al., 2019; Sehrawat et al., 2021; Yan et al., 2020). The recombinant engineering techniques for the removal of environmental pesticides have been accelerated by the evolution of advanced molecular tools which have reflected a better understanding of microbial metabolic structures and functions (Delegan et al., 2019; Iqbal, 2019; Iram et al., 2019; Marco & Abram, 2019; Puckett et al., 2020). Advanced molecular techniques that entailed directly characterizing the genome of microorganisms without using their cultured sample gave rise to omics technology (Wang et al., 2012; Yan et al., 2020; Zhang et al., 2014). Thus, the use of contemporary molecular methods such as transcriptomic, metagenomic, and proteomic generates pertinent data on gene expression in microbial populations present in polluted environments in an effort to understand the mechanism of microbial degradation. This systematic review thus addresses the biodegradation capability of microorganisms and offers insights into recent technologies and bioengineering advancement.

10.2 Microbial-assisted bioremediation

As already discussed, it is a sustainable approach using the natural ability of microbes to remove pesticides from the environment by degrading complex pollutants (Hussain et al., 2009). The use of microorganisms has proven to be an economically feasible and environmentally safe tool for the degradation and detoxification of numerous toxic pesticides (Kour et al., 2021; Parween et al., 2018). Microbes via their intrinsic metabolic processes with or without slight pathway modifications transform the pollutants into their less toxic forms by degradation/and biotransformation. There are a large number of microorganisms that possess the ability to degrade pesticides. They are also called pesticide degraders which have been isolated from various pesticide-contaminated sites.

10.2.1 Bacterial bioremediation

Among microorganisms, bacteria have been used extensively for bioremediation purposes. The bacterial classes gamma-proteobacteria (v.gr. *Pseudomonas, Aerobacter, Acinetobacter, Moraxella, Plesiomonas*), alpha-proteobacteria (*Sphingomonas*) *Actinobacteria* (*Micrococcus*), and *Flavobacteria* (*Flavobacterium*), beta-proteobacteria (*Burkholderia, Neisseria*), are considered to be active micro biodegraders (Tables 10.1 and 10.3). Oxidation of the parent compound results in the production of carbon dioxide and water during the complete biodegradation of the pesticide (Doolotkeldieva & Konurbaeva, 2018). The energy produced during this process is utilized by the microbes for their metabolism. In the soil where the natural community of microbial population is not sufficient, the external addition of pesticide degrading microflora is recommended (Doolotkeldieva, 2018).

Table 10.1 Bacterial genera with their pesticide degrading capacities.

S. no.	Bacterial species	Types of pesticides degraded	References
1.	*Flavobacterium* spp.	Diazinon, Parathion Diazinon, Pentachlorophenol	Yasouri (2006)
2.	*Sphingobium* spp.	Fenobucarb, Carbaryl	Randika et al. (2022)
3.	*Sphingobium fuliginis*	Diazinon, Parathion Buprofezin	Randika et al. (2022)
4.	*Sphingomonas* sp.	Fenvalerate, Diazinon	Yu et al. (2013)
5.	*Agrobacterium* spp.	Methyl parathion, Phoxim, Methamidophos, Chlorpyrifos, Carbofuran, Deltamethrin	Wang et al. (2012)
6.	*Pseudomonas* spp.	Profenofos Diazinon	Yasouri (2006); Malghani et al. (2009)
7.	*Arthrobacter* spp.	Atrazine	Getenga et al. (2009)
8.	*Klebsiella pneumoniae*	Endosulfan	Kwon et al. (2002)
9.	*Enterobacter* spp. *(Strain B-14)*	Chlorpyrifos, Parathion, Diazinon, Coumaphos, Isazofos	Singh et al. (2004)
10.	*Stenotrophomonas* sp.	Chlorpyrifos, Diazonin, Methyl parathion, Parathion, Phoxim, Profenofos, Triazophos	Deng et al. (2015)
11.	*Streptomyces* sp.	Atrazine, Chlordane, HCHs, Methoxychlor	Bhat et al. (2020)
12.	*Variovorax* sp.	Linuron	Bers et al. (2013)
13.	*Lactococcus* sp.	Fenpropathrin	Chen et al. (2012)
14.	*Azospirillum* sp.	Ethion	Foster et al. (2004)
15.	*Bacillus*	Acibenzolar-S-methyl, metribuzin, napropamide, propamocarb hydrochloride, thiamethoxam	Myresiotis et al. (2012)
16.	*Paracoccus* sp.	Alachlor, Acetochlor, Butachlor, Metolachlor, Propisochlor	Liu et al. (2020)
17.	*Enterobacter* sp.	Bifenthrin, Fenamiphos	Singh et al. (2006)
18.	*Nocardia* sp.	Sulfentrazone	Mielke et al. (2020)
19.	*Paenibacillus* sp.	Chlorpyrifos, Cyanophos Fenpropathrin	

(continued on next page)

Table 10.1 Bacterial genera with their pesticide degrading capacities—cont'd

S. no.	Bacterial species	Types of pesticides degraded	References
20.	*Morganella* sp.	Carbaryl	Mustapha et al. (2019)
21.	*Mesorhizobium* sp.	Chlorpyrifos	Mansotra et al. (2022)
22.	*Klebsiella* sp.	Bifenthrin, Fenpropathrin, Naphtalene, Chlorpyrifos, Fenpropathrin, Trifluralin	Bokade et al. (2022)
23.	*Herbaspirillum* sp.	Trifluralin	Lara-Moreno et al. (2022)

The most widely used organochlorine pesticides include dieldrin, aldrin, alpha-BHC, beta alpha BHC, gamma BHC (lindane), beta BHC, delta-BHC, gamma-BHC (lindane), endosulfan, methoxychlor, aroclor dichloro diphenyl trichloroethane (DDT) and heptachlor.

Majorly the pesticide contamination is mainly caused by atrazine (2-chloro-4-(aminoethyl)-6-(aminoisopropyl)-s-1,3,5-triazine). Atrazine is widely used in agriculture for the control of annual grasses and broad-leaved weeds. It is a photosynthesis-inhibitor (Eevers et al., 2017). It is known to be an endocrine disruptor and induces toxicity in amphibians. The *in situ* degradation of organochlorine pesticides was carried out by bacteria belonging to the genera *Pseudomonas, Arthrobacter and Micrococcus*, and *Bacillus* (Kumar et al., 2021). *Pseudomonas* sp. played an important role in the breakdown of dieldrin in the soil (Velázquez-Fernández, 2012). Endosulfan is another widely used agricultural pesticide (Menezes et al., 2017). Oxidation of endosulfan sulfate or endosulfan by the monooxygenase encoded by ese is caused by a bacterium species *Arthrobacter* sp. These are the metabolites formed during biodegradation of lesser toxicity-endosulfan diol, endosulfan lactone, and endosulfan ether as compared with the original compound (i.e., endosulfan) (Kafilzadeh et al., 2015). Bacterial species such as *Pseudomonas* sp., *Rhodococcus Stenotrophomonas maltophilia* CGMCC 1.1788 and *Burkholderia cepacia* strain CH-9 (Chen et al., 2008) also degrade neonicotinoids. Research studies have shown that bacterial genera are identified as carbamate degraders (Satish & Ashokrao, 2017). 37 bacteria were isolated with carbofuran degradation potential (Shin et al., 2012). These included *Rhodococcus, Spingomonas, Spingobium, Bacillus, Pseudomonas, Aeromonas*, and *Corynebacterium*, etc. Some pesticides are recalcitrant and get accumulated in the environment after their application to the soil and gain entry the food chain (Bolan, 2021). Microbial degradation is limited by the presence of anionic species in the recalcitrant pesticides. The anions like chloride, sulfate, etc. are strongly bonded to the hydrocarbon ring which prevents the microbes from attacking the ring structure. This may be due to the increased toxicity of the anionic group.

Table 10.2 Fungal species capable to degrade pesticides.

S. no.	Fungal species	Pesticides	References
1.	*Comamonas, Azospirillum, Methylobacillus,* and *Acinetobacter, Serratia marcescens*	Epoxiconazole and fludioxonil Tebuconazole	Wang et al. (2018); Alexandrino et al. (2020)
2.	*Trichoderma*	Malathion	Ghosh and Pal (2018)
3.	*Pleurotus*	Terbufos, Azinphosmethyl, Phosmet and Tribufos	Eevers et al. (2017)
4.	*Aspergillus* sp.	Pentachlorophenol	Birolli et al. (2018)
5.	*Rhodotorula* spp.	Lindane	Abdul Salam et al. (2013)
6.	*Fusarium poae* and *F. solani*	Lindane	Sagar and Singh (2011)
7.	*Bjerkandera* sp.	Azinphos methyl, Phosmet, Terbufos	Eevers et al. (2017)
8.	*Verticillium* sp.	Chlorpyrifos	Jaiswal et al. (2017)
9.	*Phanerochaete* sp.	Phosmet	Eevers et al. (2017)
10.	*Cladosporium* sp.	Pentachlorophenol	Birolli et al. (2020)
11.	*Bjerkandera* sp.	Azinphos methyl, Phosmet, Terbufos, Tribufos	Maqbool et al. (2016)
12.	*Coriolus* sp.	Aldicarb, Alachlor, Atrazine	Spina et al. (2018)
13.	*Cordyceps* sp.	Dieldrin	Xiao and Kondo (2013)
14.	*Lentinula* sp.	Difenoconazole, Pendimethalin, Terbuthylazine	Pinto et al. (2020)
15.	*Glomus* sp.	Phoxim	Wang et al. (2020)

10.2.2 Fungal bioremediation

Fungi are considered more efficient in the degradation of pesticides as they flourish in low moisture areas and low pH solutions (Deshmukh, 2016). Bioremediation technologies using fungus have given considerable attention where the bacterial species fails to degrade pesticides. Fungi make the pesticides nontoxic by inducing structural changes in them (Odukkathil & Vasudevan, 2013). Pesticide groups like dicarboximide, organophosphorus and chlorinated compounds, phenylamide, triazine and phenylurea are known to be degraded by fungal species such as *Agrocybe semiorbicularis, Coriolus versicolor Auricularia auricula, Dichomitus squalens, Flammulina velupites, Pleurotus ostreatus, Stereum hirsutum,* and *Avatha discolor Hypholoma fasciculare.* Fungal species such as *Penicillium miczynskii, Trichoderma* spp. *Pleurotus ostreatus, Aspergillus sydowii,* and *Phlebia* spp. causes the degradation of pesticides (Tables 10.2 and 10.3) (Shourie & Vijayalakshm, 2022).

P. ostreatus produces the biotransformation products of lesser toxicity, i.e., dieldrin, 9-hydroxy aldrin, 9-hydroxy dieldrin of aldrin degradation. An organochlorine compound—pentachlorophenol (PCP) has been included in the priority pollutant list of the Environmental Protection Agency (Aregbesola et al., 2020). Studies showed that degradation of PCP is carried out by many fungal species such as *Byssochlamys nivea, Scopulariopsis brumptii, Rhizopus oryzae Trichoderma harzianum, Phlebia acanthocystis*, and *Anthracophyllum discolor* (Kumar et al., 2021; Ruiz-Lara et al., 2020). Another group of pesticides—polychlorinated biphenyl (PCB) and polycyclic aromatic hydrocarbons (PAH) are degraded by a fungus—*Phanerochaete chrysosporium*. Fungal microbes including *Lentinula edodes, Penicillium brevicompactum*, and *Lecanicillium saksena* causes the degradation of difenoconazole, terbuthylazine, and pendimethalin. Rot fungi isolated from contaminated soil have the capability to degrade methomyl and diazinon pesticides. The phytopathogenic fungus easily grows on organophosphonate herbicides resulting in their degradation (Javaid et al., 2016). The higher rate of degradation can be enhanced by using a mixture of fungal strains. Pirimicarb is degraded by *Trichoderma viride* and *T. harzianum* (Kharat et al., 2018). The latter fungus was isolated from a marine plant *Didemnun ligulum* and had the capacity to degrade 50 mg/L phoxim in a liquid medium in 7 days (Wang et al., 2022). 17 white rot fungi were tested to study pesticide degradation. Out of 17 white rot fungi; 16 degraded parathion, azinphos methyl, phosmet, and terbufos after a 4-day growth period.

10.3 Enzymatic degradation pathway of pesticides by microbes

The conventional physical, chemical, and physical–chemical degradation, all of which essentially resulted in secondary pollution. Microbial degradation has been more popular in recent years as a bioremediation agent (Abatenh et al., 2017; Bharagava et al., 2017; Iqbal, 2019; Pande et al., 2020; Roy et al., 2022). The pesticides get slowly degraded in nature through mineralization and co-metabolism by naturally occurring microbial communities. The process, referred to as an enzymatic reaction, involved the compound (i.e., pesticide) entering the microbe system followed by a series of physiological and biochemical processes aided by different enzymes, finally resulting in the complete degradation of pesticide into small molecular compounds with lower or nontoxic toxicity (Singh, 2008; Singh et al., 2014; Verma et al., 2014).

Microorganisms have the ability to break down several chemical pesticides, which were analogs of natural chemicals via enzymatic processes. The pesticide serves as a source of nutrition for microorganisms which thus can be processed to be broken down by biochemical reactions into inorganic compounds, carbon dioxide, and water. The process is termed mineralization which serves as a significant way to degrade pesticides into nontoxic inorganic substances. *Archromobacter, Bacillus, Burkholderia, Chromohalobacter, Pseudomonas, Kocuria, Sphingomonas*, and other bacterial strains have demonstrated

Table 10.3 Pesticides and their degrading microorganisms.

S. no.	Pesticides	Microbes	References
1.	Organochlorine Hexachlorocyclohexane	*Microbacterium* sp., *Sphingomonas taejonensis, Sphingomonas* sp. Γ *1-7, Sphingomonas paucimobilis UT26; Sphingomonas paucimobilis Strain B90*	Martin et al. (2019)
2.	Organophosphorous insecticide Organophosphorous Fenitrothion Methyl parathion Parathion	*Achromobacter, Brucella, Clavibacter michiganense, Delftia acidovorans, Pseudaminobacter, Pseudomonas aeruginosa F10B, Alteromonas undina, Alteromonas haloplanktis* *B. diminuta and Flavobacterium* sp. *B. diminuta and Flavobacterium* sp.	Liu et al. (2005)
		Fungi: *Pleurotus ostreatus.* *Aspergillus* sp. *Penicillium* sp. *Burkholderia* *Plesiomonas* *Flavobacterium* sp. *Strain ATCC 27551*	Zhang et al. (2006b)
3.	Carbamate Carbaryl Carbofuran Thiocarbamate	*Arthrobacter oxydans, Methylobacteria organophilum, Methylomonas methylovora, Pseudomonas, Mesorhizobium, Ralstonia, Rhodococcus*	Kumar et al. (2021)
		Ochrobactrum, and *Bacillus, Achromobacter* sp. *Arthrobacter* sp. *Strain RC100* *Achromobacter* sp. *Rhodococcus erythropolis*	Zhang et al. (2022)
4.	Endosulfan	*Bacillus, Staphylococcus*	Duc et al. (2021)
5.	*Enterobacter*	*Chlorpyrifos*	Bose et al. (2021)
6.	Fitoraz	*Pseudomonas putida, Acinetobacter* sp.	Monica et al. (2016)
7.	WP 76, Decis 2.5 EC, malation	*Arthrobacter* sp.	Monica et al. (2016)
8.	chlorpyrifos and methyl parathion	*Acenetobactor* sp., *Pseudomonas* sp., *Enterobacter* sp. and *Photobacterium* sp.	Gilani et al. (2016)
9.	Pyrithroid	*Serratia, Pseudomonas* *Fungi: Aspergillus niger*	Grant et al. (2003) Liang et al. (2005)

(*continued on next page*)

Table 10.3 Pesticides and their degrading microorganisms—cont'd

S. no.	Pesticides	Microbes	References
10.	Insect growth regulators	*Azadirachtin, benzoylphenylurea, diflubenzuron, Methoxyfenozid, pyriproxyfen, spinosad, tebufenozide*	Kamal and Khater (2010)
11.	Chlorpyrifos	*Ochrobactrum* sp. *JAS2*	Abraham & Silambarasan (2016)
12.	Ochrobactrum sp. JAS2	*Bacillus subtilis*	Gangola et al. (2018)
13.	DDT	*Fomitopsis pinicola and Ralstonia pickettii*	Purnomo et al. (2020)
14.	Deltamethrin	*Streptomyces rimosus*	Khajezadeh et al. (2020)
15.	Fentopropathrin	*Rhodopseudomonas palustris PSB-S*	Raffa and Chiampo (2021)

biodegradation of lindane under axenic and anoxic conditions via dehydrochlorination, dehydrogenation, and hydroxylation reactions resulting in complete degradation or mineralization (Chen & Mulchandani, 1998; Cuozzo et al., 2017; Sharma et al., 2018; Upadhyay & Dutt, 2017; Wang et al., 2012; Zhang et al., 2014).

As most of the pesticides bear no similarity to the natural compound metabolized by the microbes thus are resistant to mineralization. An alternative strategy to combat the same is co-metabolism. It involves the integration of mineralization coupled with the addition of exogenous or iso-biomass compounds to degrade insecticides, fungicides, and herbicides from the soil environment (Li et al., 2005; Satish et al., 2017; Scott et al., 2008). The key factor in the microbial breakdown of pesticides is co-metabolism. It is used as a primary energy source and thus helps in the mineralization degradation of pesticides. Enzymes or cofactors used for the consumption of energy-yielding substrates help in the metabolism of pesticides. Co-metabolism is somehow a series' of synergistic biochemical reactions involved in a complete breakdown of pesticide. For example, the degradable pathway of β-CY by *Aspergillus niger* YAT. *Achromobacter, Arthrobacter, Azotobacter, Brevibacterium, Flavobacterium, Sphingomonas, Microbacterium, Trichoderma, Rhodococcus,* and *Aspergillus* are co-metabolism displaying strains that have shown their ability to degrade pollutants without using them as an energy source (Arora, 2018; Arora et al., 2020; Dangi et al., 2019). Another example of co-metabolic conversion is that of imidacloprid, an insecticide. Growing on two substrate types as an energy source: carbohydrates and organic acids. *P. indica* CGMCC 6648 bacterium is potentially active in degrading imidacloprid via hydroxylation and yields two metabolites namely olefin and 5-hydroxy IMI (Birolli et al., 2016; Bhatt et al., 2021).

Another mechanism in bioremediation is contaminant demobilization. Microorganisms have the ability to demobilize toxins in a variety of methods, including sorption (accumulation of organic pollutants by microbial biomass) followed by precipitation (conversion of toxic components to reduced or oxidized forms) and polymerization (fusion of organic pollutants with each other or with naturally occurring chemicals in the environment).

Three steps contribute to the entire degrading mechanism. The first involves the dynamic equilibrium process known as target adsorption, which is executed on the cell membrane's surface of a potent organism. Secondly, the target is channelized to the cell through the cell membrane's surface, and the rate and effectiveness of penetration were dependent on the molecular makeup of the targeted stereoisomer. Thirdly, xenobiotic targets quickly stimulate the enzymatic reaction at the membrane interface. Specific enzymes that hydrolyze P-C, P-O, P-F, and P-S bonds accelerate each step of the degradation process. Since all biochemical transformations rely on the presence of enzymes like hydrolases, peroxidases, oxygenases, and others, enzymatic activities play a significant role (Bhatt et al., 2021; Chen & Mulchandani, 1998; Chen et al., 2019; Singh, 2008; Sharma et al., 2018; Upadhyay & Dutt, 2017).

The biochemical reactions occurring in enzymatic degradation pathways operating during the bioremediation comprise oxidation (aliphatic hydroxylation, aromatic hydroxylation, N-hydroxylation, N-P-S-oxidation, oxidative deamination, oxidative dealkylation, oxidative dehalogenation, and epoxidation), reduction of nitro and quinone, and reductive dehalogenation, and esters hydrolysis such as thiocarbamate and thiophosphate), decarboxylation, dehydrogenation, dehalogenation, condensation, synthesis reactions (Chen et al., 2015; Dangi et al., 2019; Deng., 2015).

Several studies have reported organophosphate (OP) degrading enzymes, such as carboxylesterases and phosphodiesterase for biodegradation of OP compounds. The main class of enzymes frequently used is the hydrolases. These enzymes catalyze the hydrolysis of numerous key biochemical groups like esters, peptide linkages, carbon halide bonds, tri-esters, urea, and other similar structures. They make excellent candidates for bioremediation techniques as they are functional even without redox cofactors.

One of the most well-known multicopper enzymes is the laccase (benzenediol oxygen oxidoreductase, EC 1.10.3.2). The same has been isolated in bacteria, fungi, plants, and insects. Their potential to oxidize a wide range of aromatic and nonaromatic chemicals, which are used as hydrogen donors, is a result of the fact that they are primarily extracellular and nonspecific. Due to their biodegradability, affordability, and environmental friendliness, laccases have fascinated significant interest in the bioremediation of agricultural soil. Many pesticides' intricate structures can be successfully broken down by these enzymes. Only a few microorganisms have been found to have laccase enzymatic activity. *P. desmolyticum NCIM 2112, P. fluorescens GB1, P. putida GB1,*

P. maltophila, *P. syringae*, and *P. fluorescens* GB1 all included some of them. These bacterial laccases may execute the action in critical situations such as high salt concentrations and alkaline pH values. By breaking down organic macromolecules into smaller, nontoxic components, bioremediation prevents secondary pollution (Mishra et al., 2021; Zhang et al., 2014).

The ability of microbes to break down pesticide residues is subjected to a variety of intrinsic and external environmental conditions. The former involves the chemistry of the pesticide and that of microbes. In addition to this, microbial species, metabolic activity, and adaptability of the microbial strain directly regulate the pesticide degradation and transformation. Research has revealed that microbe species or strains of the same species behave differently to a specified organic pollutant and that these microorganisms have strongly adapted themselves accordingly (Mishra et al., 2021; Naik & Duraphe, 2012).

The rate and effectiveness of pesticide microbial detoxification were influenced by the pesticides properties, like their molecular mass, spatial configuration, quantity and type of substituent, substituted attributes, and position. Bioremediation is further related to the bioavailability of the pesticide. Generally, the lower molecular weight pesticides were more biodegradable than the polymer complex (Mihai et al., 2021; Niti et al., 2013).

Additionally, retention phenomena are essential to how pesticides behave in soils, especially in terms of their persistence. They mostly correspond to adsorption, the transfer of a component from the liquid or gaseous phase of the soil to the solid phase, and desorption. Depending on a variety of factors, adsorption is a behavior that is somewhat reversible. A significant aspect affecting the retention of pesticides in the solid soil phase is the amount of organic matter in the soil (Arora et al., 2020; Birolli et al., 2016).

Retention and deterioration are intertwined concepts. Retention lessens a compound's accessibility to soil microbes, which limits its degradability. It is possible to determine the material mobility thanks to this coupling retention/degradation. There are two mechanisms involved in the production of nonextractable pesticide residues. The first involves the creation of covalent connections between molecules and organic material. The second involves the sequestration of molecules or the breakdown of products. Environmental parameters that affect the remediation process carried out by bacteria include temperature, pH, salinity, humidity, available nutrients, substrate concentration needed, availability of surfactants, carbon dioxide, and aerobic or anaerobic conditions, soil structure, and redox potential (Madhavi & Mohini, 2012).

pH: It influences the adsorption/desorption of pesticides on soil matrix and hence bioavailability and biodegradation. The breakdown of a xenobiotic compound depends on the enzymes secreted by the microorganisms. These enzymes are pH dependent and work optimally between the pH range varying from 6.5 to 8.5 (Raffa, 2021). Microbial metabolic activity is largely affected by even the slightest changes in pH. The measurement of pH in the soil is an indication of the potential for microbial growth

(Asir & Enim, 2013). Singh et al. (2006) and Hong et al. (2007) also reported that the biodegradation of OP pesticides in soil was slower in lower pH soils in comparison with neutral and alkaline soils. Optimum soil pH is an essential criterion to regulate microbial biomass and enzyme activity and boost hydrocarbon biodegradation (Imron et al., 2020). Adsorption of weakly acidic pesticides is increased by lower soil pH.

Moisture content: Soil moisture acts as the transport medium for soil nutrients and the removal of bacterial metabolic waste products in the soil particles (Jacoby, 2017). Excess soil moisture content has adverse effects on biodegradation agents. Microbial growth requires an optimum presence of water in the environmental matrix. For optimum growth and proliferation, microorganisms require 12%–25% of moisture (Adams et al., 2015) for optimum growth and proliferation. Some workers also quantified the effect of soil moisture was noticed on the aerobic microbial mineralization of selected pesticides (isoproturon, benzoin-ethyl, and glyphosphate) in different soils. They found a linear correlation ($p < 0.0001$) between increasing soil moisture (within a soil water potential range of -20 and -0.015 MPa) and increased relative pesticide mineralization. The hydrocarbon contaminants in the soil decrease water holding capacity and porosity of the soil resulting in decreased microbial activity. Therefore providing proper moisture (water availability) in hydrocarbon-contaminated sites can improve microbial activities for hydrocarbon degradation (Huang et al.,2019; Wu et al., 2017).

Temperature: Among the physical factors temperature highly influences microbial physiological properties. Temperature ranging between 15°C and 40°C is considered favorable for the degradation of pesticides by isolated pesticide-degrading bacteria. It has a strong influence on microbial metabolism and consequently microbial activity in the environment. A linear relationship between the biodegradation rate with increasing temperature has been found and slows with decreasing temperature. The transport channels within the microbial cells are closed or may even freeze the entire cytoplasm at the sub-zero temperature of water thus, rendering most oleophilic microbes metabolically inactive (Sharma et al., 2022). The optimum temperature enhances the rate of petroleum hydrocarbon-degrading bacteria in deep and surface waters of the Northern Gulf of Mexico as reported by Liu et al. (2017).

Redox potential and concentration of oxygen: Redox reactions produce energy that is required for cell maintenance, biological functions, and reproduction of microorganisms. For example, bacteria metabolize hydrocarbon pollutants by catalyzing the transfer of electrons from electron donors to electron acceptors in aerobic or anaerobic bioremediation. This degradation of hydrocarbon-derived pollutants depends on redox potential (Truskewycz et al., 2019). The presence of dominant electron acceptors in hydrocarbon-contaminated sites is an important factor that affects the rate of biodegradation. The presence of electron acceptors such as nitrate, manganese oxides, iron oxides, and sulfate in the soil largely affects redox potential. Biological degradation is carried out in aerobic and anaerobic conditions because oxygen is a gaseous requirement for most living organisms. For example, the degradation of HCH in soil takes place under an anoxic environment (Kumar, 2018; Wang, 2019).

Nutrients: Microorganisms are dependent on nutrient accessibility for their metabolism and growth (Kuppusamy et al., 2016). The optimum level of organic matter in the soil is a rich source of nutrients that enhances the properties of soil a favors the growth and activities of soil microbes. Macronutrients—N, P, and K are added in the contaminated soil and are helpful in the distribution and degradation of hydrocarbon pollutants (Hamzah et al., 2017).

Salinity: Excess level of soil salinity decreases the rate and extent of hydrocarbon biodegradation. At 0.1–2M NaCl, significant hydrocarbon degradation was detected but it was found maximum at 0.4 M. Natural and anthropogenic activities cause the deposition and accumulation of salt which increases the stress (halophile or halotolerant) in the environment. The increased salt concentration causes high osmotic pressure which declines various microbial metabolic activities eventually hindering the transportation of essential ions (Abdulrasheed & Zakaria, 2020) resulting in a lack of oxygen for microbes (Imron et al., 2020) thus indirectly affecting pesticide remediation. Soil factors like the amount and type of clay, cation exchange capacity, and metal oxides, total iron content exchangeable calcium, and sodium percentage play an important role in microbial bioremediation. Amondham et al. (2006) investigated paraquat adsorption and degradation in the tropical soils of Thailand was studied by Amondham et al. (2006). It was found that the paraquat adsorption rate was higher in clay soil and sandy soil as compared to the sandy loam soil with KF values of 787 and 18, respectively (Nollet & Rathore, 2016).

10.4 Bioremediation processes

Microorganisms like bacteria, fungi, and yeast are highly adaptable and can survive under the divergent environment and produce metabolites whereby pesticides are broken down into less toxic compounds (Thakur et al., 2019). The microbial bioremediation strategies can be either ex-situ or in-situ. In situ, bioremediation involves the transfer of pollutants from contaminated sites to another site of treatment and the latter involves the treatment of the contaminated material at the site itself.

10.4.1 In situ bioremediation

It involves the activity of microorganisms to detoxify the contaminants at their place. These techniques are more desirable due to less cost and disturbance because the natural conditions at contaminated sites provide all the optimum conditions (nutrients, electron acceptors, donors, etc.) for the microorganisms required for biodegradation. Sufficient amounts of oxygen and nutrients are made available by the circulation of aqueous solutions through contaminated soils. It stimulates naturally occurring bacteria to degrade organic contaminants (Barupal et al., 2019a,b). Some in situ bioremediation methods are:

10.4.1.1 Biostimulation

It is a remediation technique where rate-limiting nutrients like phosphorus, nitrogen, oxygen, and electron donors are added to polluted sites to stimulate the existing bacteria which has the capability to degrade the hazardous and toxic contaminants (Tyagi et al., 2010). It has been reported that the detoxification of the petroleum-contaminated site can be accelerated by the process of biostimulation (Adams et al., 2015; Tyagi et al., 2010). Various contaminants—sulphate and polyester polyurethanes other than hydrocarbons can be remediated biologically through the process of biostimulation (Miao et al., 2021).

10.4.1.1.1 Bioaugmentation

It is the addition of pollutant-degrading microorganisms (natural/genetically engineered/exotic) to increase their biodegradative capacity (Mrozik & Piotrowska-Seget, 2010). Populations of naturally occurring microorganisms are collected from the remediation site, separately cultured, genetically modified, and returned to the site (Shukla et al., 2010). This process detoxifies soil and water contaminated with chlorinated ethenes, such as tetrachloroethylene and trichloroethylene. This process also detoxifies halogenated compounds such as pesticides and fungicides. Xu and other co-workers (2015) reported the biodegradation efficiency of *Burkholderia pickettii* against quinoline. The indigenous organisms of the contaminated site were incapable of degrading quinoline but these organisms in cooperation with the newly introduced quinoline degrader *Burkholderia pickettii* degraded quinoline (concentration of 1 mg/g of soil), within 6 and 8 hours.

10.4.1.2 Bioattenuation

It reduces the contamination in soil and groundwater via biological processes (aerobic and anaerobic biodegradation), physical phenomena (advection, dilution, diffusion, volatilization, sorption/desorption, and dispersion) and chemical reactions ion exchange, complexation, abiotic transformation by naturally occurring organisms. For example, phenoxy herbicides in groundwater are detoxified via bioattenuation (Janniche et al., 2010).

10.4.1.3 Bioeventing

Bioventing uses low airflow rates to provide sufficient oxygen to sustain microbial activity. Oxygen is supplied through direct air injection into residual contamination thus supplying the necessary amount of oxygen for biodegradation. In this way, it minimizes the release of volatile contaminants into the atmosphere. Bioventing is a prominent approach for the treatment of hydrocarbon-contaminated soil. It is used for the detoxification of those hydrocarbons which are deep under the surface soil (Harekrushna & Kumar, 2011). For example, effective bioremediation of petroleum-contaminated soil using bioventing is on record (Agarry & Latinwo, 2015).

10.4.1.4 Biosparging

Similar to bioventing, the air is injected into the soil subsurface to stimulate microbial activities to decontaminate the pollutants from polluted sites. However, unlike bioventing, the air is injected into the saturated zone, which causes the upward movement of volatile organic compounds to the unsaturated zone promoting biological degradation by naturally occurring microorganisms (Maitra, 2018). Mixing in the saturated zone is increased so that more contact is made between soil and groundwater (Harekrushna & Kumar, 2011). For instance, more than 70% of BTEX (benzene, toluene, ethylbenzene, and xylene) is decreased in BTEX-contaminated groundwater by biosparging (Kao et al., 2008).

10.4.2 Ex-situ remediation

This technique involves the transfer of pollutants and subsequently transporting them to another site for treatment (Philp & Atlas, 2015).

10.4.2.1 Landfarming

Also referred to as land treatment or land application (Tyagi & Kumar, 2021) it employs passive aeration by tilling or ploughing the contaminated soil in order to reduce contaminant levels. In this, the contaminated soil is spread over a prepared bed and periodically tilled until pollutants are degraded (Vidali, 2001). Here the soil is decontaminated with polycyclic aromatic hydrocarbons (PAHs) (Lukic, 2017).

10.4.2.2 Composting

It is a biological process where the diverse microbial communities present in composting materials are responsible for either complete mineralization/metabolism, co-metabolism, or nonspecific extracellular oxidation for the biodegradation of recalcitrant organic contaminants (Lin et al., 2021). This technology is used to treat recalcitrant organic contaminants, including polycyclic aromatic hydrocarbons (PAHs), total petroleum hydrocarbons (TPHs), diesel, phthalate-based plasticizers, organochlorine pesticides, polychlorinated dibenzo-p-dioxins and furans (PCDD/Fs), and polychlorinated biphenyls (PCBs). Bioremediation via composting assist in the degradation of recalcitrant organic contaminants. High temperatures during the composting (thermophilic phase) also increase the rate of degradation by making the contaminants less viscous and more bioavailable (Lin et al., 2021).

10.4.3 Biopiles

Biopiles (also called biocells, bioheaps, biomounds, and compost piles) are a way of excavating soil contaminated with aerobically remediable hydrocarbons. It reduces concentrations of petroleum pollutants in excavated soils during the time of biodegradation (Lin et al., 2021). It is much more efficient compared to composting and landfarming in terms of mass transfer of air, nutrients, and water, thus providing a better strategy for the removal of pollutants (Kuppusamy et al., 2016).

10.4.4 Bioreactors

Slurry bioreactors are the most important ex-situ technique. The soils and sediments polluted by recalcitrant pollutants are treated in slurry bioreactors for bioremediation under controlled environmental conditions (Lin et al., 2021). Here the contaminated soils are combined with water or wastewater depending on the concentration of pollutants and other additives and mixed in the bioreactor-engineered system in the form of closed vats, tanks, or columns) in order to maintain contact between microorganisms and pollutants present in the soil under optimum conditions for biodegradation (Bhardwaj & Kapley, 2015).

10.5 Recent advances and biotechnological implications

Microbial populations are highly adaptable and exhibit tremendous potential for the remediation of any contaminated environment (Bastida et al., 2016; Bhatt & Barh, 2018; Chen et al., 2015). However, effective implementation of microbial remediation strategies with advanced technical approaches should be adopted to provide an in-depth understanding of the dynamics aspects of microbial activity and survival under a stressed environment (Lima-Morales et al., 2016; Mao et al., 2019). Molecular technologies in the fields of molecular, biotechnology, bioinformatics, and system biology have provided new insights into microbial metabolic and functions that improve the bioremediation of pesticides (Marco & Abram, 2019). Advanced molecular techniques like metagenomic, transcriptomic, and proteomic unravel the mechanism of microbial degradation and the relevant information on genes and protein expression levels in microbial communities growing under contaminated environments (Marco & Abram, 2019). The genetically engineered microorganism thus developed can overcome the limitations of conventional bio-remediation techniques because they are way more efficient, quicker, and accurate (Mishra et al., 2020).

Recombinant DNA technology is employed to produce genetically engineered microorganisms (GEMs) which give better bioremediation efficiency (Table 10.4) (Abateneh, 2017). Genetically engineered microorganisms (GEMs) are considered as "designer biocatalysts," where desirable biodegradation pathways or enzymes from different

Table 10.4 Examples of recombinant microbial strains for detoxification of pesticides.

S. no.	Transgenic strain	Gene source	Pesticides	References
1.	*Fusarium solani*	Fusarium solani	DDT	Miguel-Ferrer & Romero-Arenas (2021)
2.	*Escherchia coli*	*Sphingomonas paucimobilis*	Lindane	Chaurasia et al. (2013)
3.	*Escherchia coli*	Moraxella sp.	Coumaphos	Kim et al. (2014)
4.	*Escherchia coli*	*Pseudomonas diminuata*	Methyl parathion	Zhao et al. (2021)

organisms are brought together in a single host. The genes are located on the extrachromosomal part of the plasmid for the detoxification of contaminated soil. The plasmids were isolated and inserted into a single strain of *Pseudomonas*. This strain enhanced the rate of degradation by 10–100 times faster than the wild type (Banu & Prasad, 2017). Another recombinant strain—*Sphingomonas* sp. CDS-1 was developed by Ali and coworkers (2021) for the degradation of OP and carbamate. Using the shotgun method, the methyl parathion hydrolase encoding gene (mpd) was cloned with the cognate regulator of aethyl parathion (MP)—degrading strain *Pseudomonas putida* DLL-1 using the shotgun method. A genetically engineered bacterium, *Pseudomonas* in a laboratory-scale sewage plant fed with a mixture of chlorophenols and 4-methyl phenol showed an increased rate of degradation of phenolic compounds (Aken et al., 2010). Genetically modified strain of a bacterium namely *P. fluorescens* strain HK44 was used to conduct field tests on bioremediation. This strain was designed to sense an environmental contaminant through bioluminescence signaling. This type of bioremediation can be used for in situ monitoring of the bioremediation process. Using recombinant DNA technology the genes encoding *s*-triazine hydrolase for dechlorination and deamination of de-ethylatrazine and de-isopropylatrazine were transferred from *R. corallinus* to *Rhodococcus* TE1 (Malik et al., 2022). Recombinant *Rhodococcus* completely degrades atrazine to cyanuric acid. Trichloroethylene (TCE) was efficiently degraded by recombinant *Escherichia coli* by harboring and expressing the phenol hydroxylase gene from *Pseudomonas putida* BH (Kumar et al., 2020). *Sphingobium chlorophenolicum* ATCC 39723 by genomic shuffling increases the degradation of the pesticide up to tenfold in comparison to the wild-type strain (Kumar et al., 2013).

These genetically modified microbes help in the speedy recovery of waste-polluted sites and enhance the rate of substrate degradation by displaying a high catalytic or utilization capacity with a small amount of cell mass by decontamination or neutralizing any harmful substances including pesticides (Abatenh, 2017).

Molecular probes have been developed to identify, enumerate and isolate microorganisms with degradative potential is an important tool to degrade xenobiotics (Fagervold et al., 2021). DNA probes for PCB-degrading genotypes have been constructed from

PCB-degrading bacteria (Dua et al., 2002). These have been employed for the detection of specific PCB-degrading bacteria, abundance of PCB-degrading genotypes and genotypic diversity among PCB-degrading bacteria in toxic, chemical-polluted environments (Walia et al., 1990). Bacterial genes are cloned in the cosmid vector pCP13 which detoxifies chlorinated biphenyls (Treviño-Quintanilla & Galán-Wong, 2002). These catabolic genes (cbpABCD) are incorporated into the genotypes of microbial communities to construct gene probes for the detection of PCB degradation genotypes (Focht, 2021).

Naturally occurring enzymes are not sufficient to work on an industrial scale. With the increasing demand for enzymes DNA shuffling has led to the production of novel enzyme variants to efficiently biodegrade pollutants (Zhang et al., 2014). DNA shuffling, also known as molecular breeding, is an in vitro technique that enables the generation of large libraries of novel genes from which to encode proteins with improved or unique functionality.

10.6 Conclusion

Numerous benefits come with microbe-mediated bioremediation of pesticides, including minimal input costs, lengthy elimination times, straightforward equipment and space requirements, and cost-effective environmental protection. It can play a bigger part in reducing the potential damage that pesticides bring to the natural ecosystem while maintaining agricultural productivity and output. Many research projects have been carried out globally in this context. Nevertheless, the harm pesticides do to the environment and human health continues to be a major concern. This may be because diverse microorganisms that can be employed to detoxify a wide range of pesticides have not yet been identified. Additionally, genetically altered microorganisms can provide fresh promise in this area. There may also be a large number of unidentified bacteria that have different pesticide-detoxifying capacities. Therefore, it is critical to do additional research in order to find new microbes with these qualities in various global places.

References

Abatenh, E., Gizaw, B., Tsegaye, Z., & Wassie, M. (2017). The role of microorganisms in bioremediation: A review. *Open Journal of Environmental Biology, 2*(1), 038–046.

Abdul Salam, J., Lakshmi, V., Das, D., & Das, N. (2013). Biodegradation of lindane using a novel yeast strain, Rhodotorula sp. VITJzN03 isolated from agricultural soil. *World Journal of Microbiology and Biotechnology, 29*(3), 475–487.

Abdulrasheed, M., Zakaria, N. N., Roslee, A. F. A., Shukor, M. Y., Zulkharnain, A., Napis, S., ..., & Ahmad, S. A. (2020). Biodegradation of diesel oil by cold-adapted bacterial strains of *Arthrobacter* spp. from Antarctica. *Antarctic Science, 32*(5), 341–353.

Abraham, J., & Silambarasan, S. (2016). Biodegradation of chlorpyrifos and its hydrolysis product 3, 5, 6-trichloro-2-pyridinol using a novel bacterium *Ochrobactrum* sp. JAS2: A proposal of its metabolic pathway. *Pesticide Biochemistry and Physiology, 126*, 13–21.

Adams, G. O., Fufeyin, P. T., Okoro, S. E., & Ehinomen, I. (2015). Bioremediation, biostimulation and bioaugmention: A review. *International Journal of Environmental Bioremediation & Biodegradation, 3*(1), 28–39.

Agarry, S., & Latinwo, G. K. (2015). Biodegradation of diesel oil in soil and its enhancement by application of bioventing and amendment with brewery waste effluents as biostimulation-bioaugmentation agents. *Journal of Ecological Engineering, 16*(2), 82–91.

Aken, B. V., Correa, P. A., & Schnoor, J. L. (2010). Phytoremediation of polychlorinated biphenyls: New trends and promises. *Environmental Science & Technology, 44*(8), 2767–2776.

Alexandrino, D. A., Mucha, A. P., Almeida, C. M. R., & Carvalho, M. F. (2020). Microbial degradation of two highly persistent fluorinated fungicides-epoxiconazole and fludioxonil. *Journal of Hazardous Materials, 394*, 122545.

Ali, S. A., Mittal, D., & Kaur, G. (2021). In-situ monitoring of xenobiotics using genetically engineered whole-cell-based microbial biosensors: Recent advances and outlook. *World Journal of Microbiology and Biotechnology, 37*(5), 1–24.

Amondham, W., Parkpian, P., Polprasert, C., Delaune, R. D., & Jugsujinda, A. (2006). Paraquat adsorption, degradation, and remobilization in tropical soils of Thailand. *Journal of Environmental Science and Health Part B, 41*(5), 485–507.

Aregbesola, O. A., Mokoena, M. P., & Olaniran, A. O. (2020). Biotransformation of pentachlorophenol by an indigenous Bacillus cereus AOA-CPS1 isolated from wastewater effluent in Durban, South Africa. *Biodegradation, 31*(4), 369–383.

Arora, P. K. (2020). Bacilli-mediated degradation of xenobiotic compounds and heavy metals. *Frontiers in Bioengineering and Biotechnology, 8*, 570307.

Arora, P. K., Srivastava, A., Garg, S. K., & Singh, V. P. (2018). Recent advances in degradation of chloronitrophenols. *Bioresource Technology, 250*, 902–909.

Asira, E. E. (2013). Factors that determine bioremediation of organic compounds in the soil. *Academic Journal of Interdisciplinary Studies, 2*(13) 125-125.

Banu, H., & Prasad, K. P. (2017). Role of plasmids in microbiology. *Journal of Aquaculture Research & Development, 8*(466), 2.

Barupal, T., Chittora, D., & Meena, M. (2019a). Solid waste: Characterization, assessment, monitoring, and remediation. In R. P. Singh, V. Prasad, & B. Vaish (Eds.), *Advances in waste-to-energy technologies* (pp. 1–19). Taylor & Francis and CRC Press. https://doi.org/10.1201/9780429423376-1.

Barupal, T., Meena, M., & Sharma, K. (2019b). Inhibitory effects of leaf extract of *Lawsonia inermis* on *Curvularia lunata* and characterization of novel inhibitory compounds by GC–MS analysis. *Biotechnology Reports, 23*, e00335. https://doi.org/10.1016/j.btre.2019.e00335.

Bastida, F., Torres, I. F., Moreno, J. L., Baldrian, P., Ondoño, S., Ruiz-Navarro, A., …, & Jehmlich, N. (2016). The active microbial diversity drives ecosystem multifunctionality and is physiologically related to carbon availability in Mediterranean semi-arid soils. *Molecular Ecology, 25*(18), 4660–4673.

Bers, K., Batisson, I., Proost, P., Wattiez, R., De Mot, R., & Springael, D. (2013). HylA, an alternative hydrolase for initiation of catabolism of the *phenylurea herbicide* linuron in *Variovorax* sp. strains. *Applied and Environmental Microbiology, 79*(17), 5258–5263.

Bharagava, R. N., Chowdhary, P., & Saxena, G. (2017). Bioremediation: An eco-sustainable green technology: Its applications and limitations. In *Environmental pollutants and their bioremediation approaches* (pp. 1–22). CRC Press.

Bhardwaj, P., Sharma, A., Sagarkar, S., & Kapley, A. (2015). Mapping atrazine and phenol degradation genes in *Pseudomonas* sp. EGD-AKN5. *Biochemical Engineering Journal, 102*, 125–134.

Bhat, M. P., Nayaka, S., & Kumar, R. S. (2022). A swamp forest *Streptomyces* sp. strain KF15 with broad spectrum antifungal activity against chilli pathogens exhibits anticancer activity on HeLa cells. *Archives of Microbiology, 204*(9), 1–17.

Bhatt, P., & Barh, A. (2018). Bioinformatic tools to study the soil microorganisms: An in silico approach for sustainable agriculture. In *silico approach for sustainable agriculture* (pp. 169–182). Springer.

Bhatt, P., Bhatt, K., Sharma, A., Zhang, W., Mishra, S., & Chen, S. (2021). Biotechnological basis of microbial consortia for the removal of pesticides from the environment. *Critical Reviews in Biotechnology, 41*(3), 317–338.

Bhatt, P., Huang, Y., Zhan, H., & Chen, S. (2019). Insight into microbial applications for the biodegradation of pyrethroid insecticides. *Frontiers in Microbiology, 10*, 1778.

Birolli, W. G., Alvarenga, N., Seleghim, M. H., & Porto, A. L. (2016). Biodegradation of the pyrethroid pesticide esfenvalerate by marine-derived fungi. *Marine Biotechnology, 18*(4), 511–520.

Birolli, W. G., da Silva, B. F., & Rodrigues-Filho, E. (2020). Biodegradation of the fungicide *Pyraclostrobin* by bacteria from orange cultivation plots. *Science of the Total Environment, 746*, 140968.

Birolli, W. G., Vacondio, B., Alvarenga, N., Seleghim, M. H., & Porto, A. L. (2018). Enantioselective biodegradation of the pyrethroid (±)-lambda-cyhalothrin by marine-derived fungi. *Chemosphere, 197*, 651–660.

Bokade, P., Gaur, V. K., Tripathi, V., Bobate, S., Manickam, N., & Bajaj, A. (2022). Bacterial remediation of pesticide polluted soils: Exploring the feasibility of site restoration. *Journal of Hazardous Materials, 441*, 129906.

Bolan, N. S., Yan, Y., Lia, Q., & Kirkham, M. B. (2021). Compost-assisted bioremediation of polycyclic aromatic hydrocarbons. In *Bioremediation science from theory to practice* (pp. 202–211). CRC Press.

Bose, S., Kumar, P. S., & Vo, D. V. N. (2021). A review on the microbial degradation of chlorpyrifos and its metabolite TCP. *Chemosphere, 283*, 131447.

Chaudhary, D. K. (2016). Bioremediation: An eco-friendly approach for polluted agricultural soil. *Emergent Life Sciences Research, 2*(1), 73–75.

Chaurasia, A. K., Adhya, T. K., & Apte, S. K. (2013). Engineering bacteria for bioremediation of persistent organochlorine pesticide lindane (γ-hexachlorocyclohexane). *Bioresource Technology, 149*, 439–445.

Chen, T, Dai, YJ, Ding, JF, Yuan, S, & Ni, JP (2008). N-demethylation of neonicotinoid insecticide acetamiprid by bacterium *Stenotrophomonas maltophilia* CGMCC 1.1788. *Biodegradation, 19*(5), 651–658.

Chen, M., Xu, P., Zeng, G., Yang, C., Huang, D., & Zhang, J. (2015). Bioremediation of soils contaminated with polycyclic aromatic hydrocarbons, petroleum, pesticides, chlorophenols and heavy metals by composting: Applications, microbes and future research needs. *Biotechnology Advances, 33*(6), 745–755.

Chen, S., & Zhan, H. (2019). Biodegradation of synthetic pyrethroid insecticides. In *Microbial metabolism of xenobiotic compounds* (pp. 229–244). Springer.

Chen, W., & Mulchandani, A. (1998). The use of live biocatalysts for pesticide detoxification. *Trends in Biotechnology, 16*(2), 71–76.

Chen, Y. P., Hsiao, P. J., Hong, W. S., Dai, T. Y., & Chen, M. J. (2012). *Lactobacillus kefiranofaciens* M1 isolated from milk kefir grains ameliorates experimental colitis in vitro and in vivo. *Journal of Dairy Science, 95*(1), 63–74.

Cuozzo, S. A., Sineli, P. E., Davila Costa, J., & Tortella, G. (2018). *Streptomyces* sp. is a powerful biotechnological tool for the biodegradation of HCH isomers: Biochemical and molecular basis. *Critical Reviews in Biotechnology, 38*(5), 719–728.

Dangi, A. K., Sharma, B., Hill, R. T., & Shukla, P. (2019). Bioremediation through microbes: Systems biology and metabolic engineering approach. *Critical Reviews in Biotechnology, 39*(1), 79–98.

Delegan, Y. A., Valentovich, L. N., Shafieva, S. M., Ganbarov, K. G., Filonov, A. E., & Vainstein, M. B. (2019). Characterization and genomic analysis of highly efficient thermotolerant oil-degrading bacterium *Gordonia* sp. 1D. *Folia Microbiologica, 64*(1), 41–48.

Deng, W., Lin, D., Yao, K., Yuan, H., Wang, Z., Li, J., & Liu, S. (2015). Characterization of a novel β-cypermethrin-degrading *Aspergillus niger* YAT strain and the biochemical degradation pathway of β-cypermethrin. *Applied Microbiology and Biotechnology, 99*(19), 8187–8198.

Deshmukh, R., Khardenavis, A. A., & Purohit, H. J. (2016). Diverse metabolic capacities of fungi for bioremediation. *Indian Journal of Microbiology, 56*(3), 247–264.

Doolotkeldieva, T., Konurbaeva, M., & Bobusheva, S. (2018). Microbial communities in pesticide-contaminated soils in Kyrgyzstan and bioremediation possibilities. *Environmental Science and Pollution Research, 25*(32), 31848–31862.

Dua, M., Singh, A., Sethunathan, N., & Johri, A. (2002). Biotechnology and bioremediation: Successes and limitations. *Applied Microbiology and Biotechnology, 59*(2), 143–152.

Duc, H. D., Hung, N. V., & Oanh, N. T. (2021). anaerobic Degradation of endosulfans by a mixed culture of *Pseudomonas* sp. and *Staphylococcus* sp. *Applied Biochemistry and Microbiology, 57*(3), 327–334.

Eevers, N., White, J. C., Vangronsveld, J., & Weyens, N. (2017). Bio-and phytoremediation of pesticide-contaminated environments: A review. *Advances in Botanical Research, 83*, 277–318.

Fagervold, S. K., Rohée, C., Rodrigues, A. M. S., Stien, D., & Lebaron, P. (2021). Efficient degradation of the organic UV filter benzophenone-3 by *Sphingomonas wittichii* strain BP14P isolated from WWTP sludge. *Science of the Total Environment, 758*, 143674.

Focht, D. D. (2021). Microbial degradation of chlorinated biphenyls. In *Soil biochemistry* (pp. 341–407). CRC Press.

Gangola, S., Bhatt, P., Chaudhary, P., Khati, P., Kumar, N., & Sharma, A. (2018). Bioremediation of industrial waste using microbial metabolic diversity. In *Microbial biotechnology in environmental monitoring and cleanup* (pp. 1–27). IGI Global.

Getenga, Z., Dörfler, U., Iwobi, A., Schmid, M., & Schroll, R. (2009). Atrazine and terbuthylazine mineralization by an *Arthrobacter* sp. isolated from a sugarcane-cultivated soil in Kenya. *Chemosphere, 77*(4), 534–539.

Ghosh, S. K., & Pal, S. (2016). Entomopathogenic potential of Trichoderma longibrachiatum and its comparative evaluation with malathion against the insect pest Leucinodes orbonalis. *Environmental Monitoring and Assessment, 188*(1), 1–7.

Gilani, R. A., Rafique, M., Rehman, A., Munis, M. F. H., Rehman, S. U., & Chaudhary, H. J. (2016). Biodegradation of chlorpyrifos by bacterial genus *Pseudomonas*. *Journal of Basic Microbiology, 56*(2), 105–119.

Gilden, R. C., Huffling, K., & Sattler, B. (2010). Pesticides and health risks. *Journal of Obstetric, Gynecologic, and Neonatal Nursing, 39*, 103–110. https://doi.org/10.1111/j.1552-6909.2009.01092.x.

Grant, R. J., & Betts, W. B. (2003). Biodegradation of the synthetic pyrethroid cypermethrin in used sheep dip. *Letters in Applied Microbiology, 36*(3), 173–176.

Hamzah, A., Hapsari, R. I., & Priyadarshini, R. (2017). The influence of rice husk and tobacco waste biochars on soil quality. *Journal of Degraded and Mining Lands Management, 5*(1), 1001.

Harekrushna, S., & Kumar, D. C. (2012). A review on: Bioremediation. *International Journal of Research in Chemistry and Environment, 2*(1), 13–21.

Hong, Q., Zhang, Z., Hong, Y., & Li, S. (2007). A microcosm study on bioremediation of fenitrothion-contaminated soil using *Burkholderia* sp. FDS-1. *International Biodeterioration & Biodegradation, 59*(1), 55–61.

Huang, Y., Pan, H., Wang, Q., Ge, Y., Liu, W., & Christie, P. (2019). Enrichment of the soil microbial community in the bioremediation of a petroleum-contaminated soil amended with rice straw or sawdust. *Chemosphere, 224*, 265–271.

Hussain, S., Siddique, T., Saleem, M., Arshad, M., & Khalid, A. (2009). Impact of pesticides on soil microbial diversity, enzymes, and biochemical reactions. *Advances in Agronomy, 102*, 159–200.

Imron, M. F., Kurniawan, S. B., Ismail, N. I., & Abdullah, S. R. S. (2020). Future challenges in diesel biodegradation by bacteria isolates: A review. *Journal of Cleaner Production, 251*, 119716.

Iqbal, A. (2019). Bacterial mediated remediation: A strategy to combat pesticide residues in agricultural soil. In *Soil microenvironment for bioremediation and polymer production* (pp. 35–53). https://doi.10.1002/9781119592129.ch3.

Iram, D., Riaz, R., & Iqbal, R. K. (2019). Usage of potential micro-organisms for degradation of plastics. *Open Journal of Environmental Biology, 4*(1), 007–015.

Jacoby, R., Peukert, M., Succurro, A., Koprivova, A., & Kopriva, S. (2017). The role of soil microorganisms in plant mineral nutrition: Current knowledge and future directions. *Frontiers in Plant Science, 8*, 1617.

Jaiswal, D. K., Verma, J. P., & Yadav, J. (2017). Microbe induced degradation of pesticides in agricultural soils. In *Microbe-induced degradation of pesticides* (pp. 167–189). Springer.

Jaiswal, S., & Shukla, P. (2020). Alternative strategies for microbial remediation of pollutants via synthetic biology. *Frontiers in Microbiology, 11*, 808.

Janniche, G. S., Lindberg, E., Mouvet, C., & Albrechtsen, H. J. (2010). Mineralization of isoproturon, mecoprop and acetochlor in a deep unsaturated limestone and sandy aquifer. *Chemosphere, 81*(7), 823–831.

Javaid, M. K., Ashiq, M., & Tahir, M. (2016). Potential of biological agents in decontamination of agricultural soil. *Scientifica (Cairo), 2016*, 1598325. https://doi.org/10.1155/2016/1598325.

Kafilzadeh, F., Ebrahimnezhad, M., & Tahery, Y. (2015). Isolation and identification of endosulfan-degrading bacteria and evaluation of their bioremediation in Kor River, Iran. *Osong Public Health and Research Perspectives, 6*(1), 39–46.

Kamal, H. A., & Khater, E. I. (2010). The biological effects of the insect growth regulators; pyriproxyfen and diflubenzuron on the mosquito *Aedes aegypti*. *Journal of the Egyptian Society of Parasitology, 40*(3), 565–574.

Kao, C. M., Chen, C. Y., Chen, S. C., Chien, H. Y., & Chen, Y. L. (2008). Application of in situ biosparging to remediate a petroleum-hydrocarbon spill site: Field and microbial evaluation. *Chemosphere, 70*(8), 1492–1499.

Khajezadeh, M., Abbaszadeh-Goudarzi, K., Pourghadamyari, H., & Kafilzadeh, F. (2020). A newly isolated *Streptomyces rimosus* strain capable of degrading deltamethrin as a pesticide in agricultural soil. *Journal of Basic Microbiology, 60*(5), 435–443.

Kharat, A. S., Parte, S. G., & More, N. S. (2018). Microbes ecofriendly tools for bioremediation of pesticide-contaminated environments. In *Recent advances in environmental management* (pp. 391–416). CRC Press.

Kim, C. S., Seo, J. H., Kang, D. G., & Cha, H. J. (2014). Engineered whole-cell biocatalyst-based detoxification and detection of neurotoxic organophosphate compounds. *Biotechnology Advances, 32*(3), 652–662.

Kour, D., Khan, S. S., Kour, H., Kaur, T., Devi, R., Rai, P. K., …, & Yadav, A. N. (2022). Microbe-mediated bioremediation: Current research and future challenges. *Journal of Applied Biology and Biotechnology, 10*(2), 6–24.

Kumar, A., Kumar, A., Singh, R., Singh, R., Pandey, S., Rai, A., …, & Rahul, B. (2020). Genetically engineered bacteria for the degradation of dye and other organic compounds. In *Abatement of environmental pollutants* (pp. 331–350). Elsevier.

Kumar, D. (2018). Biodegradation of γ-Hexachlorocyclohexane by *Burkholderia* sp. IPL04. *Biocatalysis and Agricultural Biotechnology, 16*, 331–339.

Kumar, M., Yadav, A. N., Saxena, R., Paul, D., & Tomar, R. S. (2021). Biodiversity of pesticides degrading microbial communities and their environmental impact. *Biocatalysis and Agricultural Biotechnology, 31*, 101883.

Kumar, S., Dagar, V. K., Khasa, Y. P., & Kuhad, R. C. (2013). Genetically modified microorganisms (GMOs) for bioremediation. In *Biotechnology for environmental management and resource recovery* (pp. 191–218). Springer.

Kumar, V., Shahi, S. K., & Singh, S. (2018). Bioremediation: An eco-sustainable approach for restoration of contaminated sites. In *Microbial bioprospecting for sustainable development* (pp. 115–136). Springer.

Kuppusamy, S., Palanisami, T., Megharaj, M., Venkateswarlu, K., & Naidu, R. (2016). In-situ remediation approaches for the management of contaminated sites: A comprehensive overview. *Reviews of Environmental Contamination and Toxicology Volume, 236*, 1–115.

Kwon, G. S., Kim, J. E., Kim, T. K., Sohn, H. Y., Koh, S. C., Shin, K. S., & Kim, D. G. (2002). *Klebsiella pneumoniae* KE-1 degrades endosulfan without formation of the toxic metabolite, endosulfan sulfate. *FEMS Microbiology Letters, 215*(2), 255–259.

Lara-Moreno, A., Morillo, E., Merchán, F., Madrid, F., & Villaverde, J. (2022). Bioremediation of a trifluralin contaminated soil using bioaugmentation with novel isolated bacterial strains and cyclodextrin. *Science of The Total Environment, 840*, 156695.

Li, J., Lin, D., Ji, R., Yao, K., Deng, W. Q., Yuan, H., Wu, Q., Jia, Q., Luo, P., Zhou, K., Li, H., Zou, L., & Liu, S. (2005). Simultaneous determination of β-cypermethrin and its metabolite 3-phenoxybenzoic acid in microbial degradation systems by HPLC–UV. *Journal of Chromatographic Science, 54*(9), 1584–1592.

Liang, W. Q., Wang, Z. Y., Li, H., Wu, P. C., Hu, J. M., Luo, N., …, & Liu, Y. H. (2005). Purification and characterization of a novel pyrethroid hydrolase from *Aspergillus niger* ZD11. *Journal of Agricultural and Food Chemistry, 53*(19), 7415–7420.

Lima-Morales, D., Jáuregui, R., Camarinha-Silva, A., Geffers, R., Pieper, D. H., & Vilchez-Vargas, R. (2016). Linking microbial community and catabolic gene structures during the adaptation of three contaminated soils under continuous long-term pollutant stress. *Applied and Environmental Microbiology, 82*(7), 2227–2237.

Lin, C., Cheruiyot, N. K., Bui, X. T., & Ngo, H. H. (2022). Composting and its application in bioremediation of organic contaminants. *Bioengineered, 13*(1), 1073–1089.

Liu, J., Zhang, X., Xu, J., Qiu, J., Zhu, J., Cao, H., & He, J. (2020). Anaerobic biodegradation of acetochlor by acclimated sludge and its anaerobic catabolic pathway. *Science of the Total Environment, 748*, 141122.

Liu, M., Hashi, Y., Song, Y., & Lin, J. M. (2005). Simultaneous determination of carbamate and organophosphorus pesticides in fruits and vegetables by liquid chromatography–mass spectrometry. *Journal of Chromatography A, 1097*(1-2), 183–187.

Liu, S. H., Zeng, G. M., Niu, Q. Y., Liu, Y., Zhou, L., Jiang, L. H., ..., & Cheng, M. (2017). Bioremediation mechanisms of combined pollution of PAHs and heavy metals by bacteria and fungi: A mini review. *Bioresource Technology, 224*, 25–33.

Lukić, B., Panico, A., Huguenot, D., Fabbricino, M., van Hullebusch, E. D., & Esposito, G. (2017). A review on the efficiency of landfarming integrated with composting as a soil remediation treatment. *Environmental Technology Reviews, 6*(1), 94–116.

Madhavi, GN, & Mohini, DD (2012). Review paper on–parameters affecting bioremediation. *International Journal of Life science and Pharma Research, 2*(3), 77–80.

Mahmood, I., Imadi, S. R., Shazadi, K., Gul, A., & Hakeem, K. R. (2016). Plant, soil and microbes. *Journal of Implications in Crop Science, 1*, 103–116.

Maitra, S. (2018). *In situ bioremediation: An overview*. Responsibility of Life Science Informatics Publications.

Malghani, S., Chatterjee, N., Yu, H. X., & Luo, Z. (2009). Isolation and identification of profenofos degrading bacteria. *Brazilian Journal of Microbiology, 40*, 893–900.

Malik, J. A., Rahman, M. M., & Rais, N. (2022). Microbial biotechnology: An introduction. In *Microbes and microbial biotechnology for green remediation* (pp. 3–11). Elsevier.

Mansotra, P., Sharma, P., Sirari, A., & Aggarwal, N. (2022). Ecological performance of multifunctional pesticide tolerant strains of *Mesorhizobium* sp. in chickpea with recommended pendimethalin, ready-mix of pendimethalin and imazethpyr, carbendazim and chlorpyrifos application. *Archives of Microbiology, 204*(1), 1–25.

Mao, X., Stenuit, B., Tremblay, J., Yu, K., Tringe, S. G., & Alvarez-Cohen, L. (2019). Structural dynamics and transcriptomic analysis of *Dehalococcoides mccartyi* within a TCE-Dechlorinating community in a completely mixed flow reactor. *Water Research, 158*, 146–156.

Maqbool, Z., Hussain, S., Imran, M., Mahmood, F., Shahzad, T., Ahmed, Z., ..., & Muzammil, S. (2016). Perspectives of using fungi as bioresource for bioremediation of pesticides in the environment: A critical review. *Environmental Science and Pollution Research, 23*(17), 16904–16925.

Marco, D. E., & Abram, F. (2019). Using genomics, metagenomics and other "Omics" to assess valuable microbial ecosystem services and novel biotechnological applications. *Frontiers in Microbiology, 10*, 151.

Martin, T. J., Gabure, S., Maise, J., Snipes, S., Peete, M., & Whalen, M. M. (2019). The organochlorine pesticides pentachlorophenol and dichlorodiphenyltrichloroethane increase secretion and production of interleukin 6 by human immune cells. *Environmental Toxicology and Pharmacology, 72*, 103263.

Menezes, R. G., Qadir, T. F., Moin, A., Fatima, H., Hussain, S. A., Madadin, M., Pasha, S. B., Al Rubaish, F. A., & Senthilkumaran, S. (2017). Endosulfan poisoning: an overview. *Journal of Forensic and Legal Medicine, 51*, 27–33.

Miao, Y., Heintz, M. B., Bell, C. H., Johnson, N. W., Polasko, A. L., Favero, D., & Mahendra, S. (2021). Profiling microbial community structures and functions in bioremediation strategies for treating 1, 4-dioxane-contaminated groundwater. *Journal of Hazardous Materials, 408*, 124457.

Mielke, K. C., Bertuani, R. R., Pires, F. R., Cotta, A. J. B., Egreja Filho, F. B., & Madalão, J. C. (2020). Does Canavalia ensiformis inoculation with *Bradyrhizobium* sp. enhance phytoremediation of sulfentrazone-contaminated soil? *Chemosphere, 255*, 127033.

Miguel-Ferrer, L., Romero-Arenas, O., Andrade-Hoyos, P., Sánchez-Morales, P., Rivera-Tapia, J. A., & Fernández-Pavía, S. P. (2021). Antifungal activity of *Trichoderma harzianum* and *T. koningiopsis* against *Fusarium solani* in seed germination and vigor of *Miahuateco chili* seedlings. *Revista mexicana de fitopatología, 39*(2), 228–247.

Mihai, F. C., Plana, R., Taherzadeh, M. J., Aswathi, M. K., & Ezeah, C. (2021). Bioremediation of organic contaminants based on biowaste composting practices. In *Handbook of bioremediation* (pp. 701–714). Academic Press.

Mishra, A., Ntihuga, J. N., Molitor, B., & Angenent, L. T. (2020). Power-to-protein: Carbon fixation with renewable electric power to feed the world. *Joule, 4*(6), 1142–1147.

Mishra, R., Lone, R., Manzoor, D., & Shuab, R. (2018). Imbalance due to pesticide contamination in different ecosystems. *International Journal of Theoretical and Applied Sciences, 10*, 239–246.

Mishra, S., Lin, Z., Pang, S., Zhang, W., Bhatt, P., & Chen, S. (2021). Recent advanced technologies for the characterization of xenobiotic-degrading microorganisms and microbial communities. *Frontiers in Bioengineering and Biotechnology, 9*, 632059.

Mónica, P., Darwin, R. O., Manjunatha, B., Zúñiga, J. J., Diego, R., Bryan, R. B., …, & Maddela, N. R. (2016). Evaluation of various pesticides-degrading pure bacterial cultures isolated from pesticide-contaminated soils in Ecuador. *African Journal of Biotechnology, 15*(40), 2224–2233.

Mrozik, A., & Piotrowska-Seget, Z. (2010). Bioaugmentation as a strategy for cleaning up of soils contaminated with aromatic compounds. *Microbiological Research, 165*(5), 363–375.

Mustapha, M. U., Halimoon, N., Johar, W. L. W., Shukor, Abd, & M. , Y (2019). An overview on biodegradation of carbamate pesticides by soil bacteria. *Pertanika Journal of Science & Technology, 27*(2).

Myresiotis, C. K., Vryzas, Z., & Papadopoulou-Mourkidou, E. (2012). Biodegradation of soil-applied pesticides by selected strains of plant growth-promoting rhizobacteria (PGPR) and their effects on bacterial growth. *Biodegradation, 23*(2), 297–310.

Naik, M. G., & Duraphe, M. D. (2012). Parameters affecting bioremediation. *Advance Research in Pharmaceuticals and Biologicals, 2*(3).

Niti, C., Sunita, S., Kamlesh, K., & Rakesh, K. (2013). Bioremediation: An emerging technology for remediation of pesticides. *Research Journal of Chemistry and Environment, 17*, 4.

Nkongolo, K. K., & Narendrula-Kotha, R. (2020). Advances in monitoring soil microbial community dynamic and function. *Journal of Applied Genetics, 61*(2), 249–263.

Nollet, L. M., & Rathore, H. S. (2016). *Handbook of pesticides: Methods of pesticide residues analysis*. CRC Press.

Odukkathil, G., & Vasudevan, N. (2013). Toxicity and bioremediation of pesticides in agricultural soil. *Reviews in Environmental Science and Bio/Technology, 12*(4), 421–444.

Pande, V., Pandey, S. C., Sati, D., Pande, V., & Samant, M. (2020). Bioremediation: An emerging effective approach towards environment restoration. *Environmental Sustainability, 3*(1), 91–103.

Panigrahi, S., Velraj, P., & Rao, T. S. (2019). Functional microbial diversity in contaminated environment and application in bioremediation. In *Microbial diversity in the genomic era* (pp. 359–385). Academic Press.

Parween, T., Bhandari, P., Sharma, R., Jan, S., Siddiqui, Z. H., & Patanjali, P. K. (2018). Bioremediation: A sustainable tool to prevent pesticide pollution. In *Modern age environmental problems and their remediation* (pp. 215–227). Springer.

Paul, D., Pandey, G., Pandey, J., & Jain, R. K. (2005). Accessing microbial diversity for bioremediation and environmental restoration. *TRENDS in Biotechnology, 23*(3), 135–142.

Pereira, J. L., Antunes, S. C., Castro, B. B., Marques, C. R., Gonçalves, A. M., Gonçalves, F., & Pereira, R. (2009). Toxicity evaluation of three pesticides on non-target aquatic and soil organisms: Commercial formulation versus active ingredient. *Ecotoxicology (London, England), 18*(4), 455–463.

Philp, J. C., & Atlas, R. M. (2005). Bioremediation of contaminated soils and aquifers. In Atlas, R., & Philp, J. (Eds.), Bioremediation: Applied microbial solutions for real-world environmental cleanup (1st ed., pp. 139—236). American Society for Microbiology.

Pinto, A. P., Teixeira, D. M., Caldeira, A. T., & Rodrigues, S. C. (2020). Biodegradation of pesticides by adapted fungi: Potential use on biopurification systems? In *Agrochemicals detection, treatment and remediation* (pp. 1–23). Butterworth-Heinemann.

Prabha, R., Singh, D. P., & Verma, M. K. (2017). Microbial interactions and perspectives for bioremediation of pesticides in the soils. In *Plant-microbe interactions in agro-ecological perspectives* (pp. 649–671). Springer.

Puckett, S. P., Samples, R. M., Schloss, P. D., & Balunas, M. J. (2020). 7. 25-metabolomics and the microbiome: Characterizing molecular diversity in complex microbial communities. In H.-W. Liu, & T. P. Begley (Eds.), *Comprehensive natural products III* (pp. 502–518). Elsevier.

Purnomo, H., Okarda, B., Dermawan, A., Ilham, Q. P., Pacheco, P., Nurfatriani, F., & Suhendang, E. (2020). Reconciling oil palm economic development and environmental conservation in Indonesia: A value chain dynamic approach. *Forest Policy and Economics, 111*, 102089.

Foster, R., L. , J., Kwan, B. H., & Vancov, T (2004). Microbial degradation of the organophosphate pesticide, Ethion. *FEMS Microbiology Letters, 240*(1), 49–53.

Raffa, C. M., & Chiampo, F. (2021). Bioremediation of agricultural soils polluted with pesticides: A review. *Bioengineering, 8*(7), 92.

Rajmohan, K. S., Chandrasekaran, R., & Varjani, S. (2020). A review on occurrence of pesticides in environment and current technologies for their remediation and management. *Indian Journal of Microbiology, 60*(2), 125–138.

Randika, J. L. P. C., Bandara, P. K. G. S. S., Soysa, H. S. M., Ruwandeepika, H. A. D., & Gunatilake, S. K. (2022). Bioremediation of pesticide-contaminated soil: A review on indispensable role of soil bacteria. *Journal of Agricultural Sciences–Sri Lanka, 17*(1), 19–43.

Roy, A., Roy, M., Alghamdi, S., Dablool, A. S., Almakki, A. A., Ali, I. H., ... Cabral-Pinto (2022). Role of microbes and nanomaterials in the removal of pesticides from wastewater. *Hindawi International Journal of Photoenergy, 2022*. Article ID 2131583. 12 pages. https://doi.org/10.1155/2022/2131583.

Ruiz-Lara, A., Fierro, F., Carrasco, U., Oria, J. A., & Tomasini, A. (2020). Proteomic analysis of the response of *Rhizopus oryzae* ENHE to pentachlorophenol: Understanding the mechanisms for tolerance and degradation of this toxic compound. *Process Biochemistry, 95*, 242–250.

Sagar, V., & Singh, D. P. (2011). Biodegradation of lindane pesticide by non white-rots soil fungus *Fusarium* sp. *World Journal of Microbiology and Biotechnology, 27*(8), 1747–1754.

Satish, G. P., Ashokrao, D. M., & Arun, S. K. (2017). Microbial degradation of pesticide: A review. *African Journal of Microbiology Research, 11*(24), 992–1012.

Scott, C., Pandey, G., Hartley, C. J., Jackson, C. J., Cheesman, M. J., Taylor, M. C., & Oakeshott, J. G. (2008). The enzymatic basis for pesticide bioremediation. *Indian Journal of Microbiology, 48*(1), 65–79.

Sehrawat, A., Phour, M., Kumar, R., & Sindhu, S. S. (2021). Bioremediation of pesticides: An eco-friendly approach for environment sustainability. In *Microbial rejuvenation of polluted environment* (pp. 23–84). Springer.

Sharma, A., Kumar, V., Shahzad, B., Tanveer, M., Sidhu, G. P. S., Handa, N., Kohli, S. K., Yadav, P., Bali, A. S., Parihar, R. D., & Dar, O. I. (2019). Worldwide pesticide usage and its impacts on ecosystem. *SN Applied Sciences, 1*, 1–16.

Sharma, B., Dangi, A. K., & Shukla, P. (2018). Contemporary enzyme based technologies for bioremediation: A review. *Journal of Environmental Management, 210*, 10–22.

Sharma, N., & Singhvi, R. (2017). Effects of chemical fertilizers and pesticides on human health and environment: A review. *International Journal of Agriculture, Environment and Biotechnology, 10*(6), 675–680.

Sharma, P., Singh, S. P., Parakh, S. K., & Tong, Y. W. (2022). Health hazards of hexavalent chromium (Cr (VI)) and its microbial reduction. *Bioengineered, 13*(3), 4923–4938.

Shekhar, S. K., Godheja, J., & Modi, D. R. (2020). Molecular technologies for assessment of bioremediation and characterization of microbial communities at pollutant-contaminated sites. In *Bioremediation of industrial waste for environmental safety* (pp. 437–474). Springer.

Shin, D. H., Kim, D. U., Seong, C. N., Song, H. G., & Ka, J. O. (2012). Genetic and phenotypic diversity of carbofuran-degrading bacteria isolated from agricultural soils. *Journal of Microbiology and Biotechnology, 22*(4), 448–456.

Shourie, A., & Vijayalakshmi, U. (2022). Fungal diversity and its role in mycoremediation. *Geomicrobiology Journal, 39*(3-5), 426–444.

Shukla, A. K., Singh, R. S., Upadhyay, S. N., & Dubey, S. K. (2010). Kinetics of bio-filtration of trichloroethylene by methanotrophs in presence of methanol. *Bioresource Technology, 101*(21), 8119–8126.

Singh, B. K., Walker, A., Morgan, J. A. W., & Wright, D. J. (2004). Biodegradation of chlorpyrifos by Enterobacter strain B-14 and its use in bioremediation of contaminated soils. *Applied and Environmental Microbiology, 70*(8), 4855–4863.

Singh, B., Kaur, J., & Singh, K. (2014). Microbial degradation of an organophosphate pesticide, malathion. *Critical Reviews in Microbiology, 40*(2), 146–154.

Singh, D. K. (2008). Biodegradation and bioremediation of pesticide in soil: Concept, method and recent developments. *Indian Journal of Microbiology, 48*(1), 35–40.

Singh, R., Paul, D., & Jain, R. K. (2006). Biofilms: Implications in bioremediation. *Trends in Microbiology, 14*(9), 389–397.

Spina, F., Cecchi, G., Landinez-Torres, A., Pecoraro, L., Russo, F., Wu, B., Cai, L., Liu, X. Z., Tosi, S., Varese, G. C., & Zotti, M. (2018). Fungi as a toolbox for sustainable bioremediation of pesticides in soil and water. *Plant Biosystems-An International Journal Dealing with all Aspects of Plant Biology, 152*(3), 474–488.

Thakur, M., Medintz, I. L., & Walper, S. A. (2019). Enzymatic bioremediation of organophosphate compounds—progress and remaining challenges. *Frontiers in Bioengineering and Biotechnology., 7*, 289. https://doi.org/10.3389/fbioe.2019.00289.

Treviño-Quintanilla, L., Galán-Wong, L., Rodríguez-Uribe, B., & Soberón-Chávez, G. (2002). Cloning and characterization of a FAD-monooxygenase gene (cadA) involved in degradation of chloranilic acid (2, 5-dichloro-3, 6-dihydroxybenzo-1, 4-quinone) in *Pseudomonas putida* TQ07. *Applied Microbiology and Biotechnology, 59*(4), 545–550.

Truskewycz, A., Gundry, T. D., Khudur, L. S., Kolobaric, A., Taha, M., Aburto-Medina, A., …, & Shahsavari, E. (2019). Petroleum hydrocarbon contamination in terrestrial ecosystems: Fate and microbial responses. *Molecules (Basel, Switzerland), 24*(18), 3400.

Tyagi, B., & Kumar, N. (2021). Bioremediation: Principles and applications in environmental management. In *Bioremediation for environmental sustainability* (pp. 3–28). Elsevier.

Tyagi, M., da Fonseca, M. M. R., & de Carvalho, C. C. (2011). Bioaugmentation and biostimulation strategies to improve the effectiveness of bioremediation processes. *Biodegradation, 22*(2), 231–241.

Upadhyay, L. S., & Dutt, A. (2017). Microbial detoxification of residual organophosphate pesticides in agricultural practices. In *Microbial biotechnology* (pp. 225–242). Springer.

Velázquez-Fernández, J. B., Martínez-Rizo, A. B., Ramírez-Sandoval, M., & Domínguez-Ojeda, D. (2012). Biodegradation and bioremediation of organic pesticides. In *Pesticides-Recent Trends in Pesticide Residue Assay* (pp. 253–272). IntechOpen.

Verma, J. P., Jaiswal, D. K., & Sagar, R. (2014). Pesticide relevance and their microbial degradation: A-state-of-art. *Reviews in Environmental Science and Bio/Technology, 13*(4), 429–466.

Vidali, M. (2001). Bioremediation: An overview. *Pure and Applied Chemistry, 73*(7), 1163–1172.

Walia, S., Khan, A., & Rosenthal, N. (1990). Construction and applications of DNA probes for detection of polychlorinated biphenyl-degrading genotypes in toxic organic-contaminated soil environments. *Applied and Environmental Microbiology, 56*(1), 254–259.

Wang, F., Adams, C. A., Yang, W., Sun, Y., & Shi, Z. (2020). Benefits of arbuscular mycorrhizal fungi in reducing organic contaminant residues in crops: Implications for cleaner agricultural production. *Critical Reviews in Environmental Science and Technology, 50*(15), 1580–1612.

Wang, J., Chen, X., Zhang, S., Wang, Y., Shao, X., & Wu, D. (2022). Analysis of raw materials and products characteristics from composting and anaerobic digestion in rural areas. *Journal of Cleaner Production, 338*, 130455.

Wang, S., Zhang, C., & Yan, Y. (2012). Biodegradation of methyl parathion and p-nitrophenol by a newly isolated *Agrobacterium* sp. strain Yw12. *Biodegradation, 23*(1), 107–116.

Wang, S., Zhang, C., Lv, Z., Huang, H., Cao, X., Song, Z., & Shao, M. (2019). Degradation of 3, 5, 6-trichloro-2-pyridinol by a microbial consortium in dryland soil with anaerobic incubation. *Biodegradation, 30*(2), 161–171.

Wang, X., Hou, X., Liang, S., Lu, Z., Hou, Z., Zhao, X., …, & Zhang, H. (2018). Biodegradation of fungicide *Tebuconazole* by *Serratia marcescens* strain B1 and its application in bioremediation of contaminated soil. *International Biodeterioration & Biodegradation, 127*, 185–191.

Wu, Z., Wong, H. S., & Buenfeld, N. R. (2017). Transport properties of concrete after drying-wetting regimes to elucidate the effects of moisture content, hysteresis and microcracking. *Cement and Concrete Research, 98*, 136–154.

Xiao, P. F., & Kondo, R. (2013). Biodegradation of dieldrin by *Cordyceps* fungi and detection of metabolites. In *Applied mechanics and materials: 295* (pp. 30–34). Trans Tech Publications Ltd.

Xu, X., Sun, Y., Sun, Y., & Li, Y. (2022). Bioaugmentation improves batch psychrophilic anaerobic co-digestion of cattle manure and corn straw. *Bioresource Technology, 343*, 126118.

Yan, C., Wang, F., Geng, H., Liu, H., Pu, S., Tian, Z., Chen, H., Zhou, B., Yuan, R., & Yao, J. (2020). Integrating high-throughput sequencing and metagenome analysis to reveal the characteristic and resistance mechanism of microbial community in metal contaminated sediments. *Science of The Total Environment, 707*, 136116.

Yasouri, F. (2006). Plasmid mediated degradation of diazinon by three bacterial strains *Pseudomonas* sp., *Flavobacterium* sp. and *Agrobacterium* sp. *Asian Journal of Chemistry, 18*, 2437.

Yu, F. B., Shan, S. D., Luo, L. P., Guan, L. B., & Qin, H. (2013). Isolation and characterization of a *Sphingomonas* sp. strain F-7 degrading fenvalerate and its use in bioremediation of contaminated soil. *Journal of Environmental Science and Health, Part B, 48*(3), 198–207.

Zhang, M. Y., Ying, T. E. N. G., Ye, Z. H. U., Jun, W. A. N. G., Yong-Ming, L. U. O., Christie, P., Zhen-Gao, L. I., & Udeigwe, T. K. (2014). Isolation and characterization of chlorothalonil-degrading bacterial strain H4 and its potential for remediation of contaminated soil. *Pedosphere, 24*(6), 799–807.

Zhang, M., Bai, X., Li, Q., Zhang, L., Zhu, Q., Gao, S., Ke, Z., Jiang, M., Hu, J., Qiu, J., & Hong, Q. (2022). Functional analysis, diversity, and distribution of carbendazim hydrolases MheI and CbmA, responsible for the initial step in carbendazim degradation. *Environmental Microbiology, 24*(10), 4803–4817.

Zhang, R., Cui, Z., Zhang, X., Jiang, J., Gu, J. D., & Li, S. (2006). Cloning of the organophosphorus pesticide hydrolase gene clusters of seven degradative bacteria isolated from a methyl parathion contaminated site and evidence of their horizontal gene transfer. *Biodegradation, 17*(5), 465–472.

Zhang, Y. F., Liu, S. Y., Du, Y. H., Feng, W. J., Liu, J. H., & Qiao, J. J. (2014). Genome shuffling of *Lactococcus lactis* subspecies lactis YF11 for improving nisin Z production and comparative analysis. *Journal of Dairy Science, 97*, 2528–2541. https://doi.org/10.3168/jds.2013-7238.

Zhao, S., Xu, W., Zhang, W., Wu, H., Guang, C., & Mu, W. (2021). In-depth biochemical identification of a novel methyl parathion hydrolase from *Azohydromonas australica* and its high effectiveness in the degradation of various organophosphorus pesticides. *Bioresource Technology, 323*, 124641.

CHAPTER 11

Impact of nanopesticides in the environment: Solutions, threats, and opportunities

Romica Verma[a] and Shveta Saroop[b]
[a]Department of Botany, University of Jammu, Jammu, Jammu and Kashmir, India
[b]Department of Botany, Government Degree College, Kathua, Jammu and Kashmir, India

11.1 Introduction

A new nanoparticle venture in the form of plant protection products known as "nanopesticides" has sparked interest in nanotechnology in the pesticide industry. Nanopesticides as the name signifies are very small pesticide particles of an agrochemical active ingredient, particles abbreviated as "Ais" or any other micro-sized particles with effective pesticidal potential. In contrast to conventional pesticides, nanomaterials with the potential to encapsulate and transport pesticide active ingredients (AIs) in a regulated, targeted, synchronized, and systematic manner present a new initiative in pesticidal efficacy and efficiency. These nanopesticides have revolutionized the agricultural system, driving higher agricultural yields and hence contributing to sustainable agriculture and world food security (Fraceto et al., 2016; Kah et al., 2014; Prasad et al., 2017; Yadav et al., 2020).

Nanopesticides employ targeted delivery of agrochemicals with the objective of achieving their full biological effectiveness without any overdosage thereby minimizing environmental toxicity (Fraceto et al., 2016; Prasad et al., 2017; Yadav et al., 2020). Nanopesticides offer advantages owing to their greater solubility, delivery/transportation, and persistence; the potential to minimize the number of active chemicals utilized; and the ability to use compounds that impose fewer toxic effects on organisms that are not their targets, which checks the emergence of resistance (Huang et al., 2018; Kah et al., 2014; Kuhlbusch et al., 2018; Yadav et al., 2020).

To analyze potential concerns linked with nanoagriculture, extensive scientific insights about biosafety issues, detrimental effects, trajectory, and evolved biological sensitivity of nanopesticides in the environment and ecological complexes need to be initiated (Huang et al., 2018; Yadav et al., 2020). Toxicological research should monitor and assess risks and thresholds of exposure throughout the life-cycle of nanopesticides, as well as physicochemical factors modulating nanomaterial toxicity, and potential interactions

Pesticides in the Environment: Impact, Assessment, and Remediation
DOI: https://doi.org/10.1016/B978-0-323-99427-9.00010-0

with agricultural regulators. Overall, the analysis emphasizes proper risk management, plans for pesticide company employees as well as farmers, occupational safety policies, and procedures as well as creating a competent policy framework for the utilization of nanotechnology in agriculture (Balaure et al., 2017; Khan et al., 2019).

Despite the potential advantages of nanopesticide formulations, questions about their use in agriculture have surfaced, notably with regard to the biosafety of pesticides and their impacts on the ecosystem and their complexes. The unregulated and unsystematic usage of these smart agrichemical has intensified the rate of pathogen resistance, low rates of nitrogen fixation, biodiversity issues, and their bioaccumulation potential in agricultural, livestock products, and in trophic levels of a food chain existing in the soil environment, which is posing a significant threat to the ecosystem and its sustainability (Chaud et al., 2021; Grillo et al., 2021; Prasad et al., 2017; Rajpoot, 2021). Nanopesticides from agricultural and industrial wastewater runoff enter the water supply via the soil by precipitation, penetrating followed by leaching events. By altering the quality and quantity of the ecosystem matrix thereby, increases human exposure time thereby raising health issues in the ecosystem. Long-term (acute) and short-term (chronic) negative effects of intentional or unintentional absorption of pesticide particles via food and water shorten the span and quality of life of any organisms (Chaud et al., 2021; Grillo et al., 2021; Iavicoli et al., 2017; Kuhlbusch et al., 2018; Yadav et al., 2020).

Scientific research must recognize the intricate interactions between environmental factors and pesticides applied to reveal physicochemical attributes vis-a-vis the biological reactivity of the pesticides and the agrosystem environment. Improvements in nanopesticides' solubility and bioavailability may throw light into the dynamics and pathways of the pesticides once they are absorbed by organisms, in terms of toxicokinetics. Consequently, a thorough toxicological evaluation and risk assessment of nanopesticides—both as nanoformulations of current active ingredients and as nanomaterials with pesticidal activity should be undertaken to minimize the adverse effect of nanopesticide (Chaud et al., 2021; Grillo et al., 2021; Kah, 2015; Koul, 2019; Rajpoot, 2021; Yadav et al., 2020).

This chapter puts forth insights into nanopesticides, their applications on agricultural production and ecosystem issues, regulated release strategies, and stimuli-responsive devices for gradual, prolonged, and genetic material delivery are highlighted in the present communication.

Nanotechnology has the potential to address a wide range of agricultural and environmental issues, including urbanization, power, and source restraints, the continual utility of assets, run-off, and amassing of pesticides as well as fertilizer (Chen & Yada, 2011; Ditta, 2012; Parisi et al., 2015). Amid an ever-increasing population, the need for elevated agricultural production and efficient methods of optimizing agricultural operations increases; hence, exercising nanoscale matter for farming research is growing (Gogos et al., 2012). Actuality, nanotechniques are the potential enough to dramatically increase sustainable agriculture and precision farming. The ultimate goal of environmental monitoring and targeted action is to maximize agricultural output while minimizing the input of pesticides, fertilizers, and herbicides (Fraceto et al., 2016; Servin et al., 2015). Such a segment is very interested in nanopromoted solutions because of their capacity to ameliorate germination of seed, growth, and protection of plants by controlling the discharge of agrochemicals, resulting in less chemical products used and reduced nutrient loss during the use of fertilizers. Besides, nanotechniques used in farming might lead to novel approaches to shield and rehabilitate soils as well as water, hence, improving comprehensive yield without compromising food value and being ecologically friendly at the same (Biswal et al., 2012; Ditta, 2012; Khot et al., 2012; Prasad et al., 2014; Sekhon, 2014; Sonkaria et al., 2012). Currently, the entire agricultural chemical firms are focusing their research on nanotechnology for agricultural uses (DeRosa et al., 2010). For the past 10 years, certain businesses have submitted patent applications covering a variety of methods for developing and delivering formulations of nanopesticides (Peters et al., 2016). Nevertheless, the current Food Safety Authority of Europe "inventory of nanotechnology utilization in the agricultural, feed, and food industry" indicates that only a small number of formulations for encapsulated nanopesticides are now accessible (Peters et al., 2014). Other nanoagrochemical applications, such as nanocomposites, nanoencapsulates, nanoclays, and silver and silica nanomaterials are in the early stages of progress for pesticidal and biocidal appliance and is likely to join the marketplace in the forthcoming period (Peters et al., 2014).

Nevertheless, the growing use of nanotechnology in sectors of the global economy like agriculture and many others raises worries concerning the well-being of mankind and the environment (Kah, 2015; Scott & Chen, 2013). Indeed, the intentional use of nanoelements in farming may unanticipated health repercussions (Gogos et al., 2012; Kah et al., 2013). As a result of probable accumulation in the surroundings and food chains, such a scenario calls for increased exposure of humans and the environment to the residues of nanomaterials in crops as well as soil. As a result, the projected novel and increased actions of nanopromoted uses may provide new advantages as well as dangers to the well-being of people and the environment. The author believes that the need for "green nanotechnology"—a theoretical approach, for balancing the profits offered by nanoproducts in resolving ecological issues along with assessing and managing the health as well as security threats related to the environment that nanoscale products may pose, coincides with the aim of accomplishing sustainable farming. Furthermore, risks associated with exposure to nanomaterial ought to be handled utilizing a situation-specific life-cycle strategy. It includes undertaking all stages in which solutions of nanoparticles might be discovered. It accounts for applications into agricultural land to incorporation into foodstuffs, to clearance or re-use of the end product. It also includes the potential pressure put forth by exclusive agro-based system circumstances, which might influence the dangerous properties of nanomaterials and their menace characterization.

The first nanomaterial exposures might be for agricultural employees, who are highly and persistently in touch with the as-yet-undiscovered xenobiotics during conducting everyday job duties. From the perspective of occupational health, this looks to be an even more urgent issue to examine. Concerns regarding possible occupational hazards from exposure to nanoscale materials, measuring, explaining, and managing these risks for regulatory reasons, have grown in response to rising attention to the utilization of nano-based technology in agronomy (Kookana et al., 2014). The toxicological effects of a broad array of nanosubstances are still unknown once they are discrete into physicochemically varying environments. Their exclusive means of employment in agro-fields impose nanofocused as well as highly unambiguous agricultural-directed strategies for professional risk evaluation and executive procedure.

One of the most promising current technologies is nanotechnology. The quantum incarceration of electronic states at the nanoscale lead to bizarre properties, easing the challenges associated with traditional bulk materials (iGate research, 2018). The global nanotechnology industry is expected to reach US$ 125 billion by 2020 (FAOSTAT), because it continues to pose an impact on virtually every area in the globalized world, including electronics, cosmetics, information and communication technologies, power, agriculture, and so on. Being a farming country with a considerably greater agriculture scale, India is poised to be a nation having a higher proportion demanding technologically sophisticated farming goods. Concerning India's agricultural industry, the use of nanopesticides might have a direct impact on production and food security. Pesticide misuse, as well as the harmful impacts on physical conditions and the atmosphere, insists for scientifically superior and secure alternatives against existing agro-formulations. In these situations, applying smart nanotechnology solutions might help minimize the negative health effects connected to the misuse of traditional pesticides. Currently, nanoparticle-based pesticides are created as (1) nanoscale manufactured nanostructures containing the pesticide as an active ingredient or as and (2) nanosized elements of the organic and inorganic component as functional constituents in nanopesticide formulations. Although the area is relatively young, nanoscale forms are viewed as promising candidates for overcoming problems with massive pesticides, particularly as having maximum efficiency and reduced usage of huge amounts of pesticides. Future widespread application of these nanopesticides will be hampered by their toxicity to people and other nontarget animals as well as possible environmental consequences. As a result, care is advised while thinking about the widespread usage of nanopesticides. More importantly, the scientific community must significantly increase its research effort in order to develop a vigorous regulative methods for scheming functional conventions by analyzing the probable threats linked to the products, as failure to do so could pose a serious threat to universal food safety. We require a variety of designs that are safer and technologically equipped with advanced agrochemical systems that are globally applicable to get beyond the obstacles and serious roadblocks that prevent large-scale usage.

11.1.1 Nanopesticides in agriculture and agricultural areas

Nanopesticide research is thriving among commercial, scientific, and government scientists. Climate change and agriculture have a direct link. Current agricultural practices, in a pernicious negative cycle, aggravate factors that pessimistically affect climatic changes. However, both the quality and quantity of food produced get affected by increases in environmental demands as well as by the influence of biophysical conditions. This causes an increase in ration insecurity which in turn hinders adequate nutrition and equal food distribution. Excellent medical imaging devices and medicine delivery systems have been made possible by nanotechnology; comparable advancements are also being used in intellectual agrochemical deliverance known as "nanopesticides." A recent technological breakthrough, i.e., nanopesticides, employs nanotechnological approach to generate smart pesticides based on nanoformulations for agricultural usage (Shekhar et al., 2021). To examine pesticide efficacy in innovative nanocarrier formulations based on carbon, metals, ceramics, polymers, copolymers, silica, lipids, and other materials, the form and distinguishing characteristics of nanoscale material are generated. These include the process of pesticide conjugation, encapsulation, and adsorption by using nanomaterials for increasing active ingredient dispersal, permeability, constancy, and then biodegradability. Insecticides encapsulated in nanoparticles avoid early breakdown and the risks of direct active component exposure to people. Nanopesticides, unlike standard pesticides, have a large surface area, which upgrades their aptitude for interaction with aimed pests at minimum quantity. Because traditional pesticides and nanopesticides clearly operate differently in the environment, extensive study governing the nanopesticides is needed to establish that they meet the regulatory laws and recommendations (Kah et al., 2018a, b). Nanopesticides have a critical function in reducing the regular impact of pesticides on the environment. There are two categories of nanopesticides: (1) pesticides with nanoscale articulate components, often including a dispersant suspension of vigorous nanopesticides (2) pesticides ladened, encapsulated, doped, or encrusted with nanomaterials. Pesticides of the sort described above often combine the potential advantages of traditional pesticides along with shielding, conveying, and controlling pesticide contact (Chaudhary et al., 2018; Jiang et al., 2012; Jindal et al., 2017). Contrary to nanopesticides, conformist pesticides have meager steadiness, reduced dispersibility, coarse units for carrying drugs, inferior biological action, and employment at a rate of less than 30% (Tong et al., 2018; Sun et al., 2019). Furthermore, nanopesticide nanocarriers not only increase pesticide dispersibility but also speed up the conveyance of pesticides' valuable ingredients to the chosen species thus boosting their bioavailability (Sharma et al., 2017). Because of these advantages, nanopesticides have been widely used to overcome the limitations of traditional pesticides' limited effectiveness and high dosages. Environmental Protection Agency based in the United States attributes the negative health effects of nanopesticides to: (1) surface absorption due to their tiny size and capacity to traverse cell membranes,

(2) ability to cross the blood-brain barrier and enter the lungs, (3) the immediate prospective as well as durability of environmentally concerned nanomaterials, and (4) deficiency of knowledge regarding the measurement of environmental exposures. The nanopesticides being developed include reformulated active components with fungicidal, insecticidal, and herbicidal characteristics (Walker et al., 2018). The majority of active components are organic molecules; however, few functional compounds are inorganic, such as Cu, being utilized for its fungicidal ability. Although hard nanoparticles like silica, graphene oxide, and carbon nanotube particles are also examples, soft nanoparticles like polymers, solid lipids, and others are frequently used as nanocarriers (Cao et al., 2017; Wibowo et al., 2014). Pesticides used at high doses do not fulfill their objectives, and their use usually leads to the pollution of terrestrial and marine areas. The primary motivation for implementing nanotechnology is to minimize the threshold quantity to protect crops, which can be accomplished in a variety of ways, such as improving the evident solubility, bioavailability, targeted release, controlled release, stability, and adhesion of leaf of functional components in surrounding. Restricted pesticide administration minimizes its quantity and inhibits the development of challenging pests caused by overexposure of chemicals as well as pollution of water and soil. Unlike traditional formulations, controlled administration guarantees that the active ingredient is released at a predefined pace to achieve the targeted effectiveness lifetime for the nanopesticides in the field. To build a suitable analysis of its comparative benefits, the commercialized nanoformulation should be compared with a conventional formulation, serving as a standard, practically in in-vivo conditions using the label instructions for a cropping system pressed by the Australian Registration Scheme. The business is dominated by a marker publication on "diyarex gold" nanopesticides created utilizing nanotechnology. Numerous plants have utilized it as an antibacterial against rust and powdery mildew (Battacharayya et al., 2016).

Traditional agrochemicals and pesticides are widely used in global agrofarming, resulting in a rise in the production of food and crops to meet the nutritive needs of the world's fast-rising population (Gianessi, 2013; Linquist et al., 2013). Crop yield increases can be attributable to improved pest management using pesticides instead of the usage of high-yielding crops (Popp et al., 2013). Pesticides enable greater food safety (against fungus, weeds, and insects), decreased Mn force and power usage, and better life in addition to increased agricultural yields (Cooper & Dobson, 2007). Agriculture in the 21st century has various obstacles in sustaining the trend of growing yields. Concerns with pesticide formulations include AIs' low solubility in water, nonselective behavior, and unrestrained discharge into the environment (Beggel et al., 2010; Birnbaum, 2013; Grant et al., 2011). Many studies have been conducted in recent years to determine the relevance of the increase of pesticide remains in the atmosphere, as they have produced considerable pollution in the environment while harming and/or changing the dietary constituents of crops as well as foods (Annamalai & Namasivayam, 2015; Cutulle et al., 2018;

Green, 2014; Hakeme et al., 2018). Merely 0.1% of pesticides are effective against the intended pests, whereas 99.9% escape into the surrounding. This inefficiency of the pesticides and their release in environment led in a number of negative consequences, including increased pest and disease resistance, water pollution, biodiversity loss, soil contamination, and the extinction of critical species (e.g., bees) (Goswami et al., 2017; Goulson et al., 2015). Another difficulty to be addressed is the intricacy of interactions between pesticide mixtures, which might result in unforeseen hazard consequences (Rizzati et al., 2016). The release of Rachel Carson's Silent Spring in the early 1960s sparked widespread awareness about the environmental dangers of pesticides (Kohler & Triebskorn, 2013). Pesticides called organochlorine are known to have an estrogenic or androgenic impact in animals, including mammals, birds, reptiles, and amphibians (terrestrial and aquatic) (Hatcher et al., 2008; Krief et al., 2017). Further, organophosphates, carbamates, thiocarbamates, triazoles, pyrethroids, and triazines are the most common sources of unremitting poisoning in rodents, amphibians, birds, and fish (Colborn, 2002). These chemicals are called endocrine-disrupting chemicals (EDCs) and have been linked to a variety of epigenetic and transgenerational consequences in water-loving organisms (Bhandari et al., 2015). Malathion and chlorpyrifos are two common organo-phosphates harmful to fauna. Malathion and chlorpyrifos remains can contribute to 50% of all deployed organophosphate pesticides, thus, harming the neurons of the inhabiting fauna (Wang et al., 2016b). Some acute impacts of pesticides on world pollinators (particularly bees) were observed with respect to their homing capability (Henry et al., 2012), reproductive process (Rundlof et al., 2015), and foraging behavior (Feltham et al., 2014). The impacts may be analyzed with respect to nonfocal as well as focal sources of agricultural pollen on exposure of honeybees to pesticides (McArt et al., 2017). Stanley and coworkers (2015) observed that neonicotinoid exposure to bumble colonies resulted in reduced visiting rates and pollen collecting from apple trees. Finally, the consequence of decreased pollination service delivery may be evident in apples produced with 36% fewer seeds. Thus chemical exposure reduced bee pollination capacity and disruption of consistent agricultural yields (Bommarco et al., 2013; Vanbergen, 2013; Woodcock et al., 2017). Notwithstanding the profits obtained by pesticides in the protection of crops, improper pesticide use may lead to a number of negative outcomes (like toxic residues) which may pose potential health hazards along with degradation of the environment. In accordance to market assessment, it has been found that individuals are also at risk of unintended pesticide poisoning, which is directly linked to their incorrect utility and exploitive contact to the surrounding (Jones et al., 2014). Pesticides reach people through a variety of channels (1) direct routes, such as oral, cutaneous, and breathing and (2) indirect routes, such as professional contact and intake of foodstuff (Smith et al., 2017). Pesticide exposure can cause obesity, allergic wheezing, cancer, endocrine difficulties, neurological abnormalities, and many other health issues in people (Androutsopoulos et al., 2013; George & Shukla, 2017; Hoppin et al., 2017; Lammoglia et al., 2017; Richendrfer & Creton, 2015;

Rossignol et al., 2014). Pesticides have the ability to modulate the neuroendocrine system by antagonistic and/or agnoistic actions on the oestrogen binding receptor (ER) (Hernandez et al., 2013). Along with agricultural laborers, nearby inhabitants may be affected by hazardous health concerns produced by the presence of AIs in the surrounding as a result of AI loss via volatilization and drift (Brouwer et al., 2018; Jiang et al., 2016). Larsen and colleagues (2017) found that increased exposure to agricultural pesticides to pregnant women had a negative influence on the birth rate, resulting in a 5%–9% increase in dreadful births. Quite a few number of factors influence residue levels, including (1) the qualities and formulations of AIs, (2) location circumstances and local agriculture, (3) the timespan between application of pesticide and production, (4) crop type, (5) infestation of pest, and (6) plant fitness. A poorly developed crop is more likely to have large residue levels. Lozowicka et al. (2015) discovered that some AIs contained in apples, including captan, iprodione, folpet, and procymidone are carcinogenic. Organo-chlorine pesticides pose a significant threat to cognitive damage. This damage is roughly three times the maximum in elderly aged people (70 years age), though weight reduction is taken into account (Lee et al., 2016). Losing weight can also make it more likely for organochlorine toxins to enter brain neurons. Several international standards regulate the "maximum residue levels" (MRLs) of pesticides to maintain acceptable levels of food safety, such as MRLs issued by Codex Alimentarious, a cooperative organization of the Food Agriculture Organization (FAO) and World Health Organization (WHO) (http://www.fao.org/fao-who-codexalimentarius/standards/pestres/pesticides/en[Accessed 30 November 2018]), the European Union (EU) (http://ec.europa.eu/food/plant/pesticides/eu-pesticides-database/public/?event=homepage&language=EN[Accessed 30 November 2018]), and other organizations (http://www.who.int/foodsafety/databases/en[Accessed 30 November 2018]). To lessen health risks, efforts were done throughout the years to either prohibit detrimental complexes or lessen MRLs in foodstuffs. These initiatives involve the creation of more efficient insect targeting by active chemicals with fewer side effects, as well as nonchemical pest control strategies (Polajnar et al., 2015). Exposure of pesticides, however, has not been reduced satisfactorily because of a number of factors, including the introduction of novel compounds and the occurrence of inheritance chemicals in the atmosphere due to the growing frequent use of AI in cultivation (Bahlai et al., 2010). Many studies have been conducted to identify relations between biogenic and the eco-community configuration in order to minimize unintended pollution of ecosystems (Borgan & Relyea, 2014, 2017). Exposure and environmental destiny models must be built for combining information about sources of chemicals in the surroundings, chemical and/or pesticide degradation behavior, and environmental scenarios like the nanoquantitative structure-activity or structure-property relationship technique (Domercq et al., 2018; Guardo et al., 2018; Villaverde et al., 2018). Consideration of several pesticide release mechanisms in the field can aid in establishing the principal means of exposure for both marine and terrestrial species (Dang et al., 2016).

Considering the physiochemical characters or dose of the nanomaterials, nanopesticides can alter the biochemical or photochemical stages by affecting photosynthetic cycles and thus crop production. The mechanical impact of nanoparticles on plants is associated with their form and size, limiting the nanomaterials to certain places on surfaces of organ or whole plants (Shekhar et al., 2021). The impacts, however, may be felt in the entire plant body, resulting in either suppression or encouragement of plant development. Nanopesticides interact with chloroplasts in the cytoplasm of the cell, influencing photosynthetic processes at particular places in plants by attaching to the photosynthetic system and inhibiting its activity. Metallic nanoparticles have been proven in studies to be harmful to photosynthesis by causing functional and structural damage (Tighe-Neira et al., 2018). Improved knowledge of nanomaterial electron transport inhibition, the antioxidant defence system in plants, and photoinduced oxidative stress, may result from knowing the reciprocal connection of nanopesticides and the photosynthetic process. For instance, because of the impact of seed water content on the entrance, 3 nm carbon nanotubes (CNTs) might affect the germination of tomatoes (*Lycopersicon esculentum* Mill (Khodakovskaya et al., 2009). Contrarily, nanoparticles based on carbon can impede development by mechanical harm brought on by their perforating action on organelles or cell membrane, containing the fundamental cytotoxic process (Chen & Bothun, 2014; Hu et al., 2014; Zaytseva & Neumann, 2016). Additionally, clay-based colloidal suspensions or titanium-based nanoparticles may affect the hydraulic conductivity of corn (*Zea mays* L.), lowering transpiration and growth (Asli & Peter, 2009).

This part will summarize the possible implications of nanomaterials in agronomy to boost crop yield and soil health. Pesticide's pervasiveness has led to an increase in resistance to weeds, pests, and diseases in the ecological community against pesticides (Rai & Ingle, 2012). Nanomaterials are commonly used to remove contaminants from polluted soils. Because of their high aspect ratio, tiny size and great reactivity, nanoparticles are increasingly being used to clean polluted soil by chemical oxidation, absorption, and reduction (Guerra et al., 2018). Furthermore, the recovery of nanoparticles for reuse gives a chance to eliminate contaminants by lowering the cost of cleanup. As a result, substantial emphasis must be paid to the development of stable and efficient nanomaterials which can sustain their effectiveness through several ministration cycles. As a result of their strong magnetism, magnetic nanoparticles separate easily and quickly in the reaction media on applying a magnetic field, thus, allowing renaissance or recycling for subsequent therapeutic cycles (Ajmal et al., 2018; Fu et al., 2018; Usman et al., 2014). Several nanoparticles, including CNTs, graphene, silver, and copper, have antibacterial and pest management capabilities. The increased interest in nanobiotechnology has resulted in the use of biotic approaches for the production of nanoparticles, such as fungi, algae, and bacteria, plus some particular flora molecules, like proteins, lipids, carbohydrates, and extracts of the plant, among others, by utilizing cheap and readily accessible resources with ensuring less energy consumption and least production of byproducts

(Kaur, 2018). Pursuit of eco-friendly technology highlights the competence given by green nanomaterial synthesis and the ramifications of these materials in a variety of disciplines, including food, environment, agriculture, energy, building, and biomedicine, etc. (Barabadi et al., 2019; Lu & Ozcan, 2015; Khan et al., 2018; Saratale et al., 2018; Tan et al., 2019). Nanoparticles have different physical, catalytic, and optoelectronic capabilities that promote increased light reaction in plants, plant development, and plant resilience to abiotic and biotic stress. Lipophilic pesticides can benefit from nanoformulations in terms of dispersion and solubility in water. By using appropriate surface-active agents, stable water nanodispersions of lipophilic pesticides may be generated, therefore eliminating the need for hazardous organic solvents and increasing pesticide bioavailability on the soil. Among the various nanopesticide-based formulations, one is "imidacloprid"—an extremely efficient pesticide inhibiting the growth of a variety of sucking pests. Additionally, its nanoderivative is photosensitive, which provides environmental advantages over the regular one. Certainly, nanoencapsulation provides the option to adjust pesticide liberation rates in order to retain effective pesticide concentrations for a longer length of time. Furthermore, nanoencapsulation can prevent pesticide protection from deteriorating prematurely. Without a doubt, smart nanosized deliverance methods, such as versatile dispatch systems, allow for additional prudent pesticide use. This is because the quantity of pesticides required for effectual control of pests is significantly reduced, the cost allied with pesticide manufacture and residual managing is reduced, and the possibility of developing resistance also reduces. The consistent release of active substances is a long-term goal of nanoformulations; or else, they will be unbalanced and portable on land. During such circumstances, releasing active components from nanocarriers is a critical strategy for controlling the environmental destiny of nanopesticides (Kookana et al., 2014). Carbon-based nanomaterials offer a variety of features that make them suitable for use in insecticides. Carbon dots, in particular, exhibit self-fluorescence, allowing them to be used for stress detection rather than commercial biochar. Immobilization by plants has recently received a lot of interest as a low-cost and simple strategy for remediating polluted soil (Hamid et al., 2019). By limiting their exchangeable sections, nanohydroxyapatite particles can store the contents of metal in infected soils, lowering their concentration in porous water (Dong et al., 2016; Zhang et al., 2010). Additionally, employing an enhanced oxidation technique and a variety of oxidants, nanoparticles have been used in contaminated soil to mineralize organic contaminants and catalyze their breakdown. Iron nanoparticles and chelating agents were used in conjunction with modified Fenton oxidation to remediate pyrene in polluted soil (Jorfi et al., 2013). The oxidation efficiency varies depending on the type of chelating agent, with sodium pyrophosphate providing the highest efficiency. Biotic stressors, such as herbivore assault and pathogen infection, are critical variables influencing crop protection. Pesticide use, on the other hand, has increased agricultural output and production risk. Considerable attention has been given to silver nanoparticles in the realm of nanopesticides due to their broad spectrum of antibacterial properties. Ocsoy et al. (2013) created DNA-based silver nanoparticles grafted onto graphene oxide, that reduced *Xanthomonas perforans* culturing activity. At a concentration of 16 mg/L, this pathogenic bacterium causes bacterial spots on tomatoes, resulting in a 10%–50% loss in production. Cromwell et al. created Ag nanoparticles under comparable circumstances that were effective against nematodes, a soil-borne organism. About 99% of the tested *Meloidogyne* spp. died after 6 days of exposure to doses of silver nanoparticles of 30–150 mg/L. On days 2 and 4, it was demonstrated that 150 mg/L of Ag nanoparticles decreased the nematode concentration by 82% and 92%, respectively (Cromwell et al., 2014). For other chemical insecticides, using greener methods to produce nanoparticles is ecologically friendly. Numerous metabolites found in bacterial or plant extracts are critical for capping and act as reducing agents in the formation of silver nanoparticles. The same created by Ali et al. (2015) from *Artemisia absinthium* extract, inhibited *Phytophthora* infection and improved plant life. In a different work, Mishra and colleagues created Ag nanoparticles using a biomimetic approach and tested against the wheat disease-causing organism *Bipolaris sorokiniana*, having outstanding antifungal efficacy (Mishra et al., 2014). Despite producing effective outcomes in agricultural contexts, concerns have been raised about the possible toxic effects of Ag nanoparticles in crop plants. Photochemically active metal oxides include ZnO and TiO_2 (zinc and titanium dioxide). TiO_2 has excellent antibacterial capabilities. On exposure to light, electrons get excited and arise due to the presence of O_2, resulting in the creation of superoxide radicals via electron transfer. Paret et al. (2013) studied its photocatalytic and antibacterial activities against *X. perforans*. At a concentration of 500–800 mg/L, ZnO/TiO_2 nanoparticles successfully reduced bacterial spots in a greenhouse setting, compared to controls treated and untreated with copper (Paret et al., 2013). Adisa et al. (2018), did a greenhouse investigation for the inhibition of *Fusarium* tilt in tomato plants and used cerium-based nanoparticles more recently. Seedlings of tomatoes were treated with cerium dioxide nanoparticles at doses of 50 mg/L and 250 mg/L via the foliar and root channels. The results showed that using 250 mg/L of CeO_2 nanoparticles reduced the infectivity by 53% and 57%, respectively (Adisa et al., 2018). Copper nanoparticles are well known for their antifungal and antibacterial activities. Bacterial pathogens such as *E. coli* and *Bacillus subtilis*, as well as fungal pathogens of plants such as *Fusarium oxysporum, P. destructiva, C. lunata,* and *A. alternate* can be treated with $Cu(OH)_2$ nanoparticles. These are important components in the commercial pesticide Kocide 3000 (Yoon et al., 2007; Kanhed et al., 2014). Copper nanoparticles were shown to be insecticidal on both usual and Bt-transgenic cotton, according to (Le Van et al., 2016). The resulting nanoparticles increased the expression of foreign gene encoding the Bt toxin in cotton plant tissues, enhancing resistance. Contrarily, nanoparticles based on silica were used for effective and long-lasting pesticide delivery, and the key is figuring out the right dosage for getting the preferred result. The goal of Prado and colleagues was to develop a nanosized pesticide delivery vector.

Picloram was used, which contains a number of active groups acting as attachment areas for pesticides. Interaction of chloroacetic acid with 3-amnipropyltrimethoxysilane produced novel silylating compounds having carboxyl moieties. They asserted that mesoporous silica nanoparticles produced from carboxyl may restrict herbicide release (Prado et al., 2011). Another study examined how silica nanoparticles, which encourage the up-regulation of phenolic compounds and boost a plant's resistance to *Aspergillus* species, affect the phytopathogen resistance of maize (Zhang et al., 2017). As elevated loading enables a decrease in the number of nanocarriers needed for encapsulating numerous functional components, the loading capacity of nanocarriers is essential to their performance and efficiency (Jiang et al., 2012).

11.1.2 Environmental risks

Despite advancements in plant science, there is still a lack of knowledge about how nanomaterials interact with plants in terms of bioavailability, bioaccumulation, flow, toxicity, and the security of agroecosystems, resulting in difficulty in assessing the advantages and disadvantages of AcI nanotechnology in crops (Sanzari et al., 2019). Nanopesticides are now being used in redeveloping the registered AcI, with the goal of improving performance over nontechnological AcI and offsetting the shortcomings of recent agrochemical stuffs. Perceiving the response of plants to designing nanopesticide revelation and determining toxicology requires extensive research that addresses key obstacles and features that are yet underutilized (Chaud et al., 2021). Nanocarriers are one of the most targeted techniques for overcoming key modern agricultural issues, particularly those connected to boosting output vs. environmental implications and pesticide application rate reductions (Kah & Hofmann, 2014; Kah et al., 2018; Prasad et al., 2017). According to previous research, half of the pesticides used saturate numerous microhabitats surrounding crops, contaminating the rivers, water table, and lakes via either leakage or airflow. Nevertheless, achieving all fruitful agricultural targets with no comprehensive understanding of the obstacles appears to increase the hazards when nano-based particles and AcI are highly damaging to well-being and the environment (Perez-Iglesias et al., 2019).

To effectively combat pests while reducing environmental pollution, bioaccumulation, and risks to ecosystems and human health, it is preferable to release nanopesticides in a targeted and controlled manner depending on the trigger. In recent years, nanotechnology has evolved to address the negative consequences and problems linked to wetting nanoparticles, surfactants, solvents, anionic nanoparticles and particles toxifying and genotoxifying nanoparticles in humans. Nevertheless, when AcI and nanotechnology are coupled, the toxicological risk of accomplishing these aims seems to rise (Mustafa & Hussein, 2020; Yadav et al., 2020). Materials employed in nanoparticle manufacturing can be harmful to plants, humans, and other vertebrate groups (Sehnoor et al., 2018).

Formulations based on lipids such as liposomes, micelles, and liquid crystals are primarily made of naturally occurring cholesterol and phospholipids which are utilized to transport AcI with limited aqua solubility. This necessitates the addition of surfactants, which can contribute to pesticide toxicity (Bang et al., 2011; Mustafa & Hussein, 2020). Because of the complex nature of nanostructures, size, reactivity, electric charge, and shape, it is tricky to analyze, globalize, and anticipate critical features of cyto- and genotoxicity. As a wide-spectrum nonselective organophosphate, glyphosate's toxicity has been a concern for producers, staff, and the safety of the food supply. Aminomethylphosphonic acid, a metabolite of glyphosate, has been related to harmful effects on both terrestrial and aquatic life. Humans have had allergic responses, cardiac and respiratory issues, and endocrine disturbance when exposed to the human placental JEG3-cell line. The addition of surface active agents, albumin, and solvents such as 1,4-dioxane derivates enhances the noxiousness of nanoparticles (Chen et al., 2018; Richard et al., 2005; Souza et al., 2013; Wang et al., 2016a, 2018b). When pesticides like the herbicide clomazone are used, they drift off fields and pose serious toxicity risks to both people and animals. Researchers have examined the effects of this particular herbicide both in an isolated state and when enclosed in polymeric mix nanoparticles created via ionic binding (De Oliveira et al., 2016; Patel et al., 2018). In tadpoles of *L. catesbeianus* subjected to minor exposure, the hepatic effects of nanoparticles and clomazone were investigated to sublethal levels in both forms. The results showed that eosinophil accumulation, lipidosis, and the rate of macrophage aggregation (melano-macrophage centers) increased in experimental groups exposed to clomazone (De Oliveira et al., 2016). In this investigation, the polymeric nanoparticles mix had no effect on tadpole liver protection. Placental cytotoxicity, in general, causes reproductive issues. As an insect pest management technique, dsRNA forwarded to target certain pest genes provides insect defense via plant-mediated interference RNA. As a result, specialized gene selection and dsRNA fragment design have the greatest potential for avoiding interaction with a nonselective marked individual. Conversely, while being vigorously supported, this idea was not tested. Although the RNA endogenous interference pathway's precise sequence is effective at suppressing gene targets and enabling insecticides with longer half-lives, it still has some gaps in terms of its delivery mechanism, environmental stability, toxicity toward nontarget insects, risk of insect resistance following repeated exposure to RNAi plants, and environmental destination (Fletcher et al., 2020; Yu et al., 2016). After overcoming these difficulties, RNAi-based solutions will successfully and safely change agricultural practise management and environmental protection in a short period of time. Atrazine, which is widely utilized in maize and sugarcane fields, is an environmental perseverance pesticide that has a negative influence on soil and aquatic habitats. In hydrographic basins in Brazil, levels nine times higher than permissible (2 g/L) have frequently been recorded (De Paiva et al., 2017; Nogueira et al., 2012). The toxic nature and possible dangers of atrazine to aquatic animals were examined by measuring the sensitiveness of *P. mesopotamicus* fingerlings to an atrazine dosage comparable to the LC50. The liver was found to have the most significant damage, according to the findings. Despite the fact that nuclear anomalies and the amount of micronuclei were stable, the atrazine dose affected the kidney and the liver (De Paiva et al., 2017). The possible toxic effects of atrazine in Asian clam (*C. flumineausing*) were investigated; the findings revealed that this herbicide impacted the process of biotransformation and oxidation and caused damage to DNA, without demonstrating any significant change in antioxidant defences (Dos Santos & Martinez, 2014). The chemical reaction of DNA with purine bases by means of an intercalation mechanism and the production of adducts, is thought to be the source of the genotoxic effect on *C. flumineausing* (Dos Santos & Martinez, 2014; Kushwaha et al., 2011). A member of the imidazolinones class of herbicides, imazethapyr is produced from nicotinic acid. Yet, identified as a hazardous AcI which effects the atmosphere as well as an irritates the eyes and respiratory system. Imazethapyr sensitivity varies between insects and earthworms; annelids and bees are very susceptible, whereas earthworms are more or less resistant. The oxidative DNA damage technique was used to assess the effect of imazethapyr on genotoxicity (Kushwaha et al., 2011). The findings demonstrated that imazethapyr (0.39–1.17 mg/L) caused anomalies in the nucleus of tadpoles after 96 hours of exposure, along with a substantial rise in genetic damage index (GDI) past 48 hours. People provided facts about imazethapyr's capacity to threaten nontarget exposed animal species when disseminated over water or air (Perez-Iglesias et al., 2017, 2019). Though imazethapyr is familiar with its function in controlling weeds and the harm done by it to the environment, no research employing nanostructured systems to overcome this dilemma has been discovered in scholarly literature. Fipronil is a phenylpyrazole insecticide that is used in cultivation, and medicines for humans and animals. Fipronil depolarizes the insect by acting on its aminobutyric acid receptors. Though less poisonous to vertebrates than insects; yet, has aftereffects such as changes in liver fattiness and sexual cycles, DNA impairment, and sperm death. In vitro studies show that fipronil inhibits the growth of preimplantated embryos in rats, increases cell death and decreases the proliferation of cells. Experiments in nature supported the laboratory findings and demonstrated that fipronil is hazardous for reproduction in vertebrates (Sefcikova et al., 2018). Long-term exposure to harmful agrochemicals promotes oil bioaccumulation in liver tissues, resulting in infectious diseases (hepatitis and fibrosis) (Wasef et al., 2021). Reactive oxygen species (ROS) buildup plays a vital function in the liver and is responsible for inflammation, which causes hepatic tissue damage (Mittal et al., 2014).

With nanotechnology developing so quickly, concerns have been raised about how it may affect the food chain and the following buildup of artificial nanomaterials. On the other hand, it is still unclear how hazardous pesticides and nanomaterials are to living things (Shekhar et al., 2021). As a result, we must study these concerns since the potential consequences for people and biota are transformative. Virulency is a critical factor in pesticide efficiency and is connected to how pesticides act on target organisms. Toxicity results from the use of stabilizers, reductants plus organic fluids to avoid the aggregation of unwanted suspensions. The use of nanomaterials is not inherently dangerous; for instance, conventional meals contain several nanoscale components such as milk proteins, carbohydrates, mayonnaise having fat globules, DNA, and so on. Nevertheless, certain nanoscale materials may be harmful to the environment and human health when used in water, agriculture and food (Mushtaq et al., 2020). In order to hasten the application of nanoscale compounds in agriculture, it is necessary to innovate and create procedures for their safe, effective, and financially viable manufacture. In contrast to standard formulations, several investigations have found higher toxicity of pesticides at enormously small doses of these pesticides. Although analyses possibly will not be comparable across a mass of organisms, it was indicated that the dangers caused by nano-based pesticides were substantially elevated, by a degree of figures raised, on comparison to traditional formulas (Pankaj et al., 2012; Saini et al., 2014). Sasson and colleagues (2018) showed that nanopesticides were more hazardous to target species than commercial treatments, emphasizing the need of accurately comparing nanopesticides to actual current conventional counterparts. The variety in toxicity across traditional formulations is not unusual, but it usually remains a mystery, owing to the confidentiality and private nature of the formulations and component procedures. Another research compared the activity of nanopesticides to that of pristine active components in sale products. Such studies are necessary to identify the consequences of nano vs traditional formulations by taking fungicide and pesticide toxicity into account. Since a result, products designed to regulate the release behavior of active components must be studied over an extended period of time, as the time length of the examination has been found to impact their activity (Saini et al., 2018).

The key drivers of nanopesticides' effects on human health and ecology are their behavior and destiny during application in the environment. The active components must be released gradually in nanoformulations in order to prevent them from becoming too unbalanced and portable in the ploughing field. The mechanism that regulates the destiny of nanopesticides in the environment under the aforementioned circumstances is the discharge of functional components from nanocarrier schemes. Another Gordian knot is the ineffectiveness of pesticide application and distribution. It has been shown that 10%–75% of pesticides miss their intended target. The problems linked with the usage of pesticides include their persistence, the emergence of resistance, their being toxic to untargeted creatures, as well as buildup in environmental areas. The lessened environmental strain brought on by agricultural production might result in enormous linked advantages if this technology could improve yield nutrition, fertilization, and irrigation effectiveness. To properly compare the impacts of nanopesticides to those of conventional treatments, it is required to quantify the unbound dissolved ions of latent toxic micronutrients. These days, it is believed that such research is essential for the environmental risk evaluation of nanomaterials. A revolution in farming is urgently needed to fulfill the rising food demand while eliminating its negative effects on the biotic community. It is crucial that vital research hard work is put forth to discover novel and highly competitive goods that may increase the sustainability of agricultural production in the future. Positively, the proper application of nanotechniques may have a significant function in achieving this goal.

11.2 Health impacts

Pesticide usage has a socioeconomic influence on the economy as a whole. They are undoubtedly beneficial to overall crop output; according to FAO data, pesticide consumption is steadily increasing and present consumption is 4.19 million tons per year (Food and Agriculture Organisation of the United Nations). On the other hand, the well-being menace linked to its usage must not be overlooked. A worldwide comparative review of pesticide use through time in connection to its gradual effect on population health might conclude that, across the entire globe, increased pesticide use has led to health declines, with a rise in the percentage of people with cancer or other neoplasms over time. Pesticide usage has clearly had a toll on the health of the people in nations such as China and Russia, according to data. To wit, the proportion of people being effected by onco-diseases or having tumors has been increasing in various places throughout the years. In such cases, the application of apt technology involving nanoparticles might prove beneficial in reducing the negative effects on physical conditions connected with pesticide use. Increasing pesticide use has had a significant detrimental impact on public health in developed nations like the European Union and the United States. The percentage of people with cancer in the former region has been continuously rising since 2005, although it has leveled off ever since 2015. With regard to the latter, since 2003, the fraction of people who have onco-diseases has been steadily rising. However, in a country like Australia, rising pesticide usage is unlikely to have impacted the country's population's health. Pesticide use was highest in a growing economy like largely rural India during the constructional renovative phase of the 1990s and steadily decreased over the millennium decennary beginning since 2000.

11.3 Emerging trends and future perspectives

11.3.1 Nanocarriers for pesticide formulations

The possible applications of nanotechnology for the integrative management of pests include the regulated administration of AIs with increased action at minimum concentrations and the well-organized checking of interactions of pesticide with the surrounding (Khandelwal et al., 2016). Nanocarriers are extremely beneficial for protecting insecticides from hostile environments. Nanocarriers or nanoencapsulation are used to:

(1) shield artificial intelligence (AI) molecules before their release to certain plant tissues; (2) boost the AIs' solubility and penetration of that tissue; and (3) alter the AIs' discharge functions in the atmosphere. Such aspects are mostly determined by the encapsulating material's unique properties (Cooper, 2010; Flores-Cespedes et al., 2015). AI encapsulation increases their method of action while decreasing toxicity (Maruyama et al., 2016). Controlled-release formulations are extremely desired for maximizing pesticide usage efficiency. For instance, a nanoformulation based on responses to stimulus can release AIs to effectively control pests by intelligently reacting to the stimulus generated by the insect and/or the atmosphere. Particularly nanoparticles may reach an elevated degree of solubility and wide coating of plant/leaf surfaces because to their large surface area (Bordes et al., 2009; Kah et al., 2013). Geitner et al. (2017) investigated the adsorptive relationship between nanoparticles and pesticides using discrete molecular dynamics (DMD). Through the binding and transport of pesticides, physiological morphology, and antioxidant systems in plants, this kind of interaction should have a noteworthy influence on adsorption spots. Li and colleagues (2018) found that the common postemergence herbicide Diclofop-methyl and Ag nanoparticles had an antagonistic connection, with the herbicide's presence influencing or lowering the silver ions from the nanoparticles. Such kind of contact encourages the development of *Arabidopsis thaliana*. Furthermore, lowering pesticide doses is critical to overcoming nontargeted organism toxicity and the possibility of contamination in the environment. Pesticide nanoformulations such as (1) nanosuspensions (Cui et al., 2016), (2) nanoemulsions (Du et al., 2016), and (3) nanocapsules have been suggested and created (Cao et al., 2016). The nanocapsule is made up of a case or membranous structure that houses the AIs inside. Given that prolonged AI release is advantageous for plants, nanoencapsulation resources are an efficient pesticide discharge method for restricted discharge. The following methods of controlling the release of AIs may be programmed into these nanomaterials: (1) steady, (2) fast, (3) specific, (4) humidity, (5) temperature, (6) pH, (7) ultrasonic, (8) magnetic, and (9) nuclear (DNA) profile. In addition to determining the release profile of AIs, the encapsulating material's design is crucial for shielding AIs from the environment. The ultraviolet (UV) protective properties of porous hollow silica nanoparticles were described by Li and coworkers (2007) (i.e., avermectin). In order to prolong the effects of pesticidal nanoformulations on their intended pests, it is important that they have remarkable photostability. In addition to these benefits, nanoformulations help herbicides work more effectively. For instance, several cuticular lipids are used by insects to avoid desiccation death succeeded by protection from their aqueous hurdles. However, when applied to the surface of plants, nanoformulations (such as nanosilica) have the capacity to physiosorb into cuticular lipids and cause insect death (Li et al., 2006). Here, we provide a detailed description of the properties of various nanomaterials for the encapsulating AIs, such as polymeric nanomaterials, nanoclays, lipid nanoparticles, metal-organic frameworks (MOFs), and another environmentally friendly nanoformulations.

11.3.1.1 Encapsulation based on polymers

Nanoparticulates are the principal often employed materials for AI encapsulation, owing to their environmental friendliness and biodegradability. These nanoparticles do not produce any potentially hazardous byproducts (Kumar et al., 2017; Ramasamy et al., 2017). For AI encapsulation, a number of artificial and natural polymers have been created, including as polyethylene glycol (PEG), polycaprolactone (PCL), alginate-gelatin, and cellulose (Heydarifard et al., 2018; Kulkarni et al., 2000; Kumar et al., 2018; Rami et al., 2017; Trinell & Shasha, 1988). Various nanopolymeric formulations are known to give a considerable decline in AI use while preserving benign behavior, cheap cost, and biodegradability. Meyer and colleagues (2015) explained how to encapsulate imidacloprid against phytopathogen vectors at around 200 times concentrations less than marketable compounds using serviceable nanodispensers (i.e., PLA). Polymer-based nanoformulations have the potential to improve optimum pesticide application while minimizing negative environmental and human health impacts. The inclusion of solvents (mostly organic) and surface active agents in pesticide formulations are primarily responsible for these effects (Bystrzejewska-Piotrouska et al., 2009). As a result, research efforts have been devoted to increasing pesticide efficiency while limiting the negative impacts of appurtenants (Guo et al., 2015; Li et al., 2016). The kind of polymeric nanoformulation, which includes granule-based, micro-emulsion formulations, and water microcapsule, has a major influence on the drug-release profile. Kumar et al. (2015) used ionic pregelation and polyelectrolyte complexation to create alginate-chitosan nanocapsules. The nanocapsules demonstrated pH-dependent acetamiprid release behavior with minimal adverse effects. Furthermore, the choice of surfactant had a significant impact on the dimensions and drug dispersion of the polymeric nanoformulation, which in turn impacted the drug-ensnarement competence (Edueng et al., 2017; Wang et al., 2017). To improve pesticide effectiveness and assist extend the drug's controlled effects in the environment, a progressive drug-release profile is desirable. Poly (N-isopropylacrylamide) (PNIPAm)-capped polydopamine (PDA) microspheres with a high imidacloprid-loading capacity and exceptional NIR-light and temperature-sensitive discharge profiles were disclosed by Xu and colleagues (2017). At 15°C, there is no medication release; at 25°C, there is 20.5% drug release within 5 hours; and at 40°C, there is 64.3% drug release within 5 hours. This profile of drug release dependent on temperature may be used to identify the optimal surrounding environmental factors that should be present when these nanoformulations are used in the field. After 0.5 hours of NIR light irradiation, the collective discharge of drug was 15.7% versus 5.2% without photoeriod. Medication liberation rates should be slow to maintain the pesticide nanoformulations' long-term toxicity to the targeted insect. Attapulgite NH_4HCO_3 glyphosate/amino silicon oil-poly(vinyl alcohol) (PVA) nanocomposites with temperature-sensitive glyphosate herbicide release were made by Chi and coworkers (2017). In order to minimize pesticide loss during simulated rains and, as a consequence, increase control efficacy on the intended weeds, it is essential to use a temperature-controlled pesticide release profile. It is crucial to choose the right polymeric nanoformulation for a certain pesticide; in other words, not all nanoformulations are appropriate for use in all varieties of pesticide encapsulations (Tong et al., 2017). Solubility and particle size stability are two examples of polymeric nanoformulation factors that impact pesticide composition. These variables can have a significant impact on the presentation of weakly hydro dissouble insecticides that involve cosolvents and/or solvents for nanoformulation (Yu et al., 2018; Zhu et al., 2017). Because of their excellent stability, even in diluted media, nanogels are potential nanocarriers (Wang et al., 2018b; Zan et al., 2014). However, the quantities of cross-linker and nanofiller used had a significant impact on the swelling behavior of the hydrogels (Karimi et al., 2017). Micelles have a crucial role in the regulated release of insecticides and other bioactive compounds. The hydrophobic cores of micelles serve as reservoirs for a variety of bioactive agents, whereas the hydrophilic outer shell aids with steric stability, water solubility, and AI inactivation throughout the delivery process (Perez Quinone et al., 2018; Sarkar & Singh, 2017). The photoresponsive properties of micelles, among the various drug discharge triggers, are particularly significant because they can affect the structural and time-dependent parameters for the controlled release of AIs in response to outer signals (Jiang et al., 2006). Ye and colleagues revealed the light-regulated constant release behavior of diuron-loaded 2-nitrobenzyl succinate-carboxymethyl chitosan (NBS-CMCS) (2015). The focus of the study has been on the selection of certain nanofillers in the bio-originated polymer medium to coalesce cutting-edge properties (Chevillard et al., 2014; Salgueiro et al., 2013). Chevillard and group (2012) created a montmorillonite (MMT) resistant wheat gluten milieu for encapsulating ethofumesate. The nano-based clays might be effective in altering the transfer characteristics of AIs in polymers. Exploration of the interaction of the polymer medium having pesticides and nanofillers can assist in managing the medication release pattern with high efficiency.

11.3.1.2 Encapsulations based on lipid nanomaterial

Lipid nanoparticles have been discovered as suitable colloidal nanocarriers for controlled drug-delivery systems because of their beneficial properties such as strong physiochemical storage stability, nontoxicity, high loading capacity, usability, and target-oriented release profile (Pan et al., 2016; Zheng et al., 2013). The most often employed nanoformulations of lipid nanocarriers are (1) nanostructured (NLC), (2) solid (SLC), and (3) nanoemulsions (Gaber et al., 2017; Singh et al., 2017). Because of their encapsulation stability and food system safety, NLCs are accounted for nutraceutical deliverance (Tamjidi et al., 2013). However, just a few reports of the use of pesticide nanocarriers have been published. Lipoid nanocarriers were produced to prevent pesticide photo decay with no usage of UV absorbers (Nguyen et al., 2012). SLNs have various advantages over other drug discharge means (e.g., polymeric nanoparticles, liposomes, and emulsions), together with the capacity to contain aquaphobic as well as aquaphilic AIs, they decreased the deterioration of AIs by chemicals, and can be feasibly and extensively manufactured (Cacciatore et al., 2016; Kumar et al., 2016; Li et al., 2018; Wei et al., 2017). The physical condition of nanocarriers' lipid matrix influences their incursion into plants through roots and thereby controls their shift to the shoot for targeting the pests entering organisms. Nguyen and workers (2016) used a mix of heat homogenization and sonication procedures to create three distinct lipid formulations using nanocarrier. These were (1) lipid-based nanoemulsions, (2) NLC, and (3) SLN (as solid lipid). The nanoemulsions took just 1 day to reach the roots as opposed to 3 days for NLC and SLN. These techniques were efficient in extending the pest protection period by limiting pesticide loss under a variety of environmental circumstances. However, greater research into the communications between stem cells of plant lipids and nanocarriers is required. Actually, SLN had no phytotoxic effects on *Pisum satium, Z. mays* or *Brassica rapa* throughout germination or seedling development. For polymers like chitosan/tripolyphosphate nanoparticles, such patterns were no longer visible (Nakasato et al., 2017). The SLN loading capacity of AIs is usually determined by numerous parameters, which include the physiochemical configuration of the concrete lipid medium, the dispersible nature of AIs in the liquor lipid, and the polymeric condition of the lipids. The inclusion of *Eucalyptus citriodora* oil inside nanoliposomes demonstrated antibacterial efficacy against *Staphylococcus aureus* (Lin et al., 2015). Concerns about SLNs, such as inadequate payload, drug-release kinetics, and photo deterioration can be resolved by NLCs produced from maize oil and beeswax. After being exposed to UV light for 12 hours, these NLCs significantly increased the photoprotective encapsulation efficacy (83%) and flexibility for modulating AI release, giving deltamethrin 1.79 times better photoprotection efficiency (Ngugen et al., 2012). The diverse nature of their resources, particularly recyclable ones such as raspberry seed and rice bran oil, can explain at least some of the current trends in lipid nanocarriers (Niculae et al., 2014). These nanocarriers have numerous appealing qualities, including low aftereffects, self-antioxidizing capabilities, and the capacity to co-encapsulate and liberate two AIs at the same time, similar to nanocarriers made up of polymers. NLC was known to be manufactured utilizing propolis wax, beeswax, and pomegranate seed oil (Soleimanian et al., 2018). By a mean size of 71–350 nm, these nanoparticle NLCs demonstrated outstanding physical stability and minimal polydispersity. The nanocarriers (liquid as well as solid lipid) resist the dispersion of environment-damaging chemicals and it is critical to the stability of AIs. Some simulation studies validated the impacts of numerous parameters on the dissemination of degrading materials into nanopesticide formulations, including nanoparticle size, domain size, and solid domain percentage (Nelson et al., 2017). The communication between lipid nanocarriers and chemical pesticides, on the other hand, needs further toxicological investigations to appropriately characterize the hazardous summary of such nanopesticide formulations.

11.3.1.3 Clay nanomaterial-derived encapsulation

Nanoformulations derived from clay were designed for increasing neutral and hydrophobic AI adsorption and gradual release. Nanoclays are regarded to be economically viable and biocompatible materials with significant promise in this industry due to their prospective function as nanocarriers in materialistic science and agricultural employment. Clay nanoparticles have been modified to give them with useful qualities (like water-loving nature increased surface area, charge density, and porosity) (Ianchis et al., 2017; Zhang et al., 2014).

Clay was modified by using positively charged organic particles to improve its compatibility for the adsorption of aqua-phobic AIs like acetochlor, alachlor, metolachlor, and metolachlor nor flurazon (de Sousa Rodrigues et al., 2013). Cao et al. (2017) used a postgrafting approach to create positively charged mesoporous silica nanoparticles to load anionic pesticides such as 2,4-D sodium salt. In comparison to pristine conformations (1.5%), the ionic character of the silica-nanoparticles provided a considerable enhancement in pesticide loading (21.7%). Furthermore, to provide regulated anionic AI release, polymerized clay nanocomposites were produced (Lvov et al., 2016). Cabrera and colleagues (2016) described the usage of biopolymer chitosan to modify a naturally occurring smectite for encapsulating imazamox. The mentioned nanocomplexes released this pesticide into the water at a significantly slower rate (from 27% to 75%) than commercial imazamox (86%). These nanoformulations also decreased overall soil-leaching losses by 15%. Photo-thermal instability can have a negative impact on achieving optimum biological activity at the target location. Fan et al. (2017) used specific disulfide bonds to create pectin-crosslinked silica microcapsules. These silica microcapsules increased photo-thermal stability and carrying efficacy (20% w/w) for kasugamycin. According to the Riger–Peppas concept, the above-described microcapsules provided a mechanism for releasing drugs that involved pectinase and glutathione (Peppas & Sahlin, 1989). The enzymatic-sensitive mesoporous silica nanoparticles demonstrated good chlorantraniliprole loading ability (42% w/w) (Kaziem et al., 2017). These nanoformulations demonstrated outstanding AI resistance with exceptional *Plutella xylostella* larval mortality even after 14 days. Several characteristics, such as interlayer spacing and the occurrence of H_2O molecules between exchangeable cations and interlayers, might influence AI intercalation inside clay (Turker et al., 2017). Since clayey nanocarriers have appealing qualities, thus, they are also helpful for the regulated delivery of bioactives in addition to chemical molecules (Rani et al., 2014). Recent research has concentrated on the production of sustainable and low-cost silica nanoparticles that rely heavily on flora (Mattos et al., 2016; Soltani et al., 2015). On a side note, the relationships of silica nanocomposites and extracts of plants) is still being researched. Mattos and colleagues (2017) used biological silica nanostructures for proficient and regulated neem oil release. In comparison to natural neem bark extracts, strong bipolar bonding of neem extracts and silica provided better water constancy, strong heat stability, high antioxidant capacity, gradual discharge profile of drugs, leading to 100% death of *Acromyrmex crassispinus*. However, the silica nanoparticles' uncontrolled release profile, which includes release rate and duration, made it difficult to use them in agricultural applications. To obtain controlled and intelligent AI release patterns, great effort has been made to develop stimuli-responsive silica microcapsules (Chen et al., 2017a; Huayao et al., 2016). Enzymes are particularly appealing among stimuli because to their precision, selectivity, and efficiency for a wide range of compounds (Qin et al., 2016). Liang et al. (2017) used isocyanate-operative silica articulated micro-capsules to provide urease-responsive pendimethalin delivery. The herbicidal action of these microcapsules against weeds, *Amaranthus retroflexus* L. (RPW) and *Echinochloa cruss-galli* L., lasted longer (BYG). To create a prolonged drug-release profile, the outer sheath of pendimethalin capsules proved useful in preventing the first burst release of pharmaceuticals. Regardless of these advances in nanocarrier uses, the occurrence of clear imperfections in clay nanoparticles remains a key problem. To optimize their application in pesticide delivery, proficient blending processes and appropriate facade moderation techniques must be developed.

11.3.1.4 Encapsulation based on a metal-organic framework

Metal-organic frameworks have received a much interest in recent years due to their unusual architecture-property associations. Because of their excellent drug adsorption in the pore architectures, these are described as competent nanobased carriers for aimed drug deliverance (Orellana-Tavra et al., 2015). They can be utilized to address various drawbacks of traditional medicinal drugs, which can induce cardiac toxicity and injury to a normal and strong group of cells of the same functionality (He et al., 2015; Pinto et al., 2017). The MOFs have a extensive outer sheath areas, lofty pore volumes, configurable pore dimensions, effectual facede chemistry, excellent thermic steadiness, and various topologies (Chen et al., 2017b; Nehra et al., 2019; Vellingiri et al., 2017). Other porous matrices besides MOFs include adsorption resin (Zhang et al., 2016), activated clay (Alromeed et al., 2015), activated alumina (Jeon et al., 2011), and activated carbon (Garrido-Herreae et al., 2016). AIs' effective lives have been extended by these formulations. However, the nonbiodegradability of these materials remains a serious concern that must be addressed (Glaser, 2015). MOFs (particularly green MOFs) have been proposed as an alternate approach for encapsulating AI to reduce its explosive nature. These Ca- and Fe-based green MOFs break down quickly in H_2O to their constituents (calcium and iron), which act as supplements and minimize soil contamination (Yang et al., 2017). Metal clusters or ions that serve as coordination centers are the building blocks of MOFs, which are joined by organic ligands. For delivering dynamic species, MOFs can offer a variety of active sites with various strengths. These have high adsorption capacity, good ion exchange capabilities (Rozek and Dincă, 2014), and outstanding electronic characteristics (Xia et al., 2015). To guarantee the flourishing encapsulation of AIs in a permeable medium and to minimize AI dripping via apertures, encapsulated species should be bigger than the holes of the matrix (Juan-Alcaniz et al., 2012). To support the requirements of encapsulation, two synthetic techniques must be used: (1) putting AIs together inside the MOFs' pores and (2) putting the hollow MOFs together around the AIs. Both encapsulating techniques are dependent on the sort of AIs to be encapsulated as well as the chemistry of the support. The usage of MOFs for green applications may be investigated further with reference to the employment of biodegradable metal ions (Xie et al., 2017) or the liberation of natural modules to address the matter of virulence (Liu et al., 2017). Yang et al. (2017) observed acceptable drug-release kinetics in porous structures (430 and 160 m^2/g) loaded with the fumigant cis-1,3-dichloropropene dependent upon calcium ions and lactic. The release kinetics of these sorption-based formulations were enhanced, with MOF-1201 releasing 100 times more slowly than MOF-1203. However, further research is required to assess the drug-loading capability, lethality to untargeted plants, and residual content in the end product of MOF-based encapsulation.

11.3.1.5 Greener encapsulations

Nanomaterials are one of several novel environmental stressors that are being applied to soils (e.g., TiO_2) (Conway et al., 2015). These new contaminants are fast becoming more prevalent in our ecological unit and farming lands. Simonin et al. (2016) investigated the reaction of the microbial population to TiO_2 nanoparticles in agricultural soil over 90 days. The advent of nanoparticles had a significant detrimental influence on both ammonia-oxidizing microorganism populations and nitrification enzyme activity. The harmful impact of nanoparticles necessitates more study on how to use these pollutants to improve the health of the soil and the functioning of the ecosystem. Because of their biocompatibility and biodegradability, harmless materials must be investigated to avoid these concerns (Chevillard et al., 2012; Narayanam et al., 2015; Anitha et al., 2014). Numerous research efforts have been conducted in this regard to construct starch-based nanocontainers to aid in the transport of crop protection compounds, supplements, and stimulants into the tissues of plants (Wilpiszewska et al., 2016). The potential applicability of starch-based nanocontainers to guard flora from dangerous disorder has also been identified (El-Hamshary et al., 2015; Shukla et al., 1991). Promoting more environmentally friendly manufacture of these nano-based formulations is essential. For instance, the production of starch nanoparticles by hydrolyzing starch granules using acid has several drawbacks, such as less output, extensive extracting time, and ecologically hazardous leftovers (e.g., toxic wastes) (Oum & Rhim, 2016). To address these difficulties, attempts should be devoted toward the growth of environmentally friendly techniques (Chen et al., 2014; Mauricio et al., 2015). Effective AI trapping can be achieved by surface functionalizing as-produced nanoparticles to change their physicochemical and biological properties chemically (Masina et al., 2017). Microbes (like cyanobacteria) were reported as nanocarriers in addition to biodegradable polymers due to their large quantity, compatibility with the environment and microalgal surface with a variety of working groups that can hold loading of pesticides (Prasanna et al., 2013; Giessen & Silver, 2016). Numerous algae and cyanobacteria species have been found to be capable of adsorbing and removing heavy metal ions and/or organic molecules (Anastopoulous & Kyzas, 2015; Markou et al., 2015). These microorganisms are environmentally safe and may be employed to effectively fix atmospheric nitrogen, particularly in *Oryza sativa* fields (Srivastava et al., 2017). Cyanobacteria have also been discovered to have an impact on the physical, chemical, and biological characteristics of soil as well as the intersection of soil and water (Schulz et al., 2016). Injecting cyanobacteria in soil was reported to increase soil solidity owing to lipid and polysaccharide excretion (Delattra et al., 2016). However, the positive benefits of these microbes have not been explored as thoroughly as herbicides for weed suppression. Yan et al. (2013) showed that soil seeded with blue-green algae as nanocarriers for the administration of avermectin in a regulated manner has stronger photostability than nonbound avermectin. Isopropanol revealed avermectin's high binding affinity for cyanobacteria. The use of blue–green algae extended the release discharge of avermectin in a solution of ethanol and water (4:1; v/v) combination. It may be additionally regulated by covering avermectin-loaded cyanobacteria with carbopol resin. Avermectin must travel a longer distance to disperse into solvents because a thick layer of carbopol resin offers more hydrogen binding sites for avermectin. The total procedure provides gradual release rates with minimal pesticide release amounts. Substitute to nanomanufactured particles; there are numerous naturally occurring viral nanomaterials based on plants that could be used for AI packing and distribution (Czapar & Steinmetz, 2017). It has also been observed that tobacco mild green mosaic virus (TMGMV) can transport plant parasitic nematodes (Chariou & Steinmetz, 2017). These flora nanoparticles are cost-effective, resistant to extreme ecological circumstances, proficient, have high soil mobility, and are eco-friendly. Tarkowski and Vereecke (2014) evaluated the use of plant pathogenic bacteria to increase plant productivity and yield (e.g., *Rhodococcus fascians, Agrobacterium tumefaciens, A. rhizogenes, Pseudomonas syringae*, and scab-inducing *Streptomyces*). As a result of the aforementioned benefits, greener nanoformulations for pesticide application will present substantial prospects in the near future.

11.4 Agricultural applications of stimuli responsive systems

Creation of smart delivery systems which are responsive to stimuli adds to the objective of attaining sustainable agriculture practises. Stimulus-sensitive nanoformulations can more precisely liberate their dynamic components while responding to surrounding stimuli and organic needs via aimed discharge or regulated liberation mechanisms (Camara et al., 2019). Crop growth and production are hampered by a variety of factors which include

(1) biotic facets that causes stresses due to pests, nematodes, pathogenic diseases, and weed competitors and (2) abiotic facets like intense temperatures, infertile soils, nutrient unavailability or meagre uptake, metal contaminated soils, drought, flooding, and high salinity (Lowery et al., 2019; Majeed et al., 2018). As a result, stimuli-responsive materials for the discharge of agrochemicals in response to pH, redox, temperature conditions, light, enzymes and have been investigated. Stimuli-responsive nanocarrier instances used to control insects and nematodes pH-sensitive and pH-responsive systems, can be created for the management of key farming pests because of their wide array of acidic and basic pH levels that allow the discharge mechanism to be activated after intaking nanocarriers (Kaziem et al., 2018; Wang et al., 2018c). For the herbicide cypermethrin, Patel et al. (2018) produced an alginate nanocarrier system. The high degree of ionization of the alginate polymer at acidic pH caused greater interaction with calcium ions, which improved cross-linking of the polymer mesh and increased the internal hydrophobicity of the nanoparticles, restricting pesticide release. As the pH climbed, the ionic connections weakened, enabling water to enter the system and, as a result, the spilling of the active chemical. Other pesticide methods responsive to pH have been developed. For the delivery of the insecticide abamectin, Gao et al. (2019) developed an arrangement of silica nanoparticles layered and functionalized with polystyrene and (trimethoxysilyl) propyl methacrylate. The findings showed that at pHs 5 and 7, there was a 14%–15% discharge after 15 days, indicating that the system prevented the pesticide from being released too early. However, at pH 10, the insecticide's release reached 39% after 5 hours and 87% after 15 days.

Insecticide acetamiprid was added to an alginate and chitosan system that Kumar et al. (2015) created. At pH 10, 50% of the pesticide was released after 24 hours, while at pH 7 and 4, the same amount was released after 36 hours. In the same test, the researchers examined the insecticide's release in alkaline soils (pH 8.3) and discovered that the release profile was identical to that in a buffer solution with a pH of 10. Likewise, technologies responsive to pH are capable to be utilized to combat other agricultural pests such as worms. For the transport of the nematicide avermectin, Liang et al. (2018a) developed a nanoparticle system made of polyglutamic acid and chitosan. The system showed a 20% initial liberation regardless of pH, then a 69.5% release at pH 8.5, a 60.4% release at pH 7, and a 57.5% release at pH 5.5. The employed nanocarrier system showed high stability at pH 5.5 due to the ionic interactions between chitosan and the carboxyl groups of polyglutamic acid. With the rise in pH, this interconnection decreased, enhancing the nematicide discharge. pH fluctuations occur in agricultural systems in topsoil, different organs of plants, during metabolic processes of the plants plus during fruit maturity, as well as owing to the occurrence of diseases or farming pests (Choudhary et al., 2017; Gao et al., 2019). Because the release kinetics is controlled by pH variations, nanocarrier systems capable of releasing active chemicals in reaction to pH changes constitute a key method for the expansion of novel nanopesticides.

11.5 Temperature-sensitive systems

Zhang et al. (2019) used nanocarriers responsive to temperature made of nanomycetes to create a release mechanism for the pesticide pyrethrin. The release was tested at three different temperatures (13°C, 18°C, and 26°C), which simulated the warmth of the growth stages in mosquito in H_2O. Twelve percent of the pyrethrin was released in the system during the first 16 hours at low temperatures, with no additional changes after that. The emission increased with time at increasing temperatures, reaching 31.9% and 49.7% respectively at 18°C and 26°C. Liang et al. (2018b) Silica nanoparticles coated with chitosan (temperature-responsive) increased the temperature resulting in higher avermectin release. The release was 18.82% and 34.21% at 25°C and 50°C, which demonstrates that chitosan coating was a fine choice for changing the release profile in accordance with temperature. The enhanced diffusion of the insecticide through the chitosan was induced by alterations in the thermal behavior of the molecules at higher temperatures. Enzyme-responsive polymers have been investigated for pesticide release in the existence of enzymes present in the abdomen of pests and larvae. Guo et al. (2015) used silica cross-linked with epichlorohydrin-modified carboxymethylcellulose to create microcapsules containing emamectin benzoate. Cellulase was used to break the cellulose wall into tiny particles, allowing the functional substance to be released from the microcapsules. After 1 hour of cellulase exposure, almost 30% of the emamectin was released, with around 80% liberated after 30 hours. After 30 hours in the absence of enzymes, less than 20% was liberated. When compared to emamectin benzoate alone, the usage of silica-epichlorohydrincarboxymethylcellulose microcapsules lowered the active agent's genotoxicity and increased its insecticidal effectiveness against *Myzus persicae.* Kaziem et al. (2018) investigated the use of an amylase-responsive carrier in a high-tech delivery system that reacted to internal insect stimuli and started the release of a biocide. The developed system was built on cyclodextrin, which was attached to hollow, avermectin-loaded mesoporous silica. The nano-based particles responded strongly to the presence of amylase, releasing 41.64% of the avermectin until day 17 in comparison to just 5.88% release when amylase is absent for equal time period. Even after 14 days, larvae mortality remained above 80%, whereas the efficacy of a saleable formulation of avermectin dropped to 43.4%. This nanoformulation persists 44 days post-treatment, resulting in a 6.63% mortality of larvae. Gao et al. (2017) and Xu et al. (2017) studied coumarin as a light inducer for pesticide delivery (2018). Coumarin has the capacity to increase the stability of molecules, as well as pesticidal action, strong fluorescence, quick rates of release, a broad range of absorption wavelength and fine ecocompatibility (Atta et al., 2010; Gao et al., 2017). Foremost, Gao et al. (2017), described a photo-caged pesticide made of fpronil and coumarin; bonded by a covalent bond, this pesticide produced an off/on set-up with light-dependent pesticide discharge. On being exposed to blue wavelengths (LC50 = 0.56 mol/L) and sunshine (LC50 = 0.37 mol/L), the set-up was capable to offer effective management of *Aedes larvae*. Xu and workers (2018) created a caged pesticide with covalent bonding utilizing coumarin and spirotetramat enol. The release elicited by light was tested by subjecting the device to light of 420 nm wavelength and sunshine. This demonstrated effective spirotetramat enol release utilizing both irradiation sources. The caged pesticide had little insecticidal efficacy (LC50 = 1 mmol/L) against *Aphis craccivora* in the dark, but blue radiations dramatically boosted death rates (LC50 = 0.11 mol/L).

11.6 Conclusion and future perspectives

For creating a dynamic equilibrium between agricultural output and ecological sustainability, nanotechnology has emerged as a cutting-edge framework. Regardless of the type, nanopesticide maximizes pesticide efficiency by enhancing AI stability, prolonging its effective lifetime, and reducing pesticide cost to the environment. Nanopesticides provide target-specific pest control, minimize the harmful effects of chemical-based pesticides, and aids in the development of cognitive nanosystems for mitigating issues of environmental instability, global food and nutrition security, and thus ensuring sustainable agriculture. They are efficient for long-term usage and provide an alternative to environmental issues such as unregulated nutrient-rich aquatic ecosystems and bioaccumulation and biomagnification of nonbiodegradable and toxic chemical pesticide residues in any food chain as a consequence of the controlled release of biologically active chemicals.

However, due to unprecedented and unregulated usage of these nanotechnology, aided pesticides have emerged in the form of new environmental toxins. The first line of defense to ensure environmental safety and human health prior to widespread application is a reliable and in-depth risk assessment of nanopesticides. Their reactivity, retention time, bioaccumulation index, biodegradation potential, toxicity index, and leaching and drifting potential need to be addressed before their commercial usage. Strategies and innovations in terms of novel technologies to minimize the cost of generating nanopesticides toxins, ecotoxicological monitoring and assessment of nanopesticides, and legislative and regulatory interventions for safer nanopesticide application in agriculture and other linked domains need to be undertaken.

Utilizing green chemistry and eco-friendly practices can be formulated which can improve the efficacy of nanopesticide. The smart nanopesticide formulations will enable agrochemical firms to address a wide range of issues, including the solubility of the operational components, stability, constrained liberation, and distribution of active components in the ecosystem and their complexes to remodulate agriculture while ensuring sustainable ecosystem. Further, policies and interventions regarding the use of nanomaterials in agriculture need to be strengthened to mitigate the cost associated with nanopesticides.

References

Adisa, I. O., Reddy, P., Rawat, S., Hernandez-Viezcas, J. A., Dimkpa, C. O., ... Gardea-Torresdey, J. L. (2018). Role of cerium compounds in Fusarium wilt suppression and growth enhancement in tomato (Solanum lycopersicum). *Journal of Agricultural and Food Chemistry, 66*(24), 5959–5970.

Ajmal, Z., Muhmood, A., Usman, M., Kizito, S., Lu, J., Dong, R., & Wu, S. (2018). Phosphate removal from aqueous solution using iron oxides: adsorption, desorption and regeneration characteristics. *Journal of Colloid and Interface Science, 528*, 145–155.

Ali, M., Kim, B., Belfield, K. D., Norman, D., Brennan, M., & Ali, G. S. (2015). Inhibition of Phytophthora parasitica and P. capsici by silver nanoparticles synthesized using aqueous extract of Artemisia absinthium. *Phytopathology, 105*(9), 1183–1190.

Alromeed, A. A., Scrano, L., Bufo, S. A., & Undabeytia, T. (2015). Slow-release formulations of the herbicide MCPA by using clay–protein composites. *Pest Management Science, 71*(9), 1303–1310.

Anastopoulos, I., & Kyzas, G. Z. (2015). Progress in batch biosorption of heavy metals onto algae. *Journal of Molecular Liquids, 209*, 77–86.

Androutsopoulos, V. P., Hernandez, A. F., Liesivuori, J., & Tsatsakis, A. M. (2013). A mechanistic overview of health associated effects of low levels of organochlorine and organophosphorous pesticides. *Toxicology, 307*, 89–94.

Anitha, A., Sowmya, S., Kumar, P. S., Deepthi, S., Chennazhi, K. P., Ehrlich, H., ... Jayakumar, R. (2014). Chitin and chitosan in selected biomedical applications. *Progress in Polymer Science, 39*(9), 1644–1667.

Annamalai, J., & Namasivayam, V. (2015). Endocrine disrupting chemicals in the atmosphere: Their effects on humans and wildlife. *Environment International, 76*, 78–97.

Asli, S., & Neumann, P. M. (2009). Colloidal suspensions of clay or titanium dioxide nanoparticles can inhibit leaf growth and transpiration via physical effects on root water transport. *Plant, Cell & Environment, 32*(5), 577–584.

Atta, S., Jana, A., Ananthakirshnan, R., & Narayana Dhuleep, P. S. (2010). Fluorescent caged compounds of 2, 4-dichlorophenoxyacetic acid (2, 4-D): Photorelease technology for controlled release of 2, 4-D. *Journal of Agricultural and Food Chemistry, 58*(22), 11844–11851.

Bahlai, C. A, Xue, Y., McCreary, C. M., Schaafsma, A. W., & Hallett, R. H. (2010). Choosing organic pesticides over synthetic pesticides may not effectively mitigate environmental risk in soybeans. *PLoS ONE, 5*(6), e11250.

Balaure, P. C., Gudovan, D., & Gudovan, I. (2017). Nanopesticides: A new paradigm in crop protection. In *New pesticides and soil sensors* (pp. 129–192). Academic Press.

Bang, S. H., Hwang, I. C., Yu, Y. M., Kwon, H. R., Kim, D. H., & Park, H. J. (2011). Influence of chitosan coating on the liposomal surface on physicochemical properties and the release profile of nanocarrier systems. *Journal of Microencapsulation, 28*(7), 595–604.

Barabadi, H., Tajani, B., Moradi, M., Damavandi Kamali, K., Meena, R., Honary, S., ... Saravanan, M. (2019). Penicillium family as emerging nanofactory for biosynthesis of green nanomaterials: a journey into the world of microorganisms. *Journal of Cluster Science, 30*, 843–856.

Beggel, S., Werner, I., Connon, R. E., & Geist, J. P. (2010). Sublethal toxicity of commercial insecticide formulations and their active ingredients to larval fathead minnow (*Pimephales promelas*). *Science of the Total Environment, 408*(16), 3169–3175.

Bhandari, R. K., Deem, S. L., Holliday, D. K., Jandegian, C. M., Kassotis, C. D., Nagel, S. C., ..., & Rosenfeld, C. S. (2015). Effects of the environmental estrogenic contaminants bisphenol A and 17α-ethinyl estradiol on sexual development and adult behaviors in aquatic wildlife species. *General and Comparative Endocrinology, 214*, 195–219.

Bhattacharyya, A., Duraisamy, P., Govindarajan, M., Buhroo, A. A., & Prasad, R. (2016). Nano-biofungicides: Emerging trend in insect pest control. In *Advances and applications through fungal nanobiotechnology* (pp. 307–319). Springer, Cham.

Birnbaum, L. S. (2013). When environmental chemicals act like uncontrolled medicine. *Trends in Endocrinology & Metabolism, 24*(7), 321–323.

Biswal, S. K., Nayak, A. K., Parida, U. K., & Nayak, P. L. (2012). Applications of nanotechnology in agriculture and food sciences. *International Journal of Science Innovation and Discovery, 2*(1), 21–36.

Bommarco, R., Kleijn, D., & Potts, S. G. (2013). Ecological intensification: harnessing ecosystem services for food security. *Trends in ecology & evolution, 28*(4), 230–238.

Bordes, P., Pollet, E., & Avérous, L. (2009). Nano-biocomposites: Biodegradable polyester/nanoclay systems. *Progress in Polymer Science, 34*(2), 125–155.

Brogan, W. R., III, & Relyea, R. A. (2014). A new mechanism of macrophyte mitigation: How submerged plants reduce malathion's acute toxicity to aquatic animals. *Chemosphere, 108*, 405–410.

Brogan, W. R., III, & Relyea, R. A. (2017). Multiple mitigation mechanisms: Effects of submerged plants on the toxicity of nine insecticides to aquatic animals. *Environmental Pollution, 220*, 688–695.

Brouwer, M., Kromhout, H., Vermeulen, R., Duyzer, J., Kramer, H., Hazeu, G., …, & Huss, A. (2018). Assessment of residential environmental exposure to pesticides from agricultural fields in the Netherlands. *Journal of Exposure Science & Environmental Epidemiology, 28*(2), 173–181.

Bystrzejewska-Piotrowska, G., Golimowski, J., & Urban, P. L. (2009). Nanoparticles: Their potential toxicity, waste and environmental management. *Waste Management, 29*(9), 2587–2595.

Cabrera, A., Celis, R., & Hermosín, M. C. (2016). Imazamox–clay complexes with chitosan-and iron (III)-modified smectites and their use in nanoformulations. *Pest Management Science, 72*(7), 1285–1294.

Cacciatore, I., Ciulla, M., Fornasari, E., Marinelli, L., & Di Stefano, A. (2016). Solid lipid nanoparticles as a drug delivery system for the treatment of neurodegenerative diseases. *Expert Opinion on Drug Delivery, 13*(8), 1121–1131.

Camara, M. C., Campos, E. V. R., Monteiro, R. A., do Espirito Santo Pereira, A., de Freitas Proença, & Fraceto, L. F. (2019). Development of stimuli-responsive nano-based pesticides: emerging opportunities for agriculture. *Journal of Nanobiotechnology, 17*(1), 1–19.

Cao, L., Zhang, H., Cao, C., Zhang, J., Li, F., & Huang, Q. (2016). Quaternized chitosan-capped mesoporous silica nanoparticles as nanocarriers for controlled pesticide release. *Nanomaterials, 6*(7), 126.

Cao, L., Zhou, Z., Niu, S., Cao, C., Li, X., Shan, Y., & Huang, Q. (2017). Positive-charge functionalized mesoporous silica nanoparticles as nanocarriers for controlled 2, 4-dichlorophenoxy acetic acid sodium salt release. *Journal of Agricultural and Food Chemistry, 66*(26), 6594–6603.

Chariou, P. L., & Steinmetz, N. F. (2017). Delivery of pesticides to plant parasitic nematodes using tobacco mild green mosaic virus as a nanocarrier. *ACS Nano, 11*(5), 4719–4730.

Chaud, M., Souto, E. B., Zielinska, A., Severino, P., Batain, F., Oliveira-Junior, J., & Alves, T. (2021). Nanopesticides in agriculture: Benefits and challenges in agricultural productivity, toxicological risks to human health and environment. *Toxics, 9*(6), 131.

Chaudhry, N., Dwivedi, S., Chaudhry, V., Singh, A., Saquib, Q., Azam, A., & Musarrat, J. (2018). Bio-inspired nanomaterials in agriculture and food: Current status, foreseen applications and challenges. *Microbial Pathogenesis, 123*, 196–200.

Chen, K. L., & Bothun, G. D. (2014). Nanoparticles meet cell membranes: Probing nonspecific interactions using model membranes. *Environmental Science & Technology* 2014, *48*(2), 873–880.

Chen, Y., Hong, S., Fu, C. W., Hoang, T., Li, X., Valencia, V, & Ma, S. (2017b). Investigation of the mesoporous metal–organic framework as a new platform to study the transport phenomena of biomolecules. *ACS Applied Materials & Interfaces, 9*(12), 10874–10881.

Chen, H. M., Huang, Q., Fu, X., & Luo, F. X. (2014). Ultrasonic effect on the octenyl succinate starch synthesis and substitution patterns in starch granules. *Food Hydrocolloids, 35*, 636–643.

Chen, H., & Yada, R. (2011). Nanotechnologies in agriculture: new tools for sustainable development. *Trends in Food Science & Technology, 22*(11), 585–594.

Chen, K., Yu, G., He, F., Zhou, Q., Xiao, D., Li, J., & Feng, Y. (2017a). A pH-responsive emulsion stabilized by alginate-grafted anisotropic silica and its application in the controlled release of λ-cyhalothrin. *Carbohydrate Polymers, 176*, 203–213.

Chen, C., Zhang, G., Dai, Z., Xiang, Y., Liu, B., Bian, P., …, & Cai, D. (2018). Fabrication of light-responsively controlled-release herbicide using a nanocomposite. *Chemical Engineering Journal, 349*, 101–110.

Chevillard, A., Angellier-Coussy, H., Guillard, V., Bertrand, C., Gontard, N., & Gastaldi, E. (2014). Biodegradable herbicide delivery systems with slow diffusion in soil and UV protection properties. *Pest Management Science, 70*(11), 1697–1705.

Chevillard, A., Angellier-Coussy, H., Guillard, V., Gontard, N., & Gastaldi, E. (2012). Investigating the biodegradation pattern of an ecofriendly pesticide delivery system based on wheat gluten and organically modified montmorillonites. *Polymer Degradation and Stability, 97*(10), 2060–2068.

Chevillard, A., Angellier-Coussy, H., Guillard, V., Gontard, N., & Gastaldi, E. (2012). Controlling pesticide release via structuring agropolymer and nanoclays based materials. *Journal of Hazardous Materials, 205*, 32–39.

Chi, Y., Zhang, G., Xiang, Y., Cai, D., & Wu, Z. (2017). Fabrication of a temperature-controlledrelease herbicide using a nanocomposite. *ACS Sustainable Chemistry & Engineering, 5*, 4969–4975.

Choudhary, R. C., Kumaraswamy, R. V., Kumari, S., Sharma, S. S., Pal, A., Raliya, R., …, & Saharan, V. (2017). Cu-chitosan nanoparticle boost defense responses and plant growth in maize (*Zea mays* L.). *Scientific Reports, 7*(1), 9754.

Colborn, T. (2002). Clues from wildlife to create an assay for thyroid system disruption. *Environmental Health Perspectives, 110*(suppl. 3), 363–367.

Conway, J. R., Beaulieu, A. L., Beaulieu, N. L., Mazer, S. J., & Keller, A. A. (2015). Environmental stresses increase photosynthetic disruption by metal oxide nanomaterials in a soil-grown plant. *ACS Nano, 9*(12), 11737–11749.

Cooper, E. R. (2010). Nanoparticles: A personal experience for formulating poorly water soluble drugs. *Journal of Controlled Release, 141*(3), 300–302.

Cooper, J., & Dobson, H. (2007). The benefits of pesticides to mankind and the environment. *Crop Protection, 26*(9), 1337–1348.

Cromwell, W. A., Yang, J., Starr, J. L., & Jo, Y. K. (2014). Nematicidal effects of silver nanoparticles on root-knot nematode in bermudagrass. *Journal of Nematology, 46*(3), 261.

Cui, B., Feng, L., Wang, C., Yang, D., Yu, M., Zeng, Z., …, & Cui, H. (2016). Stability and biological activity evaluation of chlorantraniliprole solid nanodispersions prepared by high pressure homogenization. *PLoS ONE, 11*(8), e0160877.

Cutulle, M. A., Armel, G. R., Kopsell, D. A., Wilson, H. P., Brosnan, J. T., Vargas, J. J., …, & Koepke-Hill, R. M. (2018). Several pesticides influence the nutritional content of sweet corn. *Journal of Agricultural and Food Chemistry, 66*(12), 3086–3092.

Czapar, A. E., & Steinmetz, N. F. (2017). Plant viruses and bacteriophages for drug delivery in medicine and biotechnology. *Current Opinion in Chemical Biology, 38*, 108–116.

Dang, V. D., Kroll, K. J., Supowit, S. D., Halden, R. U., & Denslow, N. D. (2016). Tissue distribution of organochlorine pesticides in largemouth bass (*Micropterus salmoides*) from laboratory exposure and a contaminated lake. *Environmental Pollution, 216*, 877–883.

de Oliveira, C. R., Fraceto, L. F., Rizzi, G. M., Salla, R. F., Abdalla, F. C., Costa, M. J., & Silva-Zacarin, E. C. M. (2016). Hepatic effects of the clomazone herbicide in both its free form and associated with chitosan-alginate nanoparticles in bullfrog tadpoles. *Chemosphere, 149*, 304–313.

de Paiva, P. P., Delcorso, M. C., Matheus, V. A., de Queiroz, S. C. D. N., Collares-Buzato, C. B., & Arana, S. (2017). Acute toxicity of commercial atrazine in *Piaractus mesopotamicus*: Histopathological, ultrastructural, molecular, and genotoxic evaluation. *Veterinary World, 10*(9), 1008.

de Sousa Rodrigues, L. A., Figueiras, A., Veiga, F., de Freitas, R. M., Nunes, L. C. C., da Silva Filho, E. C., & da Silva Leite, C. M. (2013). The systems containing clays and clay minerals from modified drug release: A review. *Colloids and Surfaces B: Biointerfaces, 103*, 642–651.

Delattre, C., Pierre, G., Laroche, C., & Michaud, P. (2016). Production, extraction and characterization of microalgal and cyanobacterial exopolysaccharides. *Biotechnology Advances, 34*(7), 1159–1179.

DeRosa, M. C., Monreal, C., Schnitzer, M., Walsh, R., & Sultan, Y. (2010). Nanotechnology in fertilizers. *Nature Nanotechnology, 5*(2), 91.

Di Guardo, A., Gouin, T., MacLeod, M., & Scheringer, M. (2018). Environmental fate and exposure models: Advances and challenges in 21st century chemical risk assessment. *Environmental Science: Processes & Impacts, 20*(1), 58–71.

Ditta, A. (2012). How helpful is nanotechnology in agriculture? *Advances in Natural Sciences: Nanoscience and Nanotechnology, 3*(3), 033002.

Domercq, P., Praetorius, A., & Boxall, A. B. (2018). Emission and fate modelling framework for engineered nanoparticles in urban aquatic systems at high spatial and temporal resolution. *Environmental Science: Nano, 5*(2), 533–543.

Dong, A., Ye, X., Li, H., Zhang, Y., & Wang, G. (2016). Micro/nanostructured hydroxyapatite structurally enhances the immobilization for Cu and Cd in contaminated soil. *Journal of Soils and Sediments, 16*, 2030–2040.

dos Santos, K. C., & Martinez, C. B. (2014). Genotoxic and biochemical effects of atrazine and Roundup®, alone and in combination, on the Asian clam *Corbicula fluminea*. *Ecotoxicology and Environmental Safety, 100*, 7–14.

Du, Z., Wang, C., Tai, X., Wang, G., & Liu, X. (2016). Optimization and characterization of biocompatible oil-in-water nanoemulsion for pesticide delivery. *ACS Sustainable Chemistry & Engineering, 4*(3), 983–991.

Edueng, K., Mahlin, D., Larsson, P., & Bergström, C. A. (2017). Mechanism-based selection of stabilization strategy for amorphous formulations: Insights into crystallization pathways. *Journal of Controlled Release, 256*, 193–202.

El-Hamshary, H., Fouda, M. M., Moydeen, M., El-Newehy, M. H., Al-Deyab, S. S., & Abdel-Megeed, A. (2015). Synthesis and antibacterial of carboxymethyl starch-grafted poly (vinyl imidazole) against some plant pathogens. *International Journal of Biological Macromolecules, 72*, 1466–1472.

Fan, C., Guo, M., Liang, Y., Dong, H., Ding, G., Zhang, W., …, & Cao, Y. (2017). Pectin-conjugated silica microcapsules as dual-responsive carriers for increasing the stability and antimicrobial efficacy of kasugamycin. *Carbohydrate Polymers, 172*, 322–331.

Feltham, H., Park, K., & Goulson, D. (2014). Field realistic doses of pesticide imidacloprid reduce bumblebee pollen foraging efficiency. *Ecotoxicology (London, England), 23*, 317–323.

Fletcher, S. J., Reeves, P. T., Hoang, B. T., & Mitter, N. (2020). A perspective on RNAi-based biopesticides. *Frontiers in Plant Science, 11*, 51.

Flores-Cespedes, F., Martinez-Dominguez, G. P., Villafranca-Sanchez, M., & Fernandez-Perez, M. (2015). Preparation and characterization of azadirachtin alginate-biosorbent based formulations: Water release kinetics and photodegradation study. *Journal of Agricultural and Food Chemistry, 63*(38), 8391–8398.

Fraceto, L. F., Grillo, R., de Medeiros, G. A., Scognamiglio, V., Rea, G., & Bartolucci, C. (2016). Nanotechnology in agriculture: Which innovation potential does it have? *Frontiers in Environmental Science, 4*, 20.

Fu, H., Yang, Y., Zhu, R., Liu, J., Usman, M., Chen, Q., & He, H. (2018). Superior adsorption of phosphate by ferrihydrite-coated and lanthanum-decorated magnetite. *Journal of Colloid and Interface Science, 530*, 704–713.

Gaber, M., Medhat, W., Hany, M., Saher, N., Fang, J. Y., & Elzoghby, A. (2017). Protein-lipid nanohybrids as emerging platforms for drug and gene delivery: Challenges and outcomes. *Journal of Controlled Release, 254*, 75–91.

Gao, Y., Zhang, Y., He, S., Xiao, Y., Qin, X., Zhang, Y., …, & Li, J. (2019). Fabrication of a hollow mesoporous silica hybrid to improve the targeting of a pesticide. *Chemical Engineering Journal, 364*, 361–369.

Gao, Z., Yuan, P., Wang, D., Xu, Z., Li, Z., & Shao, X. (2017). Photo-controlled release of fipronil from a coumarin triggered precursor. *Bioorganic & Medicinal Chemistry Letters, 27*(11), 2528–2535.

Garrido-Herrera, F. J., Gonzalez-Pradas, E., & Fernández-Pérez, M. (2006). Controlled release of isoproturon, imidacloprid, and cyromazine from alginate– bentonite-activated carbon formulations. *Journal of Agricultural and Food Chemistry, 54*(26), 10053–10060.

Geitner, N. K., Zhao, W., Ding, F., Chen, W., & Wiesner, M. R. (2017). Mechanistic insights from discrete molecular dynamics simulations of pesticide–nanoparticle interactions. *Environmental Science & Technology, 51*(15), 8396–8404.

George, J., & Shukla, Y. (2011). Pesticides and cancer: Insights into toxicoproteomic-based findings. *Journal of Proteomics, 74*(12), 2713–2722.

Gianessi, L. P. (2013). The increasing importance of herbicides in worldwide crop production. *Pest Management Science, 69*(10), 1099–1105.

Giessen, T. W., & Silver, P. A. (2016). Encapsulation as a strategy for the design of biological compartmentalization. *Journal of Molecular Biology, 428*(5), 916–927.

Glaser, J. A. (2015). Microplastics in the environment. *Clean Technologies and Environmental Policy, 17*, 1383–1391.

Gogos, A., Knauer, K., & Bucheli, T. D. (2012). Nanomaterials in plant protection and fertilization: current state, foreseen applications, and research priorities. *Journal of Agricultural and Food Chemistry, 60*(39), 9781–9792.

Goswami, L., Kim, K. H., Deep, A., Das, P., Bhattacharya, S. S., Kumar, S., & Adelodun, A. A. (2017). Engineered nano particles: Nature, behavior, and effect on the environment. *Journal of Environmental Management, 196*, 297–315.

Goulson, D., Nicholls, E., Botías, C., & Rotheray, E. L. (2015). Bee declines driven by combined stress from parasites, pesticides, and lack of flowers. *Science, 347*(6229), 1255957.

Grant, S., Mortimer, M., Stevenson, G., Malcolm, D., & Gaus, C. (2011). Facilitated transport of dioxins in soil following unintentional release of pesticide-surfactant formulations. *Environmental Science & Technology, 45*(2), 406–411.

Green, J. M. (2014). Current state of herbicides in herbicide-resistant crops. *Pest Management Science, 70*(9), 1351–1357.

Grillo, R., Fraceto, L. F., Amorim, M. J., Scott-Fordsmand, J. J., Schoonjans, R., & Chaudhry, Q. (2021). Ecotoxicological and regulatory aspects of environmental sustainability of nanopesticides. *Journal of Hazardous Materials, 404*, 124148.

Guerra, F. D., Attia, M. F., Whitehead, D. C., & Alexis, F. (2018). Nanotechnology for environmental remediation: materials and applications. *Molecules, 23*(7), 1760.

Guo, M., Zhang, W., Ding, G., Guo, D., Zhu, J., Wang, B., …, & Cao, Y. (2015). Preparation and characterization of enzyme-responsive emamectin benzoate microcapsules based on a copolymer matrix of silica–epichlorohydrin–carboxymethylcellulose. *RSC Advances, 5*(113), 93170–93179.

Hakme, E., Lozano, A., Ferrer, C., Díaz-Galiano, F. J., & Fernández-Alba, A. R. (2018). Analysis of pesticide residues in olive oil and other vegetable oils. *TrAC Trends in Analytical Chemistry, 100*, 167–179.

Hamid, Y., Tang, L., Sohail, M. I., Cao, X., Hussain, B., Aziz, M. Z., … Yang, X. (2019). An explanation of soil amendments to reduce cadmium phytoavailability and transfer to food chain. *Science of The Total Environment, 660*, 80–96.

Hatcher, J. M., Delea, K. C., Richardson, J. R., Pennell, K. D., & Miller, G. W. (2008). Disruption of dopamine transport by DDT and its metabolites. *Neurotoxicology, 29*(4), 682–690.

He, C., Liu, D., & Lin, W. (2015). Nanomedicine applications of hybrid nanomaterials built from metal–ligand coordination bonds: Nanoscale metal–organic frameworks and nanoscale coordination polymers. *Chemical Reviews, 115*(19), 11079–11108.

Henry, M., Beguin, M., Requier, F., Rollin, O., Odoux, J. F., Aupinel, P., …, & Decourtye, A. (2012). A common pesticide decreases foraging success and survival in honey bees. *Science, 336*(6079), 348–350.

Hernández, A. F., Parrón, T., Tsatsakis, A. M., Requena, M., Alarcón, R., & López-Guarnido, O. (2013). Toxic effects of pesticide mixtures at a molecular level: Their relevance to human health. *Toxicology, 307*, 136–145.

Heydarifard, S., Taneja, K., Bhanjana, G., Dilbaghi, N., Nazhad, M. M., Kim, K. H., & Kumar, S. (2018). Modification of cellulose foam paper for use as a high-quality biocide disinfectant filter for drinking water. *Carbohydrate Polymers, 181*, 1086–1092.

Hoppin, J. A., Umbach, D. M., Long, S., London, S. J., Henneberger, P. K., Blair, A., …, & Sandler, D. P. (2017). Pesticides are associated with allergic and non-allergic wheeze among male farmers. *Environmental Health Perspectives, 125*(4), 535–543.

http://ec.europa.eu/food/plant/pesticides/eu-pesticides-database/public/?event=homepage&language=EN. (Accessed 30 November 2018).

http://www.fao.org/fao-who-codexalimentarius/standards/pestres/pesticides/en. (Accessed 30 November 2018).

http://www.who.int/foodsafety/databases/en. (Accessed 30 November 2018).

Hu, X., Lu, K., Mu, L., Kang, J., & Zhou, Q. (2014). Interactions between graphene oxide and plant cells: Regulation of cell morphology, uptake, organelle damage, oxidative effects and metabolic disorders. *Carbon, 80*, 665–676.

Huang, B., Chen, F., Shen, Y., Qian, K., Wang, Y., Sun, C., …, & Cui, H. (2018). Advances in targeted pesticides with environmentally responsive controlled release by nanotechnology. *Nanomaterials, 8*(2), 102.

Huayao, C., Yueshun, L., Hongjun, Z., Xinhua, Z., Sheng, G., & Hua, X. (2016). Highly efficient alginate sodium encapsulated chlorpyrifos/copper (II) Schiff base mesoporous silica sustained release system with pH and ion response for pesticide delivery. *RSC Advances, 6*(115), 114714–114721.

Ianchis, R., Ninciuleanu, C. M., Gifu, I. C., Alexandrescu, E., Somoghi, R., Gabor, A. R., …, & Roseanu, A. M. (2017). Novel hydrogel-advanced modified clay nanocomposites as possible vehicles for drug delivery and controlled release. *Nanomaterials, 7*(12), 443.

Iavicoli, I., Leso, V., Beezhold, D. H., & Shvedova, A. A. (2017). Nanotechnology in agriculture: Opportunities, toxicological implications, and occupational risks. *Toxicology and Applied Pharmacology, 329*, 96–111.

iGate Research (2018). Global nanotechnology (component and applications) markets 2018–2024. Report by Research and Markets pp. 3–191.

Jeon, G., Yang, S. Y., Byun, J., & Kim, J. K. (2011). Electrically actuatable smart nanoporous membrane for pulsatile drug release. *Nano Letters, 11*(3), 1284–1288.

Jiang, J., Tong, X., Morris, D., & Zhao, Y. (2006). Toward photocontrolled release using light-dissociable block copolymer micelles. *Macromolecules, 39*(13), 4633–4640.

Jiang, L. C., Basri, M., Omar, D., Rahman, M. B. A., Salleh, A. B., Rahman, R. N. Z. R. A., & Selamat, A. (2012). Green nano-emulsion intervention for water-soluble glyphosate isopropylamine (IPA) formulations in controlling Eleusine indica (*E. indica*). *Pesticide Biochemistry and Physiology, 102*(1), 19–29.

Jiang, W., Conkle, J. L., Luo, Y., Li, J., Xu, K., & Gan, J. (2016). Occurrence, distribution, and accumulation of pesticides in exterior residential areas. *Environmental Science & Technology, 50*(23), 12592–12601.

Jindal, A. B., Bachhav, S. S., & Devarajan, P. V. (2017). In situ hybrid nano drug delivery system (IHN-DDS) of antiretroviral drug for simultaneous targeting to multiple viral reservoirs: An in vivo proof of concept. *International Journal of Pharmaceutics, 521*(1-2), 196–203.

Jones, K., Everard, M., & Harding, A. H. (2014). Investigation of gastrointestinal effects of organophosphate and carbamate pesticide residues on young children. *International Journal of Hygiene and Environmental Health, 217*(2-3), 392–398.

Jorfi, S., Rezaee, A., Moheb-Ali, G. A., & Jaafarzadeh, N. A. (2013). Pyrene removal from contaminated soils by modified Fenton oxidation using iron nano particles. *Journal of Environmental Health Science and Engineering, 11*, 1–8.

Juan-Alcaniz, J., Gascon, J., & Kapteijn, F. (2012). Metal–organic frameworks as scaffolds for the encapsulation of active species: State of the art and future perspectives. *Journal of Materials Chemistry, 22*(20), 10102–10118.

Kah, M. (2015). Nanopesticides and nanofertilizers: Emerging contaminants or opportunities for risk mitigation? *Frontiers in Chemistry, 3*, 64.

Kah, M., & Hofmann, T. (2014). Nanopesticide research: Current trends and future priorities. *Environment International, 63*, 224–235.

Kah, M., Beulke, S., Tiede, K., & Hofmann, T. (2013). Nanopesticides: State of knowledge, environmental fate, and exposure modeling. *Critical Reviews in Environmental Science and Technology, 43*, 1823–1867.

Kah, M., Kookana, R. S., Gogos, A., & Bucheli, T. D. (2018). A critical evaluation of nanopesticides and nanofertilizers against their conventional analogues. *Nature Nanotechnology, 13*(8), 677–684.

Kah, M., Walch, H., & Hofmann, T. (2018). Environmental fate of nanopesticides: Durability, sorption and photodegradation of nanoformulated clothianidin. *Environmental Science: Nano, 5*(4), 882–889.

Kanhed, P., Birla, S., Gaikwad, S., Gade, A., Seabra, A. B., Rubilar, O., … Rai, M. (2014). In vitro antifungal efficacy of copper nanoparticles against selected crop pathogenic fungi. *Materials Letters, 115*, 13–17.

Karimi, A. R., Tarighatjoo, M., & Nikravesh, G. (2017). 1, 3, 5-Triazine-2, 4, 6-tribenzaldehyde derivative as a new crosslinking agent for synthesis of pH-thermo dual responsive chitosan hydrogels and their nanocomposites: Swelling properties and drug release behavior. *International Journal of Biological Macromolecules, 105*, 1088–1095.

Kaur, P. (2018). Biosynthesis of nanoparticles using eco-friendly factories and their role in plant pathogenicity: a review. *Biotechnology Research and Innovation, 2*(1), 63–73.

Kaziem, A. E., Gao, Y., He, S., & Li, J. (2017). Synthesis and insecticidal activity of enzyme-triggered functionalized hollow mesoporous silica for controlled release. *Journal of Agricultural and Food Chemistry, 65*(36), 7854–7864.

Kaziem, A. E., Gao, Y., Zhang, Y., Qin, X., Xiao, Y., Zhang, Y., …, & He, S. (2018). α-Amylase triggered carriers based on cyclodextrin anchored hollow mesoporous silica for enhancing insecticidal activity of avermectin against Plutella xylostella. *Journal of Hazardous Materials, 359*, 213–221.

Kennedy, M. C., Glass, C. R., Bokkers, B., Hart, A. D., Hamey, P. Y., Kruisselbrink, J. W., … van Klaveren, J. D. (2015). A European model and case studies for aggregate exposure assessment of pesticides. *Food and Chemical Toxicology, 79*, 32–44.

Khan, A., Wen, Y., Huq, T., & Ni, Y. (2018). Cellulosic nanomaterials in food and nutraceutical applications: a review. *Journal of Agricultural and Food Chemistry, 66*(1), 8–19.

Khan, I., Saeed, K., & Khan, I. (2019). Nanoparticles: Properties, applications and toxicities. *Arabian Journal of Chemistry, 12*(7), 908–931.

Khandelwal, N., Barbole, R. S., Banerjee, S. S., Chate, G. P., Biradar, A. V., Khandare, J. J., & Giri, A. P. (2016). Budding trends in integrated pest management using advanced micro-and nano-materials: Challenges and perspectives. *Journal of Environmental Management, 184*, 157–169.

Khodakovskaya, M., Dervishi, E., Mahmood, M., Xu, Y., Li, Z., Watanabe, F., & Biris, A. S. (2009). Carbon nanotubes are able to penetrate plant seed coat and dramatically affect seed germination and plant growth. *ACS Nano, 3*(10), 3221–3227.

Khot, L. R., Sankaran, S., Maja, J. M., Ehsani, R., & Schuster, E. W. (2012). Applications of nanomaterials in agricultural production and crop protection: a review. *Crop Protection, 35*, 64–70.

Köhler, H. R., & Triebskorn, R. (2013). Wildlife ecotoxicology of pesticides: Can we track effects to the population level and beyond? *Science, 341*(6147), 759–765.

Kookana, R. S., Boxall, A. B., Reeves, P. T., Ashauer, R., Beulke, S., Chaudhry, Q., … Van den Brink, P. J. (2014). Nanopesticides: guiding principles for regulatory evaluation of environmental risks. *Journal of Agricultural and Food Chemistry, 62*(19), 4227–4240.

Koul, O. (2019). *Nano-biopesticides today and future perspectives*. Academic Press.

Krief, S., Berny, P., Gumisiriza, F., Gross, R., Demeneix, B., Fini, J. B., …, & Wasswa, J. (2017). Agricultural expansion as risk to endangered wildlife: Pesticide exposure in wild chimpanzees and baboons displaying facial dysplasia. *Science of the Total Environment, 598*, 647–656.

Kuhlbusch, T. A., Wijnhoven, S. W., & Haase, A. (2018). Nanomaterial exposures for worker, consumer and the general public. *NanoImpact, 10*, 11–25.

Kulkarni, A. R., Soppimath, K. S., Aminabhavi, T. M., Dave, A. M., & Mehta, M. H. (2000). Glutaraldehyde crosslinked sodium alginate beads containing liquid pesticide for soil application. *Journal of Controlled Release, 63*(1-2), 97–105.

Kumar, S., Bhanjana, G., Kumar, A., Taneja, K., Dilbaghi, N., & Kim, K. H. (2016). Synthesis and optimization of ceftriaxone-loaded solid lipid nanocarriers. *Chemistry and Physics of Lipids, 200*, 126–132.

Kumar, S., Bhanjana, G., Sharma, A., Dilbaghi, N., Sidhu, M. C., & Kim, K. H. (2017). Development of nanoformulation approaches for the control of weeds. *Science of the Total Environment, 586*, 1272–1278.

Kumar, S., Chauhan, N., Gopal, M., Kumar, R., & Dilbaghi, N. (2015). Development and evaluation of alginate–chitosan nanocapsules for controlled release of acetamiprid. *International Journal of Biological Macromolecules, 81*, 631–637.

Kumar, S., Nehra, M., Dilbaghi, N., Tankeshwar, K., & Kim, K. H. (2018). Recent advances and remaining challenges for polymeric nanocomposites in healthcare applications. *Progress in Polymer Science, 80*, 1–38.

Kushwaha, S, Vikram, A., Trivedi, P. P., & Jena, G. B. (2011). Alkaline, Endo III and FPG modified comet assay as biomarkers for the detection of oxidative DNA damage in rats with experimentally induced diabetes. *Mutation Research/Genetic Toxicology and Environmental Mutagenesis, 726*(2), 242–250.

Lammoglia, S. K., Kennedy, M. C., Barriuso, E., Alletto, L., Justes, E., Munier-Jolain, N., & Mamy, L. (2017). Assessing human health risks from pesticide use in conventional and innovative cropping systems with the BROWSE model. *Environment International, 105*, 66–78.

Larsen, A. E., Gaines, S. D., & Deschênes, O. (2017). Agricultural pesticide use and adverse birth outcomes in the San Joaquin Valley of California. *Nature Communications, 8*(1), 302.

Le Van, N., Ma, C., Shang, J., Rui, Y., Liu, S., & Xing, B. (2016). Effects of CuO nanoparticles on insecticidal activity and phytotoxicity in conventional and transgenic cotton. *Chemosphere, 144*, 661–670.

Lee, D. H., Lind, P. M., Jacobs, D. R., Jr, Salihovic, S., van Bavel, B., & Lind, L. (2016). Association between background exposure to organochlorine pesticides and the risk of cognitive impairment: A prospective study that accounts for weight change. *Environment International, 89*, 179–184.

Li, J., Fan, T., Xu, Y., & Wu, X. (2016). Ionic liquids as modulators of physicochemical properties and nanostructures of sodium dodecyl sulfate in aqueous solutions and potential application in pesticide microemulsions. *Physical Chemistry Chemical Physics, 18*(43), 29797–29807.

Li, T., Cipolla, D., Rades, T., & Boyd, B. J. (2018). Drug nanocrystallisation within liposomes. *Journal of Controlled Release, 288*, 96–110.

Li, X., Ke, M., Zhang, M., Peijnenburg, W. J. G. M., Fan, X., Xu, J., …, & Qian, H. (2018). The interactive effects of diclofop-methyl and silver nanoparticles on *Arabidopsis thaliana*: Growth, photosynthesis and antioxidant system. *Environmental Pollution, 232*, 212–219.

Li, Z. Z., Chen, J. F., Liu, F., Liu, A. Q., Wang, Q., Sun, H. Y., & Wen, L. X. (2007). Study of UV-shielding properties of novel porous hollow silica nanoparticle carriers for avermectin. *Pest Management Science: Formerly Pesticide Science, 63*(3), 241–246.

Li, Z. Z., Xu, S. A., Wen, L. X., Liu, F., Liu, A. Q., Wang, Q., …, & Chen, J. F. (2006). Controlled release of avermectin from porous hollow silica nanoparticles: Influence of shell thickness on loading efficiency, UV-shielding property and release. *Journal of Controlled Release, 111*(1-2), 81–88.

Liang, W., Yu, A., Wang, G., Zheng, F., Jia, J., & Xu, H. (2018). Chitosan-based nanoparticles of avermectin to control pine wood nematodes. *International Journal of Biological Macromolecules, 112*, 258–263.

Liang, Y., Fan, C., Dong, H., Zhang, W., Tang, G., Yang, J., …, & Cao, Y. (2018). Preparation of MSNs-Chitosan@Prochloraz nanoparticles for reducing toxicity and improving release properties of prochloraz. *ACS Sustainable Chemistry & Engineering, 6*(8), 10211–10220.

Liang, Y., Guo, M., Fan, C., Dong, H., Ding, G., Zhang, W., …, & Cao, Y. (2017). Development of novel urease-responsive pendimethalin microcapsules using silica-IPTS-PEI as controlled release carrier materials. *ACS Sustainable Chemistry & Engineering, 5*(6), 4802–4810.

Lin, L., Cui, H., Zhou, H., Zhang, X., Bortolini, C., Chen, M., …, & Dong, M. (2015). Nanoliposomes containing Eucalyptus citriodora as antibiotic with specific antimicrobial activity. *Chemical Communications, 51*(13), 2653–2655.

Linquist, B. A., Liu, L., van Kessel, C., & van Groenigen, K. J. (2013). Enhanced efficiency nitrogen fertilizers for rice systems: Meta-analysis of yield and nitrogen uptake. *Field Crops Research, 154*, 246–254.

Liu, J., Bao, T. Y., Yang, X. Y., Zhu, P. P., Wu, L. H., Sha, J. Q., …, & Lan, Y. Q. (2017). Controllable porosity conversion of metal-organic frameworks composed of natural ingredients for drug delivery. *Chemical Communications, 53*(55), 7804–7807.

Lowry, G. V., Avellan, A., & Gilbertson, L. M. (2019). Opportunities and challenges for nanotechnology in the agri-tech revolution. *Nature Nanotechnology, 14*(6), 517–522.

Lozowicka, B. (2015). Health risk for children and adults consuming apples with pesticide residue. *Science of the Total Environment, 502*, 184–198.

Lu, Y., & Ozcan, S. (2015). Green nanomaterials: On track for a sustainable future. *Nano Today, 10*(4), 417–420.

Lvov, Y. M., DeVilliers, M. M., & Fakhrullin, R. F. (2016). The application of halloysite tubule nanoclay in drug delivery. *Expert Opinion on Drug Delivery, 13*(7), 977–986.

Majeed, A., Muhammad, Z., & Ahmad, H. (2018). Plant growth promoting bacteria: Role in soil improvement, abiotic and biotic stress management of crops. *Plant Cell Reports, 37*, 1599–1609.

Markou, G., Mitrogiannis, D., Çelekli, A., Bozkurt, H., Georgakakis, D., & Chrysikopoulos, C. V. (2015). Biosorption of Cu^{2+} and Ni^{2+} by Arthrospira platensis with different biochemical compositions. *Chemical Engineering Journal, 259*, 806–813.

Maruyama, C. R., Guilger, M., Pascoli, M., Bileshy-José, N., Abhilash, P. C., Fraceto, L. F., & De Lima, R. (2016). Nanoparticles based on chitosan as carriers for the combined herbicides imazapic and imazapyr. *Scientific Reports, 6*(1), 1–15.

Masina, N., Choonara, Y. E., Kumar, P., du Toit, L. C., Govender, M., Indermun, S., & Pillay, V. (2017). A review of the chemical modification techniques of starch. *Carbohydrate Polymers, 157*, 1226–1236.

Mattos, B. D., Rojas, O. J., & Magalhães, W. L. (2016). Biogenic SiO_2 in colloidal dispersions via ball milling and ultrasonication. *Powder Technology, 301*, 58–64.

Mattos, B. D., Rojas, O. J., & Magalhães, W. L. (2017). Biogenic silica nanoparticles loaded with neem bark extract as green, slow-release biocide. *Journal of Cleaner Production, 142*, 4206–4213.

Mauricio, M. R., da Costa, P. G., Haraguchi, S. K., Guilherme, M. R., Muniz, E. C., & Rubira, A. F. (2015). Synthesis of a microhydrogel composite from cellulose nanowhiskers and starch for drug delivery. *Carbohydrate Polymers, 115*, 715–722.

McArt, S. H., Fersch, A. A., Milano, N. J., Truitt, L. L., & Böröczky, K. (2017). High pesticide risk to honey bees despite low focal crop pollen collection during pollination of a mass blooming crop. *Scientific Reports, 7*(1), 1–10.

Meyer, W. L., Gurman, P., Stelinski, L. L., & Elman, N. M. (2015). Functional nano-dispensers (FNDs) for delivery of insecticides against phytopathogen vectors. *Green Chemistry, 17*(8), 4173–4177.

Mishra, S., Singh, B. R., Singh, A., Keswani, C., Naqvi, A. H., & Singh, H. B. (2014). Biofabricated silver nanoparticles act as a strong fungicide against Bipolaris sorokiniana causing spot blotch disease in wheat. *PLoS ONE, 9*(5), e97881.

Mittal, M., Siddiqui, M. R., Tran, K., Reddy, S. P., & Malik, A. B. (2014). Reactive oxygen species in inflammation and tissue injury. *Antioxidants & Redox Signaling, 20*(7), 1126–1167.

Mushtaq, W., Shakeel, A., Fazili, M. A., Chakrabartty, I., & Sevindik, M. (2020). Pros and cons of nanotechnology. *Nanobiotechnology in Agriculture. An Approach Towards Sustainability*, 207–222.

Mustafa, I. F., & Hussein, M. Z. (2020). Synthesis and technology of nanoemulsion-based pesticide formulation. *Nanomaterials, 10*(8), 1608.

Nakasato, D. Y., Pereira, A. E., Oliveira, J. L., Oliveira, H. C., & Fraceto, L. F. (2017). Evaluation of the effects of polymeric chitosan/tripolyphosphate and solid lipid nanoparticles on germination of *Zea mays, Brassica rapa* and *Pisum sativum*. *Ecotoxicology and Environmental Safety, 142*, 369–374.

Narayanan, D., Nair, S., & Menon, D. (2015). A systematic evaluation of hydroxyethyl starch as a potential nanocarrier for parenteral drug delivery. *International Journal of Biological Macromolecules, 74*, 575–584.

Nehra, M., Dilbaghi, N., Singhal, N. K., Hassan, A. A., Kim, K. H., & Kumar, S. (2019). Metal organic frameworks MIL-100 (Fe) as an efficient adsorptive material for phosphate management. *Environmental Research, 169*, 229–236.

Nelson, D. D., Pan, Y., Tikekar, R. V., Dan, N., & Nitin, N. (2017). Compound stability in nanoparticles: The effect of solid phase fraction on diffusion of degradation agents into nanostructured lipid carriers. *Langmuir, 33*(49), 14115–14122.

Nguyen, H. M., Hwang, I. C., Park, J. W., & Park, H. J. (2012). Photoprotection for deltamethrin using chitosan-coated beeswax solid lipid nanoparticles. *Pest Management Science, 68*(7), 1062–1068.

Nguyen, M. H., Hwang, I. C., Bui, C. B., & Park, H. J. (2016). Effects of the physical state of nanocarriers on their penetration into the root and upward transportation to the stem of soybean plants using confocal laser scanning microscopy. *Crop Protection, 87*, 25–30.

Niculae, G., Lacatusu, I., Badea, N., Stan, R., Vasile, B. S., & Meghea, A. (2014). Rice bran and raspberry seed oil-based nanocarriers with self-antioxidative properties as safe photoprotective formulations. *Photochemical & Photobiological Sciences, 13*, 703–716.

Nogueira, E. N., Dores, E. F., Pinto, A. A., Amorim, R. S., Ribeiro, M. L., & Lourencetti, C. (2012). Currently used pesticides in water matrices in Central-Western Brazil. *Journal of the Brazilian Chemical Society, 23*, 1476–1487.

Ocsoy, I., Paret, M. L., Ocsoy, M. A., Kunwar, S., Chen, T., You, M., & Tan, W. (2013). Nanotechnology in plant disease management: DNA-directed silver nanoparticles on graphene oxide as an antibacterial against Xanthomonas perforans. *ACS Nano, 7*(10), 8972–8980.

Orellana-Tavra, C., Baxter, E. F., Tian, T., Bennett, T. D., Slater, N. K., Cheetham, A. K., & Fairen-Jimenez, D. (2015). Amorphous metal–organic frameworks for drug delivery. *Chemical Communications, 51*(73), 13878–13881.

Oun, A. A., & Rhim, J. W. (2016). Isolation of cellulose nanocrystals from grain straws and their use for the preparation of carboxymethyl cellulose-based nanocomposite films. *Carbohydrate Polymers, 150*, 187–200.

Pan, Y., Tikekar, R. V., & Nitin, N. (2016). Distribution of a model bioactive within solid lipid nanoparticles and nanostructured lipid carriers influences its loading efficiency and oxidative stability. *International Journal of Pharmaceutics, 511*(1), 322–330.

Pankaj, Shakil, N. , A., Kumar, J., Singh, M. K., & Singh, K. (2012). Bioefficacy evaluation of controlled release formulations based on amphiphilic nano-polymer of carbofuran against *Meloidogyne incognita* infecting tomato. *Journal of Environmental Science and Health, Part B, 47*(6), 520–528.

Paret, M. L., Vallad, G. E., Averett, D. R., Jones, J. B., & Olson, S. M. (2013). Photocatalysis: effect of light-activated nanoscale formulations of TiO_2 on Xanthomonas perforans and control of bacterial spot of tomato. *Phytopathology, 103*(3), 228–236.

Parisi, C., Vigani, M., & Rodríguez-Cerezo, E. (2015). Agricultural nanotechnologies: what are the current possibilities? *Nano Today, 10*(2), 124–127.

Patel, S., Bajpai, J., Saini, R., Bajpai, A. K., & Acharya, S. (2018). Sustained release of pesticide (Cypermethrin) from nanocarriers: An effective technique for environmental and crop protection. *Process Safety and Environmental Protection, 117*, 315–325.

Peppas, N. A., & Sahlin, J. J. (1989). A simple equation for the description of solute release: III. Coupling of diffusion and relaxation. *International Journal of Pharmaceutics, 57*(2), 169–172.

Pérez Quiñones, J., Brüggemann, O., Kjems, J., Shahavi, M. H., & Peniche Covas, C. (2018). Novel brassinosteroid-modified polyethylene glycol micelles for controlled release of agrochemicals. *Journal of Agricultural and Food Chemistry, 66*(7), 1612–1619.

Pérez-Iglesias, J. M., de Arcaute, C. R., Natale, G. S., Soloneski, S., & Larramendy, M. L. (2017). Evaluation of imazethapyr-induced DNA oxidative damage by alkaline Endo III-and Fpg-modified single-cell gel electrophoresis assay in *Hypsiboas pulchellus* tadpoles (*Anura, Hylidae*). *Ecotoxicology and Environmental Safety, 142*, 503–508.

Pérez-Iglesias, J. M., Franco-Belussi, L., Natale, G. S., & de Oliveira, C. (2019). Biomarkers at different levels of organisation after atrazine formulation (SIPTRAN 500SC®) exposure in *Rhinella schineideri* (Anura: Bufonidae) neotropical tadpoles. *Environmental Pollution, 244*, 733–746.

Peters, R., Brandhoff, P., Weigel, S., Marvin, H., Bouwmeester, H., Aschberger, K. Rauscher, H., Amenta, V., Arena, M., Botelho Moniz, F., & Gottardo, S. (2014). Inventory of Nanotechnology applications in the agricultural, feed and food sector. *EFSA Supporting Publications, 11*(7), 621E.

Peters, R. J., Bouwmeester, H., Gottardo, S., Amenta, V., Arena, M., Brandhoff, P., … Rauscher, H. (2016). Nanomaterials for products and application in agriculture, feed and food. *Trends in Food Science & Technology, 54*, 155–164.

Pinto, R. V., Antunes, F., Pires, J., Graca, V., Brandao, P., & Pinto, M. L. (2017). Vitamin B3 metal-organic frameworks as potential delivery vehicles for therapeutic nitric oxide. *Acta Biomaterialia, 51*, 66–74.

Polajnar, J., Eriksson, A., Lucchi, A., Anfora, G., Virant-Doberlet, M., & Mazzoni, V. (2015). Manipulating behaviour with substrate-borne vibrations–potential for insect pest control. *Pest Management Science, 71*(1), 15–23.

Popp, J., Pető, K., & Nagy, J. (2013). Pesticide productivity and food security: A review. *Agronomy for Sustainable Development, 33*, 243–255.

Prado, A. G., Moura, A. O., & Nunes, A. R. (2011). Nanosized silica modified with carboxylic acid as support for controlled release of herbicides. *Journal of Agricultural and Food Chemistry, 59*(16), 8847–8852.

Prasad, R., Bhattacharyya, A., & Nguyen, Q. D. (2017). Nanotechnology in sustainable agriculture: Recent developments, challenges, and perspectives. *Frontiers in Microbiology, 8*, 1014.

Prasad, R., Kumar, V., & Prasad, K. S. (2014). Nanotechnology in sustainable agriculture: present concerns and future aspects. *African Journal of Biotechnology, 13*(6), 705–713.

Prasanna, R., Chaudhary, V., Gupta, V., Babu, S., Kumar, A., Singh, R., …, & Nain, L. (2013). Cyanobacteria mediated plant growth promotion and bioprotection against *Fusarium wilt* in tomato. *European Journal of Plant Pathology, 136*, 337–353.

Qin, X., Xiang, X., Sun, X., Ni, H., & Li, L. (2016). Preparation of nanoscale Bacillus thuringiensis chitinases using silica nanoparticles for nematicide delivery. *International Journal of Biological Macromolecules, 82*, 13–21.

Rajpoot, S. (2021). Impact of nanotechnology on environment: A review. *International Journal for Modern Trends in Science and Technology*, 7(2021), 159–164.

Rai, M., & Ingle, A. (2012). Role of nanotechnology in agriculture with special reference to management of insect pests. *Applied Microbiology and Biotechnology, 94*, 287–293.

Ramasamy, T., Ruttala, H. B., Gupta, B., Poudel, B. K., Choi, H. G., Yong, C. S., & Kim, J. O. (2017). Smart chemistry-based nanosized drug delivery systems for systemic applications: A comprehensive review. *Journal of Controlled Release, 258*, 226–253.

Rani, R., Dahiya, S., Dhingra, D., Dilbaghi, N., Kim, K. H., & Kumar, S. (2017). Evaluation of anti-diabetic activity of glycyrrhizin-loaded nanoparticles in nicotinamide-streptozotocin-induced diabetic rats. *European Journal of Pharmaceutical Sciences, 106*, 220–230.

Richard, S., Moslemi, S., Sipahutar, H., Benachour, N., & Seralini, G. E. (2005). Differential effects of glyphosate and roundup on human placental cells and aromatase. *Environmental Health Perspectives, 113*(6), 716–720.

Richendrfer, H., & Creton, R. (2015). Chlorpyrifos and malathion have opposite effects on behaviors and brain size that are not correlated to changes in AChE activity. *Neurotoxicology, 49*, 50–58.

Rizzati, V., Briand, O., Guillou, H., & Gamet-Payrastre, L. (2016). Effects of pesticide mixtures in human and animal models: An update of the recent literature. *Chemico-Biological Interactions, 254*, 231–246.

Rossignol, D. A., Genuis, S. J., & Frye, R. E. (2014). Environmental toxicants and autism spectrum disorders: A systematic review. *Translational Psychiatry, 4*(2) e360–e360.

Rozek, C. K., & Dincă, M. (2014). Cation exchange at the secondary building units of metal–organic frameworks. *Chemical Society Reviews, 43*(16), 5456–5467.
Rundlöf, M., Andersson, G. K., Bommarco, R., Fries, I., Hederström, V., Herbertsson, L., …, & Smith, H. G. (2015). Seed coating with a neonicotinoid insecticide negatively affects wild bees. *Nature, 521*(7550), 77–80.
Saini, P., Gopal, M., Kumar, R., & Srivastava, C. (2014). Development of pyridalyl nanocapsule suspension for efficient management of tomato fruit and shoot borer (*Helicoverpa armigera*). *Journal of Environmental Science and Health, Part B, 49*(5), 344–351.
Saini, P., Kumar, S., Sharma, V. K., & Chobhe K. A. (2018). *Nanopesticides: Manage food security and environmental pollution.* https://www.biotecharticles.com/Agriculture-Article/Nanopesticides-ManageFood-Security-and-Environmental-Pollution-4391.html#:B:text=Nanopesticides%20has%20been%20visualized%20as,materials%20to%20deliver%20pesticides (Accessed on 28-12-2022).
Salgueiro, A. M., Daniel-da-Silva, A. L., Fateixa, S., & Trindade, T. (2013). κ-Carrageenan hydrogel nanocomposites with release behavior mediated by morphological distinct Au nanofillers. *Carbohydrate Polymers, 91*(1), 100–109.
Sanzari, I., Leone, A., & Ambrosone, A. (2019). Nanotechnology in plant science: To make a long story short. *Frontiers in Bioengineering and Biotechnology, 7*, 120.
Saratale, R. G., Karuppusamy, I., Saratale, G. D., Pugazhendhi, A., Kumar, G., Park, Y., … Shin, H. S. (2018). A comprehensive review on green nanomaterials using biological systems: Recent perception and their future applications. *Colloids and Surfaces B: Biointerfaces, 170*, 20–35.
Sarkar, D. J., & Singh, A. (2017). Base triggered release of insecticide from bentonite reinforced citric acid crosslinked carboxymethyl cellulose hydrogel composites. *Carbohydrate Polymers, 156*, 303–311.
Sasson, Y., Levy-Ruso, G., Toledano, O., & Ishaaya, I. (2007). Nanosuspensions: Emerging novel agrochemical formulations. In *Insecticides design using advanced technologies* (pp. 1–39). Berlin, Heidelberg: Springer Berlin Heidelberg.
Schnoor, B., Elhendawy, A., Joseph, S., Putman, M., Chacón-Cerdas, R., Flores-Mora, D., …, & Salvador-Morales, C. (2018). Engineering atrazine loaded poly (lactic-co-glycolic acid) nanoparticles to ameliorate environmental challenges. *Journal of Agricultural and Food Chemistry, 66*(30), 7889–7898.
Schulz, K., Mikhailyuk, T., Dreßler, M., Leinweber, P., & Karsten, U. (2016). Biological soil crusts from coastal dunes at the Baltic Sea: Cyanobacterial and algal biodiversity and related soil properties. *Microbial Ecology, 71*, 178–193.
Scott, N., & Chen, H. (2013). Nanoscale science and engineering for agriculture and food systems. *Industrial Biotechnology, 9*(1), 17–18.
Šefčíková, Z., Babeľová, J., Čikoš, Š., Kovaříková, V., Burkuš, J., Špirková, A., …, & Fabian, D. (2018). Fipronil causes toxicity in mouse preimplantation embryos. *Toxicology, 410*, 214–221.
Sekhon, B. S. (2014). Nanotechnology in agri-food production: an overview. *Nanotechnology, science and applications*, 31–53.
Servin, A., Elmer, W., Mukherjee, A., De la Torre-Roche, R., Hamdi, H., White, J. C., … Dimkpa, C. (2015). A review of the use of engineered nanomaterials to suppress plant disease and enhance crop yield. *Journal of Nanoparticle Research, 17*, 1–21.
Sharma, S., Singh, S., Ganguli, A. K., & Shanmugam, V. (2017). Anti-drift nano-stickers made of graphene oxide for targeted pesticide delivery and crop pest control. *Carbon, 115*, 781–790.
Shekhar, S., Sharma, S., Kumar, A., Taneja, A., & Sharma, B. (2021). The framework of nanopesticides: A paradigm in biodiversity. *Materials Advances, 2*(20), 6569–6588.
Shukla, P. G., Rajagopalan, N., Bhaskar, C., & Sivaram, S. (1991). Crosslinked starch-urea formaldehyde (St-UF) as a hydrophilic matrix for encapsulation: Studies in swelling and release of carbofuran. *Journal of Controlled Release, 15*(2), 153–165.
Simonin, M., Richaume, A., Guyonnet, J. P., Dubost, A., Martins, J. M., & Pommier, T. (2016). Titanium dioxide nanoparticles strongly impact soil microbial function by affecting archaeal nitrifiers. *Scientific Reports, 6*(1), 33643.
Singh, Y., Meher, J. G., Raval, K., Khan, F. A., Chaurasia, M., Jain, N. K., & Chourasia, M. K. (2017). Nanoemulsion: Concepts, development and applications in drug delivery. *Journal of Controlled Release, 252*, 28–49.

Smith, M. N., Workman, T., McDonald, K. M., Vredevoogd, M. A., Vigoren, E. M., Griffith, W. C., …, & Faustman, E. M. (2017). Seasonal and occupational trends of five organophosphate pesticides in house dust. *Journal of Exposure Science & Environmental Epidemiology, 27*(4), 372–378.

Soleimanian, Y., Goli, S. A. H., Varshosaz, J., & Sahafi, S. M. (2018). Formulation and characterization of novel nanostructured lipid carriers made from beeswax, propolis wax and pomegranate seed oil. *Food Chemistry, 244*, 83–92.

Soltani, N., Bahrami, A., Pech-Canul, M. I., & González, L. A. (2015). Review on the physicochemical treatments of rice husk for production of advanced materials. *Chemical Engineering Journal, 264*, 899–935.

Sonkaria, S., Ahn, S. H., & Khare, V. (2012). Nanotechnology and its impact on food and nutrition: a review. *Recent Patents on Food, Nutrition & Agriculture, 4*(1), 8–18.

Souza, D. R. D., Trovó, A. G., Antoniosi Filho, N. R., Silva, M. A., & Machado, A. E. (2013). Degradation of the commercial herbicide glyphosate by photo-fenton process: Evaluation of kinetic parameters and toxicity. *Journal of the Brazilian Chemical Society, 24*, 1451–1460.

Srivastava, A., Chun, S. J., Ko, S. R., Kim, J., Ahn, C. Y., & Oh, H. M. (2017). Floating rice-culture system for nutrient remediation and feed production in a eutrophic lake. *Journal of Environmental Management, 203*, 342–348.

Stanley, D. A., Garratt, M. P., Wickens, J. B., Wickens, V. J., Potts, S. G., & Raine, N. E. (2015). Neonicotinoid pesticide exposure impairs crop pollination services provided by bumblebees. *Nature, 528*(7583), 548–550.

Sun, Y., Liang, J., Tang, L., Li, H., Zhu, Y., Jiang, D., …, & Zeng, G. (2019). Nano-pesticides: A great challenge for biodiversity? *Nano Today, 28*, 100757.

Tamjidi, F., Shahedi, M., Varshosaz, J., & Nasirpour, A. (2013). Nanostructured lipid carriers (NLC): A potential delivery system for bioactive food molecules. *Innovative Food Science & Emerging Technologies, 19*, 29–43.

Tan, K., Heo, S., Foo, M., Chew, I. M., & Yoo, C. (2019). An insight into nanocellulose as soft condensed matter: Challenge and future prospective toward environmental sustainability. *Science of the Total Environment, 650*, 1309–1326.

Tarkowski, P., & Vereecke, D. (2014). Threats and opportunities of plant pathogenic bacteria. *Biotechnology Advances, 32*(1), 215–229.

Tighe-Neira, R., Carmora, E., Recio, G., Nunes-Nesi, A., Reyes-Diaz, M., Alberdi, M., …, & Inostroza-Blancheteau, C. (2018). Metallic nanoparticles influence the structure and function of the photosynthetic apparatus in plants. *Plant Physiology and Biochemistry, 130*, 408–417.

Tong, Y., Shao, L., Li, X., Lu, J., Sun, H., Xiang, S., …, & Wu, X. (2018). Adhesive and stimulus-responsive polydopamine-coated graphene oxide system for pesticide-loss control. *Journal of Agricultural and Food Chemistry, 66*(11), 2616–2622.

Tong, Y., Wu, Y., Zhao, C., Xu, Y., Lu, J., Xiang, S., …, & Wu, X. (2017). Polymeric nanoparticles as a metolachlor carrier: Water-based formulation for hydrophobic pesticides and absorption by plants. *Journal of Agricultural and Food Chemistry, 65*(34), 7371–7378.

Trimnell, D., & Shasha, B. S. (1988). Autoencapsulation: A new method for entrapping pesticides within starch. *Journal of Controlled Release, 7*(1), 25–31.

Türker, S., Yarza, F., Sánchez, R. T., & Yapar, S. (2017). Surface and interface properties of benzethonium chloride-montmorillonite. *Colloids and Surfaces A: Physicochemical and Engineering Aspects, 520*, 817–825.

Usha Rani, P., Madhusudhanamurthy, J., & Sreedhar, B. (2014). Dynamic adsorption of α-pinene and linalool on silica nanoparticles for enhanced antifeedant activity against agricultural pests. *Journal of Pest Science, 87*, 191–200.

Usman, M., Martin, S., Cimetiere, N., Giraudet, S., Chatain, V., & Hanna, K. (2014). Sorption of nalidixic acid onto micrometric and nanometric magnetites: Experimental study and modeling. *Applied Surface Science, 299*, 136–145.

Vanbergen, A. J., & Initiative, T. I. P. (2013). Threats to an ecosystem service: Pressures on pollinators. *Frontiers in Ecology and the Environment, 11*(5), 251–259.

Vellingiri, K., Philip, L., & Kim, K. H. (2017). Metal–organic frameworks as media for the catalytic degradation of chemical warfare agents. *Coordination Chemistry Reviews, 353*, 159–179.

Villaverde, J. J., Sevilla-Morán, B., López-Goti, C., Alonso-Prados, J. L., & Sandín-España, P. (2018). Considerations of nano-QSAR/QSPR models for nanopesticide risk assessment within the European legislative framework. *Science of the Total Environment, 634*, 1530–1539.

Walker, G. W., Kookana, R. S., Smith, N. E., Kah, M., Doolette, C. L., Reeves, P. T., …, & Navarro, D. A. (2017). Ecological risk assessment of nano-enabled pesticides: A perspective on problem formulation. *Journal of Agricultural and Food Chemistry, 66*(26), 6480–6486.

Wang, L., Espinoza, H. M., MacDonald, J. W., Bammler, T. K., Williams, C. R., Yeh, A., …, & Gallagher, E. P. (2016b). Olfactory transcriptional analysis of salmon exposed to mixtures of chlorpyrifos and malathion reveal novel molecular pathways of neurobehavioral injury. *Toxicological Sciences, 149*(1), 145–157.

Wang, T., Ren, J., Qu, G., Liang, D., & Hu, S. (2016a). Glyphosate contaminated soil remediation by atmospheric pressure dielectric barrier discharge plasma and its residual toxicity evaluation. *Journal of Hazardous Materials, 320*, 539–546.

Wang, Y., Wang, A., Wang, C., Cui, B., Sun, C., Zhao, X., …, & Cui, H. (2017). Synthesis and characterization of emamectin-benzoate slow-release microspheres with different surfactants. *Scientific Reports, 7*(1), 12761.

Wang, Q., Wu, J., Hao, L., Wu, Q., Wang, C., & Wang, Z. (2018b). Magnetic solid-phase extraction of benzoylurea insecticides by Fe_3O_4 nanoparticles decorated with a hyper-cross-linked porous organic polymer. *Journal of Separation Science, 41*(16), 3285–3293.

Wang, G., Xiao, Y., Xu, H., Hu, P., Liang, W., Xie, L., & Jia, J. (2018c). Development of multifunctional avermectin poly (succinimide) nanoparticles to improve bioactivity and transportation in rice. *Journal of Agricultural and Food Chemistry, 66*(43), 11244–11253.

Wang, Y., Zhang, S., & Benoit, D. S. (2018a). Degradable poly (ethylene glycol)(PEG)-based hydrogels for spatiotemporal control of siRNA/nanoparticle delivery. *Journal of Controlled Release, 287*, 58–66.

Wasef, L., Nassar, A. M., El-Sayed, Y. S., Samak, D., Noreldin, A., Elshony, N., …, & Shaheen, H. M. (2021). The potential ameliorative impacts of cerium oxide nanoparticles against fipronil-induced hepatic steatosis. *Scientific Reports, 11*(1), 1310.

Wei, W., Lu, X., Wang, Z., Pérez, B., Liu, J., Wu, C., …, & Guo, Z. (2017). Single-component solid lipid nanocarriers prepared with ultra-long chain amphiphilic lipids. *Journal of Colloid and Interface Science, 505*, 392–401.

Wibowo, D., Zhao, C. X., Peters, B. C., & Middelberg, A. P. (2014). Sustained release of fipronil insecticide in vitro and in vivo from biocompatible silica nanocapsules. *Journal of Agricultural and Food Chemistry, 62*(52), 12504–12511.

Wilpiszewska, K., Spychaj, T., & Paździoch, W. (2016). Carboxymethyl starch/montmorillonite composite microparticles: Properties and controlled release of isoproturon. *Carbohydrate Polymers, 136*, 101–106.

Woodcock, B. A., Bullock, J. M., Shore, R. F., Heard, M. S., Pereira, M. G., Redhead, J., …, & Pywell, R. F. (2017). Country-specific effects of neonicotinoid pesticides on honey bees and wild bees. *Science, 356*(6345), 1393–1395.

Xia, W., Mahmood, A., Zou, R., & Xu, Q. (2015). Metal–organic frameworks and their derived nanostructures for electrochemical energy storage and conversion. *Energy & Environmental Science, 8*(7), 1837–1866.

Xie, Q., Li, Y., Lv, Z., Zhou, H., Yang, X., Chen, J., & Guo, H. (2017). Effective adsorption and removal of phosphate from aqueous solutions and eutrophic water by Fe-based MOFs of MIL-101. *Scientific Reports, 7*(1), 1–15.

Xu, X., Bai, B., Wang, H., & Suo, Y. (2017). A near-infrared and temperature-responsive pesticide release platform through core–shell polydopamine@PNIPAm nanocomposites. *ACS Applied Materials & Interfaces, 9*(7), 6424–6432.

Xu, Z., Gao, Z., & Shao, X. (2018). Light-triggered release of insecticidally active spirotetramat-enol. *Chinese Chemical Letters, 29*(11), 1648–1650.

Yadav, R. K., Singh, N. B., Singh, A., Yadav, V., Bano, C., & Khare, S. (2020). Expanding the horizons of nanotechnology in agriculture: Recent advances, challenges and future perspectives. *Vegetos, 33*(2), 203–221.

Yan, Y., Hou, H., Ren, T., Xu, Y., Wang, Q., & Xu, W. (2013). Utilization of environmental waste cyanobacteria as a pesticide carrier: Studies on controlled release and photostability of avermectin. *Colloids and Surfaces B: Biointerfaces, 102*, 341–347.

Yang, J., Trickett, C. A., Alahmadi, S. B., Alshammari, A. S., & Yaghi, O. M. (2017). Calcium L-lactate frameworks as naturally degradable carriers for pesticides. *Journal of the American Chemical Society, 139*(24), 8118–8121.

Ye, Z., Guo, J., Wu, D., Tan, M., Xiong, X., Yin, Y., & He, G. (2015). Photo-responsive shell cross-linked micelles based on carboxymethyl chitosan and their application in controlled release of pesticide. *Carbohydrate Polymers, 132*, 520–528.

Yoon, K. Y., Byeon, J. H., Park, J. H., & Hwang, J. (2007). Susceptibility constants of Escherichia coli and Bacillus subtilis to silver and copper nanoparticles. *Science of the Total Environment, 373*(2–3), 572–575.

Yu, D. G., Li, J. J., Williams, G. R., & Zhao, M. (2018). Electrospun amorphous solid dispersions of poorly water-soluble drugs: A review. *Journal of Controlled Release, 292*, 91–110.

Yu, X. D., Liu, Z. C., Huang, S. L., Chen, Z. Q., Sun, Y. W., Duan, P. F., …, & Xia, L. Q. (2016). RNAi-mediated plant protection against aphids. *Pest Management Science, 72*(6), 1090–1098.

Zan, M., Li, J., Luo, S., & Ge, Z. (2014). Dual pH-triggered multistage drug delivery systems based on host–guest interaction-associated polymeric nanogels. *Chemical Communications, 50*(58), 7824–7827.

Zaytseva, O., & Neumann, G. (2016). Carbon nanomaterials: Production, impact on plant development, agricultural and environmental applications. *Chemical and Biological Technologies in Agriculture, 3*(1), 1–26.

Zhang, Y., Chen, W., Jing, M., Liu, S., Feng, J., Wu, H., …, & Ma, Z. (2019). Self-assembled mixed micelle loaded with natural pyrethrins as an intelligent nano-insecticide with a novel temperature-responsive release mode. *Chemical Engineering Journal, 361*, 1381–1391.

Zhang, W., He, S., Liu, Y., Geng, Q., Ding, G., Guo, M., …, & Cao, Y. (2014). Preparation and characterization of novel functionalized prochloraz microcapsules using silica–alginate–elements as controlled release carrier materials. *ACS Applied Materials & Interfaces, 6*(14), 11783–11790.

Zhang, Z., Li, M., Chen, W., Zhu, S., Liu, N., & Zhu, L. (2010). Immobilization of lead and cadmium from aqueous solution and contaminated sediment using nano-hydroxyapatite. *Environmental Pollution, 158*(2), 514–519.

Zhang, D. X., Li, B. X., Zhang, X. P., Zhang, Z. Q., Wang, W. C., & Liu, F. (2016). Phoxim microcapsules prepared with polyurea and urea–formaldehyde resins differ in photostability and insecticidal activity. *Journal of Agricultural and Food Chemistry, 64*(14), 2841–2846.

Zhang, Q., Liu, M., & Ruan, J. (2017). Metabolomics analysis reveals the metabolic and functional roles of flavonoids in light-sensitive tea leaves. *BMC Plant Biology, 17*(1), 1–10.

Zheng, M., Falkeborg, M., Zheng, Y., Yang, T., & Xu, X. (2013). Formulation and characterization of nanostructured lipid carriers containing a mixed lipids core. *Colloids and Surfaces A: Physicochemical and Engineering Aspects, 430*, 76–84.

Zhu, Y., Yang, B., Chen, S., & Du, J. (2017). Polymer vesicles: Mechanism, preparation, application, and responsive behavior. *Progress in Polymer Science, 64*, 1–22.

CHAPTER 12

Plant hormones (brassinosteroids, salicylates, and jasmonates) as potential mitigators of pesticide and herbicide stress in plants

Barket Ali
Department of Botany, Government Degree College, Kilhotran, Doda, Jammu and Kashmir, India

12.1 Introduction

The human population is growing at an exponential rate and has recently crossed eight billion mark. The abrupt population explosion has enhanced the demand for more and more food, shelter, and many other requirements. Feeding this ever-increasing population is one of the biggest challenges faced by human. To overcome the challenge, attempts are being made to increase the productivity of many agricultural and horticultural crops by adopting different techniques. The pre- and post-harvest protection of these crops from diverse pests is also equally important, which may be achieved by using different pesticides, insecticides, fungicides, etc. Although these chemicals protect the plants effectively, the side effects are quite serious. Pesticides are generally applied to the plant via the presowing seed treatment, soil application, or foliar spray. One of the most serious problems arising out of the excessive use of pesticides is the disturbance in the soil ecosystem and microflora (Gomes et al., 2017a, b; Xu et al., 2017). The pesticide uptake by the plant is determined by diverse factors such as plant type, pesticide type, stage of plant growth, and other environmental factors (Führ, 1991). Once making entry into a plant via the root or leaf, these chemicals are translocated to the different parts of the plant via the apoplast, xylem, or phloem, accompanied by water or photoassimilate transport (Bell et al., 2011; Shimabukuro et al., 1982). Although plants exhibit a system comprised of enzymes and specific metabolites to counter these toxic compounds (xenobiots), they accumulate in a significant concentration in plants, which results in their biomagnifications (Mwevura et al., 2002; Zhou et al., 2015). The enzymes that facilitate pesticide detoxification include oxidases (e.g., phenol oxidases, peroxidases), reductases (e.g., nitroreductase), esterases, glutathione-S-transferases and uridine 5-diphospho-glucuronosyltransferase (UDPG) transferase (Sharma et al., 2020). It also affects root-microbe interaction, nodulation, nitrogen fixation, plant metabolism, dry matter accumulation, and yield in legumes

Pesticides in the Environment: Impact, Assessment, and Remediation
DOI: https://doi.org/10.1016/B978-0-323-99427-9.00009-4

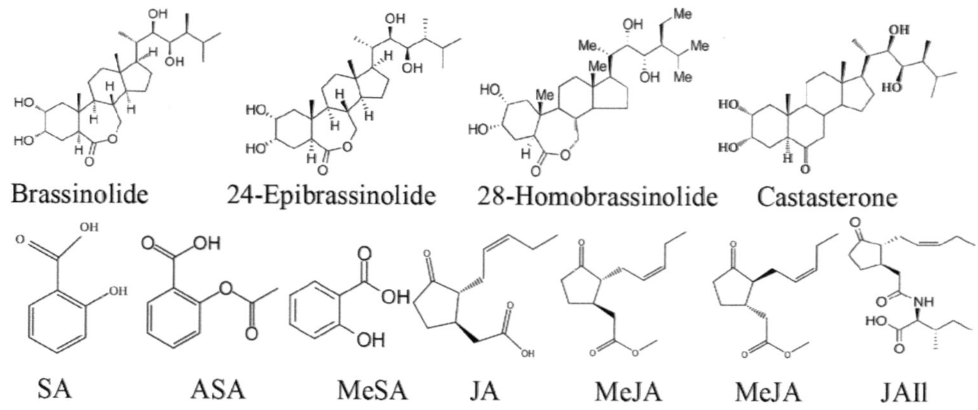

Figure 12.1 Chemical structures of the brassinosteroid (BS) analogues; Brassinolide, 24-Epibrassinolide, 28-Homobrassinolide and Castasterone, Salicylate analogues; Salicylic acid (SA), Acetyl salicylic acid (ASA) and Methylsalicylic acid (MeSA), and Jasmonate analogues; Jasmonic acid (JA), Methyl jasmonates (MeJA) Jasmonoyl isoleucine (JAII).

(Shahid et al., 2020). Chemicals such as insecticides, herbicides, fungicides are reported to decrease seed germination (Dubey & Fulekar, 2011), pigment concentration (Ekmekci & Terzioglu, 2005), photosynthesis (Nemat Alla et al., 2007), amino acid accumulation (Jaleel et al., 2007) plant growth and dry matter accumulation (Huang & Xiong, 2009; Mishra et al., 2009), and oxidative stress and antioxidants (Chagas et al., 2008). All these altered aspects inflict heavy losses on the productivity of the crops.

Plants are bestowed with a wide variety of biochemicals, including plant hormones or phytohormones. Phytohormones are biochemical substances, which are produced at one site and transported to another site in a plant, where at a very low concentration; they exert a profound influence on a physiological response (Taiz & Zeiger, 2010). Earlier five classical groups of phytohormone namely auxins, gibberellins, cytokinins, abscisic acid, and ethylene have been reported. Some more groups namely brassinosteroids (BS), salicylates, jasmonates, and polyamines have also been added to the list of plant hormones during the past couple of decades (Yamaguchi et al., 2010). The recently added classes of plant hormones are represented by important and active analogues viz brassinosteroids; brassinolide, 24-epibrassinolide, 28-homobrassinolide and castasterone, Salicylates; salicylic acid (SA), acetyl salicylic acid (ASA) and methylsalicylic acid (MeSA), and Jasmonates; jasmonic acid (JA), methyl jasmonates (MeJA) and jasmonoyl isoleucine (Fig. 12.1). These plant hormones play an indispensable role in the regulation of plant growth and development at molecular, cellular and whole plant levels. The important phenomena regulated by them include, seed germination, root initiation, and elongation, shoot initiation and elongation, cell division and differentiation, flowering, reproduction, sex determination, fruiting, fruit ripening, seed set, seed dormancy, senescence, etc. (Taiz & Zeiger, 2010). They also enhance the yield of many agricultural and horticultural crops (Ali, 2017a, 2019, 2022) besides enhancing the secondary metabolite production in many others (Ali, 2021a, b; Ali et al., 2021). Some of these hormones have been reported to play an important role in abiotic stress tolerance in plants such as salt stress (Ali et al., 2007, 2008b), water stress (Fariduddin et al., 2009), drought stress, heavy metals (Ali, 2017b; Ali et al., 2008a, b; Hasan et al., 2008; Hayat et al., 2007). In addition to this, they also exert a promising impact on the amelioration of pesticide toxicity in different plants. However, these studies, predominantly related to brassinosteroids, salicylates, and jasmonates are dispersed over a wide range of journals and books. These studies are comprehensively presented here to propose a possible mechanism involved in the plant hormone mediate mitigation of the pesticide and herbicide stress in plants.

12.2 Effect of brassinosteroids on the herbicide and pesticide stress

There is an immense and immediate need to increase food productivity, as the population is increasing rapidly. This can be accomplished by maintaining soil fertility, using high-yielding cultivars, and preventing the losses inflicted by different pests. One of the measures which can help in maintaining soil fertility is the prevention of the growth of weeds/plants that can otherwise exploit the soil nutrients, thereby affecting the desired crop yield. Therefore, herbicides are used to curb these unwanted plants/herbs, which can act as a potential threat to crop productivity. However, the excessive use of these agrochemicals induces oxidative stress (Jabłonska-Trypuc, 2017) and also shows deleterious effects on diverse physiological processes, and the growth and development of the plants (Jan et al., 2020; Singh & Tiwari, 2020).

The induction of oxidative stress is evident from the increased concentration of proline and malondialdehyde in the presence of nicosulfuron herbicide in maize (Liu et al., 2019) and glyphosate herbicide in winter wheat (Shopova et al., 2021b). However, phytohormones including BS show ameliorative effects on the plants receiving herbicides (Varshney et al., 2012). BS application ameliorated the herbicide toxicity via the activation of the antioxidant system; catalase, peroxidase, superoxide dismutase, glutathione S-transferase, and glutathione reductase (Liu et al., 2019; Shopova et al., 2021b). Brassinolide application also improved the performance of foxtail millet (*Setaria italica*) exposed to Sigma Broad herbicide. The antioxidant-mediated mitigation of the stress was visible in the improved photosynthesis-related parameters; chlorophyll content, chlorophyll fuorescence, electron transport, photochemical efficiency of PS II and PSI, net photosynthetic rate (P_N), and plant growth (Yuan et al., 2017). However, BS-induced amelioration of herbicide stress was manifested in the increased biomass and photosynthesis together with the overexpression of glutathione biosynthesis genes (*GST1, ABC-2, ALS1*) in maize plants (Liu et al., 2019).

The use of pesticides in agriculture to protect the crop and the produce from different kinds of pests is unavoidable. However, excessive usage of pesticides results in their accumulation in vegetables, food grains, and fruits, thereby posing various health hazards as well as polluting the environment (Jan et al., 2020). However, plants activate certain enzymes such as methyltransferases, glycosyltransferases, laccases, and ATP-binding cassette transporters to metabolize and detoxify their deleterious effects (Zhang & Yang, 2021). The phytohormones SA, JA, and BS also show some ameliorative responses in plants facing pesticide stress (Jan et al., 2020; Zhang & Yang, 2021). Exogenous application of EBL strengthened the antioxidant system to counter the oxidative stress induced by pesticides imidacloprid in Indian mustard (Sharma et al., 2016, 2017a) and by chlorothalonil in grapevine (Wang et al., 2017a). Similarly, brassinolide (Wang et al., 2017b) and castasterone (Sharma et al., 2019) also enhanced the antioxidative enzymes to attenuate the toxicity generated by the pesticides polychlorinated biphenyls and imidacloprid, respectively (Sharma et al., 2019). The pesticide detoxification was facilitated via the enhanced activities of enzymes such as cytochrome P450 monooxygenase, glutathione synthase, carboxylesterase, glutathione transporter-1 (Sharma et al., 2017b, 2019), together with those associated with pigment metabolism, carbon fixation, respiration, antioxidative enzymes, and phenolic biosynthesis (Sharma et al., 2019). EBL treatment also enhanced the pesticide metabolism by inducing the expression of the *RESPIRATORY BURST OXIDASE HOMOLOG1* (*RBOH1*) gene to attenuate the toxic effects of chlorothalonil pesticide (Hou et al., 2018). Moreover, BS exhibited crosstalk with nitric oxide (NO) and mitogen-activated protein kinase (MPK) (Yin et al., 2016) and mild oxidative stress (Hou et al., 2018, 2019) to increase the expression of genes associated with the metabolism and detoxification of pesticide. It is pertinent to mention here that EBL enhanced the expression of the glutathion redoxin gene (*GRXS25*), which subsequently enhanced the expression and activity of the glutathione S-transferases (*GSTs*) gene and chlorothalonil pesticide metabolism. Here the role of the *GRXS25* gene in pesticide metabolism was confirmed by silencing the *GRXS25* gene which led to the increased accumulation of the pesticide (Hou et al., 2019).

12.3 Effect of salicylates on the herbicide and pesticide stress

Salicylates constitute a class of well-recognized plant hormones, which includes important analogs; salicylic acid, methyl salicylic acid, and acetylsalicylic acid (Fig. 12.1) (Hayat et al., 2007). They are primarily known for their role in systemic acquired resistance (Ding & Ding, 2020). They are known to play diverse physiological roles in plants such as thermogenesis, flowering (Hayat et al., 2007), productivity of horticultural crops (Ali, 2022), secondary metabolite production (Ali, 2021), and abiotic stress tolerance (Ali, 2017; Hayat et al., 2007). One of the most important aspects affected by SA under biotic and abiotic stress conditions is their interaction with the reactive oxygen species ROS signaling (Saleem et al., 2021), besides the regulation of the synthesis and accumulation of proline (Matysik et al., 2002), an important metabolite under stress situations.

In conformity with these trends, salicylates play an important role in pesticide and herbicide resistance in different crop plants. SA treatment of *Brassica napus* plants exposed to Napropamide declined the oxidative stress generated by it. The SA treatment recovered the plant growth that was accompanied by the enhanced antioxidant system (Cui et al., 2010). The SA-mediated enhanced antioxidant system also reduced the oxidative stress in *Vicia faba* plants pre-exposed to alphamethrin and endosulfan insecticides (Singh et al., 2013). Here, the SA-mediated insecticide stress mitigation was reflected in the reduced mitotic index and chromosomal abnormalities besides an increase in the concentration of carotenoids. However, proline concentration exhibited a decline in this study (Singh et al., 2013). In another study with *Vicia faba* plants, a fungicide namely Ridomil MZ was used to counter *Botrytis fabae* fungus. Although the fungicide countered the fungal infection efficiently, it had many deleterious effects on the plant, which were reversed by the SA treatment. The fungicide detoxification was reflected in the improvement in the growth and yield of the stressed plant (Soliman, 2015). Likewise, the barley plants exposed to the pesticide (paraquat) stress showed a deleterious impact on the antioxidant enzyme except for catalase which was elevated by the stress. However, the post-stress treatment of the plants with SA improved the overall performance of the plants, in addition to the antioxidant enzymes (Ananieva et al., 2002, 2004). Likewise, SA application before the seed sowing detoxified the impact of pesticides and herbicides via the elevated antioxidant system in greengram (Fatma et al., 2018). The stress mitigation was manifested in the improvement in the concentration of photosynthetic pigments and other biochemical attributes. A similar pattern of the response was recorded in sunflower plants subjected to the herbicide (aryloxyphenoxy propionate) stress (Bayram et al., 2015). Here SA induced herbicide resistance was accompanied by enhanced antioxidant and photosynthetic pigments beside phenolic compounds. Besides fungicides and herbicides, SA treatment also decreased the toxicity inflicted by some insecticides in cotton plants (Ali, 2008). In this study, SA not only mitigated the insecticide toxicity but also improved the anti-insect potential of the insecticides to counter the *Spodoptera littoralis* larva. The beneficial effect of the SA treatment was evident from the improved growth and yield in the treated plants.

Glyphosate is an important herbicide/weedicide. However, it also exerts some deleterious effects on different crops (Fiori et al., 2012). It exerts multifaceted deleterious effects on different crop plants such as aromatic amino acid biosynthesis via blockade of the shikmic acid pathway (Basantani et al., 2011; Hernandez et al., 1999), photosynthesis and enzyme actives (Ahsan et al., 2008), plant growth and mineral nutrition (Bott et al., 2008), and triggers oxidative stress (Singh et al., 2017). However, exogenous application of SA has been found to control the damage inflicted by the herbicide in maize (Akbulut et al., 2015), tomato (Singh et al., 2017), and wheat (Shopova et al., 2021a). In maize plants, the SA (0.5 mM) treatment mitigated the herbicide toxicity by increasing the activities of antioxidant enzymes (Fig. 12.2) (Akbulut et al., 2015). The SA treatment

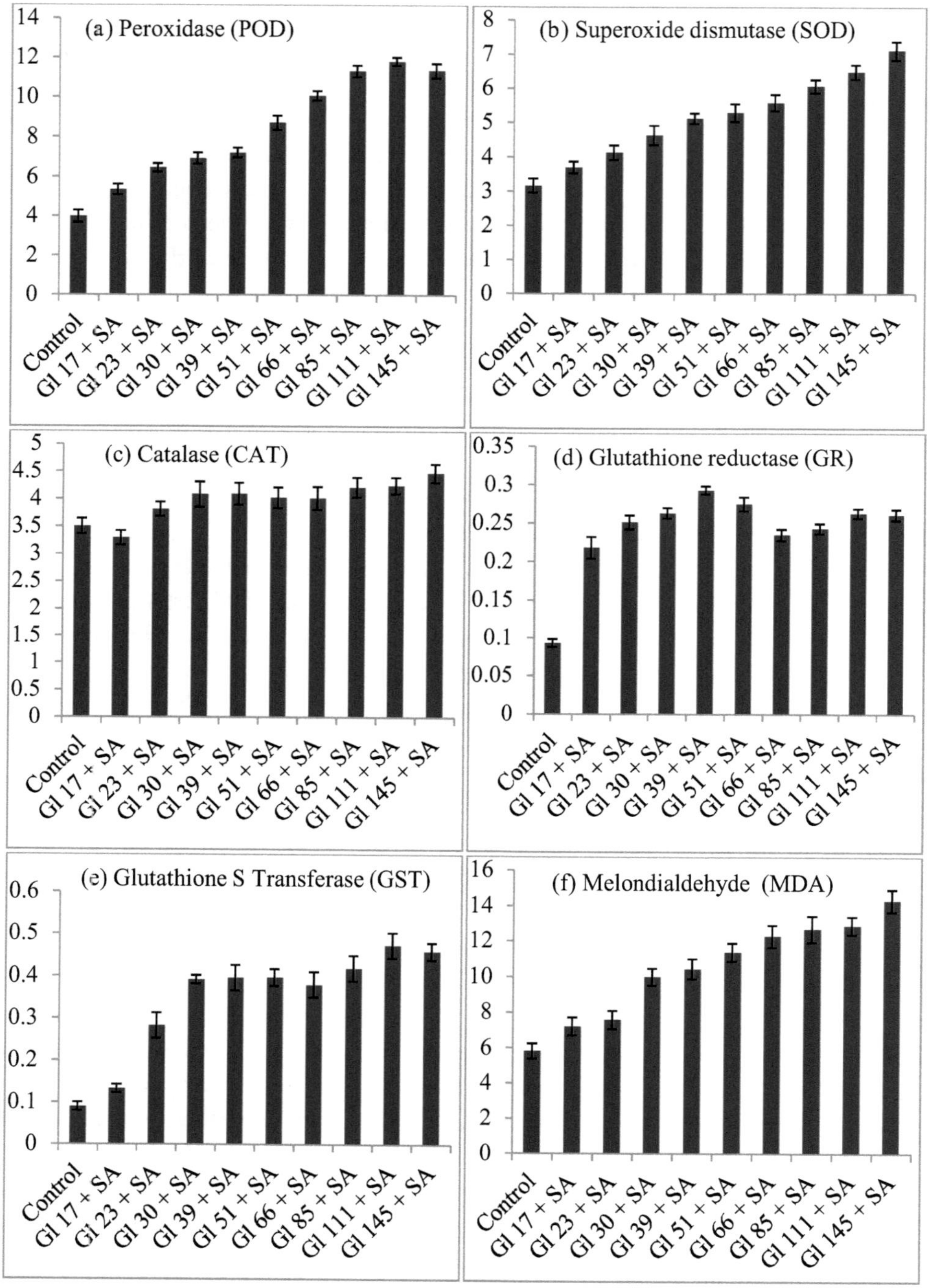

Figure 12.2 Interactive effect of SA and Glyphosate (Gl) in the antioxidant system (a-f) in maize (Akbulut et al., 2015).

also improved the activities of antioxidant enzymes in tomato plants and also declined the phytotoxicity of the herbicide (Singh et al., 2017). Similarly, in wheat plants, a sharp decline in the stress markers was noticed in response to the 1 mM SA treatment. It was denoted as the amelioration of the stress induced by the herbicide. The stress mitigation was evident from the restoration of the plant growth (Shopova et al., 2021a).

Exogenous SA application also ameliorated the toxic effects of other herbicides like clethodim (Radwan, 2012), flurochloridone (Kaya & Yigit, 2014), terbuthylazine, ametryn and atrazine (Yu et al., 2021), and mesosulfuron-methyl + Iodosulfuron-methyl (Khatami et al., 2022). Exposure of maize plants to clethodim herbicide triggered oxidative burst that was evident from the increased concentrations of hydrogen peroxide and melondialdehyde. However, the subsequent treatment of the plants with 1 mM SA mitigated the stress by stimulating the antioxidant enzymes (Radwan, 2012). SA treatment also enhanced the activities of antioxidant enzymes in sunflower plants exposed to flurochloridone herbicide. Moreover, the SA pretreatment also declined the level of residual herbicide, compared to the untreated plants (Kaya & Yigit, 2014). Likewise, the damage inflicted by mesosulfuron-methyl + iodosulfuron-methyl herbicide was reduced by the SA treatment in wheat plants, and subsequently the physiological functions were restored (Khatami et al., 2022). Moreover, the toxic effects of quinclorac herbicide were also neutralized by SA treatment in rice seedlings, which was accomplished by the enhanced activities of antioxidant enzymes and the subsequent decline in oxidative stress (Wang et al., 2020).

12.4 Effect of jasmonates on herbicide and pesticide stress

Like salicylates, jasmonates also comprise a class of naturally occurring plant hormones, represented by jasmonic acid (JA), methyl jasmonate (Me-JA), and jasmonoyl isoleucine (JA-Il) analogs (Fig. 12.1) (Taiz & Zeiger, 2010). They regulate diverse physiological responses in plants, both under normal and stress conditions. However, they are basically known for their role in the mitigation of abiotic stress (salinity, drought, temperature, water logging, and heavy metal) (Avanci et al., 2010; Taiz & Zeiger, 2010) and secondary metabolite elicitation (Ali, 2021b). They also play a promising role in the mitigation of herbicide and pesticide stress which has been demonstrated in a large number of studies.

Jasmonates themselves are efficient pesticides and counter different pathogenic attacks (Wasternack, 2007), including nematodes (Du et al., 2021), fungi (Ávila et al., 2019), and other pathogens (Creelman & Mullet, 1995). They counter pathogenic attacks by activating the specific defence-related genes (Rosahl & Feussner, 2005), which are also quite helpful in repelling various herbivores (Doares et al., 1995; Howe & Jander, 2008; Wasternack & Parthier, 1997). Not only do endogenous jasmonates repel herbivores, even exogenous application of JA helps in protecting the plants from herbivores. In tomato plants, exogenous JS application prevented the possible damage inflicted by the *Helicoverpa zea* herbivore, mediated via the enhanced activities of some important defence-related enzymes including polyphenol oxidase (Thaler et al., 1996). Likewise, the exogenous application of JA also countered the attack of many other herbivores like larvae, caterpillars, aphids, thrips, etc. in tomato plants, besides enhancing crop productivity (Thaler et al., 2001). Similarly, in chilli plants, comparing the effect of jasmonates and insecticides on disease control, the former was more effective than the latter (Awang et al., 2015). Methyl jasmonate (MeJA) has also been reported to counter the herbivore attack mediated via the expression of defense-related genes and subsequent enzymes in *Arabidopsis* (McConn et al., 1997) and spinach (Schmelz et al., 2002). The other possible aspect of the mechanism behind this effect is the stimulation of the biosynthesis and accumulation of JA, besides secondary metabolites (terpenes) and some important enzymes like polyphenol oxidase and peroxidase (Zebelo et al., 2014). Therefore, they can act as an efficient and eco-friendly alternative to chemical pesticides.

Besides acting as endogenous and exogenous herbivore repellents, jasmonates are also demonstrated to mitigate the toxicity generated by different pesticides and herbicides. In *Brassica juncea* plants, the treatment of the seeds with JA mitigated the toxic effects of the pesticide imidacloprid. The ameliorative response was visible from the improvement in the activities of some important enzymes including Ribulose-1,5-bisphosphate carboxylase/oxygenase mediated via the enhanced expression of the respective genes corresponding to them (Sharma et al., 2018). JA also facilitated the degradation of the pesticide chlorothalonil residues mediated via glutathione and nitric oxide, in addition to the enhancement in the activities of antioxidant enzymes and detoxification-related enzymes (Yu et al., 2022). Likewise, the exogenous application of JA detoxified the negative effects of the herbicide imazapic in *Nicotiana tabacum*, which was mediated via the antioxidant enzymes. The stress mitigation was evident from the increase in the pigment concentration and hormonal level, and a sharp decline in the concentrations of MDA and herbicide residue (Kaya & Doganlar, 2016).

12.5 Conclusion

Plant hormones are indispensable biochemicals that regulate the entire growth and development of a plant starting from the seed germination till the formation of new seed and the death of a plant. They also play an important role in the plants exposed to different kinds of biotic and abiotic stress factors, including that of pesticides and herbicides. As an innate response to the oxidative stress triggered by pesticides and herbicides, plants invoke the antioxidant system comprised enzymes and some metabolites such as proline to rescue the plants from stressful situations. Phytohormones significantly enhance the expression of many resistance-related genes such as *GST1, ABC-2, ALS, and GRXS25, RESPIRATORY BURST OXIDASE HOMOLOG1* (*RBOH1*), etc. which are quite helpful in the pesticide and herbicide stress mitigation. The stress mitigation by plant

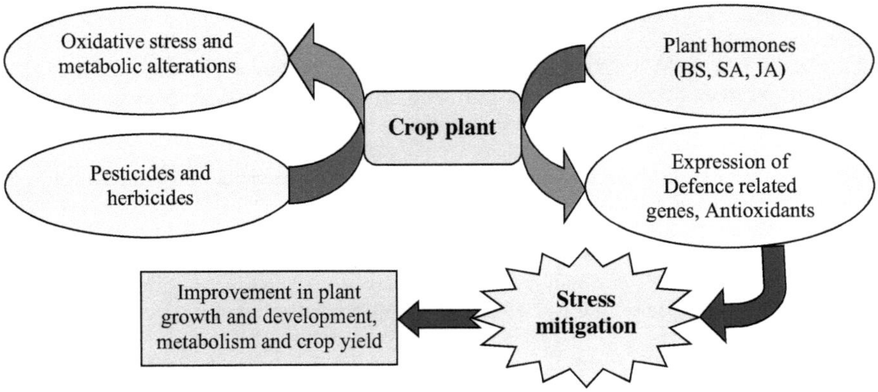

Figure 12.3 Flow diagram depicting the mechanism of the phytohormone regulated mitigation of pesticide and herbicide stress in plants.

hormones is evident from the improvement in different physiological parameters as well as the yield of the crop (Fig. 12.3). The most important aspect of the plant hormones as the pesticide and herbicide stress mitigators are their ecological feasibility and affordability.

References

Akbulut, G. B., Yigit, E., & Bayram, D. (2015). Investigation of the effects of salicylic acid on some biochemical parameters in *Zea mays* to glyphosate herbicide. *Journal of Environmental & Analytical Toxicology, 5*, 271. https://doi.org/10.4172/2161-0525.1000271.

Ali, B. (2017a). Practical applications of brassinosteroids in horticulture: Some field perspectives. *Scientia Horticulturae, 225*, 15–21. https://doi.org/10.1016/j.scienta.2017.06.051.

Ali, B. (2017b). Salicylic acid induced antioxidant system enhances the tolerance to aluminium in mung bean (*Vigna radiate* L. Wilczek) plants. *Journal of Plant Physiology, 22*(2), 178–189. https://doi.org/10.1007/s40502-017-0292-1.

Ali, B. (2019). Brassinosteroids: The promising plant growth regulators in horticulture. In S. Hayat, M. Yusuf, R. Bhardwaj, & A. Bajguz (Eds.), *Brassinosteroids: Plant growth and development* (pp. 349–365). Springer. https://doi.org/10.1007/978-981-13-6058-9_12.

Ali, B. (2022). Exploiting the recuperative potential of brassinosteroids in agriculture. In A. Krishana, & G. Sirhindi (Eds.), *Jasmonates and rassinosteroids in lants: Metabolism, signaling and applications* (pp. 177–185). Tailor & Francis Group. https://doi.org/10.1201/9781003110651-16.

Ali, B. (2021a). Salicylic acid: An efficient elicitor of secondary metabolite production in plants. *Biocatalysis and Agricultural Biotechnology, 31*, 101884. https://doi.org/10.1016/j.bcab.2020.101884.

Ali, B. (2021b). Practical applications of jasmonates in the biosynthesis and accumulation of secondary metabolites in plants. *Biocatalysis and Agricultural Biotechnology, 38*, 102205. https://doi.org/10.1016/j.bcab.2021.102205.

Ali, B. (2022). Role of salicylic acid in pre- and post-harvest attributes in horticulture. *Salicylic acid: A versatile plant hormone*, 47–64. https://doi. org/10. 1007/978-3-030-79229-9_4.

Ali, B., Hasan, S. A., Hayat, S., Hayat, Q., Yadav, S., Fariduddin, Q., & Ahmad, A. (2008a). A role for brassinosteroids in the amelioration of aluminium stress through antioxidant system in mung bean (*Vigna radiate* L. Wilczek). *Environmental and Experimental Botany, 62*, 153–159.

Ali, B., Hayat, S., & Ahmad, A. (2007). 28-Homobrassinolide ameliorates the saline stress in chickpea (*Cicer arietinum* L). *Environmental and Experimental Botany, 59*, 217–223.

Ali, B., Hayat, S., Fariduddin, Q., & Ahmad, A. (2008b). 24-Epibrassinolide protects against the stress generated by salinity and nickel in *Brassica juncea*. *Chemosphere, 72*, 1387–1392.

Ali, B., Wani, Z. A., & Dar, M. I. (2021). Brassinosteroids- the phytohormones with potential to enhance the secondary metabolite production in plants. In M. Shahnawaz (Ed.), *Biotechnological approach to enhance plant secondary metabolites: Recent trends and future prospects* (pp. 125–131). Tailor & Francis Group. https://doi.org/10.1201/9781003034957-8.

Ali, O. S. (2008). Effect of salicylic acid and its mixtures with three insecticides on some cotton insect pests. *Doctoral dissertation, Faculty of Agriculture,* Tanta University.

Ananieva, E. A., Alexieva, V. S., & Popova, L. P. (2002). Treatment with salicylic acid decreases the effects of paraquat on photosynthesis. *Journal of Plant Physiology, 159*, 685–693.

Ananieva, E. A., Christov, K. N., & Popova, L. P. (2004). Exogenous treatment with salicylic acid leads to increased antioxidant capacity in leaves of barley plants exposed to paraquat. *Journal of Plant Physiology, 161*, 319–328.

Ashan, N., Lee, D. G., Lee, K. W., Alam, I., Lee, S. H., Bahk, J. D., & Lee, B. H. (2008). Glyphosate induced oxidative stress in rice leaves revealed by proteomic approach. *Plant Physiology and Biochemistry, 46*, 1062–1070.

Avanci, N. C., Luche, D. D., Goldman, G. H., & Goldman, M. H. S. (2010). Jasmonates are Jasmonates are phytohormones with multiple functions, including plant defense and reproduction. *Genetics and Molecular Research (Electronic Resource), 9*, 484–505.

Ávila, A. C., Ochoa, J., Proaño, K., & Martínez, K. C. (2019). Jasmonic acid and nitric oxide protects naranjilla (*Solanum quitoense*) against infection by *Fusarium oxysporum* f. sp. Quitoense by eliciting plant defense responses. *Physiological and Molecular Plant Pathology, 106*, 129–136. https://doi.org/10.1016/j.pmpp.2019.01.002.

Awang, N. A. A., Ismail, M. R., Omar, D., & Islam, M. R. (2015). Comparative study of the application of jasmonic acid and pesticide in chilli: Effects on physiological activities, yield and viruses control. *Journal of Biosciences, 31*, 672–681.

Basantani, W. M. I., Srivastava, A., & Sen, S. (2011). Elevated antioxidant response and induction of tau-class glutathione S-transferase after glyphosate treatment in *Vigna radiata* (L.). *Pesticide Biochemistry and Physiology, 99*, 111–117.

Bayram, D. D., Yigit, E., & Akbulut, G. B. (2015). The effects of salicylic acid on *Helianthus annuus* L. exposed to Quizalofop-P-Ethyl. *American Journal of Plant Sciences, 6*, 2412.

Bell, J. L., Burke, I. C., & Prather, T. S. (2011). Uptake, translocation and metabolism of aminocyclopyrachlor in prickly lettuce, rush skeleton weed and yellow star thistle. *Pest Management Science, 67*, 1338–1348.

Bott, S., Tesfamariam, T., Candan, H., Cakmak, I., Römheld, V., & Neumann, G. (2008). Glyphosate-induced impairment of plant growth and micronutrient status in glyphosate-resistant soybean (*Glycine max* L.). *Plant and Soil, 312*, 185–194.

Chagas, R. M., Silveira, J. A. G., Ribeiro, R. V., Vitorello, V. A., & Carrer, H. (2008). Photochemical damage and comparative performance of superoxide dismutase and ascorbate peroxidase in sugarcane leaves exposed to paraquat-induced oxidative stress. *Pesticide Biochemistry and Physiology, 90*, 181–188.

Creelman, R. A., & Mullet, J. E. (1995). Jasmonic acid distribution and action in plants: Regulation during development and response to biotic and abiotic stress. *Proceedings of the National Academy of Sciences of the United States of America, 92*, 4114–4119.

Cui, J., Zhang, R., Wu, G. L., Zhu, H. M., & Yang, H. (2010). Salicylic acid reduces napropamide toxicity by preventing its accumulation in rapeseed (*Brassica napus* L.). *Archives of Environmental Contamination and Toxicology, 59*, 100–108.

Ding, P., & Ding, Y. (2020). Stories of salicylic acid: A plant defense hormone. *Trends in Plant Science, 25*, 549–565. https://doi.org/10.1016/j.tplants.2020.01.004.

Doares, S. H., Syrovets, T., Weiler, E. W., & Ryan, C. A. (1995). Oligogalacturonides and chitosan activate plant defensive genes through the octadecanoid pathway. *Proceedings of the National Academy of Sciences of the United States of America, 92*, 4095–4098.

Du, C., Shen, F., Li, Y., Zhao, Z., Xu, X., Jiang, J., & Li, J. (2021). Effects of salicylic acid, jasmonic acid and reactive oxygen species on the resistance of *Solanum peruvianum* to *Meloidogyne incognita*. *Scientia Horticulturae, 275*, 109649. https://doi.org/10.1016/j.scienta.2020.109649.

Dubey, K. K., & Fulekar, M. H. (2013). Investigation of potential rhizospheric isolate for cypermethrin degradation. *3 Biotech, 3*, 33–43.

Ekmekci, Y., & Terzioglu, S. (2005). Effects of oxidative stress induced by paraquat on wild and cultivated wheat. *Pesticide Biochemistry and Physiology, 83*, 69–81.

Fariduddin, Q., Khanam, S., Hasan, S. A., Ali, B., Hayat, S., & Ahmad, A. (2009). Effect of 28-homobrassinolide on the drought stress-induced changes in photosynthesis and antioxidant system of *Brassica juncea* L. *Acta Physiologiae Plantarum, 31*(5), 141–146.

Fatma, F., Kamal, A., & Srivastava, A. (2018). Exogenous application of salicylic acid mitigates the toxic effect of pesticides in *Vigna radiata* (L.) Wilczek. *Journal of Plant Growth Regulation, 37*, 1185–1194. https://doi.org/10.1007/s00344-018-9819-6.

Fiori, E., Di, H., Pizarro, M., Dos Santos, A., & Cataldo, D. (2012). Impact of the invasive mussel *Limno perna* fortunei on glyphosate concentration in water. *Ecotoxicology and Environmental Safety, 81*, 106–113.

Führ, F. (1991). Radiotracers in pesticide studies-advantages and limitations. *Science Cult, 43*, 211–216.

Gomes, Y. C. B., Dalchiavon, F. C., & Valadão, F. C. D. A. (2017a). Joint use of fungicides, insecticides and inoculants in the treatment of soybean seeds. *Revista Ceres, 64*(3), 258–265.

Gomes, A. R., Justino, C., Rocha-Santos, T., Freitas, A. C., Duarte, A. C., & Pereira, R. (2017b). Review of the ecotoxicological effects of emerging contaminants on soil biota. *Journal of Environmental Science and Health, Part A, 52*, 992–1007. https://doi.org/10.1080/10934529.2017.1328946.

Hasan, S. A., Hayat, S., Ali, B., & Ahmad, A. (2008). 28-homobrassinolide protects chickpea (*Cicer arietinum*) from cadmium toxicity by stimulating antioxidants. *Environmental Pollution, 151*, 60–66. https://doi.org/10.1016/j.envpol.2007.03.006.

Hayat, S., Ali, B., & Ahmad, A (2007). Salicylic acid: Biosynthesis, metabolism and physiological in plant. In S. Hayat, & A. Ahmad (Eds.), *Salicylic acid: A plant hormone* (pp. 1–14). Dordrecht Heidelberg. https://doi.org/10.1007/1-4020-5184-0_1.

Hayat, S., Ali, B., Hasan, S. A., & Ahmad, A. (2007). Brassinosteroid enhanced the level of antioxidants under cadmium stress in *Brassica juncea*. *Environmental and Experimental Botany, 60*, 33–41.

Hernandez, A., Garcia-Plazaola, J. I., & Becerril, J. M. (1999). Glyphosate effects on phenolic metabolism of nodulated soybean (*Glycine max* L. Merr.). *Journal of Agricultural and Food Chemistry, 47*, 2920–2925.

Hou, J. Y., Sun, Q., Li, J., Ahammed, G. J., Yu, J. Q., Fang, H., & Xia, X. J. (2019). Glutaredoxin S25 and its interacting TGACG motif-binding factor TGA2 mediate brassinosteroid-induced chlorothalonil metabolism in tomato plants. *Environmental Pollution, 255*, 113256.

Hou, J. Y., Zhang, Q. H., Zhou, Y., Ahammed, G. J., Zhou, Y. H., Yu, J. Q., Fang, H., & Xia, X. J. (2018). Glutaredoxin *GRXS16* mediates brassinosteroid-induced apoplastic H_2O_2 production to promote pesticide metabolism in tomato. *Environmental Pollution, 240*, 227–234.

Howe, G. A., & Jander, G. (2008). Plant immunity to insect herbivores. *Annual Review of Plant Biology, 59*, 41–66.

Huang, H., & Xiong, Z. T. (2009). Toxic effects of cadmium, acetochlor and bensulfuron-methyl on nitrogen metabolism and plant growth in rice seedlings. *Pesticide Biochemistry and Physiology, 94*, 64–67.

Jabłonska-Trypuc, A. (2017). Pesticides as inducers of oxidative stress. *Reactive Oxygen Species, 3*(8), 96–110.

Jaleel, C. A., Gopi, R., & Pannerselvam, R. (2007). Alterations in lipid peroxidation electrolyte leakage, and proline metabolism in *Catharanthus roseusunder* treatment with triadimefon, a systemic fungicide. *Comptes Rendus Biologies, 330*, 905–912.

Jan, S., Singh, R., Bhardwaj, R., Ahmad, P., & Kapoor, D. (2020). Plant growth regulators: A sustainable approach to combat pesticide toxicity. *3 Biotechnol, 10*, 466.

Kaya, A., & Doganlar, Z. B. (2016). Exogenous jasmonic acid induces stress tolerance in tobacco (*Nicotiana tabacum*) exposed to imazapic. *Ecotoxicology and Environmental Safety, 124*, 470–479. https://doi.org/10.1016/j.ecoenv.2015.11.026.

Kaya, A., & Yigit, E. (2014). The physiological and biochemical effects of salicylic acid on sunflowers (*Helianthus annuus*) exposed toflurochloridone. *Ecotoxicology and Environmental Safety, 106*, 232–238. https://doi.org/10.1016/j.ecoenv.2014.04.041.

Khatami, S. A., Barmaki, M., Alebrahim, M. T., & Bajwa, A. A. (2022). Salicylic acid pre-treatment reduces the physiological damage caused by the herbicide *Mesosulfuron-methyl + Iodosulfuron-methyl* in wheat (*Triticum aestivum*). *Agronomy, 12*, 3053. https://doi.org/10.3390/agronomy12123053.

Liu, S., He, Y., Tian, H., Yu, C., Tan, H., Li, Z., & Duan, L. (2019). Application of brassinosteroid mimetics improves growth and tolerance of maize to nicosulfuron toxicity. *Journal of Plant Growth Regulation, 38*, 701–712.

Matysik, J., Alia, B. B., & Mohanty, P. (2002). Molecular mechanisms of quenching of reactive oxygen species by proline under stress in plants. *Current Science, 82*, 525–532.

McConn, M., Creelman, R. A., Bell, E., Millet, J. E., & Browse, J. (1997). Jasmonate is essential for insect defense in Arabidopsis. *Proceedings of the National Academy of Sciences of the United States of America, 93*, 5473–5477.

Mishra, V., Srivastava, G., & Prasad, S. M. (2009). Antioxidant response of bitter gourd (*Momordica charantia* L.) seedlings to interactive effect of dimethoate and UV-B irradiation. *Scientia Horticulturae, 120*, 373–378.

Mwevura, H., Othman, O. C., & Mhehe, G. L. (2002). Organochlorine pesticide residues in sediments and biota from the coastal area of Dar-es-Salaam city. *Tanzania Marine Pollution Bulletin, 45*, 262–267.

Nemat Alla, M. M., Hassan, N. M., & El-Bastawisy, Z. M. (2007). Differential influence of herbicide treatments on activity and kinetic parameters of C_4 photosynthetic enzymes from *Zea mays*. *Pesticide Biochemistry and Physiology, 89*, 198–205.

Radwan, D. E. M. (2012). Salicylic acid induced alleviation of oxidative stress caused by clethodim in maize (*Zea mays* L.) leaves. *Pesticide Biochemistry and Physiology, 102*, 182–188. https://doi.org/10.1016/j.pestbp.2012.01.002.

Rosahl, S., & Feussner, I. (2005). Oxylipins. In D. J. Murphy (Ed.), *Plant lipids: Biology, utilization, and manipulation* (pp. 329–354). Blackwell Publishing.

Saleem, M., Fariduddin, Q., & Castroverde, C. D. M. (2021). Salicylic acid: A key regulator of redox signalling and plant immunity. *Plant Physiology and Biochemistry, 168*, 381–397. https://doi.org/10.1016/j.plaphy.2021.10.011.

Schmelz, E. A., Grebenok, R. J., Ohnmeiss, T. E., & Bowers, W. S. (2002). Interactions between *Spinacia oleracea* and *Bradysia impatiens*: A role for phytoecdysteroids. *Archives of Insect Biochemistry and Physiology, 51*, 204–221. https://doi.org/10.1002/arch.10062.

Shahid, M., Khan, M. S., & Zaidi, A. (2020). Fungicide toxicity to legumes and its microbial remediation: A current perspective. In P. K. Srivastava, V. P. Singh, A. Singh, D. K. Tripathi, S. Singh, S. M. Prasad, & D. K. Chauhan (Eds.), *Pesticides in crop production: Physiological and biochemical action* (1st ed., pp. 15–34). John Wiley & Sons Ltd.

Sharma, A., Kumar, V., Yuan, H., Kanwar, M. K., Bhardwaj, R., Thukral, A. K., & Zheng, B. (2018). Jasmonic acid seed treatment stimulates insecticide detoxification in *Brassica juncea* L. *Frontiers in Plant Science, 9*, 1609. https://doi.org/10.3389/fpls.2018.01609.

Sharma, A., Kumar, V., Kanwar, M. K., Thukral, A. K., & Bhardwaj, R. (2017a). Ameliorating imidacloprid induced oxidative stress by 24-epibrassinolide in *Brassica juncea* L. *Russian Journal of Plant Physiology (Translation of Fiziologiya Rastenii (Moscow)), 64*(4), 509–517.

Sharma, A., Kumar, V., Singh, R., Thukral, A. K., & Bhardwaj, R. (2016). Effect of seed pre-soaking with 24-epibrassinolide on growth and photosynthetic parameters of *Brassica juncea* L. in imidacloprid soil. *Ecotoxicology and Environmental Safety, 133*, 195–201.

Sharma, A., Kumar, V., Kohli, S. K., Kaur, R., Kaur, T., Arora, S., Thukral, A. K., & Bhardwa, R. (2020). Pesticide metabolism in plants, insects, soil microbes and fishes: An Overview. In P. K. Srivastava, V. P. Singh, A. Singh, D. K. Tripathi, S. Singh, S. M. Prasad, & D. K. Chauhan (Eds.), *Pesticides in crop production: Physiological and biochemical action* (1st ed., pp. 35–53). John Wiley & Sons Ltd.

Sharma, A., Thakur, S., Kumar, V., Kesavan, A. K., Thukral, A. K., & Bhardwaj, R. (2017b). 24-epibrassinolide stimulates imidacloprid detoxification by modulating the gene expression of *Brassica juncea* L. *BMC Plant Biology, 17*(1), 5336.

Sharma, A., Yuan, H., Kumar, V., Ramakrishnan, M., Kohli, S. K., Kaur, R., Thukral, A. K., Bhardwaj, R., & Zheng, B. (2019). Castasterone attenuates insecticide induced phytotoxicity in mustard. *Ecotoxicology and Environmental Safety, 179*, 50–61.

Shimabukuro, R. H., Lamoureux, G. L., & Frear, D. S. (1982). Pesticide metabolism in plants reactions and mechanisms. In F. Matsumura, & C. R. K. Murti (Eds.), *Biodegradation of pesticides* (pp. 21–66). Plenum Press.

Shopova, E., Brankova, L., Katerova, Z., Dimitrova, L., Todorova, D., Sergiev, I., & Talaat, N. B. (2021a). Salicylic acid pretreatment modulates wheat responses to glyphosate. *Crops, 1*, 88–96. https://doi.org/10.3390/crops1020009.

Shopova, E., Katerova, Z., Brankova, L., Dimitrova, L., Sergiev, I., Todorova, D., & Talaat, N. B. (2021b). Modulation of physiological stress response of *Triticum aestivum* L. to glyphosate by brassinosteroid application. *Life, 11*, 1156.

Singh, A., Srivastava, A. K., & Singh, A. K. (2013). Exogenous application of salicylic acid to alleviate the toxic effects of insecticides in *Vicia faba* L. *Environmental Toxicology, 28*, 666–672.

Singh, H., Singh, N. B., Singh, A., & Hussain, H. (2017). Exogenous application of salicylic acid to alleviate glyphosate stress in *Solanum lycopersicum*. *International Journal of Vegetable Science, 23*, 552–566. https://doi.org/10.1080/19315260.2017.1347845.

Singh, S., & Tiwari, S. (2020). Responses of plants to herbicides: Recent advances and future prospective. In D. K. Tripathi, V. P. Singh, D. K. Chauhan, S. Sharma, S. M. Prasad, N. K. Dubey, & N. Ramawat (Eds.), *Plant life under changing environment* (pp. 237–250). Elsevier. https://doi.org/10.1016/B978-0-12-818204-8.00011-4.

Soliman, A. H. (2015). Shikimic acid and salicylic acid induced protection on growth vigor, seed yield and biochemical aspects of yielded seeds of *Vicia faba* plants infected by *Botrytis fabae*. *Journal of Plant Pathology & Microbiology, 6*, 2. https://doi.org/10.4172/2157-7471.1000304.

Taiz, L., & Zeiger, E. (2010). *Plant Physiology* (pp. 607–611). Sinauer Associates.

Thaler, J. S., Stout, M. J., Karban, R., & Duffey, S. S. (1996). Exogenous jasmonates simulate insect wounding in tomato plants, *Lycopersicon esculentum*, in the laboratory and field. *Journal of Chemical Ecology, 22*, 1767–1781.

Thaler, J. S., Stout, M. J., Karban, R., & Duffey, S. S. (2001). Jasmonate-mediated induced plant resistance affects a community of herbivores. *Ecological Entomology, 26*, 312–324.

Varshney, S., Hayat, H., Alyemeni, M. N., & Ahmad, A. (2012). Effects of herbicide applications in wheat fields. *Plant Signaling & Behavior,* 7(5), 570–575.

Wang, J., Zhang, C., Shi, Y., Long, M., Islam, F., Yang, C., Yang, S., He, Y., & Zhou, W. (2020). Evaluation of quinclorac toxicity and alleviation by salicylic acid in rice seedlings using ground-based visible/near-infrared hyperspectral imaging. *Plant Methods, 16*, 30. https://doi.org/10.1186/s13007-020-00576-7.

Wang, X., Teng, Y., Zhang, N., Christie, P., Li, Z., Luo, Y., & Wang, J. (2017b). Rhizobial symbiosis alleviates polychlorinated biphenyls-induced systematic oxidative stress via brassinosteroids signaling in alfalfa. *Science of the Total Environment, 592*, 68–77.

Wang, Z., Jiang, Y., Peng, X., Xu, S., Zhang, H., Gao, J., & Xi, Z. (2017a). Exogenous 24-epibrassinolide regulates antioxidant and pesticide detoxification systems in grapevine after chlorothalonil treatment. *Plant Growth Regulation, 81*, 455–466.

Wasternack, C. (2007). Jasmonates: An update on biosynthesis, signal transduction and action in plant stress response, growth and development. *Annals of Botany, 100*, 681–697.

Wasternack, C., & Parthier, B. (1997). Jasmonate-signaled plant gene expression. *Trends in Plant Science, 2*, 302–307.

Xu, Z., Huan, Z., Luo, J., & Xie, D. (2017). Simultaneous determination of eight pesticide residues in cowpeas by GC–ECD. *Journal of Chromatographic Science, 55*(1), 1–6.

Yamaguchi, I., Cohen, J. D., Culler, A. H., Quint, M., Slovin, J. P., Nakajima, M., Yamaguchi, S., Sakakibara, H., Kuroha, T., Hirai, N., Yokota, T., Ohta, H., Kobayashi, Y., Mori, H., & Sakagami, Y. (2010). Plant hormones. In H.-W. (Ben) Liu, & L. Mander (Eds.), *Comprehensive natural products II* (pp. 9–125). Elsevier. https://doi.org/10.1016/B978-008045382-8.00092-7.

Yin, Y. L., Zhou, Y., Zhou, Y. H., Shi, K., Zhou, J., Yu, Y., & Xia, X. J. (2016). Interplay between mitogen-activated protein kinase and nitric oxide in brassinosteroid-induced pesticide metabolism in *Solanum lycopersicum*. *Journal of Hazardous Materials, 316*, 221–231.

Yu, G.-B., Chen, R.-N., Chen, Q.-S., Chen, F.-Q., Liu, H.-L., Ren, C.-L., Zhang, Y. X., Yang, F.-J., & Wei, J.-P. (2022). Jasmonic acid promotes glutathione assisted degradation of chlorothalonil during tomato growth. *Ecotoxicology and Environmental Safety, 233*, 113296. https://doi.org/10.1016/j.ecoenv.2022.113296.

Yu, Q. Q., Lu, F. F., Ma, L. Y., Yang, H., & Song, N. H. (2021). Residues of reduced herbicides Terbuthylazine, Ametryn, and Atrazine and toxicology to maize and the environment through salicylic acid. *ACS Omega, 6*, 27396–27404. https://doi.org/10.1021/acsomega.1c04315.

Yuan, X.-Y., Zhang, L.-G., Huang, L., Yang, H.-J., Zhong, Y.-T., Ning, N., Wen, Y.-Y., Dong, S.-Q., Song, X.-E., Wang, H.-F., & Guo, P.-Y. (2017). Spraying brassinolide improves Sigma Broad tolerance in foxtail millet (*Setaria italic* L.) through modulation of antioxidant activity and photosynthetic capacity. *Scientific Reports, 7*, 11232.

Zebelo, S., Piorkowski, J., Disi, J., & Fadamiro, H. (2014). Secretions from the ventral eversible gland of *Spodoptera exigua* caterpillars activate defense- related genes and induce emission of volatile organic compounds in tomato, *Solanum lycopersicum*. *BMC Plant Biology, 14*, 140.

Zhang, J. J., & Yang, H. (2021). Metabolism and detoxification of pesticides in plants. *Science of the Total Environment, 790*, 148034.

Zhou, Y., Xia, X., Yu, G., Wang, J., Wu, J., Wang, M., Yang, Y., Shi, K., Yu, Y., Chen, Z., Gan, J., & Yu, J. (2015). Brassinosteroids play a critical role in the regulation of pesticide metabolism in crop plants. *Scientific Reports, 5*, 9018. https://doi.org/10.1038/srep09018.

CHAPTER 13

Alternatives to chemical pesticides: Current trends and future implications

Sumit Singh[a], Shivali Puri[b] and Satwinder Kaur Sohal[a]
[a]Department of Zoology, Guru Nanak Dev University, Amritsar, Punjab, India
[b]Department of Zoology, DAV University, Jalandhar, Punjab, India

13.1 Introduction

In the global scenario, the agricultural sector plays a major role in the growth and development of a country as plants provide the principal source of nutrition to livestock and about 80% of the total food consumed by the human population is obtained from plants (Rizzo et al., 2021). Advancement in agriculture not only improves yield and economy but indirectly influences the health of humans and other animals. Presently, our agricultural community faces two major challenges; one is production, while the other is the protection of crops. Production refers to adequately meeting the growing global demand for food, whereas, protection deals with alternatives to various conventional agricultural approaches that can prevent the deterioration of the environment in terms of air, water, and land (Koul, 2011; Verma et al., 2021). Globally, crop losses occur due to diseases, nematodes, insects, rodents, etc., and are considered as major constraints in agriculture growth (Mpandeli & Maponya, 2014). An additional factor is the worldwide decline in the population of major pollinators which results in low-quality production of crops (Cappa et al., 2022). So far, halogenated organophosphates and carbamates were widely used as synthetic pesticides to meet the global food demand (Sharma et al., 2019). Although chemical pesticides have proven successful in reducing pest populations, it is also well known that their application can have adverse effects on nontarget individuals especially on insect pollinators and crops itself creating havoc in the environment (Samada & Tambunan, 2020; Shao & Zhang, 2017). Also, excessive use for longer duration results in pest resistance to these synthetic pesticides by modulating their receptors involved in the mechanism of action (Casida & Durkin, 2013; Nawaz et al., 2016; Rattan, 2010). Further, due to their biomagnification and persistent nature, extensive pesticide usage, in the long run, can cause serious health hazards for living beings and the environment by direct or indirect contamination of air, water, soil, and the overall ecosystem (Sharma et al., 2019). Recently, Sharma et al. (2019) have thoroughly and critically reviewed the global usage of different types of synthetic pesticides, their contamination status, and their

Pesticides in the Environment: Impact, Assessment, and Remediation
DOI: https://doi.org/10.1016/B978-0-323-99427-9.00003-3

various adverse impacts on the global ecosystem. Organochlorine pesticides, amongst the different classes of chemical pesticides, are the most destructive ones due to their greater stability, slow rate of decomposition, and longer half-life (Sharma et al., 2019). Therefore, eco-friendly management of pests using other alternatives is the need of the hour for sustainable production and protection of crop plants.

Biopesticides can be defined as "naturally occurring agents or compounds derived from plants, microorganisms (including bacteria, microalgae, and cyanobacteria) and other animals that are used for the control and management of agricultural pests" (Kumar et al., 2021). As per the United States Environmental Protection Agency (EPA), biopesticides are "derived from natural materials like plants, animals, viruses, bacteria, and certain minerals." Products like genes or other metabolites obtained from such biocontrol agents can also be used for crop protection (EPA (Environmental Protection Agency) 2021; Kumar et al., 2021). Since biopesticides are plant-, animal-, and microorganism-based pesticides or their derived products as well as natural products, they can be an effective alternative against synthetic pesticides for sustainable agriculture production and protection (Beas-Catena et al., 2014; Manda et al., 2020; Mazid et al., 2011; Olson, 2015; Prabha et al., 2016; Regnault-Roger, 2012; Verma et al., 2021). Historically, studies prove that plants and plant-derived products, essential oils (EOs), and microorganisms such as viruses, fungi, bacteria, etc. have a great potential for pest management (Bateman et al., 2021; Faria & Rodrigeuz, 2021; Marwal et al., 2022; Santos et al., 2022). In comparison to synthetic chemical pesticides, biopesticides are eco-friendly, nontoxic to natural enemies/predators of pests, and can help in reducing the selection pressure for the evolution of pesticide resistance in pest populations (Kumar & Singh, 2015; Senthil-Nathan, 2015). If the target specificity/spectra and the environmental fate of a biopesticide are known, it can provide as a potential benign substitute for synthetic pesticides (Schnarr et al., 2022). All these characteristics make these pesticides appropriate for inclusion in IPM strategies.

This chapter summarizes various biopesticides with their modes of action, their current status, associated limitations and challenges for improved effect and global commercialization, and the future prospects in agricultural crop management.

13.2 Advantages of using biopesticides over conventional chemical pesticides

Presently, there are approximately 50,000 species of plant pathogens, 9000 species of insects and mites, and 8000 species of weeds that damage crops (Zhang et al., 2011), of which insect pests cause maximum crop loss (Pimentel, 2009a, b). Exclusive reliance on chemical pesticides to manage insect pests and increase the crop yield has, in the long run, deteriorated the environment and human health with adverse effects on other organisms as well (Sharma et al., 2019). Organic pesticides have now been considered as the best alternatives to chemical pesticides. They are very effective even in small quantities and decompose rapidly, which results in significantly less pollution and exposure. Nonhazardous residual properties allow their application even before harvesting stage. They are generally less toxic to other animals, cause minimum environmental pollution, highly target specific and harmless to beneficial insects than conventional pesticides (Razaq & Shah, 2022). Further, biopesticides do not consist of heavy metals and are formed entirely of biodegradable materials and hence, soil damage by biopesticides application is also negligible in comparison to chemical ones (Butu et al., 2020). The application of biological pesticides, therefore, allows the preservation and maintenance of a good state of soil and the reduction in its pollution indices, thereby, beneficial not only to human health but also to land usage and economic potential (Butu et al., 2020). Biological pest control also encourages beneficial soil micro-floral communities leading to increased crop yield (Kumari et al., 2022; Patil & Solanki, 2016). The improved performance, cost-competitiveness and reduced environmental costs of biological pesticides result in declined production of new chemical pesticides (Starnes et al., 1993). The major functional characteristic that differentiates biopesticides from synthetic pesticides is their mode of action. Although most, if not all, chemical/synthetic pesticides are neurotoxic to their target pests, biopesticides carry other modes of action as well which include mating disruption, suffocation, antifeedant and desiccation (Olson, 2015). Fig. 13.1 presents an overview of the advantages of using biopesticides over conventional synthetic/chemical pesticides.

13.3 Adverse effects of chemical pesticides on human health

There are three common routes through which the pesticides can access entry into the human body viz. skin (contact), lungs (inhalation) and mouth (ingestion) (Khursheed et al., 2022). Considering the agricultural workers to be the first ones to be vulnerable, the summary analysis of available literature published during the period May 2018 to May 2019 related to epidemiology of thorough relationship between the chemical pesticide's exposure and human health confirmed that the exposure of synthetic/chemical pesticides resulted in the development of a number of health disorders among humans, especially in the agricultural workers, such as thyroid, respiratory and metabolic effects, oxidative stress, DNA or genomic damage, neurological disorders, and even, cancer (Curl et al., 2020; Khursheed et al., 2022). The chemical/synthetic pesticides or their constituents have also been reported to act as mutagens that can alter the DNA or genomic expressions (Hassaan & Nemr, 2020). Population oriented studies have also revealed significant correlation between the adverse human health impacts and exposure to organophosphate pesticides including disorders of cardiovascular system, nervous system, male reproductive system, and proneness to develop non-Hodgkin's lymphoma (Khursheed et al., 2022). These adverse health effects are not only limited to direct pesticide exposure but also indirect exposure via contamination of water resources. Organophosphates exposure to parents

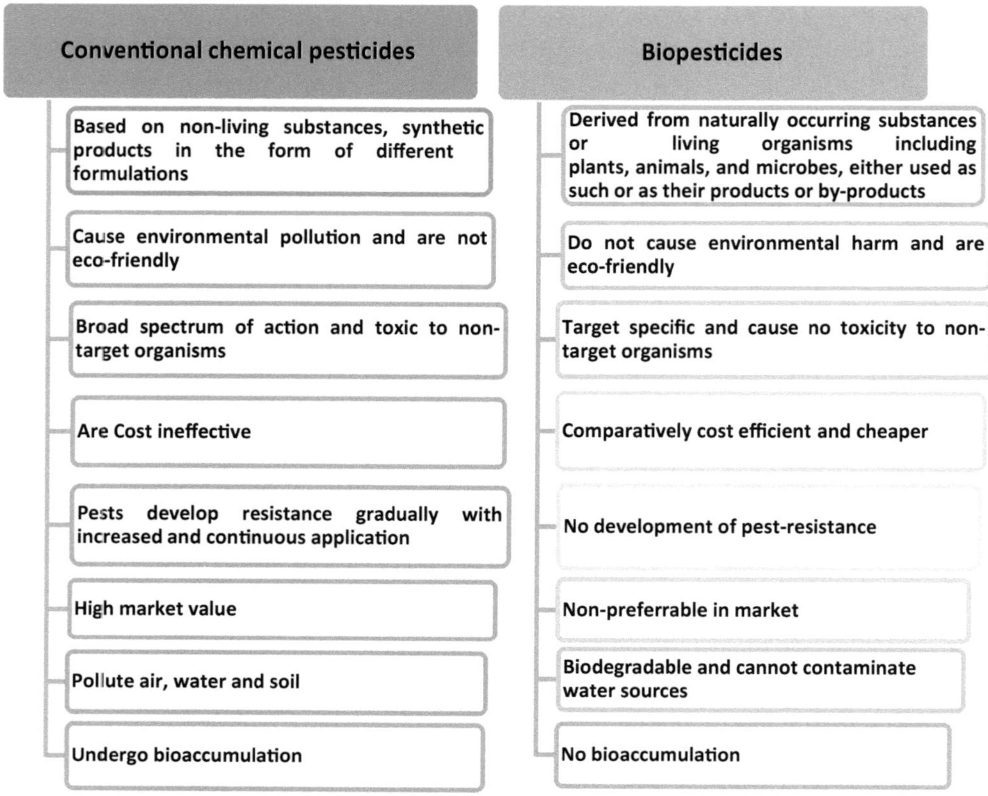

Figure 13.1 Advantages of using biopesticides over conventional chemical pesticides.

has also been linked to the occurrence of various neurological disorders in children and in the reduction of gestational periods (Eskenazi et al., 2004; Rauh et al., 2015).

13.4 Types of biopesticides and their modes of action

13.4.1 Botanical pesticides

Naturally occurring substances in plants used for controlling pest populations are called botanicals or botanical pesticides. They are categorized under herbal pesticides (Pal et al., 2013). All plant derived products if deviating from nontoxic mode of action, are not considered as botanical pesticides by EPA (Mazid et al., 2011). Around 6000 plant species and their products have been considered as environmentally safe potential insecticides. They are volatile in nature, target specific, have minimal residual activity and do not bioaccumulate and so do not contaminate the environment in contrast to chemical pesticides. They either kill pests or otherwise cause growth inhibition of pests

Table 13.1 List of some botanical pesticides and their target pests. Reproduced from Wakeil (2013).

Botanical pesticides	Target pests
Nicotine	Aphids, caterpillars, thrips
Ryania	Onion thrips, codling moths, corn earworms, potato aphids, silkworms
Rotenone	Aphids, bugs, potato beetles, carpenter ants, spider mites
Pyrethrum	Aphids, caterpillars, leafhoppers, beetles, bugs, spider mites, cabbage worms
Synthetic pyrethroids	Caterpillars, aphids, thrips
Sabadilla	Armyworms, squash bugs, grasshoppers, aphids, codling moths, cabbage loopers
Neem products	Caterpillars, aphids, thrips, armyworms, whiteflies, bollworms, cutworms, stemborers, leaf miners, leafhoppers, mites, psyllids
Essential oils	Land snails, caterpillars, aphids, cabbage worms, white flies

that damage important crop plants, shrubs, trees, and other vegetation. Several naturally occurring botanicals are available from various plants viz. neem, rotenone, ryania, nicotine, pyrethrins, grapefruit seeds, sabadilla, lantana, garlic, etc. and are used for control and management of various pests (Table 13.1).

13.4.1.1 Azadirachta indica *A. Juss. (Meliaceae) (neem)*

Azadirachta indica A. Juss. (Meliaceae) is a key botanical alternative to chemical pesticides in pest management. Neem tree possesses potent active compounds such as azadirachtin, salannin, desacetyl salannin, meliantriol, nimbin, nimbidin and desacetyl nimbin. Tetranortriterpenoid limonoid (Azadirachtin), is amongst the most promising potent active compounds extracted from neem leaves and seeds (Ahmad et al., 2013; Schmutterer, 1990). Several studies have reported its insecticidal activity against various economically important insects such as *Pieris brassicae* L., *Spodoptera* spp., *Helicoverpa. Armigera* (Hübner), *Plutella xylostella* L., aphids, leaf miners, and whiteflies (Ahmad et al., 2015; Isman, 2006; Pineda et al., 2009; Schmutterer, 1990). Azadirachtin acts by hindering the functioning of prothoracic gland, thereby inhibiting the synthesis and discharge of molting hormone (MH) leading to defective ecdysis in immature stages and hence, preventing adult formation. Besides, it also acts as a strong antifeedant, oviposition deterrent, repellent, and growth inhibitor against most insects (Ahmad et al., 2015; Hasan & Ansari, 2011; Isman, 2006; Mordue & Blackwell, 1993; Schmutterer, 1990).

13.4.1.2 Chrysanthemum cinerariifolium *(pyrethrin)*

Chrysanthemum cinerariifolium is another plant species with natural insecticidal compounds called pyrethrins. These are esters extracted from flowers of *C. cinerariifolium* and are commercially available botanical insecticides. Its main constituents are cinerin I, cinerin II, jasmolin I, jasmolin II, pyrethrin I, and pyrethrin II. Cinerin I, jasmolin I, and pyrethrin I are the chrysanthemic acid esters, while cinerin II, jasmolin II, and pyrethrin II are the pyrethric acid esters (Grdiša & Gršić, 2013). Among cinerins, jasmolins and pyrethrins, pyrethrins are the most dominant active ingredients. Pyrethrins produce "knock-down" effect in insects by opening Na^+ channels in the nerve membranes leading to hyperactivity, excitement and convulsions followed by spasms (Davies et al., 2007; Wakeil, 2013).

13.4.1.3 Rotenone (Derris spp. and Lonchocarpus spp.)

Rotenone is an isoflavonoid, considered as one of the broad-spectrum botanical pesticide obtained from stems and roots of *Lonchocarpus utilis, Lonchocarpus urucu, Tephrosia virginiana, Derris elliptica* and *D. involuta*. It is used in the form of dry powder or spray. Rotenone acts by inhibition of electron transport chain (ETC) at the mitochondrial level. The phosphorylation of ADP into ATP is blocked which adversely affects insect metabolism (Li et al., 2003). Pure rotenone, functionally, in terms of its acute toxicity to mammals, is equivalent to chemical pesticide such as dichlorodiphenyltrichloroethane (DDT) (Isman, 2006; Wakeil, 2013).

13.4.1.4 Eucalyptus camaldulensi

Eucalyptus belongs to family Myrtaceae, represented by tall, evergreen, and wonderful trees with 700 species cultivated worldwide for its oil, pulp, medicine, timber, gum, and aesthetic importance (Batish et al., 2008). Eucalyptus oil consists of sesquiterpenes, monoterpenes, aromatic phenols, alcohols, oxides, aldehydes, ketones, and ethers, collectively, leading to its broad-spectrum biological activity against insects and other pests such fungi, viruses, nematodes etc. (Batish et al., 2008). The major active ingredients are γ-terpinene, 1,8-cineole, eucamalol, citronellyl acetate and citronellol (Liu et al., 2008; Su et al., 2008; Watanabe et al., 1993). Eucalyptus oil has also been employed as an antifeedant, predominantly against insects' bite (Trigg, 1996a, b). The eucalyptus leaves extract was also found to possess toxicity against insect pests (Koul et al., 2008).

13.4.1.5 Datura stramonium

Datura or *Datura stramonium* is a popular herb and is well known for its toxic effect against insect pests. Reports are available where seed extracts show insecticidal activity against the rice weevil, *Sitophilus oryzae* (Jawalkar et al., 2016). Growing plants of *Datura* repel insects and protects nearby neighboring plants from herbivore insects (Das et al., 2012; Jawalkar et al., 2016). Alkaloids, phenols, tannins, flavonoids, steroids, saponins, and glycosides are active ingredients present in plants (Shagal et al., 2012). Ripened seeds carry the highest contents of alkaloids though they are present in all parts of plants (Jawalkar et al., 2016; Shagal et al., 2012). Alkaloids have shown growth retarding, antifeedant and antioviposition properties against insect pests (Awulachew, 2022).

13.4.1.6 *Tobacco (*Nicotiana tabacum*; Nicotine)*

In tobacco plant, active ingredients contributing in pesticidal properties are alkaloids nicotine (1, 3-(di-methyl pyrrolidine) pyridine) (C_{10} H_{14} N_2 O_4 S), anabasine (1-3(2 pyridyl) pyridine) and nornicotine (3-(2 pyrrolidinyl) pyridine) (Nayak, 2020). Nicotine and other alkaloids are essentially nonpersistent contact insecticides. Nicotine mimics neurotransmitter acetylcholine and thereby binds to acetylcholine receptors present in the nerve synaptical complexes subsequently leading to uncontrolled or impaired nerve firing. This, in turn, leads to failure of all those body processes/systems that solemnly rely on nervous input to carry out their normal and balanced functioning (Wakeil, 2013). Nicotine ingestion or absorption through cuticle kills the insect pests rapidly by causing intensive tremors and convulsions followed by paralysis (Kanmani et al., 2021).

13.4.1.7 *Ryania-based botanicals*

Ryania-based botanical insecticides (ryanodine) are prepared from grounded stem and root of American shrub, *Ryana speciosa* belonging to Flacourtiaceae. Ryania shows highly toxicity to the fruitmoths and citrus thrips. Active ingredient present in plant is ryanodine, which is a complex of polycyclic, polyhydroxylic diterpenes with diversely substituted oxapentacyclo pentadecyl pyrrole-2-carboxylate configuration (Sharma, 2019). The powdered wood carries less than 1% ryanodine, an alkaloid, which interferes with the release of calcium in muscle tissues. Organic apple growers use it to a limited extent for control and management of the codling moth, *Cydia pomonella* (Dar et al., 2017). Ryania does not cause rapid knockdown paralysis but acts as a slow-stomach poison by causing insects to stop feeding soon after ingestion. It is an efficiently synergized piperonyl butoxide (PBO) and is reportedly the most effective in hot weather conditions.

13.4.1.8 *Essential oils*

Essential oils (EOs) are derived from aromatic plants and are known to provide defense against herbivore insects (Basaid et al., 2020; Regnault-Roger et al., 2012; Walling, 2000). In recent years, out of all the botanicals used, EOs has the strongest market growth. Currently, EOs are of utmost importance to industries with wide range of potential applications in the agricultural, nutritional, pharmaceutical and cosmetic fields due to their several biological activities, viz. insecticidal, repellent, antimicrobial, analgesic, and anti-inflammatory (Asbahani et al., 2015; Bilia et al., 2014; Turek et al., 2013). EOs are allelochemicals developed during the course of coevolution as a part of chemical defense system of plants. They may affect insects' growth by acting as repellants, oviposition deterrents, development, molting and adult emergence inhibitors in insects. (Upadhyay et al., 2018). EOs are produced by aromatic plants belonging to families, viz. Myrtaceae, Lamiaceae, Asteraceae, and Lauraceae (Regnault-Roger et al., 2012). EOs mainly belong to two groups, i.e., terpenoids and phenylpropanoids containing terpenes, terpenoids (the oxygenated derivatives of terpenes), alkaloids, polyphenols, and cyanogenic glucosides as their main chemical compounds. The terpenes, characterized by the blend of isoprene units (C_5H_8), are categorized based on the number of isoprene units attached in their structure, i.e., hemiterpene with one isoprene unit, monoterpene with two isoprene units. The hydro distillation or the supercritical fluid extraction method is usually adopted for extraction of the complex mixtures of simple aromatic terpenes from plants (Ikbal & Pavela., 2019; Machalova et al., 2015).

EOs are alleged to impede basic metabolic, physiological, behavioral, and biochemical functions of insects. The mono- and sesquiterpenoid constituents of EOs interfere with various neurotransmitter receptors on nerve cells thereby acting as neurotoxins and make them fast acting insecticides. These oils act by attaching to various receptors such as octopamine receptors, gamma-aminobutyric acid-gated chloride channels receptors, and nicotinic acetylcholine receptors, as well as acetylcholine esterase receptors in case of oil terpenoids and biogenically linked phenols (Isman & Tak, 2017; Robu et al., 2015). Most of the EOs are readily absorbed by insect pests on contact which make them highly potent bioinsecticide. However, in closed spaces their volatile nature and constituents make them efficient fumigants (Isman, 2020). Plant EOs have minimum residual activity as contact insecticides and there are evidence that after their application, pests may be deterred or repelled for a prolonged time period (Isman, 2020). Various technologies, viz. nanoformulations and micro-encapsulation are being deployed in order to supply slow release of terpenoids so as to increase their efficiency under natural field conditions (Benelli et al., 2017; De Oliveira et al., 2014).

These compounds also display potentially insecticidal properties such as antifeedant, repellence, ovicidal, larvicidal, pupicidal, and oviposition deterrence (Dey & Gupta, 2016). The mechanism of action depends upon the type of EOs as reported by Chang et al. (2022). Ten EOs were studied for their acute toxic and feeding deterrent bioactivity against three stored-grain pests (*Rhyzopertha dominica, Tribolium castaneum,* and *Sitophilus oryzae*). Eugenol, carvacrol, transanethole, and thymol demonstrated differential species-specific toxicity by acting as acute toxins rather feeding deterrents (Kanda et al., 2017).

Presently, a number of insecticides produced contain eucalyptus, cinnamon, clove, thyme oils, peppermint, rosemary, etc. in combination or singly as their active ingredients (Isman, 2020) (Table 13.2).

13.4.2 Microbial pesticides

Microbial pesticides comprise any microorganism whether bacteria, fungi, bacteriophage, virus, protozoan, nematode, etc. used to curb plant diseases and pests (Yadav et al., 2017, 2020). Among all, microbial pesticides constitute the largest group and are user-friendly, nontoxic with quick degradation in environment (Kumari et al., 2014; Shukla et al., 2019). They can be applied by direct application, spraying, drenching, encapsulating seeds, root before sowing, etc. (Montesinos & Bonaterra, 2009; Thakur et al., 2020). Nowadays, a

Table 13.2 Some plant-based essential oils, their botanical sources and major constituents.

Essential oil	Botanical source (Family)	Major constituent (%)	References
Cinnamon oil	*Cinnamomum verum* (Lauraceae)	55%–76% Cinnamaldehyde, 5%–18 % Eugenol	Nollet and Rathore (2017)
Eucalyptus oil	*Eucalyptus globules* (Myrtaceae)	67%–84% 1,8-Cineole	Nollet and Rathore (2017)
Clove oil	*Syzygium aromaticum* (Myrtaceae)	89% Eugenol	Nollet and Rathore (2017)
Lemongrass oil	*Cymbopogon citrates* (Poaceae)	34%–45% Geranial, 5%–51% Neral, 9%–25% Myrcene	Nollet and Rathore (2017)
Peppermint oil	*Mentha piperita* (Lamiaceae)	7%–48% Menthol, 20%–46% Menthone	Leung (1980)
Orange oil	*Citrus sinensis* (Rutaceae)	91%–97% d-Limonene	Nollet and Rathore (2017)
Rosemary oil	*Rosmarinus officinalis* (Lamiaceae)	52% 1,8-Cineole, 10% α-Pinene, 9% Camphor, 8% α-Pinene	Isman et al. (2008)
Tea Tree oil	*Melaleuca alternifolia* (Myrtaceae)	35%–48% Terpinen-4-ol, 14%–28% α-Terpinene	International Organization for Standardization (ISO) (2017)
Thyme oil	*Thymus vulgaris* (Lamiaceae)	50% Thymol, 33% p-Cymene	Procházka et al. (2022); Tak et al. (2016)
Citronella oil	*Cymbopogon winterianus* (Poaceae)	27%–33% Citronellal, 10%–16% Citronellol, 24%–40% Geraniol	Nollet and Rathore (2017)

Table 13.3 Some important bacterial species and their target pests. Reproduced from Kunimi (2007) and Thakur et al. (2020).

Bacterial species	Target pest
Bacillus thuringiensis	*Helicoverpa armigera*, *Plutella xylostella*, *Earias* species, *Heliothis armigera*, *Spodoptera litura*, *Eutectona machaeralis*, *Cnaphalocrocis medinalis*, *Achaeae janata*, and *Hyblaea pured*
Bacillus subtilis	Powdery mildews, rust, and downy
Bacillus pumilus	Bacterial leaf blight (paddy), sigatoka (banana)
Agrobacterium adiobacter	Crown galls
Pseudomonas chlororaphis	Fungal pathogen of barley and oats
Pseudomonas cepacia	Pathogenic fungi of soil
Pseudomonas fluorescens	*Meloidogyne* species
Bacillus thuringiensis subspecies *kurstakia*	*Spodoptera exigua*, *Plutella xylostella*, *Bombyx mori*, and *Helicoverpa armigera*
Bacillus thuringiensis subspecies *aizawaia*	*Agrotis ipsilon*
Bacillus thuringiensis subspecies *japonensis*	*Cyclocephala borealis* and *Anomala orientalis*
Paenibacillus popilliae	*Popillia japonica*

variety of target-specific microbial fauna is being utilized for the management of insect pest populations. Their contact with pest tissues results in the formation of specific crystalline proteins or release of siderophores, which inhibit the availability of important minerals and ions required for the growth of pests. Microbial pesticides are gaining more attention as the continuous and indiscriminate use of chemical pesticides not only degenerates the normal basic structure of the soil but also deteriorates its vigor (Kumar et al., 2019; Singh & Yadav, 2020).

13.4.2.1 Bacteria

Among the most prominent microbial pesticides are the bacteria particularly *Bacillus thuringiensis* (Bt) derived products which have received wide attention for the control of caterpillars. Several entomopathogenic bacterial species such as *Bacillaceae, Pseudomonas, Serratia, Yersinia, Streptomyces, Burkholderia, Chromobacterium,* and *Saccharopolyspora* are effective against insect pests of orders, viz. Lepidoptera, Hymenoptera, Diptera and Coleoptera (Raymond & Federici, 2017; Ruiu, 2018; Sarwar & Aslam, 2021) (Tables 13.3).

B. thuringiensis Berliner is amongst the most successful and potent microbial pesticide, with more than 240 known holotypes of toxic crystalline protein (Cry). Presently, there are around 5 *Bt kurkstai* strains and 2 *Bt aizawai* strains commercially available for management of insect pests on brassicas, cotton, ornamentals, fruiting vegetables and a number of other commercial crops and tree fruits (Arthurs & Dara, 2019; Chakroun et al., 2016).

Mechanism of action involves the attachment of Cry proteins to insect gut receptors (Kumar, 2012). *B. thuringiensis* and other few toxic strains are endorsed for addition of parasporal bodies or d-endotoxins through sporulation to enhance the activity (Aronson & Shai, 2001). Ingested protoxins transform into active toxins within the mid-gut of insect which bind to gut receptors and after assimilation cause tissue damage, thereby, leading to the paralysis of gut. Finally, the larvae cease to feed and ultimately die of starvation due to impairment of epithelium of insect midgut (Darboux et al., 2001; Soberón et al., 2016).

Currently, Bt genes incorporated crops are produced such as maize, cotton, soybean, etc. to prevent damage caused by caterpillars (Tabashnik & Gould, 2012). According to Chakroun et al. (2016) entomopathogenic bacteria produce insecticidal proteins of which a few proteins like Cry or Cyt proteins usually accumulate in parasporal crystals or inclusion bodies while the others are discharged into the culture medium are known as vegetative insecticidal proteins (Vip proteins: Vip1, Vip 2, Vip3A, and Vip4 family). Among these, Vip1 and Vip2 type acts as binary toxins against Coleopterans and Hemipterans by binding to mid gut membrane receptors, hindering its ADP-ribosyltransferase activity against actin filaments, thereby preventing the formation of microfilaments. Vip3A shows insecticidal activity against a wide variety of Lepidopterans and its mode of action resembles Cry viz. proteolytic activation, attachment to epithelial membrane of the midgut as well as pore formation. This property suggested the enhanced effect when pooled with Cry proteins in transgenic plants (*Bt*-treated crops).

Pseudomonas spp. also constitute major groups that can be utilized as biopesticide against various insects as they inhibit growth due to secretion of antibiotics, toxins, biosurfactants and secretion of cell wall-degrading enzymes including chitinase, β-1,3-glucanase, etc. Production of metabolites during interaction triggers systemic acquired resistance induction (Kumar et al., 2017). Production of contractile phage tail-like particles by bacteria is pest specific, efficient, and promising alternative over conventional synthetic chemical pesticides (Becker et al., 2022).

13.4.2.2 Fungi

Entomopathogenic fungi (EPF) are parasitic fungi that can be used as potent biopesticide against arthropods although extracted from carcasses of arthropods and from soil (Behie & Bidochka, 2014; Litwin et al., 2020). Like entomopathogenic bacteria (EPB), they are

Table 13.4 Some important fungi species and their target pests. Reproduced from Kunimi (2007) and Thakur et al. (2020).

Fungi species	Target pest
Aschersonia aleyrodis	*Bactericera cockerelli*
Beauveria brongniartii	*Pyronota festiva*
Lecanicillium longisporum	*Frankliniella occidentalis*, *Myzus persicae*, and *Bemisia tabaci*
Metarhizium anisopliae sensulato	*Bactericera cockerelli*
Nomuraea rileyi	*Spodoptera litura*, *Helicoverpa armigera*, *Anticarsia gemmatalis*, and *Tricoplusia ni*

ecologically safer alternatives to toxic synthetic chemical pesticides (Jaihan et al., 2016; Lovett & Leger, 2017; Ríos-Moreno et al., 2016).

Entomopathogenic fungi do not form lone monophyletic group and are classified into six classes as Oomycetes, Entomophtoromycota, Chytridiomycota, Basidiomycota, Microsporidia, and the most familiar Ascomycota. So far, 12 Oomycetes, 474 Entomophtoromycota, 65 Chytridiomycota, 238 Basidiomycota, 339 Microsporidia, and 476 Ascomycota species have been reported (Araújo & Hughes, 2016; Litwin et al., 2020). Species categorized in classes Ascomycota and Entomophtoromycota exist most commonly in nature. Within the literature, species belonging to class Ascomycota and the genera *Metarhizium spp., Ophiocordyceps spp., Beauveria spp., Lecanicillium spp., Isaria spp., Torubiella spp., Cordyceps* spp., *Pochonia spp., Hirsutella spp.,* etc. have been reported (Jaihan et al., 2016; Khan et al., 2012; Litwin et al., 2020; Tkaczuk et al., 2015). Among all, Entomophthorales (e.g., *Furia, Entomophaga, Conidiobolus,* or *Erynia*) has shown highest insecticidal activity but saprotrophic fungi belonging to Ascomycota is commonly used EPF due to easy culturing (Mascarin & Jaronski, 2016). Globally, wide variety of fungal (genera *Paecilomyces, Metarhizium, Beauveria, Isaria* and *Lecanicillium*) biopesticide formulations are available carrying wide bioactive spectrum against diverse arthropods (Castro et al., 2016; Khan et al., 2012; Ríos-Moreno et al., 2016). Table 13.4 enlists some of the important EPF and their target organisms.

Unlike EPB or entomopathogenic viruses (EPV), EPF spores penetrate directly into the cuticle (Bilgo et al., 2018). The EPF spores interact with cuticle and produce enzymes and proteins known as hydrophobins (Greenfield et al., 2014; Skinner et al., 2014). In ambient conditions, spores germinate on the cuticle of insects leading to appressoria formation, which triggers the production of lytic enzymes viz., proteolytic, lipolytic, chitin lytic, etc., ultimately causing disintegration of the body wall (Lacey et al., 2015; Skinner et al., 2014). Fungal hyphae penetrate the body cavity and haemolymph and tissues forming secondary hyphae. These growing hyphae produce an array of secondary

Table 13.5 Some important viruses and their target pests. Reproduced from Kunimi (2007) and Thakur et al. (2020).

Viruse	Target pest
Cotton bollworm nuclear polyhedrosis virus (NPV) HezeSNPV	*Helicoverpa armigera*
Corn earworm NPV	*Helioth virescens, Helicoverpa zea*
Diamond back moth granulosis virus (GV)	*Plutella xylostella*
AngeMNPV	*Anticarsia gemmatalis*
BuzuNPV	*Buzura suppressaria*
AucaMNPV	*Autographa californica*

metabolites ultimately causing paralysis and disruption of host's vital physiological processes resulting in mechanical damage to host's internal organs with nutrient exhaustion (Donzelli & Krasnoff, 2016; Fan et al., 2017; Mascarin & Jaronski, 2016). Host's body, thereby, becomes stiff owing to fluid absorption by the EPF forming infective spores for further contamination (Skinner et al., 2014).

13.4.2.3 Viruses

A virus is an infectious microbe that consists of a segment of either DNA or RNA as nucleic acid enfolded inside a protein coat (mainly lipoprotein) and carries the capability to replicate inside the body of its host. Major groups of EPV include: baculovirus, entomopoxvirus, cypovirus, iridovirus, ascovirus, polydnavirus, and mesoniviridae (Sarwar & Aslam, 2021). Viruses infecting and killing a wide variety of insects are called as EPV (Soumia et al., 2021). Table 13.5 enlists some important viruses along with their target organisms. They infect economically important and destructive pests such as cabbage looper (*Trichoplusia ni*), corn earworm (*Heliothis zea*) and beet armyworm (*Spodoptera exigua*) whose control and management has become difficult owing to either development of pesticide resistance or progressive pressure to lessen pesticide residues (Arthurs & Dara, 2019; Deka et al., 2021; Sarwar & Aslam, 2021).

Baculoviruses entrenched in a matrix of protein are the most promising biopesticides universally. They are divided into two categories, namely: nucleopolyhedrovirus (NPV; HzNPV, SeNPV and HearNPV) and granulovirus (GV) (Cory, 2000; Deka et al., 2021). The occlusion bodies (OBs) assist them to stay viable outside the host cadaver (Cory, 2000). Noticeable symptoms of EPV infection in insect are sluggishness and discontinued feeding and growth. Ingested viral particles adhere to stomach receptors infecting haemocoel and vital organs, resulting in death of larvae. Insects appear to be whitish with translucent exoskeleton and lean integument which finally ruptures. Discharge of larvae, carries billions of OBs thereby spreading inoculum in the entire field (Deka et al., 2021; Granados & Williams, 1986; Sarwar & Aslam, 2021).

In some cases, virus infection depicts an inflated abdomen (due to spheroids' accumulation), swollen rectal sac, irregular in form with white colors (Sarwar & Aslam, 2021). Some members invade only epithelial tissue of the host midgut, resulting in diarrhea and vomit, cessation of feeding, inhibition of larval growth and ultimately death overcomes just before onset of pupation (Sarwar & Aslam, 2021; Tanada & Kaya, 1993). These microbial products are stable, highly target specific and compatible with parasitoids when integrated with other biological control methods (Arthurs & Dara, 2019; Raymond et al., 2006).

13.4.3 Genetically modified organism (GMO) products

Here, firstly incorporation of the genetic material is done into some crop plant, which is then subsequently utilized as a source for production of compounds carrying pesticidal properties, also called plant-incorporated protectants (PIPs). By far, Cry proteins were the first-generation PIPs with insecticidal properties to be introduced into the genetic makeup of GM crops loaded with the soil bacteria *Bt.* derived transgenes (Parween & Jan, 2019). PIPs also can be alternatives which demand the focus of the research community essential for the ongoing environmental fate appraisal of these molecules, mainly the RNAi-based PIPs (Parker et al., 2019; Kumar et al., 2021; Parween & Jan, 2019).

13.4.4 Pheromones as biochemical pesticides

Chemical communication is ubiquitous throughout the animal kingdom. Pheromones are the chemical signals or compounds secreted by some living organisms for communication purposes. They can induce a behavioral or physiological alteration either in an organism of the same species or of different species (Chandler et al., 2011; Icoz & Stotzky, 2008; Liu et al., 2021). For crop protection, the most commonly used pheromones are insect sex pheromones, a few of which are now-a-days manufactured and used commercially for control and management of pests via mass trapping method, lure-and-kill strategy and mating disruption (El-Sayed et al., 2009; Liu et al., 2021; Reddy et al., 2009). Pheromones are usually provided in the form of pheromone traps, which in turn are further associated with other kinds of traps (Butu et al., 2020). This attribute provides ecological importance to pheromone "treatments or strategies" affecting only the target species against whom these are applied and hence, the ecosystem remains unaffected (Butu et al., 2020). Based on their biological functions, pheromones have been classified as sex pheromones, dispersal pheromones, aggregation pheromones, and alarm pheromones. Resistance to such pheromones has, till date, not been documented. This may be possible due to two main reasons: first, they are present in low concentrations in the crop fields and secondly, majority of them contain a mixture of compounds posing difficulties for pests to develop resistance properties (Copping & Menn, 2000). The US EPA has already registered around 30 mating-disruption based pheromone products for controlling as many as eleven lepidopteran pest species which include: oriental fruit moth (*Grapholitha molesta* (Busck)), peach tree-borer (*Anarsia lineatella* Zeller), codling moth (*Laspeyresia pomonella* (L)), pink bollworm (*Pectinophora gossypiella* (Saunders)) and gypsy moth (*Lymantria dispar* (L)) amongst others (Copping & Menn, 2000). Considering the enormous potential of pheromones as biopesticides for pest control strategies, many researchers and users have already considered these to be the "pesticides" of the future generations (Butu et al., 2020).

13.5 Current scenario of biopesticides

At present, every year, approximately a total of 2 million tons of pesticides are consumed around the globe which include herbicides (47.5%), insecticides, (29.5%), fungicides (17.5%) and other pesticides (5.5%) (Sharma et al., 2019). The top 10 pesticide consuming countries throughout the world are China, United States, Argentina, Thailand, Brazil, Italy, France, Canada, Japan, and India (Sharma et al., 2019). A Business Communications Company, Inc., report stated that the total worth of global pesticide market in 2017 was around 61.2 billion USD which by 2022 was estimated to rise by approximately 18.1 billion USD (Chen, 2018; Khursheed et al., 2022). In other words, during 2017–2022, the global market for pesticides (synthetic and biopesticides) should expand at a compound annual growth rate (CAGR) of 5.3% from $61.2 billion in 2017 to around $79.3 billion by 2022. Of the total global pesticide/crop protection market, only a small share, i.e., 5%, worth around 3 billion USD, is contributed by biopesticides. Its share has increased from approximately 3.5% (worth $1.6 billion) of the total global pesticide market in 2009 (BCC Research, 2010) to around 5% ($3 billion) and it has been projected to be around $4.5 billion by the year 2023 (Olson et al., 2013, 2015). Around 200 products are available in the US market while only 60 analogues are there in the EU market (Kumar et al., 2021). About 45% of total biopesticides were consumed and sold in the United States, Mexico, and Canada (NAFTA countries). On the other hand, Asia consumes only 5% of the total biopesticides sold globally (Okrikata, 2021). Overall, about 225 microbial biopesticides are now being manufactured within 30 Organization for Economic Cooperation and Development (OECD) countries and around 90% of these are *Bt.* derived (Okrikata, 2021).

Nevertheless, in the past two decades, the growth rate of biopesticides market is demonstrating an upward trend. At a global scale, biopesticides consumption is increasing by around 10% per year (Markets & Markets, 2020). Many countries have made improvements in their regulatory policies for reducing the consumption of chemical pesticides and promoting biopesticides culture, though, biopesticides are still mostly regulated by the same system originally designed for chemical pesticides assessment (Kumar & Singh, 2015; Kumar et al., 2021). Compared to United States, Brazil, China, and India, a fewer number of biopesticides have reportedly been registered in EU due to a complicated and time consuming registration process (Czaja et al., 2015; Damalas & Koutroubas, 2018; Kumar et al., 2021). Biopesticides consumption is also low in Nigeria due to inferior foundations,

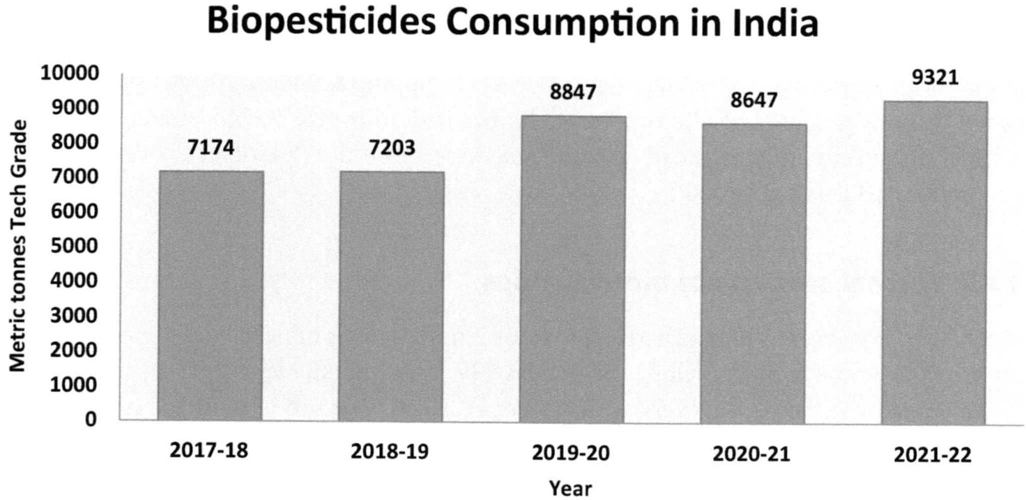

Figure 13.2 Biopesticides consumption of India for the last five years (Source: Data obtained from DPPQS, Ministry of Agriculture & Farmers Welfare, Government of India).

elevated costs, and limitations in governmental regulatory policies (Ivase et al., 2017). In India, 970 biopesticide products have reportedly been registered with the Central Insecticides Board and Registration Committee, India's leading governing body for all types of biopesticides (Rani et al., 2021). In India, as per records reported by the Directorate of Plant Protection, Quarantine, and Storage, Ministry of Agriculture and Farmers Welfare, Government of India, the total area under biopesticides use is only 11%, while, a maximum of approximately 84% area falls under the chemical pesticides application and merely 5% area falls under both chemical as well as biopesticides (Rani et al., 2021; Yadav et al., 2022). The trend shown by biopesticides consumption of India has revealed a remarkable increase over the past five years with values of 8847 metric tonnes in 2019–20 and 9321 metric tonnes in 2021–2022 (Fig. 13.2).

Further, in India, among different biopesticides' categories, the consumption of fungi based biopesticides is highest, i.e., 50% followed by 21% bacterial, 15% viral, and 14% other botanical pesticides (Rani et al., 2021).

At global level, the use of biopesticides has now been estimated to be on par with synthetic ones by the early 2050s, although, major uncertainties with respect to the rates of uptake/consumption, especially in regions like Southeast Asia and Africa, account for the majority of the flexibility in such expectations (Damalas & Koutroubas, 2018).

13.6 Limitations and challenges in the promotion, commercialization, and consumption of biopesticides

Several factors are there that hamper the potential growth of ecofriendly biopesticides in the global market. But, there are three major barriers that must be considered for successful commercialization of such natural pesticides: resource sustainability, standardization of extracts or formulations, and finally, regulatory approval (Isman, 1997; Wakeil, 2013). Further, important cost considerations are also there for each of these three key constraints. The lack of large-scale manufacturing units or facilities also affects the biopesticides market. Other limitations include the slow action of many biopesticides and also, farmers must also be encouraged to gain confidence in those pesticides which do not generate an instant "knockdown" effect and the lack of residual action for most biopesticides.

Further, several aspects of the agricultural sectors also probably hamper the development of new biopesticides (Tripathi et al., 2020). Many biopesticides show a highly selective nature such as those based on baculoviruses, like the CpGV. These, typically, are very selective for a particular pest, thereby, having stumpy profit potential due to their limited market size (Chandler et al., 2011). Moreover, the farmers also face a huge, fixed cost in comparison to those farmers who utilize conventional chemical synthetic pesticides. This is because the fixed cost price associated to biopesticides is not dispersed among a large population of farmers, and as such, early users discover it unfavorable (Chandler et al., 2011; Tripathi et al., 2020). Further, a majority of farmers who have already been using chemical pesticides for quite a long period of time have attained ample practical experience and also confidence in their potential, while a lot of uncertainties are there within the minds of farmers' when it comes to applying new products without even any prior practical knowledge (Tripathi et al., 2020).

Even, all the governments try to regulate authorization and use of many biopesticides based on various properties of their living and nonliving constituents. Such regulations are intended for safety of humans as well as environment and for ensuring the fact that only reliable products enter the open market (Chandler et al., 2011). For example, in the EU, only those biopesticides whose effectiveness has been thoroughly quantified and proved are used for pest management or crop protection. Even, in this respect, the guidelines and approval of OECD are more strict and only those biopesticides are granted authorization which pose minimal or almost zero risk.

Furthermore, sometimes, the regulatory bodies have inadequate expertise in biopesticides, and as such, they tend to delay the authorization process of a new biopesticide. Another problem is that the regulatory system in case of biopesticides is also based on conventional chemical pesticide model (Kumar & Singh, 2015; Kumar et al., 2021). Some government regulatory bodies treat and evaluate the biopesticides in the approach like chemical pesticides and are, therefore, slow in recognizing the reality that a separate regulatory process is the need of the hour for biopesticides (Chandler et al., 2011).

The efforts of many government institutions for popularization of IPM and usage of biopesticides have had modest effect. Due to lack of active promotion by the agriculture departments, the demands for such products does not increase and as such, majority of our private firms and dealers usually do not stock and sell biopesticides.

13.7 Way forward: Future prospective to be considered

Pest management in an ecofriendly approach is not a dream anymore. Biopesticides can now be formed applying the modern tools and techniques of molecular biology and biochemistry, biotechnology and nanotechnology in crop plants itself in a sustainable way. Their commercialization or branding can only be done by co-operation of both the public and private sector institutions. Discovery of new products and extensive research on their formulation and delivery through modern technical inputs together with proper training to producers and users can help facilitate the development, commercialization, promotion and overall consumption of biopesticides. Following future prospects should be considered for the same (Yadav et al., 2022):

- The governments or the regulatory authorities should keep on imposing stringent regulatory policies on conventional synthetic chemical pesticides and should give incentives to those private and public sector bodies that are focused toward manufacture and promotion of biopesticides. This will ensure a significant opportunity for the biopesticides marketing firms to assist in filling the gap and the supply of biopesticides at reasonable cost.
- Encouragement and empowerment of developing countries should be done to expand their biopesticides production as well as consumption.
- In order to incorporate biopesticides in the mainstream of our agriculture sector, focus should be directed towards a better understanding regarding their action mechanisms for enhancement of their action spectra against crop pests and other parameters including field performance, advanced delivery systems, prolonged shelf life, cheap production cost, ease of availability, promotion of awareness among concerned farmers and uncomplicated registration and regulatory policies (Fig. 13.3).
- Researchers should exploit all the available knowledge of pests' genomes together with their natural enemies and use new and advance molecular-based tools and technologies for reconstructing the evolution of various natural microbial enemies and further split the molecular origin for their pathogenicity. Further, ecological studies on the possible dynamics of disease caused in the insect population must be encouraged.
- Adequate training centers should be opened for mass awareness and training of farmers regarding storage, field application, and benefits of biopesticides so that they can gain maximum benefits.
- Although, nanotechnology research in agricultural sector is still in its early stages, but it is developing quickly. With respect to future perspective, research on the development of novel nanoparticles acting both as fertilizers as well as pesticides is also necessary (Abdollahdokht et al., 2022). Owing to their novel physico-chemical characteristics, investigations on the application of nanostructures as agrochemicals for promoting plant growth and for crop-protection must be boosted. It has been expected that the controlled application of pesticides using nanotechnologies will eventually become

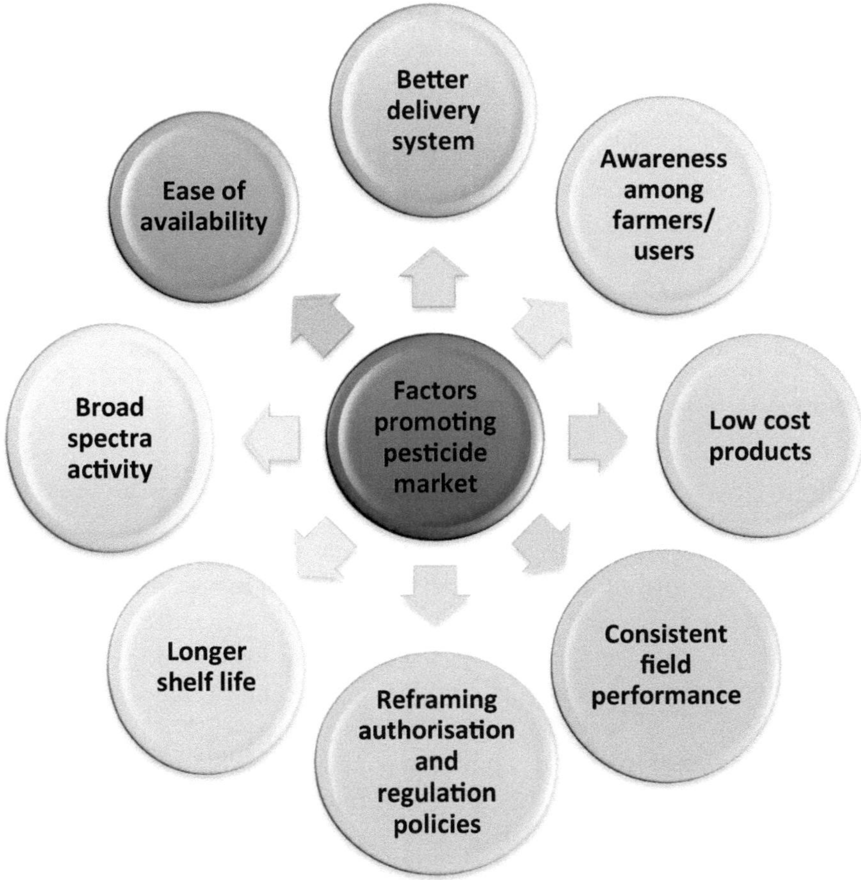

Figure 13.3 Factors that promote biopesticide's market. *(Reproduced from Yadav et al. (2022)).*

necessary in the future for promotion of cropland fertility and protection with the minimum effect on human and environmental health.

13.8 Policy recommendations

To lessen the rate of excessive consumption of hazardous chemical pesticides and to promote our biopesticides industry and the associated research and development (R&D), following policy measures have been recommended (Yadav et al., 2022):

- Agenda of sustainable agriculture should be adopted by promotion of
  - pest and disease resistant varieties,
  - judicious intercropping systems, and
  - reduction in crop intensity.
- Quality and rate of training of farmers for IPM should be improved followed by usual contact with those trained farmers both for providing maximum continuous support and for receiving feedbacks from them for further improvements.

- Agricultural institutions and universities having a decisive effect over governmental agencies should also promote biopesticides for control and management of pests.
- Regular and continuous investments should be done towards biopesticides R&D industry to have expertise in discovery, development, implementation, and delivery mechanisms of biopesticides and to support the development and registration of newer biopesticides. Together, both public and private organizations are required to educate producers, retailers, as well as the public on the usage, and benefits of biopesticides.
- Research and in situ or field trials, particularly, on area-specific and crop-specific formulations should be promoted.
- Different funding agencies should join hands for R&D of many pest-specific novel and innovative biopesticides formulations.

13.9 Conclusion

Pest management in an ecofriendly manner is not a dream anymore. Biopesticides based on plant and other animal products are now being used in many countries throughout the globe. Recently, there has been an increasing public demand for EOs and other biopesticides for pest management strategies due to their lower residual problem, cost-effectiveness, nontoxicity to nontarget organisms, accessibility, and easy-to-use, fewer human health as well as environmental concerns (Marrone, 2019). Now-a-days, an increase in pest resistance to several chemical insecticides and other health concerns has reportedly shifted the focus of our scientific community as well as farmers from synthetic insecticides to natural pesticides which are environmentally friendly. Now, biopesticides can be produced using several new and innovative approaches of different branches of sciences including molecular biology, biochemistry, biotechnology and nanotechnology in crop plants and that too in a sustainable way. They are now being seen as safer alternatives to chemical pesticides and their branding or commercialization can be done more efficiently only through proper cooperation of the concerned public and private sector institutions. Further, extensive research on the discovery, formulation, mode of action and delivery of new such substances should be done to promote the development, commercialization and consumption rates of biopesticides for a better and healthy future.

References

Abdollahdokht, D., Gao, Y., Faramarz, S., Poustforoosh, A., Abbasi, M., Asadikaram, G., & Nematollahi, M. H. (2022). Conventional agrochemicals towards nano-biopesticides: An overview on recent advances. *Chemical and Biological Technologies in Agriculture, 9*(1), 1–19.

Ahmad, S., Ansari, M. S., & Moraiet, M. A. (2013). Demographic changes in *Helicoverpa armigera* after exposure to neemazal (1% EC azadirachtin). *Crop Protection, 50*, 30–36.

Ahmad, S., Ansari, M. S., & Muslim, M. (2015). Toxic effects of neem-based insecticides on the fitness of *Helicoverpa armigera* (Hübner). *Crop Protection, 68*, 72–78.

Araújo, J. P. M., & Hughes, D. P. (2016). Diversity of entomopathogenic fungi which groups conquered the insect body? *Advances in Genetics, 94*, 1–39.

Aronson, A. I., & Shai, Y. (2001). Why *Bacillus thuringiensis* insecticidal toxins are so effective: Unique features of their mode of action. *FEMS Microbiology Letters, 195*, 1–8.

Arthurs, S., & Dara, S. K. (2019). Microbial biopesticides for invertebrate pests and their markets in the United States. *Journal of Invertebrate Pathology, 165*, 13–21.

Asbahani, A. E., Miladi, K., Badri, W., Sala, M., Addi, E. H. A., Casabianca, H., Mousadik, A. E., Hartmann, D., Jilale, A., Renaud, F. N. R., & Elaissari, A. (2015). Essential oils: From extraction to encapsulation. *International Journal of Pharmaceutics, 483*, 220–243.

Awulachew, M. T. (2022). Pharmacological, biopesticide and post-harvest loss management application of jimsonweed (*Datura stramonium*). In *Medicinal plants*. IntechOpen.

Basaid, K., Chebli, B., Bouharroud, R., Furze, J. N., de Oliveira, A. L., & Mayad, E. H. (2020). Chemical characterization of essential oils of *Senecio glaucus* ssp. *Coronopifolius* (Maire) Alexander and *Ridolfia segetum* (L.) Moris growing in Morocco. *Journal of Essential Oil-Bearing Plants, 23*(5), 918–930.

Bateman, M. L., Day, R. K., Rwomushana, I., Subramanian, S., Wilson, K., Babendreier, D., & Edgington, S. (2021). Updated assessment of potential biopesticide options for managing fall armyworm (*Spodoptera frugiperda*) in Africa. *Journal of Applied Entomology, 145*(5), 384–393.

Batish, D. R., Singh, H. P., Kohli, R. K., & Kaur, S. (2008). Eucalyptus essential oil as a natural pesticide. *Forest Ecology and Management, 256*(12), 2166–2174.

BCC Research. (2010). *Biopesticides: The global market CHM029C*. https://www.bccresearch.com/market-research/chemicals/biopesticides-market-chm029c.html.

Beas-Catena, A., Sánchez-Mirón, A., García-Camacho, F., Contreras-Gómez, A., & Molina-Grima, E. (2014). Baculovirus biopesticides: An overview. *Journal of Animal and Plant Science, 24*(2), 362–373.

Becker, Y., Patz, S., Werner, S., Hoppe, B., Feltgen, S., Berger, B., & Becker, M. (2022). Bacteria producing contractile phage tail-like particles (CPTPs) are promising alternatives to conventional pesticides. *Journal für Kultrpflanzen, 74*(3-4), 85–93.

Behie, S. W., & Bidochka, M. J. (2014). Ubifquity of insect-derived nitrogen transfer to plants by endophytic insect-pathogenic fungi: An additional branch of the soil nitrogen cycle. *Applied and Environmental Microbiology, 80*, 1553–1560. https://doi.org/10.1128/AEM.03338-13.

Benelli, G., Pavela, R., Maggi, F., Petrelli, R., & Nicoletti, M. (2017). Commentary: Making green pesticides greener? the potential of plant products for nanosynthesis and pest control. *Journal of Cluster Science, 28*, 3–10.

Bilgo, E., Lovett, B., Leger, R. J. S., Antoine, A., Debiré, R. K., & Diabate, A. (2018). Native entomopathogenic *Metarhizium* spp. from *Burkina Faso* and their virulence against the malaria vector *Anopheles coluzzii* and non-target insects. *Parasites Vectors, 11*, 11–16.

Bilia, A. R., Guccione, C., Isacchi, B., Righeschi, C., Firenzuoli, F., & Bergonzi, M. C. (2014). Essential oils loaded in nanosystems: A developing strategy for a successful therapeutic approach. *Evidence-Based Complementary and Alternative Medicine, 2014*, 1–13.

Butu, M., Stef, R., Grozea, I., Corneanu, M., & Butnariu, M. (2020). Biopesticides: Clean and viable technology for healthy environment. In *Bioremediation and biotechnology* (pp. 107–151). Springer.

Cappa, F., Baracchi, D., & Cervo, R. (2022). Biopesticides and insect pollinators: Detrimental effects, outdated guidelines, and future directions. *Science of the Total Environment, 837*, 155714.

Casida, J. E., & Durkin, K. A. (2013). Neuroactive insecticides: Targets, selectivity, resistance, and secondary effects. *Annual Review of Entomology, 58*(1), 99–117.

Castro, T., Mayerhofer, J., Enkerli, J., Eilenberg, J., Meyling, N. V., de Andrade Moral, R., Borges Demétrio, C. G., & Italo Delalibera, I., Jr. (2016). Persistence of Brazilian isolates of the entomopathogenic fungi *Metarhizium anisopliae* and *M. robertsii* in strawberry crop soil after soil drench application. *Agriculture Ecosystems and Environment, 233*, 361–369. https://doi.org/10.1016/j.agee.2016.09.031.

Chakroun, M., Banyuls, N., Bel, Y., Escriche, B., & Ferre, J. (2016). Bacterial vegetative insecticidal proteins (Vip) from entomopathogenic bacteria. *Microbiology and Molecular Biology Reviews, 80*, 329–350.

Chandler, D, Bailey, A. S., Tatchell, G. M., Davidson, G., Greaves, J., & Grant, W. P. (2011). The development, regulation, and use of biopesticides for integrated pest management. *Philosophical Transactions of the Royal Society B Biological Sciences, 366*(1573), 1987–1998.

Chang, Y., Harmon, P. F., Treadwell, D. D., Carrillo, D., Sarkhosh, A., & Brecht, J. K. (2022). Biocontrol potential of essential oils in organic horticulture systems: From farm to fork. *Frontiers of Nutrition, 8*, 805138.
Chen, J. (2018). *Biopesticides: Global markets to 2022* [Report Code CHM029G]. BCC Research.
Copping, L. G., & Menn, J. J. (2000). Biopesticides: A review of their action, applications, and efficacy. *Pest Management Science, 56*(8), 651–676.
Cory, J. S. (2000). Assessing the risks of releasing genetically modified virus insecticides: Progress to date. *Crop Protection, 19*, 779–785.
Curl, C. L., Spivak, M., Phinney, R., & Montrose, L. (2020). Synthetic pesticides and health in vulnerable populations: Agricultural workers. *Current Environmental Health Reports, 7*(1), 13–29.
Czaja, K., Góralczyk, K., Strucinski, P., Hernik, A., Korcz, W., Minorczyk, M., Łyczewska, M., & Ludwicki, J. K. (2015). Biopesticides: Towards increased consumer safety in the European Union. *Pest Management Science, 71*, 3–6.
Damalas, C. A., & Koutroubas, S. D. (2018). Current status and recent developments in biopesticide use. *Agriculture, 8*, 13.
Dar, S. A., Padder, B. A., Mahendran, G., Raja, W. H., Parey, S. H., & Hassan, G. I. (2017). Codling moth *Cydia pomonella* (L.) (Lepidoptera: Tortricidae) management a potential threat to horticulture industry. *Journal of Pharmacognosy Phytochemistry, 6*(3), 312–319.
Darboux, I., Nielsen-LeRoux, C., Charles, J. F., & Pauron, D. (2001). The receptor of *Bacillus sphaericus* binary toxin in *Culex pipiens* (Diptera: Culicidae) midgut: Molecular cloning and expression. *Insect Biochemstry Molecular Biology, 31*(10), 981–990.
Das, S., Kumar, P., & Basu, S. P. (2012). Phytoconstituents and therapeutic potentials of *Datura stramonium* Linn. *Journal of Drug Delivery Therapeutics, 2*(3), 4–7.
Davies, T. G. E., Field, L. M., Usherwood, P. N. R., & Williamson, M. S. (2007). DDT, pyrethrins, pyrethroids and insect sodium channels. *IUBMB Life, 59*(3), 151–162.
De Oliveira, J. L., Ramos Campos, E. V., Bakshi, M., Abhilash, P. C., & Franceto, L. F. (2014). Application of nanotechnology for the encapsulation of botanical insecticides for sustainable agriculture: Prospects and promises. *Biotechnology Advances, 32*, 1550–1561.
Deka, B., Baruah, C., & Babu, A. (2021). Entomopathogenic microorganisms: Their role in insect pest management. *Egyptian Journal of Biological Pest Control, 31*(1), 1–8.
Dey, D., & Gupta, M. (2016). Use of essential oils for insect pest management-a review. *Innovative Farming, 1*, 21–29.
Donzelli, B. G. G., & Krasnoff, S. B. (2016). Molecular genetics of secondary chemistry in *Metarhizium* Fungi. *Advances in Genetics, 94*, 365–436.
El-Sayed, A. M., Suckling, D. M., Byers, J. A., Jang, E. B., & Wearing, C. H. (2009). Potential of "lure and kill" in long-term pest management and eradication of invasive species. *Journal of Economic Entomology, 102*(3), 815–835.
Eskenazi, B., Harley, K., Bradman, A., Weltzien, E., Jewell, N. P., Barr, D. B., & Holland, N. T. (2004). Association of in utero organophosphate pesticide exposure and fetal growth and length of gestation in an agricultural population. *Environmental Health Perspectives, 112*(10), 1116–1124.
Fan, Y., Liu, X., Keyhani, N. O., Tang, G., Pie, Y., Zhang, W., & Tong, S. (2017). Regulatory cascade and biological activity of *Beauveria bassiana* oosporein that limits bacterial growth after host death. *Proceedings National Academy of Science USA, 114*, E1578–E1586.
Faria, J. M., & Rodrigues, A. M. (2021). Essential oils as potential biopesticides in the control of the genus *Meloidogyne:* A review. *Biology and Life Science Forum, 3*(1), 26.
Granados, R. R., & Williams, K. A. (1986). In vivo infection and replication of baculoviruses. In R. R. Granados, & B. A. Federici (Eds.), *The biology of baculoviruses* (Vol. I, pp. 89–108). CRC Press.
Grdiša, M., & Gršić, K. (2013). Botanical insecticides in plant protection. *Agriculturae Conspectus Scientificus, 78*(2), 85–93.
Greenfield, B. P. J., Lord, A. M., Dudley, E., & Butt, T. M. (2014). Conidia of the insect pathogenic fungus, *Metarhizium anisopliae,* fail to adhere to mosquito larval cuticle. *Royal Society Open Science, 1*(2), 140193. https://doi.org/10.1098/rsos.140193.
Hasan, F., & Ansari, M. S. (2011). Toxic effects of neem-based insecticides on *Pieris brassica* (Linn.). *Crop Protection, 30*(4), 502–507.

Hassaan, M. A., & El Nemr, A. (2020). Pesticides pollution: Classifications, human health impact, extraction, and treatment techniques. *Egyptian Journal of Aquatic Research, 46*(3), 207–220.

Icoz, I., & Stotzky, G. (2008). Fate and effects of insect-resistant Bt crops in soil ecosystems. *Soil Biology & Biochemistry, 40*(3), 559–586.

Ikbal, C., & Pavela, R. (2019). Essential oils as active ingredients of botanical insecticides against aphids. *Journal of Pest Science, 92*(3), 971–986.

International Organization for Standardization (ISO). (2017). *Essential oil of melaleuca, terpinen-4-ol type (tea tree oil)*. ISO 4730. iso.org/standard/69082.html.

Isman, M. B. (1997). Neem and other botanical insecticides: Barriers to commercialization. *Phytoparasitica., 25*, 339–344.

Isman, M. B. (2006). Botanical insecticides, deterrent and repellents in modern agriculture and an increasingly regulated world. *Annual Review of Entomology, 51*, 45–66.

Isman, M. B. (2020). Bioinsecticides based on plant essential oils: A short overview. *Zeitschrift für Naturforschung C, 75*(7-8), 179–182.

Isman, M. B., & Tak, J. H. (2017). Inhibition of acetylcholinesterase by essential oils and monoterpenoids: A relevant mode of action for insecticidal essential oils. *Biopesticides International, 13*, 71–78.

Isman, M. B., Wilson, J. A., & Bradbury, R. (2008). Insecticidal activities of commercial rosemary oils (*Rosmarinus officinalis*) against larvae of *Pseudaletia unipuncta* and *Trichoplusia ni* in relation to their chemical compositions. *Pharmaceutical Biology, 46*, 82–87.

Ivase, T. J. P., Nyakuma, B. B., Ogenyi, B. U., Balogun, A. D., & Hassan, M. N. (2017). Current status, challenges, and prospects of biopesticide utilization in Nigeria. *Acta Universitatis Sapientiae, Agriculture Environment, 9*, 95–106.

Jaihan, P., Sangdee, K., & Sangdee, A. (2016). Selection of entomopathogenic fungus for biological control of chili anthracnose disease caused by *Colletotrichum* spp. *European Journal of Plant Pathology, 146*, 551–564. https://doi.org/10.1007/s10658-016-0941-7.

Jawalkar, N., Zambare, S., & Zanke, S. (2016). Insecticidal property of *Datura stramonium* L. seed extracts against *Sitophilus oryzae* L. (Coleoptera: Curculionidae) in stored wheat grains. *Journal of Entomology Zoology Studies, 4*(6), 92–96.

Kanda, D., Kaur, S., & Koul, O. (2017). A comparative study of monoterpenoids and phenylpropanoids from essential oils against stored grain insects: Acute toxins or feeding deterrents. *Journal of Pest Science, 90*, 531–545. https://doi.org/10.1007/s10340-016-0800-5.

Kanmani, S., Kumar, L., Raveen, R., Tennyson, S., Arivoli, S., & Jayakumar, M. (2021). Toxicity of tobacco *Nicotiana tabacum Linnaeus* (Solanaceae) leaf extracts to the rice weevil *Sitophilus oryzae Linnaeus* 1763 (Coleoptera: Curculionidae). *The Journal of Basic and Applied Zoology, 82*(1), 1–12.

Khan, S., Guo, L., Maimaiti, Y., Mijit, M., & DeWen, Q. (2012). Entomopathogenic fungi as microbial biocontrol agent. *Molecular Plant Breeding, 3*(7), 63–79. https://doi.org/10.5376/mpb.2012.03.0007.

Khursheed, A., Rather, M. A., Jain, V., Rasool, S., Nazir, R., Malik, N. A., & Majid, S. A. (2022). Plant based natural products as potential ecofriendly and safer biopesticides: A comprehensive overview of their advantages over conventional pesticides, limitations, and regulatory aspects. *Microbial Pathogenesis, 173*, 105854.

Koul, O. (2011). Microbial biopesticides: Opportunities and challenges. *CAB Reviews: Perspectives in Agriculture, Veterinary Science, Nutrition and Natural Resources, 6*, 1–26.

Koul, O., Walia, S., & Dhaliwal, G. S. (2008). Essential oils as green pesticides: Potential and constraints. *Biopesticides International, 4*(1), 63–84.

Kumar, A., Verma, H., Singh, V. K., Singh, P. P., Singh, S. K., Ansari, W. A., & Pandey, K. D. (2017). Role of *Pseudomonas* sp. in sustainable agriculture and disease management. In *Agriculturally important microbes for sustainable agriculture* (pp. 195–215). Springer.

Kumar, J., Ramlal, A., Mallick, D., & Mishra, V. (2021). An overview of some biopesticides and their importance in plant protection for commercial acceptance. *Plants, 10*(6), 1185.

Kumar, M., Kour, D., Yadav, A. N., Saxena, R., Rai, P. K., Jyoti, A., & Tomar, R. S. (2019). Biodiversity of methylotrophic microbial communities and their potential role in mitigation of abiotic stresses in plants. *Biologia, 74*, 287–308. https://doi.org/10.2478/s11756-019-00190-6.

Kumar, S. (2012). Biopesticides: A need for food and environmental safety. *Journal of Biofertilizers and Biopesticides, 3*(4), 1–3.
Kumar, S., & Singh, A. (2015). Biopesticides: Present status and the future prospects. *Journal of Fertilizers and Pesticides, 6*(2), 100–129.
Kumari, I., Hussain, R., Sharma, S., & Ahmed, M. (2022). Microbial biopesticides for sustainable agricultural practices. In *Biopesticides* (pp. 301–317). Woodhead Publishing.
Kumari, S., Chourasia, S. K., Jha, M. N., kant, R., Singh, U., & Kumar, P. (2014). Microbial pesticide: A boom for sustainable agriculture. *International Journal of Scitific and Engineering Research, 5*, 1394–1397.
Kunimi, Y. (2007). Current status and prospects on microbial control in Japan. *Journal of Invertebrate Pathology, 95*, 181–186.
Lacey, L. A., Grzywacz, D., Shapiro-Ilan, D. I., Frutos, R., Brownbridge, M., & Goettel, M. S. (2015). Insect pathogens as biological control agents: Back to the future. *Journal of Invertebrate Pathology, 132*, 1–41.
Leung, A. Y. (1980). *Encyclopedia of common natural ingredients used in food, drugs, and cosmetics.* John Wiley & Sons.
Li, N., Ragheb, K., Lawler, G., Sturgis, J., Rajwa, B., Melendez, J. A., & Robinson, J. P. (2003). Mitochondrial complex I inhibitor rotenone induces apoptosis through enhancing mitochondrial reactive oxygen species production. *Journal of Biological Chemistry, 278*(10), 8516–8525.
Litwin, A., Nowak, M., & Różalska, S. (2020). Entomopathogenic fungi: Unconventional applications. *Reviews in Environmental Science and Biotechnology, 19*(1), 23–42.
Liu, X., Cao, A., Yan, D., Ouyang, C., Wang, Q., & Li, Y. (2021). Overview of mechanisms and uses of biopesticides. *International Journal of Pest Management, 67*(1), 65–72.
Liu, X., Chen, Q., Wang, Z., Xie, L., & Xu, Z. (2008). Allelopathic effects of essential oil from *Eucalyptus grandis* × *E. urophylla* on pathogenic fungi and pest insects. *Frontiers of Forestry in China, 3*(2), 232–236.
Lovett, B., & Leger, R. J. S. (2017). The insect pathogens. *Microbiological Spectrum, 5*, 925–943. https://doi.org/10.1128/microbiolspec.funk-0001-2016.
Machalova, Z., Sajfrtova, M., Pavela, R., & Topiar, M. (2015). Extraction of botanical pesticides from *Pelargonium graveolens* using supercritical carbon dioxide. *Industrial Crops and Products, 67*, 310–317.
Markets and Markets (2020). Biopesticides market: Global forecast to 2022. By type (bioinsecticides, biofungicides, bioherbicides, and bionematicides), origin (beneficial insects, microbials, plant-incorporated protectants, and biochemicals), mode of application, formulation, crop type and region. https://www.marketsandmarkets.com/Market-Reports/biopesticides-267.html.
Manda, R. R., Addanki, V. A., & Srivastava, S. (2020). Microbial bio-pesticides and botanicals as an alternative to synthetic pesticides in the sustainable agricultural production. *Plant Cell Biotechnology and Molicular Biology, 21*, 31–48.
Marrone, P. (2019). Pesticidal natural products: Status and future potential. *Pest Management Science, 75*, 2325–2340.
Marwal, A., Srivastava, A.K., & Gaur, R.K. (2022). Plant viruses as biopesticides. In H.B. Singh, & A. Vaishnav (Eds.), New and Future Developments in Microbial Biotechnology and Bioengineering (pp. 181–194), Elsevier.
Mascarin, G. M., & Jaronski, S. T. (2016). The production and uses of *Beauveria bassiana* as a microbial insecticide. *World Journal of Microbiology and Biotechnology, 32*, 1–26.
Mazid, S., Kalida, J. C., & Rajkhowa, R. C. (2011). A review on the use of biopesticides in insect pest management. *International Journal of Science and Advanced Technology, 1*, 169–178.
Montesinos, E., & Bonaterra, A. (2009). Microbial pesticides. In *Encyclopedia of microbiology* (3rd ed., pp. 110–120). Elsevier.
Mordue, A. J., & Blackwell, A. (1993). Azadirachtin: An update. *Journal of Insect Physiology, 39*(11), 903–924.
Mpandeli, S., & Maponya, P. (2014). Constraints and challenges facing the small scale farmers in Limpopo Province, South Africa. *Journal of Agricultural Science, 6*(4), 135.
Nawaz, M., Mabubu, J. I., & Hua, H. (2016). Current status and advancement of biopesticides: Microbial and botanical pesticides. *Journal of Entomology and Zoological Studies, 4*(2), 241–246.
Nayak, P. (2020). Botanical pesticides: An insecticide from plant derivatives. *Biotica Research Today, 2*(8), 727–730.
Nollet, L. M., & Rathore, H. S. (2017). *Green pesticides handbook: Essential oils for pest control.* CRC Press.

Okrikata, E. (2021). Present views, status, and updates in biopesticide usage. In O. A. Timothy, & T. V. Ralf (Eds.), *Biopesticides: Botanicals and microorganisms for improving agriculture and human health* (pp. 249–280). Berlin, Germany: Logos-Verlag Berlin.

Olson, S. (2015). An analysis of the biopesticide market now and where it is going. *Outlooks on Pest Management, 26*(5), 203–206.

Olson, S., Ranade, A., Kurkjy, N., Pang, K., & Hazekamp, C. (2013). *Green dreams or growth opportunities: Assessing the market potential for "greener" agricultural technologies*. Lux Research. https://portal.luxresearchinc.com/research/tidbit/15753.

Pal, G. K., Kumar, B., & Shahi, S. K. (2013). Antifungal activity of some common weed extracts against phytopathogenic fungi *Alternaria* spp. *International Journal of Universal Pharmacy and Life Sciences, 3*(2), 6–14.

Parker, K. M., Barragán, B. V., van Leeuwen, D. M., Lever, M. A., Mateescu, B., & Sander, M. (2019). Environmental fate of RNA interference pesticides: Adsorption and degradation of double-stranded RNA molecules in agricultural soils. *Environmental Science & Technology, 53*, 3027–3036.

Parween, T., & Jan, S. (2019). Pesticides and environmental ecology. In T. Parween, & S. Jan (Eds.), *Ecophysiology of pesticides* (pp. 1–38). Academic Press.

Patil, H. J., & Solanki, M. K. (2016). Microbial inoculant: Modern era of fertilizers and pesticides. In *Microbial inoculants in sustainable agricultural productivity* (pp. 319–343). Springer.

Pimentel, D. (2009a). Environmental and economic costs of the application of pesticides primarily in the United States. In P. Rajinder, & A. Dhawan (Eds.), *Integrated pest management: Innovation-development process* (Vol. 1, pp. 88–111). Springer.

Pimentel, D. (2009b). Pesticides and pest control. In P. Rajinder, & A. Dhawan (Eds.), *Integrated pest management: Innovation-development process* (Vol. 1, pp. 83–87). Springer.

Pineda, S., Martínez, A. M., Figueroa, J. I., Schneider, M. I., Del Estal, P., Viñuela, E., Gomez, B., Smagghe, G., & Budia, F. (2009). Influence of azadirachtin and methoxyfenozide on life parameters of *Spodoptera littoralis* (Lepidoptera: Noctuidae). *Journal of Economic Entomology, 102*(4), 1490–1496.

Prabha, S., Yadav, A., Kumar, A., Yadav, A., Yadav, H. K., Kumar, S., & Kumar, R. (2016). Biopesticides: An alternative and eco-friendly source for the control of pests in agricultural crops. *Plant Archives, 16*(2), 902–906.

Procházka, P., Řehoř, J., Vostřel, J., & Fraňková, A. (2022). Use of botanicals to protect early stage growth of hop plants against *Pseudoperonospora humuli*. *Crop Protection, 157*, 105978.

Rani, A. T., Kammar, V., Keerthi, M. C., Rani, V., Majumder, S., Pandey, K. K., & Singh, J. (2021). Biopesticides: An alternative to synthetic insecticides. In *Microbial technology for sustainable environment* (pp. 439–466). Springer.

Rattan, R. S. (2010). Mechanism of action of insecticidal secondary metabolites of plant origin. *Crop Protection, 29*(9), 913–920.

Rauh, V. A., Garcia, W. E., Whyatt, R. M., Horton, M. K., Barr, D. B., & Louis, E. D. (2015). Prenatal exposure to the organophosphate pesticide chlorpyrifos and childhood tremor. *Neurotoxicology, 51*, 80–86.

Raymond, B., & Federici, B. A. (2017). In defense of Bacillus thuringiensis, the safest and most successful microbial insecticide available to humanity: A response to EFSA. *FEMS Microbiology Ecology, 93*(7), 1–7.

Raymond, B., Sayyed, A. H., & Wright, D. J. (2006). The compatibility of a nucleopolyhedrosis virus control with resistance management for *Bacillus thuringiensis*: Co-infection and cross-resistance studies with the diamondback moth, *Plutella xylostella*. *Journal of Invertebrate Pathology, 93*(2), 114–120.

Razaq, M., & Shah, F. M. (2022). Biopesticides for management of arthropod pests and weeds. In *Biopesticides* (pp. 7–18). Woodhead Publishing.

Reddy, G. V. P., Cruz, Z. T., & Guerrero, A. (2009). Development of an efficient pheromone-based trapping method for the banana root borer *Cosmopolites sordidus*. *Journal of Chemical Ecology, 35*, 111–117.

Regnault-Roger, C. (2012). Trends for commercialization of biocontrol agent (biopesticide) products. In *Plant defence: Biological control* (pp. 139–160). Springer.

Regnault-Roger, C., Vincent, C., & Arnason, J. T. (2012). Essential oil in insect control: Low risk products in a high-stakes world. In *Annual Review of Entomology, 57*, 405–424.

Ríos-Moreno, A., Garrido-Jurado, I., Resquı´n-Romero, G., Arroyo-Manzanares, N., Arce, L., & Quesada-Moraga, E. (2016). Destruxin A production by *Metarhizium brunneum* strains during transient endophytic colonisation of *Solanum tuberosum*. *Biocontrol Science and Technology, 26*, 1574–1585.

Rizzo, D. M., Lichtveld, M., Mazet, J. A., Togami, E., & Miller, S. A. (2021). Plant health and its effects on food safety and security in a One Health framework: Four case studies. *One Health Outlook, 3*(1), 1–9.

Robu, V., Covaci, G., & Popescu, I. M. (2015). The use of essential oils in organic farming. *Research Journal of Agricultural Science, 47*(4), 134–137.

Ruiu, L. (2018). Microbial biopesticides in agroecosystems. *Agronomy, 8*(11), 235.

Samada, L. H., & Tambunan, U. S. F. (2020). Biopesticides as promising alternatives to chemical pesticides: A review of their current and future status. *Online Journal of Biological Science, 20*, 66–76.

Santos, E. N., Menezes, L. P., Dolabella, S. S., Santini, A., Severino, P., Capasso, R., & Jain, S. (2022). *Bacillus thuringiensis*: From biopesticides to anticancer agents. *Biochimie, 192*, 83–90.

Sarwar, M., & Aslam, R. (2021). Usage of entomopathogenic viruses for insect pest control. In *Biopesticides in organic farming* (pp. 99–108). CRC Press.

Schmutterer, H. (1990). Properties and potential of natural pesticides from the neem tree, *Azadirachta indica*. *Annual Review of Entomology, 35*(1), 271–297.

Schnarr, L., Segatto, M. L., Olsson, O., Zuin, V. G., & Kümmerer, K. (2022). Flavonoids as biopesticides: Systematic assessment of sources, structures, activities, and environmental fate. *Science of the Total Environment, 824*, 153781.

Senthil-Nathan, S. (2015). A review of biopesticides and their mode of action against insect pests. In P. Thangavel, & G. Sridevi (Eds.), *Environmental sustainability* (pp. 49–63). New Delhi: Springer. https://doi.org/10.1007/978-81-322-2056-5_3.

Shagal, M. H., Modibbo, U. U., & Liman, A. B. (2012). Pharmacological justification for the ethnomedical use of *Datura stramonium* stem-bark extract in treatment of diseases caused by some pathogenic bacteria. *International Research of Pharmacy and Pharmacology, 2*(1), 16–19.

Shao, H., & Zhang, Y. (2017). Non-target effects on soil microbial parameters of the synthetic pesticide carbendazim with the biopesticides cantharidin and norcantharidin. *Scientific Reports,* 7(1), 1–12.

Sharma, A., Kumar, V., Shahzad, B., Tanveer, M., Sidhu, G. P. S., Handa, N., Kohli, S. K., Yadav, P., Bali, A. S., Parihar, R. D., Dar, O. I., & Thukral, A. K. (2019). Worldwide pesticide usage and its impacts on ecosystem. *SN Applied Sciences, 1*(11), 1–16.

Sharma, R. K. (2019). Insecticide residues in foods. In *Bioactive molecules in food* (pp. 1793–1813). Springer.

Shukla, N., Singh, E. A. N. A., Kabadwa, B. C., Sharma, R., & Kumar, J. (2019). Present status and future prospects of bio-agents in agriculture. *International Journal of Current Microbiology and Applied Sciences, 8*(4), 2138–2153.

Singh, J., & Yadav, A. N. (2020). *Natural bioactive products in sustainable agriculture*. Springer.

Skinner, M., Parker, B. L., & Kim, J. S. (2014). Role of entomopathogenic fungi. In D. P. Abrol (Ed.), *Integrated pest management* (pp. 169–191). Academic Press.

Soberón, M., Monnerat, R., & Bravo, A. (2016). Mode of action of cry toxins from Bacillus thuringiensis and resistance mechanisms. *Mode of action of cry toxins from Bacillus thuringiensis and resistance mechanisms* (pp. 1–13). Microbial toxins, part of the series toxinology. Springer Science Media.

Soumia, P. S., Krishna, R., Jaiswal, D. K., Verma, J. P., Yadav, J., & Singh, M. (2021). Entomopathogenic microbes for sustainable crop protection: Future perspectives. In A. N. Yadav, J. Singh, C. Singh, & N. Yadav (Eds.), *Current trends in microbial biotechnology for sustainable agriculture*. Springer Environmental and microbial biotechnology.

Starnes, R. L., Liu, C. L., & Marrone, P. G. (1993). History, use, and future of microbial insecticides. *American Entomologist, 39*(2), 83–91.

Su, V., King, D., Woodrow, I., McFadden, G., & Gleadow, R. (2008). *Plasmodium falciparum* growth is arrested by monoterpenes from eucalyptus oil. *Flavour and Fragrance Journal, 23*(5), 315–318.

Tabashnik, B. E., & Gould, F. (2012). Delaying corn rootworm resistance to Bt corn. *Journal of Economic Entomology, 105*(3), 767–776.

Tak, J. H., Jovel, E., & Isman, M. B. (2016). Contact, fumigant and cytotoxic activities of thyme and lemongrass essential oils against larvae and an ovarian cell line of the cabbage looper, *Trichoplusia ni*. *Journal of Pest Science, 89*, 183–193.

Tanada, Y., & Kaya, H. K. (1993). *Insect pathology* (p. 666). Academic Press.

Thakur, N., Kaur, S., Tomar, P., Thakur, S., & Yadav, A. N. (2020). Microbial biopesticides: Current status and advancement for sustainable agriculture and environment. In *New and future developments in microbial biotechnology and bioengineering* (pp. 243–282). Elsevier.

Tkaczuk, C., Harasimiuk, M., Król, l. A., & Beres, P. K. (2015). The effect of selected pesticides on the growth of entomopathogenic fungi *Hirsutella nodulosa* and *Beauveria bassiana*. *Journal of Ecological Engineering, 16*, 177–183. https://doi.org/10.12911/22998993/2952.

Trigg, J. K. (1996a). Evaluation of a eucalyptus-based repellent against *Anopheles* spp. in Tanzania. *Journal of the American Mosquito Control Association, 12*(2), 243–246.

Trigg, J. K. (1996b). Evaluation of a eucalyptus-based repellent against *Culicoides impunctatus* (Diptera: Ceratopogonidae) in Scotland. *Journal of the American Mosquito Control Association, 12*(2), 329–330.

Tripathi, Y. N., Divyanshu, K., Kumar, S., Jaiswal, L. K., Khan, A., Birla, H., & Upadhyay, R. S. (2020). Biopesticides: Current status and future prospects in India. In C. Keswani, (Ed.) *Bioeconomy for sustainable development* (pp. 79–109). Singapore: Springer. https://doi.org/10.1007/978-981-13-9431-7_6.

Turek, C., & Stintzing, F. C. (2013). Stability of essential oils: A review. *Comprhensive Reviews in Food Science and Food Safety, 12*, 40–53.

Upadhyay, N., Dwivedy, A. K., Kumar, M., Prakash, B., & Dubey, N. K. (2018). Essential oils as eco-friendly alternatives to synthetic pesticides for the control of *Tribolium castaneum* (Herbst) (Coleoptera: Tenebrionidae. *Journal of Essential Oil-Bearing Plants., 21*(2), 282–297.

US EPA. Ingredients used in pesticide products: Pesticides. What are biopesticides? Retrieved May 10, 2021, from https://www.epa.gov/ingredients-used-pesticide-products/what-are-biopesticides.

Verma, D. K., Guzmán, K. N. R., Mohapatra, B., Talukdar, D., Chávez-González, M. L., Kumar, V., Srivastava, S., Singh, V., Yulianto, R., Malar, S. E., Ahmad, A., Utama, G. L., & Aguilar, C. N. (2021). Recent trends in plant-and microbe-based biopesticide for sustainable crop production and environmental security. In *Recent developments in microbial technologies* (pp. 1–37). Springer.

Wakeil, N. E. (2013). Botanical pesticides and their mode of actions. *Gesunde Pflanzen, 65*(4), 125–149.

Walling, L. L. (2000). The myriad plant responses to herbivores. *Journal of Plant Growth Regulation, 19*, 195–216.

Watanabe, K., Shono, Y., Kakimizu, A., Okada, A., Matsuo, N., Satoh, A., & Nishimura, H. (1993). New mosquito repellent from *Eucalyptus camaldulensis*. *Journal of Agricultural and Food Chemistry, 41*(11), 2164–2166.

Yadav, A. N., Rastegari, A. A., Yadav, N., & Kour, D. (2020). *Advances in plant microbiome and sustainable agriculture, volume 2: Functional annotation and future challenges*. Springer.

Yadav, A. N., Verma, P., Kour, D., Rana, K. L., Kumar, V., Singh, B., … Dhaliwal, H. S. (2017). Plant microbiomes and its beneficial multifunctional plant growth promoting attributes. *International Journal of Environmental Sciences and Natural Resources, 3*(1), 1–8.

Yadav, R., Singh, S., & Singh, A. N. (2022). Biopesticides: Current status and future prospects. *Proceedings of the International Academy of Ecology and Environmental Sciences, 12*(3), 211–233.

Zhang, W., Jiang, F., & Ou, J. (2011). Global pesticide consumption and pollution: With China as a focus. *Proceedings of the international academy of ecology and environmental sciences, 1*(2), 125.

Index

Page numbers followed by "*f*" and "*t*" indicate figures and tables respectively.

A

Abiotic factors, 93
Acetonitrile, 134
Acetylcholine esterase, 214
Adsorption
  desorption processes, 98
  of soil particles, 53
Advanced extraction and detection methods, 140
Agrochemicals
  fertilizers, 181
  pesticides, 181
Agrothoate, 188
AI encapsulation, 268
  metal-organic frameworks (MOFs), 272
Amperometric method, 137
Apoplastic
  pathway, 75
  translocation, 76
Arachis hypogaea, 19
Atrazine, 5
Azotobacter vinelandii, 56

B

B. thuringiensis, 317
  derived products, 316
Bacterial bioremediation, 205
Bactericides, 183
Baculoviruses, 319
Bavistin, 188
Beekeeper-applied pesticides, 27
Benzene hexachloride, 203
Biogeochemical cycles, 16
  mineralization processes, 16
  nitrogen fixation, 16
Biomagnification of
    dichloro-diphenyl-trichloroethane, 101*f*
Biomimetic immunoassay coupled method, 135
Biopesticides, 308
  advantages over chemical pesticides, 308
  current scenario, 321
  policy recommendations, 325
Biopiles, 238
Bioremediation processes, 235
Biotic stressors, 259
Bioventing, 236
Bipolaris sorokiniana, 259
Birth disorders, 16
Botanical pesticides, 310
Brassinolide application, 295
Brassinosteroids, 37
  effect on herbicide and pesticide stress, 295

C

Carbamates, 208, 307
  derived pesticides, 187
Carbamic acid, 208
Carbofuran, 215
Carbon
  based nanomaterials, 259
  dots, 259
  hydroxylation, 32
Central composite design, 133
Chagas disease, 212
Chemical compounds, 80
Chemical pesticides, 87, 100
Chemical pesticides, 307
  adverse effects on human health, 309
Chlorantraniliprole, 28
Chlorothalonil, 53, 55
Chlorpyrifos, 32
Chromatographic techniques, 135
Chronic toxicity, 16
Chrysanthemum cinerariifolium, 311
Clay nanomaterial-derived encapsulation, 271
Cloud point extraction, 137
Codex Alimentarius Commission, 120
Conformist pesticides, 255
Currently used pesticides, 91
Cyanobacteria, 273

D

Datura stramonium, 312

Detection techniques for pesticide residual analysis, 130
extraction techniques, 130
Matrix solid-phase dispersion, 133
solid-phase extraction, 131
solvent extraction and partitioning method, 131
Detoxification mechanism, 16
Dichlorodiphenyltrichloroethane, 157
Dichloro-diphenyl-trichloroethane, 91
Difluoroacetic acid, 135
DNA-based silver nanoparticles, 259
DNA methylation, 21
DNA shuffling, 240

E

Ecofriendly biopesticides, 322
Electrochemical techniques, 136
Endocrine disruption, 16
Endosulfan, 8
Entomopathogenic fungi (EPF), 317
Environmental Protection Agency, 2
Environmental Protection Agency (EPA), 1, 67
Enzymatic antioxidant defense system, 195
Enzymatic antioxidants, 196
Enzyme-linked immunosorbent test (ELISA), 136
Essential oils (EOs), 313
Eucalyptus camaldulensi, 312
European Union reference laboratories, 121

F

Food and Agriculture Organization, 155
Fumigant pesticides, 183
Fungal bioremediation, 209
Fungicides, 90

G

Gas chromatography, 138
Gene-level methods of bioremediation, 205
Genetic damage index (GDI), 262
Glutathione-S-transferases, 35
Glyphosate, 297
Greener encapsulations, 273
Green nanotechnology, 253
Green Revolution, 94
GRXS25 gene, in pesticide metabolism, 295

H

Halogenated organophosphates, 307
Herbicides, 5, 87, 204, 295
atrazine, 79
isoproturon, 79
Honeybees, 21
Hydrophilic interaction liquid chromatography (HILIC), 144
Hydrophobic pesticides, 76

I

Imidacloprid, 259
India Food Safety and Standard Authority of India, 120
Insecticides, 87
Insect pests, 37
Instrumental detection methods, 138
Integrated pest management (IPM), 120
International Council for Harmonization of technical, 128
Ionized pesticides, 78
Iron nanoparticles, 259

J

Jasmonic acid, 36

L

LC–based techniques, 139
Limit of detection, 130
Linearity, 128
Lipid-based nanoemulsions, 269
Lipid nanocarriers, 269
Lipid nanoparticles, 269
Lipophilicity, 77
Lipophilic pesticides, 259
Liquid chromatography, 139

M

Malathion, 8, 12, 214
Matrix interference, 138
Matrix solid-phase dispersion, 133
Mesoporous silicon nanospheres (MSN), 136
Methomyl, 215
Methyl jasmonate (MeJA), 299
Micelles, 268
Microbial-assisted bioremediation, 204
Microbial degradation, 205
Microbial metabolic activity, 234
Microbial pesticides, 314
MIP-based lab-on-paper device, 141
Molecular imprinted polymer (MIP) biosensors, 140

Molecular probes, 239
Multicopper enzymes, 232
Multiresidue methods (MRMs), 121

N

NADH-dehydrogenase, 20
Nanoagriculture, 251
Nano-based clays, 268
Nanocapsules, 268
Nanocarriers, 269
  for pesticide formulations, 266
Nanoencapsulation, 259
Nanomaterials, in agronomy, 259
Nanoparticlebased pesticides, 254
Nanoparticles, 259
Nanoparticulates, 268
Nanopesticides, 251
  human health and ecology, effects on, 265
  pesticide misuse, 255
Nanotechnology, 262
  agricultural and environmental issues, 252
  in crops, 262
Neonicotinoids, 27, 93, 106
Nerve disorders, 16
Nonenzymatic antioxidant defense system, 195
Nonenzymatic antioxidants, 196
Nonsystemic pesticides, 183, 206

O

O–glycosylation products, 35
Organic chemical, 75
Organic compounds, 204
Organochlorines, 207
  pesticides, 76, 213, 307
Organophosphates, 207

P

Persistent organic pollutants (POPs), 68
Pesticide, 266
Pesticides, 12, 18, 56, 67, 68, 94, 108, 182, 197, 212
  accumulation of, 56
  benzoic acid herbicides, 15
  bioavailability of, 2
  biology in soil, 49
  carbamate, 15
  classification of, 1, 184*f*
  -containing soil water fractions, 95
  contaminated feedstuffs, 87
  detoxification mechanism in plants, 32
  encapsulations, 268
  enzymatic degradation pathway of, 229
  exposure, 21
  flow chat, 17*f*
  induced oxidative burst, 185
  induced oxidative stress, 196
  Intensive and uncontrolled applications of, 91
  leaching ability, 57*t*
  mechanism of action, 20
  mechanism of involved neurotoxicity, 22*f*
  metabolized enzymes, 197
  movement of, 77
  organochlorine, 15
  organophosphate, 15
  overutilization of, 2
  persistence of, 50
  phenoxy herbicides, 15
  physical and chemical properties of, 76
  plant-protectant chemicals, 52
  pyrethroids, 15
  and repetitive use of, 1
  residues, 19, 96, 106, 107*f*, 122*f*, 140
  route of, 2
  in soil, 55
  storage and metabolism of, 78
  toxicity, 2, 3, 4*t*, 5, 7, 8, 21, 29, 33
  toxicity and mechanism of action in humans, 17
  toxicity and mechanism of action in plants, 17
  triazines, 15
  types, 50, 89
Pest management, 324
Phase III metabolism, 80
Phase II metabolism, 79, 80
Phase I metabolism, 78
Phase I reaction, 79
Phenazine-1-carboxamide, 60
Pheromones, 320
Phytohormones, 294
Phytoremediation, 19
Placental cytotoxicity, 262
Plant EOs, 314
Plant hormones, 36
Plant hormones, 294, 300
Plant-incorporated protectants (PIPs), 320
Plant-parasitic fungi, 93
Poisoning pesticides, 183
Polycyclic aromatic hydrocarbons (PAHs), 237

Polymer-based nanoformulations, 268
Potentiometric measurements, 137
Pseudomonas spp., 317
Pyrethroids, 209

Q

QuEChERS
  approach, 143
  method, 140
  process, 143
  technique, 134

R

Reactive oxygen species (ROS), 16, 182, 186
Recombinant DNA technology, 238
Redox reactions, 234
Redox-sensitive transcription, 199
Remediation technique, 236
Repellent pesticides, 183
Responsive nitrogen species (RNS), 20
Ridomil, 188
Riger–Peppas concept, 271
ROS-generating systems, 188
Rotenone (Derris spp. and Lonchocarpus spp.), 312
Ryana speciosa, 313

S

SA induced herbicide resistance, 297
Salicylates, effect on herbicide and pesticide stress, 296
Salicylic acid, 197
SA-mediated enhanced antioxidant system, 297
SANTE guidelines, 134
Sigma Broad herbicide, 295
Signaling transduction mechanism, 16
Silica microcapsules, 271
Silica nanostructures, 271
Slurry bioreactors, 238
Soil microbial biomass, 56
Soil microorganisms, 29
Spectrophotometric techniques, 137
Starch-based nanocontainers, 273
Stimuli-responsive nanocarrier, 274
Stimuli responsive systems, agricultural applications of, 274, 275
Stimulus-sensitive nanoformulations, 274
Stomach Toxicant, 183
SWAP-PEARL models, 50
Symplastic translocation, 76
Systemic pesticides, 183, 206

T

Temperature-sensitive systems, 276
Thin-layer chromatographic (TLC), 136
Tick-borne diseases, 212
*Tobacco (*Nicotiana tabacum*; Nicotine)*, 313
Translocation factor (TF), 77
Triticum aestivum, 19

U

United Nations Environment Programme (UNEP), 68

V

Vector-borne diseases, 212
Virulency (GDI), 264

W

Western Calabrian coast, 6
Withania somnifera, 182
World Health Organisation (WHO), 184, 205

X

Xanthomonas perforans, 259
Xenobiots, 293

Z

ZnS quantum dots (ZnSQDs), 138